Supply Chain Engineering

Models and Applications

The Operations Research Series

Series Editor: A. Ravi Ravindran
Professor, Department of Industrial and Manufacturing Engineering
The Pennsylvania State University – University Park, PA

Published Titles:

Supply Chain Engineering: Models and Applications
A. Ravi Ravindran & Donald Paul Warsing

Analysis of Queues: Methods and Applications
Natarajan Gautam

Integer Programming: Theory and Practice
John K. Karlof

Operations Research and Management Science Handbook
A. Ravi Ravindran

Operations Research Applications
A. Ravi Ravindran

Operations Research: A Practical Introduction
Michael W. Carter & Camille C. Price

Operations Research Calculations Handbook, Second Edition
Dennis Blumenfeld

Operations Research Methodologies
A. Ravi Ravindran

Probability Models in Operations Research
C. Richard Cassady & Joel A. Nachlas

Forthcoming Titles:

Introduction to Linear Optimization and Extensions with MATLAB®
Roy H. Kwon

Supply Chain Engineering

Models and Applications

A. Ravi Ravindran
Donald P. Warsing, Jr.

CRC Press
Taylor & Francis Group
Boca Raton London New York

CRC Press is an imprint of the
Taylor & Francis Group, an **informa** business

CRC Press
Taylor & Francis Group
6000 Broken Sound Parkway NW, Suite 300
Boca Raton, FL 33487-2742

First issued in paperback 2017

© 2013 by Taylor & Francis Group, LLC
CRC Press is an imprint of Taylor & Francis Group, an Informa business

No claim to original U.S. Government works

Version Date: 2012924

ISBN 13: 978-1-138-07772-0 (pbk)
ISBN 13: 978-1-4398-1198-6 (hbk)

Library of Congress Cataloging-in-Publication Data

Ravindran, A., 1944-
    Supply chain engineering : models and applications / A. Ravi Ravindran, Donald P. Warsing, Jr.
    p. cm. -- (The operations research series)
    Includes bibliographical references and index.
    ISBN 978-1-4398-1198-6 (hardback)
    1. Business logistics. I. Warsing, Donald Paul. II. Title.

HD38.5.R38 2012
658.5--dc23   2012031946

Visit the Taylor & Francis Web site at
http://www.taylorandfrancis.com

and the CRC Press Web site at
http://www.crcpress.com

To our wives

Bhuvana and Mary

Contents

Preface...xvii
Acknowledgments xxi
Authors xxiii

1. Introduction to Supply Chain Engineering 1
  1.1   Understanding Supply Chains...2
       1.1.1   Flows in Supply Chains 4
  1.2   Meaning of Supply Chain Engineering...4
  1.3   Supply Chain Decisions 5
       1.3.1   Strategic Decisions 5
       1.3.2   Tactical Decisions..6
       1.3.3   Operational Decisions..6
  1.4   Enablers and Drivers of Supply Chain Performance.....................7
       1.4.1   Supply Chain Enablers...7
       1.4.2   Supply Chain Drivers...8
             1.4.2.1   Inventory 8
             1.4.2.2   Transportation 8
             1.4.2.3   Facilities 9
             1.4.2.4   Suppliers 9
  1.5   Assessing and Managing Supply Chain Performance...................9
       1.5.1   Supply Chain Efficiency..10
       1.5.2   Supply Chain Responsiveness 12
       1.5.3   Supply Chain Risk 12
       1.5.4   Conflicting Criteria in Supply Chain Optimization.........13
  1.6   Relationship between Supply Chain Metrics and Financial
       Metrics 13
       1.6.1   Inventory Measures..13
             1.6.1.1   Inventory Turns 13
             1.6.1.2   Days of Inventory 14
             1.6.1.3   Inventory Capital...14
       1.6.2   Business Financial Measures 15
             1.6.2.1   Return on Assets 15
             1.6.2.2   Working Capital 15
             1.6.2.3   Cash-to-Cash Cycle 15
  1.7   Importance of Supply Chain Management....................................16
       1.7.1   Supply Chain Top 25 17
  1.8   Organization of the Textbook..18
       1.8.1   Chapter 2 (Planning Production in Supply Chains).........19
       1.8.2   Chapter 3 (Inventory Management Methods
             and Models) 19

	1.8.3	Chapter 4 (Transportation Decisions in Supply Chain Management)	20
	1.8.4	Chapter 5 (Location and Distribution Decisions in Supply Chains)	20
	1.8.5	Chapter 6 (Supplier Selection Models and Methods)	21
	1.8.6	Chapter 7 (Managing Risks in Supply Chains)	21
	1.8.7	Chapter 8 (Global Supply Chain Management)	21
1.9	Summary and Further Readings		22
	1.9.1	Summary	22
	1.9.2	Further Readings	22

Exercises 23
References 25

2. Planning Production in Supply Chains 27
2.1	Role of Demand Forecasting in Supply Chain Management	27
2.2	Forecasting Process	28
2.3	Qualitative Forecasting Methods	29
	2.3.1	Executive Committee Consensus	29
	2.3.2	Delphi Method	30
	2.3.3	Survey of Sales Force	30
	2.3.4	Customer Surveys	31
2.4	Quantitative Forecasting Methods	31
	2.4.1	Time Series Forecasting	31
	2.4.2	Constant Level Forecasting Methods	33
	2.4.3	Last Value Method	34
	2.4.4	Averaging Method	34
	2.4.5	Simple Moving Average Method	35
	2.4.6	Weighted Moving Average Method	35
	2.4.7	Computing Optimal Weights by Linear Programming Model	36
	2.4.8	Exponential Smoothing Method	38
2.5	Incorporating Seasonality in Forecasting	39
2.6	Incorporating Trend in Forecasting	42
	2.6.1	Simple Linear Trend Model	43
	2.6.2	Holt's Method	45
2.7	Incorporating Seasonality and Trend in Forecasting	47
	2.7.1	Method Using Static Seasonality Indices	47
	2.7.2	Winters' Method	49
2.8	Forecasting for Multiple Periods	51
	2.8.1	Multi-Period Forecasting under Constant Level	51
	2.8.2	Multi-Period Forecasting with Seasonality	52
	2.8.3	Multi-Period Forecasting with Trend	52
	2.8.4	Multi-Period Forecasting with Seasonality and Trend	53
2.9	Forecasting Errors	54
2.10	Monitoring Forecast Accuracy	57

2.11   Forecasting Software 59
       2.11.1 Types of Forecasting Software 59
       2.11.2 User Experience with Forecasting Software 61
2.12   Forecasting in Practice 61
       2.12.1 Real World Applications 61
       2.12.2 Forecasting in Practice: Survey Results 62
2.13   Production Planning Process 63
2.14   Aggregate Planning Problem 64
2.15   Linear Programming Model for Aggregate Planning................... 65
2.16   Nonlinear Programming Model for Aggregate Planning............. 70
2.17   Aggregate Planning as a Transportation Problem....................... 72
       2.17.1 Basic Transportation Problem 72
       2.17.2 Aggregate Planning as a Transportation Problem............ 75
       2.17.3 Greedy Algorithm for Aggregate Planning..................... 78
2.18   Aggregate Planning Strategies: A Comparison............................. 80
2.19   Summary and Further Readings 81
       2.19.1 Demand Forecasting: Summary.. 81
       2.19.2 ARIMA Method 81
       2.19.3 Croston's Method 82
       2.19.4 Further Readings in Forecasting 82
       2.19.5 Production Planning: Further Readings 83
       2.19.6 Managing Demand.. 83
       2.19.7 Bullwhip Effect... 84
       2.19.8 Collaborative Planning, Forecasting
               and Replenishment (CPFR) 85
Exercises 85
References 92

3. Inventory Management Methods and Models 95
3.1   Decision Framework for Inventory Management........................... 95
3.2   Some Preliminary Modeling Issues 98
       3.2.1   Two Critical Tasks 98
       3.2.2   ABC Analysis.. 99
3.3   Single-Item, Single-Period Problem: The Newsvendor 101
       3.3.1   Service Measures in Inventory Models 105
       3.3.2   Service Impact of Shortage Costs 106
       3.3.3   Safety Stock: A First Look... 108
3.4   Single-Item, Multi-Period Problems 108
       3.4.1   Continuous-Review: Reorder Point–Order Quantity
               Model 110
       3.4.2   Continuous-Review under Uncertainty 115
       3.4.3   Periodic-Review, Reorder-Point–Order-up-to Models..... 121
       3.4.4   Other Periodic-Review Inventory Models........................ 125
       3.4.5   Non-Stationary Demand: Distribution Requirements
               Planning 127

3.5   Multi-Item Inventory Models 133
3.6   Multi-Echelon Inventory Systems.. 136
       3.6.1   Centralized versus Decentralized Control 139
       3.6.2   Serial Supply Chain with Deterministic Demand
                and Fixed Ordering Costs.. 140
       3.6.3   Two-Stage Serial System under Decentralized
                Control...141
       3.6.4   Two-Stage Serial System under Centralized Control 143
       3.6.5   Serial Supply Chain with Stochastic Demand and
                Negligible Fixed Ordering Costs..................................... 146
       3.6.6   Serial Supply Chain with Fixed Costs and Stochastic
                Demand 151
3.7   Summary and Further Readings.. 152
       3.7.1   Summary 152
       3.7.2   Further Readings 153
3.A   Appendix: The Bullwhip Effect 154
Appendix References... 165
Exercises 166
References 171

4. Transportation Decisions in Supply Chain Management 175
4.1   Introduction 175
4.2   Motor Carrier Freight: Truckload Mode...................................... 176
       4.2.1   Accounting for Goods in Transit 180
4.3   Stepping Back: Freight Transportation Overview....................... 184
4.4   More General Models of Freight Rates 187
4.5   Building A Rate Model: LTL Service.. 190
       4.5.1   LTL Mode: Building the Inventory Decision Model....... 194
       4.5.2   LTL Mode: Discount from Published Tariff.................... 199
4.6   A More General Rate Model for LTL Service............................... 211
4.7   Beyond Truck Transport: Rail and Air Cargo............................... 214
4.8   Summary and Further Readings.. 219
       4.8.1   Summary.. 219
       4.8.2   Further Readings 219
Exercises 220
References 225

5. Location and Distribution Decisions in Supply Chains 229
5.1   Modeling with Binary Variables..230
       5.1.1   Capital Budgeting Problem 230
       5.1.2   Fixed Charge Problem.. 231
       5.1.3   Constraint with Multiple Right-Hand-Side
                Constants... 232
       5.1.4   Quantity Discounts 233

	5.1.5	Handling Nonlinear Integer Programs	236
	5.1.6	Set Covering and Set Partitioning Models	238
		5.1.6.1 Set Covering Problem	238
		5.1.6.2 Set Partitioning Problem	240
		5.1.6.3 Application to Warehouse Location	240
5.2	Supply Chain Network Optimization		241
	5.2.1	Warehouse Location	241
	5.2.2	Distribution Planning	242
	5.2.3	Location-Distribution Problem	244
	5.2.4	Location-Distribution with Dedicated Warehouses	247
	5.2.5	Supply Chain Network Design	249
5.3	Risk Pooling or Inventory Consolidation		253
	5.3.1	Principles of Risk Pooling	253
	5.3.2	General Risk Pooling Model	256
	5.3.3	Pros and Cons of Risk Pooling	259
	5.3.4	Risk Pooling under Demand Uncertainty	260
	5.3.5	Risk Pooling Example	263
	5.3.6	Practical Uses of Risk Pooling	265
5.4	Continuous Location Models		266
	5.4.1	Continuous Location Model: Single Facility	266
		5.4.1.1 Gravity Model	266
		5.4.1.2 Iterative Method	267
		5.4.1.3 Illustrative Example: Gravity Model	268
		5.4.1.4 Limitations of Gravity Model	271
	5.4.2	Multiple Facility Location	271
5.5	Real-World Applications		272
	5.5.1	Multi-National Consumer Products Company	272
		5.5.1.1 Case 1: Supply Chain Network Design	273
		5.5.1.2 Case 2: Distribution Planning	274
	5.5.2	Procter and Gamble (P&G)	274
	5.5.3	Ford Motor Company	275
	5.5.4	Hewlett-Packard (HP)	276
	5.5.5	BMW	276
	5.5.6	AT&T	277
	5.5.7	United Parcel Service (UPS)	277
5.6	Summary and Further Readings		279
	5.6.1	Summary	279
	5.6.2	Further Readings	279
		5.6.2.1 Multiple Criteria Models for Network Design	279
		5.6.2.2 Risk Pooling	280
		5.6.2.3 Facility Location Decisions	281
		5.6.2.4 Case Studies	282
Exercises			282
References			291

6. Supplier Selection Models and Methods 293
   6.1   Supplier Selection Problem... 293
        6.1.1   Introduction... 293
        6.1.2   Supplier Selection Process...................................... 294
        6.1.3   In-House or Outsource.. 295
        6.1.4   Chapter Overview 296
   6.2   Supplier Selection Methods... 296
        6.2.1   Sourcing Strategy.. 296
        6.2.2   Criteria for Selection.. 297
        6.2.3   Pre-Qualification of Suppliers............................. 299
        6.2.4   Final Selection 300
               6.2.4.1   Single Sourcing Methods 300
               6.2.4.2   Multiple Sourcing Methods 303
   6.3   Multi-Criteria Ranking Methods for Supplier Selection............. 308
        6.3.1   Ranking of Suppliers.. 309
               6.3.1.1   Case Study 1: Ranking of Suppliers.................. 309
        6.3.2   Use of L_p Metric for Ranking Suppliers........................... 311
               6.3.2.1   Steps of the L_2 Metric Method 312
        6.3.3   Rating (Scoring) Method... 312
        6.3.4   Borda Count.. 314
        6.3.5   Pair-Wise Comparison of Criteria 316
        6.3.6   Scaling Criteria Values... 317
               6.3.6.1   Simple Scaling..................................... 317
               6.3.6.2   Ideal Value Method 317
               6.3.6.3   Simple Linearization (Linear Normalization) 318
               6.3.6.4   Use of L_p Norm (Vector Scaling) 318
               6.3.6.5   Illustrative Example of Scaling Criteria Values 318
               6.3.6.6   Simple Scaling Illustration................. 319
               6.3.6.7   Scaling by Ideal Value Illustration...................... 320
               6.3.6.8   Simple Linearization (Linear Normalization) Illustration................................... 320
               6.3.6.9   Scaling by L_p Norm Illustration............................. 321
        6.3.7   Analytic Hierarchy Process.................................... 322
               6.3.7.1   Basic Principles of AHP...................... 322
               6.3.7.2   Steps of the AHP Model...................... 323
        6.3.8   Cluster Analysis.. 326
               6.3.8.1   Procedure for Cluster Analysis 328
        6.3.9   Group Decision Making 329
        6.3.10 Comparison of Ranking Methods.......................... 330
   6.4   Multi-Objective Supplier Allocation Model.................... 330
        6.4.1   Notations Used in the Model 331
        6.4.2   Mathematical Formulation of the Order Allocation Problem 332
        6.4.3   Goal Programming Methodology.......................... 334
               6.4.3.1   General Goal Programming Model.................. 335

6.4.4 Preemptive Goal Programming 336
6.4.5 Non-Preemptive Goal Programming 337
6.4.6 Tchebycheff (Min–Max) Goal Programming................... 338
6.4.7 Fuzzy Goal Programming...................................... 339
6.4.8 Case Study 2: Supplier Order Allocation 339
  6.4.8.1 Preemptive Goal Programming Solution 342
  6.4.8.2 Non-Preemptive Goal Programming............... 342
  6.4.8.3 Tchebycheff Goal Programming....................... 343
  6.4.8.4 Fuzzy Goal Programming 343
6.4.9 Value Path Approach.. 343
  6.4.9.1 Value Path Approach for the Supplier
       Selection Case Study.................................... 344
  6.4.9.2 Discussion of Value Path Results 345
6.5 Summary and Further Readings 346
  6.5.1 Ranking Suppliers 346
  6.5.2 Supplier Order Allocation 347
  6.5.3 Global Sourcing.. 349
  6.5.4 Supplier Risk... 351
Exercises 351
References 357

7. **Managing Risks in Supply Chain**.................................. 363
7.1 Supply Chain Risk 363
7.2 Real World Risk Events and Their Impacts......................... 364
  7.2.1 Importance of Supply Chain Risk Management............ 365
7.3 Sources of Supply Chain Risks 367
7.4 Risk Identification 368
7.5 Risk Assessment... 369
  7.5.1 Risk Mapping 370
  7.5.2 Risk Prioritization... 371
    7.5.2.1 Risk Priority Numbers 371
7.6 Risk Management 372
  7.6.1 Risk Management Strategies 373
  7.6.2 Developing a Risk Management Plan...................... 374
  7.6.3 Risk Mitigation Strategies.................................. 375
    7.6.3.1 Traditional Strategies 375
    7.6.3.2 Flexible Strategies................................. 375
7.7 Best Industry Practices in Risk Management 376
  7.7.1 Teradyne Inc. 377
  7.7.2 Hewlett-Packard (HP) 378
  7.7.3 Federal Express 378
  7.7.4 Wal-Mart 379
  7.7.5 Johnson and Johnson....................................... 380
7.8 Risk Quantification Models..................................... 380
  7.8.1 Basic Risk Quantification Models........................... 381

7.9   Value-at-Risk (VaR) Models 382
       7.9.1   VaR Type Impact Function 382
       7.9.2   Generalized Extreme Value Distribution (GEVD)
                Functions for Risk Impact.................................... 384
       7.9.3   Estimating GEVD Parameters................................ 384
       7.9.4   VaR Occurrence Functions 386
       7.9.5   VaR Disruption Risk Function............................... 387
                7.9.5.1   Simulation Approach............................. 387
                7.9.5.2   VaR Type Occurrence Function 390
                7.9.5.3   VaR Type Disruption Risk Function....... 390
7.10   Miss-the-Target (MtT) Risk Models.................................. 393
       7.10.1   MtT Type Impact Function 393
       7.10.2   MtT Type Occurrence Function 395
                7.10.2.1   Gamma Distribution for S-Type........... 395
                7.10.2.2   Beta Distribution for the L-Type 396
                7.10.2.3   Generalized Hyperbolic Distribution
                           for N-Type...................................... 396
       7.10.3   MtT Type Risk function 397
                7.10.3.1   S-Type Risk Function 397
                7.10.3.2   L-Type Risk..................................... 398
                7.10.3.3   N-Type Risk Function 398
7.11   Risk Measures.. 402
7.12   Combining VaR and MtT Type Risks.............................. 405
       7.12.1   Combining Different VaR Type or MtT Type Risks
                from the Same Supplier 405
                7.12.1.1   VaR Type Risk Combination................ 405
                7.12.1.2   MtT Type Risk Combination............... 406
       7.12.2   Combining the Same VaR Type or MtT Type Risks
                from Different Suppliers...................................... 407
                7.12.2.1   VaR Type Combination...................... 407
                7.12.2.2   MtT Type Risk Combination............... 407
       7.12.3   Combining Total VaR Type or MtT Type Risks
                from All Suppliers 408
                7.12.3.1   VaR Type Combination...................... 408
                7.12.3.2   MtT Type Combination 408
7.13   Risk Detectability and Risk Recovery............................. 409
       7.13.1   Detectability of Disruptive Events 409
                7.13.1.1   Some Basic Properties of Markov Chains........ 410
                7.13.1.2   Computing the MFPT Matrix 410
                7.13.1.3   Using MFPT in Disruption Risk
                           Quantification 411
       7.13.2   A Conceptual Model for Risk Recovery 412
       7.13.3   Illustrative Example of Risk Detectability
                and Recovery 413

7.14 Multiple Criteria Optimization Models for Supplier
       Selection Incorporating Risk 416
       7.14.1 Phase 1 Model (Short-Listing Suppliers) 417
       7.14.2 Results of Phase 1 Experiments 418
               7.14.2.1 Ranking of the Criteria 418
               7.14.2.2 Comparison across Methods
                        for the Same DM 419
               7.14.2.3 Comparison across DMs for the Same Method 419
               7.14.2.4 Individual Supplier Rankings 420
               7.14.2.5 Group Supplier Rankings 422
               7.14.2.6 Conclusions from Phase I Results 423
       7.14.3 Risk Adjusted Multi-Criteria Optimization Model
               for Supplier Sourcing (Phase 2) 423
               7.14.3.1 Model Objectives 424
               7.14.3.2 Model Constraints 425
       7.14.4 Solution Methodology 426
               7.14.4.1 Preemptive GP Model 426
               7.14.4.2 Non-Preemptive GP Model 427
               7.14.4.3 Tchebycheff (Min–Max) GP Model 428
               7.14.4.4 Fuzzy GP Model 429
       7.14.5 Data Description 429
               7.14.5.1 MtT Type Risk Calculations 430
               7.14.5.2 VaR Type Risk Calculations 431
       7.14.6 Phase 2 Model Results 432
               7.14.6.1 Preemptive GP Solution 432
               7.14.6.2 Non-Preemptive GP Solution 433
               7.14.6.3 Tchebycheff GP Solution 433
               7.14.6.4 Fuzzy GP Solution 434
       7.14.7 Comparison of Phase 2 Results 434
       7.14.8 Discussion of the Results 436
7.15 Summary and Further Readings 436
       7.15.1 Summary 436
               7.15.1.1 Extensions 437
       7.15.2 Literature on Supply Chain Risk Quantification
               and Management 438
               7.15.2.1 Mathematical Models for Supply
                        Chain Risk Quantification
                        and Management 438
               7.15.2.2 Conceptual Models for Supply Chain Risk
                        Management 439
               7.15.2.3 Surveys and Case Studies on Supply Chain
                        Risk Management 439
Exercises 443
References 445

8. Global Supply Chain Management 449
  8.1   History of Globalization..449
  8.2   Impacts of Globalization 450
       8.2.1   Changes to World Economies 450
       8.2.2   Global Products...451
       8.2.3   Impact of Globalization in U.S. Manufacturing..............453
       8.2.4   Risks in Globalization 454
  8.3   Managing Global Supply Chains 455
       8.3.1   Global Risk Factors 455
       8.3.2   Global Supply Chain Strategies 456
       8.3.3   Examples of Globalization Strategies.................457
  8.4   Global Sourcing...458
       8.4.1   Benefits and Barriers to Global Sourcing 459
            8.4.1.1   Reasons for Global Sourcing 459
            8.4.1.2   Barriers to Global Sourcing 459
       8.4.2   Issues in Global Sourcing 460
            8.4.2.1   Hidden Costs in Global Sourcing 460
       8.4.3   Factors Affecting International Supplier Selection.........461
            8.4.3.1   Financial Issues.....................................461
            8.4.3.2   Logistics Issues 463
            8.4.3.3   Manufacturing Practices 463
            8.4.3.4   Strategic Issues.....................................464
       8.4.4   Tools for Global Sourcing 465
  8.5   International Logistics...466
       8.5.1   Steady Demand 467
       8.5.2   High Demand Variability 467
  8.6   Designing a Resilient Global Supply Chain: A Case Study........467
       8.6.1   Problem Background...468
       8.6.2   Model Features..470
       8.6.3   Decision Criteria and Risk Assessment............470
       8.6.4   Model Results and Managerial Insights............474
            8.6.4.1   Results of Profit Maximization Model 474
            8.6.4.2   Multi-Criteria Analysis 475
  8.7   Summary and Further Readings 477
       8.7.1   Summary..477
       8.7.2   Further Readings 478
  Exercises 479
  References 485

Appendix A: Multiple Criteria Decision Making: An Overview...........489

Index 509

Preface

This book emphasizes a quantitative approach to solving problems related to designing and operating supply chains. Importantly, though, it is not so "micro" in its focus that the perspective on the larger business problems is lost, nor is it so "macro" in its treatment of that business context that it fails to develop students' appreciation for, and skills to solve, the tactical problems that must be addressed in effectively managing flows of goods in supply chains. Economists often speak of the need to understand "first principles" before one can understand and solve larger problems. We share that view, and we have therefore structured the book to provide a grounding in the "first principles" relevant to the broad and challenging problem of managing a supply chain that spans the globe. We feel strongly that students of supply chain engineering are best served by *first* developing a solid understanding of, and a quantitative toolkit for, tactical decision making in areas such as demand forecasting, inventory management, and transportation management—in both an intrafirm and firm-to-firm (dyadic) context—*before* making any attempt to "optimize the supply chain," a task that is clearly much easier said than done, or to optimize large swaths of any given supply chain.

Still, the idea of optimization is indeed prevalent throughout the book. This book is careful and deliberate in its approach to supply chain optimization. Indeed, the perspective taken is one that is well known to engineers of all types, namely, the perspective of *design*. Engineers design things. Some engineers design discrete physical items, and some design collections of items that operate together as systems. Engineers that design supply chains take on the latter challenge. But, in the same way that it is difficult to say that an engineer that designs automotive suspension systems that achieve a particular set of objectives is in some way "optimizing the automobile," it is difficult to say that an engineer who formulates a decision to locate a distribution center in order to achieve a particular set of objectives for the firm that owns and/or manages that distribution center is somehow "optimizing the supply chain." What that engineer is doing, however, is critically important to the function of the portion of the supply chain that is connected to that distribution center.

Thus, a devotion to mathematical precision and optimization is evident throughout the book. Each chapter is presented from this mathematical perspective, and in each chapter, specific mathematical problems are formulated and solved. In addition, in the latter half of the text, specifically in Chapters 6 through 8, we address another important issue in designing supply chains and their supporting systems, namely, the issue of *conflicting criteria*. Indeed, a key issue in designing anything—be it an automotive suspension or a network that connects sources of supply to points of final consumption—is the notion of *trade-offs*. Often, design objectives are in conflict. For example, it is generally not possible to achieve the fastest fulfillment of demand at the lowest transportation cost. This trade-off between speed and cost must be resolved in a way that identifies the best combined outcome, and this is the province of multicriteria decision making (MCDM).

Formally incorporating MCDM in supply chain design and decision making is one of the unique aspects of this book. Therefore, we include an appendix on MCDM that discusses important principles from this area of applied mathematics. This appendix serves as an important resource to Chapters 6 through 8, where we integrate MCDM into the process of designing and managing portions of the supply chain. This fresh perspective, utilizing MCDM in supply chain management and design, is particularly important to our treatment of supplier selection in Chapter 6 and supply chain risk management in Chapter 7.

Other unique aspects of the book are as follows:

- An emphasis on contemporary techniques and a focus on realism in modeling. These are evident, for example, in Chapter 4, where we extensively utilize publicly available data on truck transportation rates in building various examples to illustrate the effects of incorporating transportation cost in inventory decision models.

- Our emphasis on contemporary techniques is also evident in Chapter 5, where we make significant use of the concept of risk pooling in identifying whether more centralized or more decentralized networks are preferred, based on the relevant supply chain costs.

- We devote an entire chapter to managing risks in the supply chain, emphasizing risk quantification models and risk mitigation strategies, and presenting important problems that extend beyond the traditional treatment of supply chain management.

- We include an entire chapter on the effects of globalization on managing supply chains.

The flow of the book proceeds from a basic overview that defines supply chain engineering and establishes the book's emphasis on design, and then presents several topics addressed by nearly all books on supply chain management (forecasting, inventory, transportation, and network design), although in some unique ways, as we discussed earlier. Then, we establish the link to MDCM through a series of chapters that address topics that are not often covered in the level of depth that we devote in this book, namely, supplier selection, supply chain risk management and mitigation, and global supply chain management. Each chapter concludes with a section that presents a collection of further readings, extending from, and beyond, the concepts discussed in the chapter. This is followed by a series of end-of-chapter exercises. Each set of exercises includes 5–6 conceptual questions, 5–6 quantitative problems, and 1–2 "mini case studies." An instructor's manual, with solutions to the quantitative problems and mini case studies, is available for those adopting the book for classroom use.

The book is targeted to serve in the following contexts:

- A textbook for graduate-level and advanced undergraduate-level courses in industrial engineering
- A textbook for, or reference book to support, advanced MBA elective courses in operations management, logistics, management science, or supply chain management that emphasize quantitative analysis
- A reference for technical professionals and researchers in industrial engineering, operations management, logistics, and supply chain management

This book grew out of two sources. One is a graduate course in supply chain engineering taught in the industrial and manufacturing engineering department at Pennsylvania State University since the fall 2002 semester. The other is a comprehensive chapter on supply chain management written for the *Operations Research and Management Science Handbook* (2008, CRC Press), which itself was based on materials developed for graduate courses, one in supply chain management, taught at the Smeal College of Business at Penn State, and one in logistics management, taught at the Poole College of Management at North Carolina State University. The book was the result of a realization by the authors—one of whom wrote the comprehensive chapter while the other edited the volume in which it appears—that there was clear value in combining these two pools of content and organizing them into a targeted textbook that uses the precise tools of engineering analysis to address broad and challenging problems in supply chain management. The result, we believe, fills a gap that has resulted from various textbooks on these topics focusing only on one or the other of these perspectives.

Thus, the book is organized to present each of the elemental problems undergirding supply chain management, building up the reader's content knowledge before finally tackling broad issues related to managing across company boundaries and country boundaries. This approach has been influenced by other textbooks utilized in our teaching. Ultimately, though, we found that the books that are best at framing the important strategic issues in supply chain management fail to sufficiently build the kinds of skills in modeling and analysis that we believe are critical for effective tactical decision making, while the books that are best at presenting quantitative models for tactical decision making generally fail to place those modeling efforts in a larger context that aids in students' understanding of the important strategic issues in supply chain management. This book blends the best of those two perspectives, "bookending" the text, as it were, with an introductory discussion that lays out the strategic framework for effective design of the supply chain and its supporting policies, then studying the elemental problems one by one, and finally pulling this content knowledge together in the context of managing the global supply chain. The result is what we believe to be a comprehensive treatment of the subject that we hope will serve many students and practitioners of the science of designing effective supply chains for many years to come.

A. Ravi Ravindran
University Park, Pennsylvania

Donald P. Warsing, Jr.
Raleigh, North Carolina

Acknowledgments

First and foremost, we express our sincere appreciation to Madhana Raghavan, an industrial engineering doctoral student at Penn State University, for his outstanding help with the preparation of this book, typing several chapters and reviewing the examples used. We also acknowledge the valuable contributions of Subramanian Pazhani, another doctoral student, who joined the book project recently.

We thank our former PhD students—Ufuk Bilsel, Rodolfo Portillo, Vijay Wadhwa, and Tao Yang—for their valuable contributions to the material presented in Chapters 6 through 8. Their reviews of the chapter drafts are also much appreciated. In addition, several former and current industrial engineering graduate students at Penn State helped us by reviewing several chapters and providing valuable comments that improved the presentation. In particular, we wish to acknowledge the reviews provided by Pastor Casanova, Aixa Cintron, Nok Kungwalsong, Abraham Mendoza, Ajay Natarajan, Richard Titus, Aineth Torres, and Victor Valdebenito. Special thanks go to P. Balasubramanian of Theme Work Analytics for his careful review and helpful comments on several chapters. We also acknowledge the early typing assistance provided by Sharon Frazier.

Colleagues at North Carolina State provided significant and valuable support and insights on the material in Chapters 3 and 4. In particular, we thank Reha Uzsoy for his insights on multi-echelon base-stock models and Michael Kay for providing a fresh perspective on data sources and modeling for freight transportation. Also, we would be remiss if we did not recognize the work of Farhad Azadivar and Atul Rangarajan, whose chapter on inventory control in the *Operations Research and Management Science Handbook* (CRC Press, 2008) provided a solid basis upon which to build our discussion of the multi-echelon economic order quantity models. Of course, any errors of commission or omission in assimilating these insights into our work are ours alone.

We thank Cindy Renee Carelli, senior acquisitions editor at CRC Press, for her constant support and encouragement from inception to completion of this book. Finally, we thank our families for their support, love, understanding, and encouragement, when we were focused completely on writing this book.

Authors

A. Ravi Ravindran is a professor and former department head of industrial and manufacturing engineering at the Pennsylvania State University since 1997. Formerly, he was a faculty member in the School of Industrial Engineering at Purdue University for 13 years (1969–1982) and at the University of Oklahoma for 15 years (1982–1997). At the University of Oklahoma, he served as the director of the School of Industrial Engineering for 8 years and as the associate provost of the university for 7 years with responsibility for budget, personnel, and space for the academic area. He received his BS in electrical engineering with honors from BITS Pilani, India, and his MS and PhD in industrial engineering and operations research from the University of California, Berkeley.

Dr. Ravindran's area of specialization is operations research, with research interests in multiple criteria decision-making, financial engineering, healthcare delivery systems, and supply chain optimization. He has published five books (*Operations Research: Principles and Practice, Engineering Optimization: Methods and Applications, Handbook of Operations Research and Management Science, Operations Research Methodologies,* and *Operations Research Applications*) and over 100 journal articles in operations research. He is a Fulbright Fellow and a fellow of the Institute of Industrial Engineers (IIE). In 2001, he was recognized by IIE with the Albert G. Holzman Distinguished Educator Award for significant contributions to the industrial engineering profession by an educator. He has been a consultant to AT&T, General Motors, IBM, Kimberly Clark, General Electric, U.S. Department of Transportation, the Cellular Telecommunications Industry Association, and the U.S. Air Force. He currently serves as the operations research series editor for Taylor & Francis Group/CRC Press.

Donald P. Warsing, Jr. is an associate professor of operations and supply chain management at the Poole College of Management at North Carolina State University (NC State). Prior to joining the faculty at NC State, Dr. Warsing served on the faculty of the Smeal College of Business at Pennsylvania State University and also worked for several years at IBM Corporation in roles spanning from industrial engineering to manufacturing management. He received his PhD in operations management from the Kenan-Flagler Business School at the University of North Carolina-Chapel Hill, a master of science in management with a concentration in industrial engineering from North Carolina State University, and a bachelor of science in industrial and systems engineering from Ohio State University.

Dr. Warsing's research concerns the development of tools and policies for effectively managing inventory, logistics, and business operations and studying the ways in which various management practices contribute to improved performance outcomes. His work has been published in *Production and Operations Management*, the *Journal of Operations Management*, *Decision Sciences*, the *European Journal of Operational Research*, and the *International Journal of Logistics*. He is also the author of several book chapters on logistics and supply chain management, including one that appears in both the *Operations Research and Management Science Handbook* (CRC Press, 2008) and *Operations Research Applications* (CRC Press, 2009).

1

Introduction to Supply Chain Engineering

At its heart, this is a book about design. This stands to reason since a prominent word in its title is "engineering." Design is the province of much of engineering. While some engineers design physical products (e.g., computers, automobiles, bridges), others design systems. Having said that, we must briefly clarify the concept of a system. While the prevailing notion of a system is that it is "more than the sum of its parts," we can perhaps be a little clearer than that. One of the most concise and useful definitions of a system, in our minds, was offered some years ago by the Nobel-prize-winning economist Herb Simon, who stated that a system is comprised of a "number of parts that interact in a non-simple way." The "nonsimple-ness" (i.e., complexity) of the interaction of the parts is the hallmark of a system, and what leads to the "more than the sum of its parts" notion of how a system operates (Simon, 1962).

Indeed, the design focus of some engineers goes beyond individual, discrete products to deal with systems, which are, collections of discrete entities that interact—often in "non-simple" ways—to produce a desired outcome. In some cases, those systems are comprised of entities that are physical and tangible, like the heating and air conditioning system designed by a mechanical engineer. In some cases, some aspects of the systems being designed are more conceptual in nature. Industrial engineers (IEs) design such systems. IE-related systems are typically comprised of a mix of tangible and intangible components. For example, production systems take physical inputs (e.g., materials and labor) and conceptual inputs (e.g., projected consumer demand, short-term and long-term business plans) to achieve an output of physical products that ultimately are sold to satisfy customers. Moreover, distribution systems take as their input the products generated by the production system and, along with labor and other plans related to business goals and customer demand, coordinate the movement, storage, and transport of those products to ultimately satisfy that demand, hopefully on-time and in the right quantity. Our focus in this book is on the design of the *supply chain system*, which involves connecting many such production and distribution systems, often across wide geographic distances, in such a way that the businesses involved can ultimately satisfy consumer demand as efficiently as possible, resulting in maximum financial returns to those businesses connected to that supply chain system.

Having established that this book is about design, let us be clear about another important issue, specifically that design always involves *tradeoffs*.

Indeed, in designing any product, or any system, you cannot get everything* (e.g., the highest customer service at the lowest cost). Therefore, another important aspect of this book is that, in presenting and developing the mathematical models that form the toolkit for supply chain design, we will continually refer back to the nature of the tradeoff(s) involved in the decision at hand. In some cases, we will formalize this decision tradeoff through the use of *multi-criteria decision making* (MCDM). This is a unique aspect of this book, something that distinguishes it and makes it a valuable resource in building and extending the understanding of this topic that touches so many individuals in so many ways around the globe. Where appropriate, we return to the MCDM framework, and at various points in the text, utilize the tools of MCDM to build decision models that explicitly account for the tradeoffs between competing objectives. An introduction to MCDM is given in Appendix A.

Given this broad background on the subject of the book, let us now spend some time carefully setting the stage for the various topics that will be covered in the chapters that follow. In this chapter, we begin with the meaning of supply chain engineering and describe, at a high level, the types of decisions made in managing a supply chain. We introduce a variety of supply chain performance measures and show how they relate to a company's financial measures. We then discuss the importance of supply chain engineering and the distinguishing characteristics of those firms recognized as leaders in supply chain management. We end with a chapter-by-chapter overview of the textbook.

1.1 Understanding Supply Chains

Before we formally define *supply chain engineering*, we begin with the definition of what a *supply chain* is.

A supply chain consists of the following:

1. A series of *stages* (e.g., suppliers, manufacturers, distributors, retailers, and customers) that are physically distinct and geographically separated at which inventory is either stored or converted in form and/or in value.

2. A coordinated set of *activities* concerned with the procurement of raw materials, production of intermediate and finished products, and the distribution of these products to customers within and external to the chain.

---

* At the hearing of which, fans of rock music perhaps immediately begin to hum the Rolling Stones' "You Can't Always Get What you Want," which was written, interestingly, by another famous individual who studied economics—but was never awarded a Nobel Prize—Mick Jagger (London School of Economics—left in 1963 to pursue other, more lucrative interests).

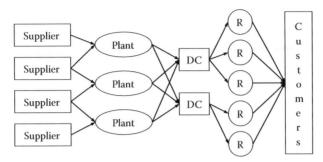

(DC = Distribution center, R = Retailer)

FIGURE 1.1
Typical supply chain network.

Thus, a supply chain includes all the partners involved in fulfilling customer demands and all the activities performed in fulfilling those demands. Figure 1.1 illustrates a typical supply chain.

It is important to recognize that the different stages of the supply chain (suppliers, plants, distribution centers (DCs), and retailers) may be located in different countries for a multi-national company with a global supply chain network. It is also possible that a firm may employ fewer supply chain stages than those represented in Figure 1.1, or perhaps more. Indeed, some business researchers (e.g., Fine, 2000) argue that the supply chains in various industries follow historical cycles that move from periods of significant vertical integration to periods of significantly less integration, where firms in the supply chain rely more on partnerships than on owning substantial portions of the value chain within a single firm. In vertical integration periods, the supply chain may employ only a few stages from raw material extraction to final production, owned primarily or exclusively by a single firm. A good example of this would be the early days of Ford Motor Company in the 1920s. In the less integrated periods, it is the horizontal, across-firm relationships that are prominent. A good example would be in the 1990s and early 2000s, when Dell led the global market for personal computers with a highly decentralized supply chain in which they served only as the final assembler and direct distributor. In this latter case, Dell not only relied heavily on its suppliers to independently manage the production and supply of components, but also simply bypassed independent distributors and retailers and dealt with the final consumers directly, without the "middle men." Interestingly, Fine's (2000) hypothesis regarding cycles of change in various industries may be coming to light in the PC industry, since Dell has recently added the retail "middle men" back to its supply chain.

1.1.1 Flows in Supply Chains

Following Chopra and Meindl (2001), the key flows in a supply chain are the following:

- *Products*: Includes raw materials, work-in-progress (WIP), sub-assemblies, and finished goods.
- *Funds*: Includes invoices, payments, and credits.
- *Information*: Includes orders, deliveries, marketing promotions, plant capacities, and inventory levels.

Thus, the flows in the supply chain are not just "goods." Tracking flows from the suppliers to the customers is called *"moving downstream"* in the supply chain. Tracking flows from the customers to the suppliers is called *"moving upstream"* in the supply chain.

1.2 Meaning of Supply Chain Engineering

Supply chain engineering encompasses the following key activities for the effective management of a supply chain:

1. Design of the supply chain network, namely the location of plants, DCs, warehouses, etc.
2. Procurement of raw materials and parts from suppliers to the manufacturing plants
3. Management of the production and inventory of finished goods to meet customer demands
4. Management of the transportation and logistics network to deliver the final products to the warehouses and retailers
5. Management of the integrity of the supply chain network by mitigating supply chain disruptions at all levels

Most of these activities involved in supply chain engineering also come under the rubric of *supply chain management* (SCM). Chopra and Meindl (2001) define supply chain management (SCM) as "the management of flows between and among supply chain stages to maximize supply chain profitability." A more complete definition of SCM by Simchi-Levi et al. (2003) states, "SCM is a set of approaches utilized to efficiently integrate suppliers, manufacturers, warehouses and stores, so that merchandise is produced and distributed at the right quantities to the right locations, and at the right time in order to minimize system-wide costs while satisfying service level requirements." Their definition brings out all the key aspects of SCM including the two conflicting objectives in SCM—minimizing supply chain costs while simultaneously maximizing customer service.

An important distinction, however, between Supply Chain Engineering (SCE) and Supply Chain Management (SCM) is the emphasis in SCE on the design of the supply chain network and the use of mathematical models and methods to determine the optimal strategies for managing the supply chain.

This emphasis on mathematical models makes sense if one returns to the introductory comments of the chapter, where we point out that our emphasis in this book is really on design, namely on the design of a relatively complex system of interconnections and flows that move both goods and information across the globe. In order to accomplish that in a systematic way, we must turn to some reasonably precise and verifiable tools of analysis. Hence, our emphasis on design—ultimately what we hope to characterize as *optimal* design—requires a hand-in-hand emphasis on mathematical models.

Without doubt, commerce has become increasingly global in scope over the past several decades. This trend toward *globalization* has resulted in supply chains whose footprint is often huge, spanning multiple countries and continents. Since products and funds now regularly flow across international boundaries, engineering a global supply chain becomes impractical—or at least ill-advised—without the use of sophisticated mathematical models. The use of the methods of operations research and applied mathematics for supply chain engineering decisions will be the primary focus of this textbook.

1.3 Supply Chain Decisions

The various decisions in supply chain engineering can be broadly grouped into three types—*strategic*, *tactical*, and *operational*.

1.3.1 Strategic Decisions

Strategic decisions deal primarily with the design of the supply chain network and the selection of partners. These decisions are not only made over a relatively long time period (usually spanning several years), and have greater impact in terms of the company's resources, but are also subject to significant uncertainty in the operating environment over this lengthy span of time. Examples of strategic decisions include the following:

1. *Network Design*: where to locate and at what capacity?
   - Number and location of plants and warehouses
   - Plant and warehouse capacity levels

2. *Production and Sourcing*: make or buy?
    - Produce internally or outsource
    - Choice of suppliers, sub-contractors, and other partners
3. *Information Technology*: how to coordinate the chain?
    - Develop software internally or purchase commercially available packages—e.g., SAP, Oracle.

1.3.2 Tactical Decisions

Tactical decisions are primarily *supply chain planning decisions* and are made in a time horizon of moderate length, generally as monthly or quarterly decisions, covering a planning horizon of one or two years. Thus, these decisions are typically made in an environment characterized by less uncertainty relative to strategic decisions, but where the effects of uncertainty still are not inconsequential. Examples of tactical decisions include the following:

- Purchasing decisions—e.g., how much to buy and when?
- Production planning decisions—e.g., how much to produce and when?
- Inventory management decisions—e.g., how much and when to hold to balance the costs of resupply with the risks of shortages?
- Transportation decisions—e.g., which modes to choose and how frequently to ship on them?
- Distribution decisions—e.g., how to coordinate DC replenishment with production schedules?

1.3.3 Operational Decisions

Operational Decisions are short-term decisions made on a daily/weekly basis, at which point much of the operational uncertainty that existed when the strategic and tactical decisions were made has been resolved. In addition, since the time scale is so short, most of these decisions involve a significantly lower expenditure of funds. Examples include the following:

- Setting delivery schedules for shipments from suppliers
- Setting due dates for customer orders
- Generating weekly or daily production schedules
- Allocating limited supply (e.g., between backorders and new customer demand)

It is important to recognize that the three types of supply chain decisions—strategic, tactical, and operational—are interrelated. For example, the number and locations of plants affect the choice of suppliers and modes to transport raw materials. Moreover, the number and locations of plants and warehouses also affect the inventory levels required at the warehouses and the delivery times of products to customers. Aggregate production planning decisions affect product movement decisions and customer fulfillment.

1.4 Enablers and Drivers of Supply Chain Performance

1.4.1 Supply Chain Enablers

Enablers make things happen and, in the case of supply chain management, are considered essential for a supply chain to perform effectively. Without the necessary enablers, the supply chain will not function smoothly. Simply having the necessary enablers, however, does not guarantee a successful supply chain performance.

Based on a survey of supply chain managers, Marien (2000) describes four enablers of effective management of the supply chain. In order of their ranked importance by the survey respondents, they are *organizational infrastructure, information technology, strategic alliances,* and *human resource management.*

- *Organizational infrastructure*: The essential issue in this case is, whether supply chain management activities internal to the firm, and across firms in the supply chain, are organized in more of a vertical orientation, or with greater decentralization. The fact that this enabler ranked first among practicing supply chain managers is clearly consistent with much that has been written in the trade press regarding the fact that intra-firm SCM processes must be in place and operating effectively before there is any hope that inter-firm collaboration on supply chain management activities is to be successful.

- *Technology*: Two types of technology are critical to success in designing and managing supply chains effectively, information technology and manufacturing technology. Although the emphasis of many consultants and software providers is often on information technology, product design can often have more impact on whether supply chain efficiencies can ultimately be achieved. Prominently, product design should account for manufacturability—for example, using modular components that utilize common interfaces with multiple final products—and allow for efficiencies in managing inventories and distribution processes.

- *Alliances*: The effectiveness of alliances is particularly important in supply chains that are more decentralized, wherein more authority is given to suppliers. In some cases, these suppliers may take on roles that go beyond just supplying components, becoming instead outsourcing partners who assume significant responsibility for product design and manufacturing.

- *Human Resources*: Two categories of employees are critical to effective supply chain management. First, technical employees assume an important role in designing networks that minimize costs while simultaneously achieving high levels of customer service performance. These employees must have a solid understanding of the types of mathematical tools we discuss in this book. Second, managerial employees must similarly have a solid conceptual grasp on the key issues addressed by the models and tools of the technical staff, and must clearly understand how such tools can be applied to ultimately allow the firm to achieve its strategic goals.

1.4.2 Supply Chain Drivers

Supply chain drivers represent the critical areas of decision making in SCE, those that ultimately generate the outcomes that impact the supply chain performance. Thus, they appear as the decision/design variables in the optimization models used in SCE decision making. The key drivers of supply chain performance are described next, and the chapters that follow are organized along the key drivers of supply chain performance.

1.4.2.1 Inventory

Companies maintain inventory of raw materials, WIP, and finished goods to protect against unpredictable demand and unreliable supply. Inventory is considered an *idle asset* to the company and is one of the major portions of supply chain costs. Maintaining large inventories increases supply chain costs but provides a higher level of customer service.

The key decision variables here are *what items to hold in inventory*, (raw materials, WIP, and finished products), *how much and when to order* (inventory policies), and *where to hold inventory* (locations). Chapter 3 discusses these issues in detail.

1.4.2.2 Transportation

Transportation is concerned with the movement of items between the supply chain stages—suppliers, plants, DCs, retailers. Use of faster transportation modes such as air and roadways incurs higher transportation costs but reduces delivery times and increases reliability. The key decision variables here are as follows:

1. Whether to outsource transportation decision making and execution to a *third party logistics* (3PL) provider
2. What transportation mode(s) to use (air, sea, road, etc.) for what items (raw materials, WIP, and finished goods)
3. Distribution options for finished goods (either shipped to customers directly or through intermediate distribution centers)

Chapter 4 discusses the transportation decisions in detail, and Chapter 5 discusses issues regarding distribution strategy and network design.

1.4.2.3 Facilities

Facilities (plants and distribution centers) play a key role in supply chain engineering. They are generally considered as strategic decisions and directly affect the performance of the supply chain. The key decision variables under facilities are the following:

1. Number of plants and their locations
2. Plant capacities and product mix allocated to plants
3. Number of DCs and their locations
4. Distribution strategies

Chapter 5 discusses the use of integer programming models to make optimal decisions regarding the number and locations of facilities.

1.4.2.4 Suppliers

Raw material cost accounts for 40%–60% of the production cost in most manufacturing industries. In fact, for the automotive industry, the cost of components and parts from outside suppliers may exceed 50% of sales (Wadhwa and Ravindran, 2007). For technology firms, component cost as a fraction of sales could be as high as 80%. Hence, the selection of suppliers for raw materials and intermediate components is considered a critical area of strategic decision making in supply chain engineering. Chapter 6 discusses supplier selection models under multiple conflicting criteria.

1.5 Assessing and Managing Supply Chain Performance

The idea that there are key drivers of supply chain performance is useful in thinking about another theme emphasized by many authors and first proposed by Fisher (1997) in an important article that advanced the notion that "one size

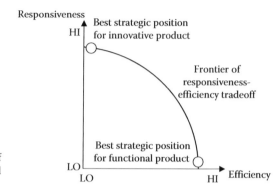

FIGURE 1.2
Responsiveness-efficiency tradeoff
frontier (based on Fisher, 1997 and
Chopra and Meindl, 2001).

fits all" is not an effective approach to managing supply chains. Fisher (1997) cogently lays out a matrix that matches product characteristics (what he describes as a dichotomy between *innovative* products like technology-based products and *functional* products like toothpaste or other staple goods) and supply chain characteristics—another dichotomy between *efficient* (cost-focused) supply chains and *responsive* (customer service-focused) supply chains. Chopra and Meindl (2001) take this conceptual model a step further, first by pointing out that Fisher's product characteristics and supply chain strategies are really continuous spectrums, and then by superimposing the Fisher model, as it were, on a frontier that represents the natural tradeoff between responsiveness and efficiency. Clearly, it stands to reason that a firm, or a supply chain, cannot maximize cost efficiency and customer responsiveness simultaneously. Some aspects of each of these objectives necessarily work at cross purposes. A combined version of Chopra and Meindl's frontier and Fisher's product dichotomy is presented in Figure 1.2.

The value of this perspective is that it clearly identifies a market-driven basis for strategic choices regarding the supply chain drivers: Should our inventory management decisions be focused more on efficiency (e.g., minimizing inventory levels) or on responsiveness (e.g., maximizing product availability)? Should our transportation choices be focused more on efficiency (e.g., minimizing transportation costs, perhaps through more extensive economies of scale) or on responsiveness (e.g., minimizing delivery lead times and maximizing reliability)? Should our facilities (network design) decisions be focused more on efficiency (e.g., minimizing the number of locations and maximizing their size and scale) or on responsiveness (e.g., seeking high levels of customer service by choosing many, focused locations closer to customers)?

Next, we discuss efficiency and responsiveness in more detail, and we also introduce *supply chain risk* as an additional criterion to consider in designing the supply chain network and its associated operating policies.

1.5.1 Supply Chain Efficiency

Generally, efficiency is measured by a ratio of the level of output generated to the level of input consumed to generate that output. This concept can be applied to both physical systems—e.g., an automobile engine that converts the energy stored in the fuel consumed by the engine into horsepower generated by the engine to drive the wheels of the vehicle—and to businesses—e.g., the conversion of dollar-valued inputs (labor, materials, and the costs of owning and/or operating physical assets like plants and warehouses) into sales revenue. Therefore, the efficiency of a given supply chain focuses on how well resources are utilized across the chain in fulfilling customer demand.

In the conceptual framework of Fisher (1997), discussed earlier, efficient supply chains are more focused on cost minimization, the idea being that a supply chain that requires less cost input to generate the same amount of sales revenue output is more efficient. Therefore, efficiency measures in SCE are often focused on costs, and include the following:

- Raw materials cost
- Manufacturing cost
- Distribution cost
- Inventory holding cost
- Facility operating costs
- Freight transportation costs
- Shortage costs

In addition, other measures that may influence the costs listed above include the following:

- *Product Cycle Time*: This is the time that elapses from the start of production of the item up to its conversion into a product that can be shipped to the customer. Clearly, longer cycle times can result in larger costs (e.g., labor costs and/or inventory holding costs).
- *Inventory Levels*: Again, higher levels of inventory can result in a number of associated costs, beyond just the cost of tying up the firm's cash in currently idle assets. Higher inventory levels generate greater needs for storage space and for labor hours and/or employee levels in order to manage these inventories as they reside in and flow between storage facilities. Later in this chapter, we will present two important measures of inventory levels, namely *inventory turns* and *days of inventory*.

Typically, supply chain optimization models focus on minimizing costs, since the decisions of supply chain managers often involve choices that directly influence costs, while revenue may often be outside the scope of the supply chain manager's decisions. Some SCE models, however, may appropriately involve maximizing profit, to the extent that it is clear that the decision at hand has both cost and revenue implications.

1.5.2 Supply Chain Responsiveness

Responsiveness refers to the extent to which customer needs and expectations are met, and also the extent to which the supply chain can flexibly accommodate changes in these needs and expectations. Thus, in the efficiency–responsiveness tradeoff introduced by the Fisher (1997) framework discussed earlier, firms whose supply chains are focused on responsiveness are willing to accept higher levels of cost (i.e., lower cost efficiency) in order to improve their ability to meet and flexibly accommodate customer requirements (i.e., higher responsiveness). Common measures of responsiveness are the following:

- Reliability and accuracy of fulfilling customer orders
- Delivery time
- Product variety
- Time to process special or unique customer requests (customization)
- Percent of customer demand filled from finished goods inventory versus built to order from raw materials or component inventories

1.5.3 Supply Chain Risk

A third supply chain criterion has gained attention in recent years. The September 11 terrorist attacks in the United States in 2001 obviously had broad and lasting impacts on society in general. From the standpoint of managing supply chains, the disruption in material flows over the days and weeks after September 11 caused companies to realize that a singular emphasis on the cost efficiency of the supply chain can actually make the chain brittle and much more susceptible to the risk of disruptions. This includes not only catastrophic disruptions like large-scale terrorist attacks, but even mundane, commonly occurring disruptions like a labor strike at a supplier. Thus, effective supply chain management no longer just involves moving products efficiently along the supply chain, but it also includes mitigating risks along the way. Supply chain risk can be broadly classified into two types as follows:

1. *Hazard Risks*: These are disruptions to the supply chain that arise from large-scale events with broad geographic impacts, such as natural disasters (e.g., hurricanes, floods, blizzards), terrorist attacks, and major political actions like wars or border closings.

2. *Operational Risks*: These are more commonly occurring disruptions whose impacts are localized (e.g., affecting only a single supplier) and resolved over a relatively short period of time. Examples include information technology disruptions (e.g., a server crash due to a computer virus infection), supplier quality problems, and temporary logistics failures (e.g., temporarily "lost" shipments).

Chapter 7 of the textbook is devoted entirely to supply chain risk management. Multi-criteria optimization models that consider profitability, customer responsiveness, and supply chain risk are discussed in this context.

1.5.4 Conflicting Criteria in Supply Chain Optimization

It is important to recognize that efficiency and responsiveness are conflicting criteria in managing supply chains. For example, customer responsiveness can be increased by having a larger inventory of several different products, but this increases inventory costs and thereby reduces efficiency. Similarly, using fewer distribution centers reduces facility costs and, as we show in later chapters, can also reduce inventory levels across the network through "pooling" effects. The downside, however, is that such a network design increases delivery time and thereby reduces responsiveness, and it also increases supply chain risk by concentrating the risk of distribution failure in fewer facilities. Thus, supply chain optimization problems are generally multiple criteria optimization models. Where appropriate in the chapters that follow, we use multiple criteria models to determine optimal solutions. An introduction to multiple criteria decision making (MCDM) models is given in Appendix A.

1.6 Relationship between Supply Chain Metrics and Financial Metrics

We will demonstrate in this section that improvement in some selected supply chain metrics also results in improvements in some important financial metrics of the firm, which should, of course, be closely correlated with its overall business performance. To illustrate this relationship, let us consider several interrelated inventory measures—*inventory turns, days of inventory,* and *inventory capital*—and how they affect some important financial measures—*return on assets, working capital,* and *cash-to-cash cycle*.

1.6.1 Inventory Measures

1.6.1.1 Inventory Turns

Inventory turns is a measure of how quickly inventory is turned over from production to sales, specifically

$$\text{Inventory turns} = \frac{\text{Annual sales}}{\text{Average inventory}} \tag{1.1}$$

For example, if the annual sales is 1200 units and the average inventory is 100 units, then the inventory turns would be $1200/100 = 12$. In other words, on the average, goods are stored in inventory for one month before they are sold. Companies prefer a higher value for inventory turns so that products reach the end customer as soon as possible after production, and, from a financial standpoint, so that the funds tied up in inventory can be freed up and converted into cash (or accounts receivable) more quickly. For many years, Dell's inventory turns exceeded 100, indicating a quick turnover of computers after assembly.

1.6.1.2 Days of Inventory

Days of inventory refers to how many days of customer demand is carried in inventory, specifically

$$\text{Days of inventory} = \frac{\text{Average inventory}}{\text{Daily sales}} \tag{1.2}$$

Using Equations 1.1 and 1.2, *days of inventory* can be written as follows:

$$\text{Days of inventory} = \frac{\text{Average inventory}}{(\text{Annual sales}/365)}$$

$$= \frac{365}{(\text{Annual sales}/\text{Average inventory})}$$

$$= \frac{365}{\text{Inventory turns}} \tag{1.3}$$

Thus for Dell, with inventory turns equal to 100, the days of inventory will be $365/100 = 3.65$ days. In other words, Dell carries less than 4 days of inventory, on average. However, Dell requires its suppliers to carry 10 days of inventory (Dell, 2004).

A Harvard Business School study in 2007 reported that the consumer goods industry carries, on average, 11 weeks (77 days) of inventory and retailers carry 7 weeks (49 days) of inventory. In spite of that, the stock-out rate in the retail industry averages nearly 10%! Note that, increasing the days of inventory is one means of attempting to increase customer responsiveness, but this increase would come at the expense of additional supply chain cost.

1.6.1.3 Inventory Capital

Inventory capital refers to the total investment in inventory, specifically

$$\text{Inventory capital} = \sum_{k=1}^{N} I_k V_k \tag{1.4}$$

where
I_k is the Average inventory of item k
V_k is the value of item k per unit
N is the total number of items held in inventory

Using Equations 1.1 through 1.4, we can state that *increasing the inventory turns* will impact the other inventory measures as follows (assuming the same annual demand):

- Average inventory decreases
- Days of inventory decreases
- Inventory capital decreases

1.6.2 Business Financial Measures

1.6.2.1 Return on Assets

Return on assets refers to the ratio of company's net income to its total assets, computed as

$$\text{Return on Assets (ROA)} = \frac{\text{Annual income (\$)}}{\text{Total assets (\$)}} \qquad (1.5)$$

ROA provides a general proxy for the overall operational efficiency of a company—that is its ability to utilize assets (input) to generate profits (output). Since inventory capital is included in the total assets, reducing inventory capital may increase ROA for the same annual income, as long as it decreases assets in total (see "Working Capital" below for details).

1.6.2.2 Working Capital

Working capital is the difference between a company's short-term assets (e.g., cash, inventories, accounts receivable) and its short-term liabilities (e.g., accounts payable, interest payments, and short-term debt). Thus, working capital—particularly the portion reflected by cash—represents the amount of flexible funds available to the company to invest in R&D and other projects. Increasing inventory turns, therefore, can shift working capital toward cash, thereby freeing up funds for immediate use in profit-generating activities and projects.

1.6.2.3 Cash-to-Cash Cycle

Cash-to-cash cycle refers to the difference in the length of time it takes for a company's *accounts receivable* to be converted into cash inflows and the length of time it takes for the company's *accounts payable* to be converted into cash outflows. Historically, it was often the case that a company paid for its raw materials, production and distribution of its products on a cycle that was shorter than the cycle by which it received payments from its customers after sales. Thus, for most companies, the cash-to-cash cycle has historically been positive. Companies would, however, like to decrease their cash-to-cash cycle in order to improve their profitability. For example, when a company increases its inventory turns, the finished goods reach the end customer sooner. Thus, the company converts purchased components into receivables more quickly, and the cash-to-cash cycle decreases.

In summary, there is a direct and significant relationship between supply chain inventory metrics and a company's financial metrics. An *increase in inventory turns* has a cascading effect on the other inventory and financial measures as follows:

- Days of inventory decreases.
- Inventory capital decreases.
- ROA may increase.*
- Working capital may increase.*
- Cash-to-cash cycle decreases.

1.7 Importance of Supply Chain Management

As shown in Section 1.6, supply chain management can have a significant impact on business performance. Based on a 2003 Accenture study, done in conjunction with Stanford University, Mulani (2005) reported the following:

- Nearly 90% of the companies surveyed said that supply chain management is critical or very important.
- 51% said that the importance of supply chain management had increased significantly in the 5 years leading up to the survey.
- Supply chain management accounted for nearly 70% of the companies' operating costs and comprised at least half of all the typical company's assets.

Mulani (2005) also reported that a significant percentage of promised synergies for many company mergers and acquisitions came from supply chain management. For example, during the HP/Compaq merger, it was estimated that the merger would result in a savings of $2.5 billion, of which $1.8 billion would be due to supply chain efficiency.

Moreover, failure to excel in supply chain management can negatively affect a company's stock price. Hendricks and Singhal (2005) found this to be true in a study of 885 supply chain disruptions reported by publicly traded companies during 1989–2000. The list of companies included small, medium, and large companies with respect to market capitalization and covered both manufacturing and IT industries. Hendricks and Singhal

---

* Recall that while inventory is part of working capital, so is cash. Thus, if an increase in inventory turns reduces inventory capital by shifting it to cash, there is no net change in assets or working capital.

found that companies suffering a supply chain disruption experienced the following effects, on average:

- A loss of over $250 million in shareholder value per disruption
- 10% reduction in stock price
- 92% reduction in return on assets
- 7% lower sales
- 11% increase in cost of doing business
- 14% increase in inventory

Supply chain management has become sufficiently important to business performance to warrant a mantra, of sorts, namely that "companies do not compete with each other, but their supply chains do."* While some might debate whether entire supply chains could literally compete with each other, there is no doubt that efficient management of the supply chain has become a competitive differentiator for many companies. In the next section, we discuss the top 25 supply chains and what supply chain characteristics make them industry leaders.

1.7.1 Supply Chain Top 25

Since 2004, AMR Research (now owned by Gartner Group) has annually ranked the "Supply Chain Top 25"—i.e., "the companies that best exemplify the demand-driven ideal for today's supply chain" (Gartner Group, 2012). The rankings are based broadly on both qualitative and quantitative measures. Half of the ranking weight is based on the company's financial measures, specifically return on assets (25%), inventory turns (15%), and revenue growth (10%). The financial data is obtained from publicly available company information. The other half of the ranking weight is based on voting by 32 AMR analysts (25%) and a peer panel of 156 senior-level supply chain executives (25%). The top 10 companies from the 2011 rankings are listed in Table 1.1.

A complete list of the Supply Chain Top 25 for 2011 and the data used in their rankings by AMR research are available in the Gartner report (Hofman et al., 2011). Based on the study of the top 25 companies over the years, AMR research reports that the supply chain leaders as a group exhibit the following attributes compared to their competitors:

- Carry 15% less inventory
- Are 60% faster-to-market
- Complete 17% more "perfect" orders
- Have 35% shorter cash-to-cash cycles
- Have 5% higher profit margins

---

* Indeed, this mantra can be found embedded in the website of the well-known logistics academic Martin Christopher (http://www.martin-christopher.info/).

TABLE 1.1

Supply Chain Top 10 for 2011

Rank	Company
1	Apple
2	Dell
3	Procter & Gamble
4	Research In Motion
5	Amazon
6	Cisco Systems
7	Wal-Mart
8	McDonald's
9	Pepsi
10	Samsung

In addition, these top 25 companies have also outperformed the S&P 500 in terms of average stock price growth. They have agile supply chain networks that can respond quickly to changes in customer demand and supply chain network disruptions.

It is important to recognize that a company is unlikely to be successful without leadership in supply chain management. However, market factors that are outside the scope of what can be achieved by effective SCM can adversely affect a company's viability. For example, although Research In Motion (RIM) appears at number 4 on this list in Table 1.1, Apple's iPhone has in recent years dramatically captured a huge chunk of the smartphone market that had previously been dominated by RIM's Blackberry.

Nonetheless, the goal of AMR's rankings of the Supply Chain Top 25 is to raise the importance of supply chain management and its potential to positively impact companies' profitability. By identifying the leaders and their best practices, the hope is that other companies can learn from them and improve their supply chain performance.

1.8 Organization of the Textbook

This book is primarily addressed to those who are interested in learning how to apply operations research (OR) models to supply chain management problems. The book assumes that the reader has a basic understanding of the elementary OR methods such as linear programming, nonlinear optimization, and applied probability. To build the reader's knowledge in supply chain engineering, the book covers first the traditional issues in operations, logistics, and supply chain management—demand forecasting and aggregate planning (Chapter 2), managing inventories (Chapter 3), managing transportation (Chapter 4), and locating facilities (Chapter 5). The book also includes a number of contemporary approaches, such as risk pooling, multi-criteria ranking methods and goal programming for addressing supplier selection problems (Chapter 6), managing risks in supply chains (Chapter 7), and global supply chain management (Chapter 8).

A chapter-by-chapter overview of the contents of the book is given in the following sections.

1.8.1 Chapter 2 (Planning Production in Supply Chains)

The success of a manufacturing supply chain depends on its ability to accurately forecast customer demands and produce in time to meet those demands. This chapter is divided into two parts. Part I (Demand Forecasting) discusses the commonly used forecasting methods (both qualitative and quantitative) used in industry. It also includes a discussion of available forecasting software and real-world applications of forecasting methods.

Part II (Aggregate Planning) discusses mathematical models for developing optimal production plans. Both linear and nonlinear programming models for aggregate planning are discussed. Modeling aggregate planning as a transportation model is also presented with a *greedy algorithm* for its solution.

1.8.2 Chapter 3 (Inventory Management Methods and Models)

We begin this chapter by presenting a framework for characterizing inventory management methods according to the product and business environment, concluding that the emphasis in this book should be on the management of what are called independent-demand items, which are subject to uncertainty in demand that is outside the control of the inventory decision maker. From there we establish the essence of inventory decision models, which require that the decision maker *describe demand* and *assess relevant costs*. Our first object lesson in doing this is an analysis of the simplest, single-period inventory decision model for uncertain demand, the newsvendor problem. Within the context of demand uncertainty, this model establishes an important cost tradeoff inherent to many inventory management problems, namely the balance between the cost of ordering too much (i.e., the cost of overage) and ordering too little (i.e., the cost of underage). Extending to multi-period problems, we introduce continuous-review policies, proceeding from the certain demand case, which establishes the basic cost tradeoffs inherent in the classical economic order quantity (EOQ) model, to the uncertain demand case, which allows us to return to the cost-service tradeoff originally implied by the newsvendor model.

We continue to build on these basic models throughout the chapter, using them to explore various inventory policies for single-item systems, policies for multi-item inventory systems, and ultimately working our way up to multi-echelon inventory systems. In all cases, however, we continue to return to the basic frameworks and tradeoffs introduced by the newsvendor and EOQ problems. We conclude the chapter by showing that the basic structure of these models persists in providing good, approximate solutions to very complicated problems, such as coordinating inventory policies across stages in the supply chain.

1.8.3 Chapter 4 (Transportation Decisions in Supply Chain Management)

This chapter presents a unique treatment of freight transportation in a way that couples it to the inventory decision-making models presented in Chapter 3. We focus mostly on truck-based motor freight, since that is the transportation mode on which the lion's share of product value moves in the United States. Moreover, truck-based freight presents the most interesting tradeoff between less frequent shipments at larger volumes (often in full truckloads) and more frequent shipments at smaller volumes, but higher costs—but not so much higher for smaller volumes as to obviate any need for decision making, as is the case with rail shipments or ocean-going freight. Thus, we can couple this low-frequency, low-freight-cost versus high-frequency, high-freight-cost tradeoff as an extension to similar tradeoffs inherent in the inventory decision models introduced in Chapter 3. Indeed, our view is that previous textbooks have unduly treated inventory decision-making and transportation choices as separate problems, whereas we view them as inextricably intertwined in any well thought-out logistics system.

We spend a good portion of our time in the chapter dealing with realistic cost values drawn from publicly available data, and we use this data as the basis for developing cost functions that can be appended to inventory cost models in a way that allows us to utilize the standard tools of nonlinear analysis to find optimal, transportation-cost-inclusive inventory decisions that can be compared across freight transportation modes. In the context of less-than-truckload motor freight, we build cost functions directly from rate tariffs that are representative of those charged by LTL carriers, and also using a recently published approximation model that can be employed more generally for shipments demonstrating a wide range of distances and shipment characteristics.

1.8.4 Chapter 5 (Location and Distribution Decisions in Supply Chains)

This chapter focuses on location and distribution strategies for designing and operating the supply chain network. It begins with an introduction to the formulation of integer programming models, in particular the use of binary (0-1) variables in modeling real-world problems.

Next we apply these integer programming models to different supply chain network optimization problems, including warehouse location, network design, and distribution problems. The topic of *risk pooling* or *inventory consolidation* is presented next. In this portion of the chapter, we discuss the pros and cons of *consolidated* and *de-consolidated* networks and how to determine the optimal level of network consolidation. The tradeoff between supply chain cost and customer service under risk pooling is also presented.

Next we present some basic results in *continuous location models* and how they relate to supply chain network design. We conclude the chapter by discussing several real-world applications of integer programming models used successfully in supply chain network design and distribution problems.

1.8.5 Chapter 6 (Supplier Selection Models and Methods)

In this chapter, we discuss supplier selection models and the multiple conflicting criteria used in supplier selection. Multi-criteria ranking methods for the *prequalification* of suppliers are discussed in detail. Next, multi-criteria optimization models are presented to determine the optimal order allocation among the shortlisted suppliers. Several variants of goal programming methods for solving multiple criteria mathematical programming models are presented with a case study.

1.8.6 Chapter 7 (Managing Risks in Supply Chains)

This chapter presents important problems in supply chain risk management that extend beyond the traditional treatment of supply chain management commonly found in textbooks. We discuss in detail the vulnerability, driven by globalization, of supply chains around the world to a disruptive event in one part of a single country. This chapter presents models and methods to answer the following questions:

- How to identify supply chain risks?
- How to classify and prioritize supply chain risks?
- How to develop appropriate risk mitigation strategies?
- How to quantify the impacts of supply chain disruptions?
- How to quantitatively integrate supply chain risks in decision making at both strategic and operational levels?

We also illustrate through real-world examples the best industry practices in supply chain risk management.

1.8.7 Chapter 8 (Global Supply Chain Management)

This chapter discusses globalization and its impacts on supply chain management. It begins with the history of globalization and illustrates the trends in globalization with examples. Key issues in managing global supply chains are discussed, including the off-shore outsourcing phenomenon.

Next, supplier criteria and vendor management under global sourcing are presented. Finally multi-objective optimization models for designing global supply chains are discussed and illustrated with actual case studies.

Every chapter ends with a *Summary and Further Readings* section. This provides additional references that extend the topics discussed in the chapter. It also includes references to other related topics, including other models and methods not discussed in this textbook.

All chapters have a set of *Exercises* which are divided into two parts. First is a set of "word questions" requiring short responses. They test the reader's basic understanding of the concepts presented in the chapter. These are followed by a set of numerical problems to apply the models and methods presented in that chapter.

1.9 Summary and Further Readings

1.9.1 Summary

In this chapter, we have defined the term "supply chain engineering" and discussed the types of decisions that are made in managing supply chains. We also demonstrated the close relationship between certain supply chain performance metrics and a company's financial performance. Finally, the importance of supply chain management was illustrated in our discussion of the Supply Chain Top 25 (Hofman et al., 2011) and their essential characteristics. An overview of the topics covered in Chapters 2 through 8 was also presented.

1.9.2 Further Readings

A good description of the supply chain design cycle is given in Warsing (2008, 2009). In addition, Warsing provides an introduction to the supply chain operations reference (SCOR) model that captures the detailed business process content encapsulated by well-functioning supply chains. The popular textbook by Chopra and Meindl (2001) contains a good discussion of the drivers and enablers of supply chain performance. The authors present two distinct perspectives on supply chain management: a *cycle view* for operational decisions and *push–pull* view for strategic decisions.

The article by Min and Zhou (2002) provides an excellent review of supply chain modeling research done prior to 2002. They identify the key challenges and opportunities in supply chain modeling. In addition, they emphasize the importance of integrated supply chain models and suggest areas for further research.

The recent textbook by Snyder and Shen (2011) contains a good treatment of the mathematical approaches for managing supply chains. It would be an excellent companion to this textbook for those who are more mathematically inclined in their approach to studying supply chain theory. Another recent textbook by Srinivasan (2010) emphasizes quantitative models in operations and supply chain management. It contains a concise treatment of several models and methods with numerical examples. Finally, the classic textbook by Ballou (2005) emphasizes business logistics in supply chain management, and would also serve as a helpful reference companion to this book, emphasizing the application of supply chain management approaches to long-standing frameworks for logistics management.

Exercises

1.1 Discuss at least two differences among the strategic, tactical, and operational decisions in a supply chain. Give two examples of each type of decision.

1.2 Describe three types of flow in a supply chain and give two examples of each.

1.3 Discuss the differences between the two key criteria of supply chain performance: Efficiency and Responsiveness. Identify three measures for each criterion.

1.4 Describe the four major drivers of supply chain performance. For each driver, identify one efficiency measure and one responsiveness measure.

1.5 What is supply chain risk? Discuss the differences between *hazard risks* and *operational risks*.

1.6 Why is supply chain management important for companies? Give three characteristics that set the "Supply Chain Top 25" apart from their competitors.

1.7 Give two examples of supply chain enablers. How do they differ from supply chain drivers?

1.8 A typical supply chain manager has to make the decisions listed below. Categorize each decision as strategic, tactical, or operational and briefly justify your choice.

(a) Number of warehouses needed and where to locate them

(b) Assignment of customer orders to inventory or production

(c) Choosing vendors to supply critical raw materials

(d) Selecting a transportation provider for shipments to customers

(e) Given the factory capacity, determining the quarterly production schedule

(f)   Choosing to produce a part internally or to outsource that part

(g)   Setting inventory policies at retail locations

(h)   Choosing a transportation mode for shipping

(i)   Reordering materials to build inventory

(j)   Allocating inventory to customer backorders

(k)   Creating the distribution plan to move finished goods inventory from warehouses to retail locations

1.9   Sales of cars at a local auto dealer total 3000 cars per year. The dealer reports inventory turns of 30 per year. Compute the following:

(a)   Average inventory in the dealer's lot

(b)   The length of time the average car sits in the dealer lot before it is sold

Suppose the dealer wants to increase his customer responsiveness. What actions should he take? How will these actions impact inventory turns, average inventory, days of inventory, and supply chain cost?

1.10 XYZ company's annual sales for its products is 5000 units and its average inventory is 500 units.

(a)   Compute the company's inventory turns.

(b)   If XYZ decides to increase its inventory turns, how will this impact (increase or decrease) the following?

(i)   Response time to customers

(ii)   Inventory capital

(iii) Days of inventory

(iv) Return on assets

(v)   Cash-to-cash cycle

(vi) Working capital

1.11 State whether each of the following statements is true or false, or whether the answer ultimately depends on some other factors (which you should identify).

(a)   Speed of delivery is an efficiency measure for the transportation driver.

(b)   Large inventory is a warning sign for poor supply chain efficiency.

(c)   Location of plants and DCs are tactical decisions.

(d)   Selection of suppliers is an operational decision.

(e)   Increasing product variety improves supply chain responsiveness.

(f)   Increasing inventory turns increases working capital.

(g)   Increasing days of inventory increases return on assets.

(h) To decrease supply chain cost, inventory turns should be increased.

(i) The Japanese earthquake and tsunami of 2011 was a hazard risk to the supply chain.

(j) Failure of IT systems is considered an operational risk to the supply chain.

(k) Operational risks have a more significant impact on supply chain performance than hazard risks.

(l) Hazard risks are uncontrollable rare events that disrupt supply chain performance.

(m) Functional products are best matched with responsive supply chains.

References

Ballou, R. H. 2005. *Business Logistics/Supply Chain Management*, 5th edn. Upper Saddle River, NJ: Prentice Hall.

Chopra S. and P. Meindl. 2001. *Supply Chain Management: Strategy, Planning and Operation*. Upper Saddle River, NJ: Prentice Hall.

Dell, M. September 3, 2004. Conversations with Dell. *Presentation at The Pennsylvania State University*.

Fine, C. H. 2000. Clock speed based strategies for supply chain design. *Production and Operations Management*. 9(3): 213–221.

Fisher, M. 1997. What is the right supply chain for your product? *Harvard Business Review*. March–April Issue. 75(2): 105–116.

Gartner Group (2012), Supply Chain Top 25, http://www.gartner.com/technology/supply-chain/top25.jsp.

Hendricks, K. B. and V. R. Singhal. 2005. Association between supply chain glitches and operating performance. *Management Science*. 51(5): 695–711.

Hofman, D., K. O'Marah, and C. Elvy. June 2011. The Gartner Supply Chain Top 25 for 2011. *Gartner Report*, ID number: G00213740, Scotsdale, AZ: Gartner Inc.

Marien, E. J. 2000. The four supply chain enablers. *Supply Chain Management Review*. March–April Issue. 4(1): 60–68.

Min, H. and G. Zhou. 2002. Supply chain modeling: Past, present and future. *Computers and Industrial Engineering*. 43: 231–249.

Mulani, N. 2005. High performance supply chains. *Production and Operations Management Society (POMS) Conference*. Chicago, IL: Production and Operations Management Society.

Simchi-Levi, D., P. Kaminisky, and E. Simchi-Levi. 2003. *Designing and Managing the Supply Chain*, 2nd edn. New York: McGraw Hill.

Simon, H. A. 1962. The architecture of complexity. *Proceedings of the American Philosophical Society*. 106(6): 467–482.

Snyder, L. V. and Z. M. Shen. 2011. *Fundamentals of Supply Chain Theory*. Hoboken, NJ: Wiley.

Srinivasan, G. 2010. *Quantitative Models in Operations and Supply Management*. New Delhi, India: Prentice Hall.

Wadhwa, V. and A. Ravindran. 2007. Vendor selection in outsourcing. *Computers and Operations Research*. 34: 3725–3737.

Warsing, D. P. 2008. Supply chain management. Chapter 22, *Operations Research and Management Science Handbook*, ed. A. R. Ravindran. Boca Raton, FL: CRC Press.

Warsing, D. P. 2009. Supply chain management. Chapter 8, *Operations Research Applications*, ed. A. R. Ravindran. Boca Raton, FL: CRC Press.

2

Planning Production in Supply Chains

The success of a manufacturing supply chain depends on its ability to accurately forecast customer demands and produce in time to meet those demands. Thus, *forecasting* is the starting point for most supply chain management decisions. Based on the demand forecasts, *aggregate production planning* is done to plan production. Distribution and allocation of resources (equipment, labor, etc.), called *material requirements planning*, generally follows an aggregate production plan. In this chapter, we will discuss commonly practiced forecasting methods in industry and quantitative models for aggregate production planning.

2.1 Role of Demand Forecasting in Supply Chain Management

Forecasts form the basis of planning production in a supply chain. Strategic plans (e.g., where to locate factories, warehouses, etc.) are based on long-term forecasts covering several years. Aggregate production plans (e.g., allocation of labor and capital resources for the next quarter's operations) are usually made a few months in advance based on medium-range forecasts. Thus, forecasting is a key activity that influences many aspects of supply chain design, planning, and operations.

Simply stated, a forecast is the best estimate of a random variable (demand, price, etc.) based on the available information. Clearly forecasting is as much an art as a science. Forecasting is different than estimating the probability distribution of demand. Since the demand distributions may change over the *forecast horizon*, forecasting handles *non-stationary* data by estimating the mean of the probability distribution at a given time. Some common features of forecasts are as follows (Nahmias, 1993; Foote and Murty, 2008, 2009):

- *Forecasts are generally wrong*—Forecasts are estimates of the mean of the random variable (demand, price, etc.) and actual outcomes may be very different.

- *Aggregate forecasts are more accurate*—Demand forecasts for the entire product family are more accurate compared to those made for each member of the product family.

- *Forecasting errors increase with the forecast horizon*—Demand forecasts for the next month will be more accurate than those for the next year.
- *Garbage in garbage out principle*—Source and accuracy of data used for forecasting are very important.

Demand forecasts are used by all business functions. Given next are a few examples:

- *Accounting*: Cost/profit estimates
- *Finance*: Cash flow and investment decisions
- *Human resources*: Hiring/recruiting/training
- *Marketing*: Sales force allocation, pricing and promotion strategy
- *Operations*: Production plans, materials requirement planning, workloads

2.2 Forecasting Process

The elements of a good forecast are as follows:

- Timeliness
- Reliability
- Accuracy
- Regular reviews
- Equal chance of being over and under
- Good documentation
- Easy to use

The major steps in the forecasting process are the following:

Step 1: Determine the purpose of the forecast

Step 2: Establish a *forecast* horizon, namely how far into the future we would like to make prediction.

Step 3: Select the appropriate forecasting technique(s).

Step 4: Get past data and analyze. Validate the chosen forecasting technique.

Step 5: Prepare the final forecast.

Step 6: Monitor the forecasts regularly and make adjustments when needed.

Step 3 is an important and difficult step in the forecasting process. There are a number of forecasting methods that are available in the literature. They can be broadly classified into two categories:

1. Qualitative or judgmental methods
2. Quantitative or statistical methods

Qualitative forecasting methods are usually based on subjective judgments and past historical data are not needed. They are most useful for new products or services. Section 2.3 discusses in detail the various qualitative forecasting methods.

Quantitative forecasting methods require historical data that are accurate and consistent. They assume that past represents the future, namely, history will tend to repeat itself. Most quantitative methods fall in the category of *time series analysis* and are discussed in detail in Section 2.4.

2.3 Qualitative Forecasting Methods

These methods are based on subjective opinions obtained from company executives and consumers. These methods are ideally suited for forecasts where there are no past data or past data is not reliable due to the changes in environment (e.g., peacetime data on spare part needs for military aircrafts are not useful for forecasts during war time). Qualitative methods are commonly used for strategic decisions where long-term forecasts are necessary. The qualitative approaches vary in sophistication from scientifically conducted consumer surveys to intuitive judgments from top executives. Quite frequently qualitative methods are also used as supplements to quantitative forecasts. In a 1994 survey of forecasting in practice, 78% of top 500 companies responded that they always or frequently used qualitative methods for forecasting (Sanders and Manrodt, 1994).

A brief review of the most commonly used qualitative methods is given in the following text. For a detailed discussion of qualitative methods, the reader is referred to the textbook by Lilien et al. (2007).

2.3.1 Executive Committee Consensus

In this approach, a group of senior executives of the company from various departments meet, discuss, and arrive at consensus forecasts for products and services. This is primarily a *top-down approach*. Once the forecasts are agreed upon by the top management, they are communicated to the various departments for use in planning their activities. Because of the face-to-face meetings used in this approach, the forecasts may be skewed due to the dominance of authoritative figures, "bandwagon" effect and persuasiveness of some individuals. The next approach, *Delphi Method*, avoids these pit falls.

2.3.2 Delphi Method

The Delphi Method was developed in 1950 at the RAND Corporation as a long-term forecasting tool (Delbecq et al., 1975). It uses a scientifically conducted anonymous survey of key experts who are knowledgeable about the item being forecasted.

The Delphi Method requires the following:

1. *A panel of experts*: The Panel members can be geographically displaced in multiple countries. They remain anonymous throughout the process. The panel members are generally top management people with different background and expertise. The selection of expert panel is key to the success of the Delphi Method.
2. *A facilitator*: Only the facilitator interacts directly with the panel members, obtains the individual responses, and presents a statistical summary of the responses to the group.

The Delphi process is conducted as follows:

Step 1: The facilitator sends a survey to the panel members asking for their expert opinions on the forecast.

Step 2: The panel's responses are analyzed by the facilitator, who then prepares a statistical summary of the forecast values, for example, the median value and the two quartiles (75th and 25th percentile values).

Step 3: The statistical summary is shared with the panel members. They will then be asked to revise their forecasts based on the group's response. The experts are also asked to provide comments on why they did or did not change their original forecasts.

Step 4: A statistical summary of the panel's response and their comments are anonymously shared with the experts again.

The process is continued until a consensus among the experts is reached. The process can take several rounds to reach consensus. In the past, Delphi process used to be very time-consuming and took several weeks. However, using email and online surveys, the Delphi process can be done in a matter of days. Because of the statistical feedback, there is no need to restrict the size of the panel.

2.3.3 Survey of Sales Force

Unlike the previous two methods, this method is a *bottom-up approach*. Here the regional sales people, who directly interact with the customers, are asked to estimate the sales for their regions. They are reviewed at the managerial level and an aggregate forecast is then developed by the company.

2.3.4 Customer Surveys

This approach uses scientifically designed customer surveys to determine their needs for products and services. The survey results are tabulated at the corporate level and forecasts are prepared. This method is referred to as the "grassroots" approach since it directly involves end users. This approach is frequently employed for estimating demands for brand new product lines or services.

2.4 Quantitative Forecasting Methods

Quantitative forecasting methods require historical data on past demand. They assume that the "conditions" that generated the past demand will generate the future demand, i.e., history will tend to repeat itself. Most of the quantitative forecasting methods fall under the category of *time series analysis*.

2.4.1 Time Series Forecasting

A *time series* is a set of values for a sequence of random variables over time. Let X_1, X_2, X_3, ..., X_n be random variables denoting demands for periods 1, 2, ..., n. The forecasting problem is to estimate the demand for period $(n + 1)$ given the observed values of demands for the last n periods, D_1, D_2, ..., D_n. If F_{n+1} is the forecast of demand for period $(n + 1)$, then F_{n+1} is the predicted mean of the random variable X_{n+1}. In other words,

$$F_{n+1} \approx E[X_{n+1}]$$

In quantitative forecasting, we assume that the time series data exhibit a *systematic component*, superimposed by a *random component* (noise). The systematic component may include the following:

- Constant level
- Constant level with seasonal fluctuations
- Constant level with trend (growth or decline)
- Constant level with seasonality and trend

Figures 2.1 through 2.4 illustrate the four different time series patterns with random fluctuations or noise.

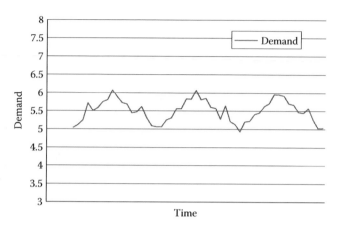

FIGURE 2.1
Constant level.

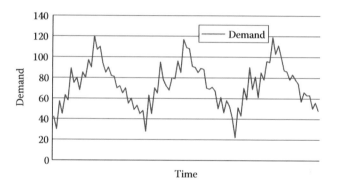

FIGURE 2.2
Constant level with seasonality.

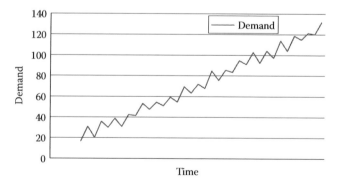

FIGURE 2.3
Constant level with trend.

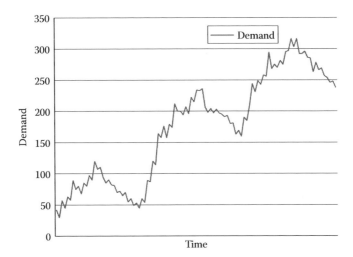

FIGURE 2.4
Constant level with seasonality and trend.

2.4.2 Constant Level Forecasting Methods

Under constant level forecasting methods, it is assumed that the random variable X_t for period t is given by

$$X_t = L_t + e_t \quad \text{for } t = 1, 2, ..., n$$

where
  L_t is the constant level
  e_t is the random fluctuation or noise

We assume e_t is normally distributed with mean zero and negligible variance. Given the values of $X_t = D_t$ for $t = 1, 2, ..., n$, the forecast for period $(n + 1)$ is given by

$$F_{n+1} = E(X_{n+1}) = L_{n+1}$$

There are five commonly used forecasting methods for estimating the value of F_{n+1} as follows:

1. Last value method
2. Simple averaging method
3. Moving average method
4. Weighted moving average method
5. Exponential smoothing method

We will discuss them in detail in the following sections. We will then incorporate seasonality and trend into the constant level forecasting methods.

2.4.3 Last Value Method

Under this method, also known as the *naïve method,* the next forecast is the last value observed for that variable. In other words,

$$F_{n+1} = D_n$$

This is not a bad method for new and innovative products whose demands are changing rapidly during the growth phase. Historical values are not relevant for fast-moving products.

Consider the demand data for a product during the past 6 months given in Table 2.1.

Under the naïve method, the forecast for month $7 = F_7 =$ Demand for month $6 = 340$.

2.4.4 Averaging Method

Here the entire historical data is used by computing the average of all the past demand for the forecast.

$$F_{n+1} = \sum_{t=1}^{n} \frac{D_t}{n}$$

For the demand data given in Table 2.1,

$$F_7 = \frac{270 + 241 + 331 + 299 + 360 + 340}{6} \approx 307$$

TABLE 2.1

Demand Data

Month	Demand
1	270
2	241
3	331
4	299
5	360
6	340

The main drawback of this method is that it assumes all the past data are relevant for the current forecast. This may not be true if the past data covers several years.

2.4.5 Simple Moving Average Method

In this method, only the most recent data is relevant for estimating future demand. For the *m-month moving average method* ($m < n$), *only* the average of the past *m* months' data is used:

$$F_{n+1} = \frac{(D_n + D_{n-1} + D_{n-2} + \cdots + D_{n-m+1})}{m}$$

Using the 3-month moving average method for the example data given in Table 2.1, we get

$$F_7 = \frac{(D_6 + D_5 + D_4)}{3} = \frac{(340 + 360 + 299)}{3} = 333$$

Note that all the demands used in the moving average method are given *equal weights*.

2.4.6 Weighted Moving Average Method

Under this method, increasing weights are given to more recent demand data. Let $W_1, W_2, \ldots, W_m$ be the weights in the *m*-period weighted moving average method, with W_1 assigned to the oldest demand and W_m assigned to the most recent demand. Then the forecast for period $(n + 1)$ is computed as follows:

$$F_{n+1} = W_m D_n + W_{m-1} D_{n-1} + \cdots + W_1 D_{n-m+1} \tag{2.1}$$

$$W_m \geq \cdots \geq W_1 \geq 0$$

$$W_1 + W_2 + \cdots + W_m = 1$$

For the example data given in Table 2.1, let the weights $(W_1, W_2, W_3) = (0.2, 0.3, 0.5)$. Then the 3 month weighted moving average forecast for month 7 will be

$$F_7 = (0.5)\,(340) + (0.3)\,(360) + (0.2)\,(299) \approx 338$$

2.4.7 Computing Optimal Weights by Linear Programming Model

To determine the best weights for the weighted moving average method, a linear programming (LP) model can be used. The objective of the LP model is to find the optimal weights that minimize the forecast error. Let e_t denote the forecast error in period t. Then, e_t is given by

$$e_t = F_t - D_t$$

where
F_t is given by Equation 2.1
D_t is the actual demand in period t

Note that e_t can be positive or negative. Define the total forecast error as

$$Z = \sum |e_t| \tag{2.2}$$

In fact, Z/n is called the *mean absolute deviation (MAD)* in forecasting and is used as a key measure of forecast error in validating the forecasting method (refer to Section 2.9). For the LP model, the unrestricted variable e_t will be replaced by the difference of two non-negative variables as follows:

$$e_t = e_t^+ - e_t^-$$

Since at most one of the non-negative variables, e_t^+ or e_t^-, can be positive in the LP optimal solution, we can write $|e_t|$ as

$$|e_t| = e_t^+ + e_t^-, \quad \text{for } t = 1, ..., n$$

Thus, the complete LP model for finding the weights for the weighted m-period moving average method will be as follows:

$$\text{Minimize } Z = \sum_{t=m+1}^{n} (e_t^+ + e_t^-)$$

Subject to

$$e_t^+ - e_t^- = (W_m D_{t-1} + W_{m-1}D_{t-2} + \cdots + W_1 D_{t-m}) - D_t \quad \text{for } t = m+1, m+2, ..., n$$

$$W_m \geq W_{m-1} \geq W_{m-2} ... \geq W_1 \geq 0$$

$$\sum_{t=1}^{m} W_t = 1$$

$$e_t^+, e_t^- \geq 0 \quad \text{for } t = m+1, m+2, \ldots, n$$

NOTE: In the objective function Z, the index t begins at $t = m + 1$, because m-demand values are necessary for the m-period moving average.

Example 2.1

For the sample data given in Table 2.1, formulate the LP model for determining the optimal weights (W_1, W_2, W_3) for the 3-month moving average. Solve the LP problem and determine the forecast for month 7.

Solution (LP model)

$$\text{Minimize } Z = \sum_{t=4}^{6} (e_t^+ + e_t^-)$$

Subject to

$$e_4^+ - e_4^- = 331W_3 + 241W_2 + 270W_1 - 299$$

$$e_5^+ - e_5^- = 299W_3 + 331W_2 + 241W_1 - 360$$

$$e_6^+ - e_6^- = 360W_3 + 299W_2 + 331W_1 - 340$$

$$W_3 \geq W_2 \geq W_1 \geq 0$$

$$W_1 + W_2 + W_3 = 1$$

$$e_t^+, e_t^- \geq 0 \quad \text{for } t = 4, 5, 6$$

The optimal solution is given by

$$W_1 = 0, \ W_2 = 0.356, \ W_3 = 0.644$$

$$e_4^+ = e_4^- = e_5^+ = e_6^+ = 0$$

$$e_5^- = 49.62, \ e_6^- = 1.69$$

$$\text{Minimum } Z = 51.31$$

Then, the forecast for month 7 is given by

$$F_7 = 340 \,(0.644) + 360 \,(0.356) + 299 \,(0) = 347$$

2.4.8 Exponential Smoothing Method

Perhaps the most popular forecasting method in practice is the *exponential smoothing method*, introduced and popularized by Brown (1959). It is basically a weighted averaging method with weights decreasing exponentially on older demands. Unlike the weighted moving average method, it uses all the data points. Under this method, given demands $D_1, D_2, \ldots, D_n$, the forecast for period $(n + 1)$ is given by

$$F_{n+1} = \alpha D_n + \alpha(1-\alpha)D_{n-1} + \alpha(1-\alpha)^2 D_{n-2} + \cdots \qquad (2.3)$$

where α is between 0 and 1 and is called the *Smoothing Constant*. Note that $\alpha > \alpha(1-\alpha) > \alpha(1-\alpha)^2 > \cdots$

Thus, the most recent demand is given the highest weight α and the weights are decreased by a factor $(1-\alpha)$ as the data gets older.

Equation 2.3 can be rewritten as follows:

$$F_{n+1} = \alpha D_n + (1-\alpha)[\alpha D_{n-1} + \alpha(1-\alpha)D_{n-2} + \alpha(1-\alpha)^2 D_{n-3}\ldots]$$

$$F_{n+1} = \alpha D_n + (1-\alpha)F_n \qquad (2.4)$$

Thus, the forecast for period $(n + 1)$ uses the forecast for period n and the actual demand for period n. The value of α is generally chosen between 0.1 and 0.4. In other words, the weights assigned to the actual demand is less than that of the forecasted demand, the reason being, the actual demands fluctuate a lot, while the forecast has smoothed the fluctuations.

Example 2.2

Using the sample data given in Table 2.1, determine the forecast for month 7 using the Exponential Smoothing method.

Solution

In order to use this method, the value of α and the initial forecast for month 1 (F_1) are necessary. The value of F_1 can be chosen by averaging all the demands or using the most recent demand value. Assuming $\alpha = 0.2$, $F_1 = 307$ (Averaging method) and using Equation 2.4, we get the following:

$$F_2 = 0.2\,(270) + 0.8\,(307) = 299.6$$

$$F_3 = 0.2\,(241) + 0.8\,(299.6) = 287.9$$

$$F_4 = 0.2\,(331) + 0.8\,(287.9) = 296.5$$

$$F_5 = 0.2\,(299) + 0.8\,(296.5) = 297$$

$$F_6 = 0.2\,(360) + 0.8\,(297) = 309.6$$

Thus, the forecast for month 7 is given by

$$F_7 = (0.2)\,(340) + (.8)\,(309.6) = 315.7 \approx 316$$

NOTE: The initial value chosen for F_1 will have negligible effect on the forecast as the number of data points increases. However, the choice of α is very important and will affect the forecast accuracy.

Choice of smoothing constant, (α)
Typically, values of α between 0.1 and 0.4 are used in practice. Smaller values of α (e.g., $\alpha = 0.1$) yield forecasts that are relatively smooth (low variance). However, it takes longer to react to changes in the demand process. Higher values of α (e.g., $\alpha = 0.4$) can react to changes in data quicker, but the forecasts have significantly higher variations, resulting in forecast errors with high variance. One disadvantage of both the method of moving averages and the exponential smoothing method is that when there is a definite trend in the demand process (either growing or falling), the forecasts obtained by them lag behind the trend. Holt (1957) developed a modification to the exponential smoothing method by incorporating trend. This will be discussed in Section 2.6.

2.5 Incorporating Seasonality in Forecasting

The constant level forecasting methods discussed so far assume that the values of demand in the various periods form a stationary time series. In some applications this series may be seasonal, that is, it has a pattern that repeats after every few periods. For example, retail sales during Christmas season are usually higher. For some retailers, sales during Christmas season may account for as much as 40% of their annual sales. For a number of companies, the majority of sales are arranged by sales agents, who operate on quarterly sales goals. In those cases, the demands for products tend to be higher during the third month of each quarter as the agents work much harder to meet their quarterly goals. It is relatively easy to incorporate seasonality in the constant level forecasting methods by computing the *seasonality index* for each period.

The basic steps to incorporate seasonality in forecasting are as follows:

Step 1: Compute the seasonality index for any period given by

$$\text{Seasonality Index} = \frac{\text{Average demand during that period}}{\text{Overall average of demand for all periods}}$$

Step 2: Seasonally adjust the actual demands in the time series by dividing by the seasonality index, to get *deseasonalized demand* data.

Step 3: Select an appropriate time series forecasting method.

Step 4: Apply the forecasting method on the *deseasonalized demand forecast*.

Step 5: Compute the *actual forecast* by multiplying the deseasonalized forecast by the seasonality index for that period.

Let us illustrate the basic steps with an example.

Example 2.3

The quarterly sales of laptop computers for 3 years (2008–2010) are given in Table 2.2.

The problem is to determine the forecast for quarter 1 of year 2011 (period 13).

Let us use the five basic steps given earlier for forecasting under seasonality.

Step 1: Compute the *seasonality* index for each quarter.

- Compute the *overall average* of quarterly demand using the demand values for 12 quarters

$$\text{Overall average} = \frac{(540 + 522 + 515 + \cdots + 550 + 629 + 785)}{12} = \frac{7236}{12} = 603$$

TABLE 2.2

Laptop Computer Sales for 2008–2010 (Example 2.3)

Year	Quarter	Period (*t*)	Actual Demand (*D_t*)
2008	1	1	540
2008	2	2	522
2008	3	3	515
2008	4	4	674
2009	1	5	574
2009	2	6	569
2009	3	7	616
2009	4	8	712
2010	1	9	550
2010	2	10	550
2010	3	11	629
2010	4	12	785

- Compute the quarterly average using the three demand values for each quarter as follows:

$$\text{Quarter 1 average} = \frac{(540 + 574 + 550)}{3} = 555$$

$$\text{Quarter 2 average} = \frac{(522 + 569 + 550)}{3} = 547$$

$$\text{Quarter 3 average} = \frac{(515 + 616 + 629)}{3} = 587$$

$$\text{Quarter 4 average} = \frac{(674 + 712 + 785)}{3} = 724$$

Note that the third quarter average is slightly higher primarily due to "Back-to-School" sales to students. The fourth quarter average is much higher due to "Christmas Sales."

- The *Seasonality Indices (SI)* are then computed as follows:

$$SI \text{ for Quarter } 1 = \frac{555}{603} = 0.92$$

$$SI \text{ for Quarter } 2 = \frac{547}{603} = 0.907$$

$$SI \text{ for Quarter } 3 = \frac{587}{603} = 0.973$$

$$SI \text{ for Quarter } 4 = \frac{724}{603} = 1.2$$

Step 2: Compute the deseasonalized sales data $\left(\overline{D_t}\right)$ by dividing the actual sales data (D_t) by its appropriate seasonality index. For example, deseasonalized sales for 2008 Quarter 1 = 540/.92 = 587

Table 2.3 gives the deseasonalized data for all quarters.

Step 3: Select any time series forecasting method. For illustration, we will use the *exponential smoothing* forecasting method with $\alpha = 0.2$. For the initial forecast for Quarter 1 of year 2008, we will use 600.

Step 4: Apply the exponential smoothing method on the *deseasonalized demand* to get *deseasonalized* forecast as shown in Table 2.3. For example, deseasonalized forecast for Quarter 1 (2011) is given by

$$\overline{F_{13}} = (0.2)\,(654.2) + (0.8)\,(610.2) = 619$$

Step 5: Compute the actual forecast for year 2011, Quarter 1 as follows:

$$F_1(2011) = \overline{F_1}(2011) \times SI_1 = (619)(0.92) = 569.5$$

Note that we could have used any of the time series forecasting method discussed in Step 3 of Section 2.4.

TABLE 2.3

Forecasting with Seasonality (Example 2.3)

Year	Quarter	Period (t)	Actual Demand (D_t)	Seasonality Index (SI)	Deseasonalized Demand $\left(\overline{D_t}\right)$	Deseasonalized Forecast $\left(\overline{F_t}\right)$	Actual Forecast (F_t)
2008	1	1	540	0.92	587.0	600.0	552.0
2008	2	2	522	0.907	575.5	597.4	541.8
2008	3	3	515	0.973	529.3	593.0	577.0
2008	4	4	674	1.2	561.7	580.3	696.3
2009	1	5	574	0.92	623.9	576.6	530.4
2009	2	6	569	0.907	627.3	586.0	531.5
2009	3	7	616	0.973	633.1	594.3	578.2
2009	4	8	712	1.2	593.3	602.0	722.5
2010	1	9	550	0.92	597.8	600.3	552.3
2010	2	10	550	0.907	606.4	599.8	544.0
2010	3	11	629	0.973	646.5	601.1	584.9
2010	4	12	785	1.2	654.2	610.2	732.2
2011	1	13		0.92		619.0	569.5

For the sake of illustration, the forecasts for the first quarter of 2011, using the Naïve method and the 4-quarter moving average method are also given in the following:

Under naïve method

F_{13} = Deseasonalized demand for Quarter 4 of 2010 = $\overline{D_{12}}$ = (654.2)

$$F_{13} = (654.2)(.92) \approx 602$$

Under 4-quarter moving average

$$\overline{F_{13}} = \frac{(597.8 + 606.4 + 646.5 + 654.2)}{4} = 626.2$$

$$F_{13} = (626.2)(.92) \approx 576$$

NOTE: Seasonality indices are not static. They can be updated as more data become available.

2.6 Incorporating Trend in Forecasting

One disadvantage of both the moving average and exponential smoothing methods is that when there is a definite trend in the demand process (either growing or falling), the forecasts obtained by them lag behind the trend. There are ways to accommodate trend in forecasting from a very *simple linear trend model* to Holt's model (Holt, 1957). Holt's method uses variations of the exponential smoothing method to track linear trends over time.

2.6.1 Simple Linear Trend Model (Srinivasan, 2010)

Here we assume that the level and trend remain constant over the forecast horizon. Thus, $X_t = a + bt + \varepsilon$ for $t = 1, 2, \ldots, n$, where X_t is the random variable denoting demand at period t, a, and b are constant level and trend and ε is the random error. Given the observed values of $X_t = D_t$ for $t = 1, \ldots, n$, the forecast for period t is given by

$$F_t = E(X_t) = a + bt$$

Then the forecast error is

$$e_t = F_t - D_t, \quad \text{for } t = 1, 2, \ldots, n$$

From the given data, we then have n linear equations relating the actual and forecasted demands. Following the approach given in Srinivasan (2010), we can use the least square regression method to estimate the level (a) and trend (b) parameters. The unconstrained optimization problem is to determine a and b such that

$$\sum_{t=1}^{n} e_t^2 = \sum_{t=1}^{n} \left[(a + bt) - D_t\right]^2 \tag{2.5}$$

is minimized.

Note that the minimization function given by Equation 2.5 is a convex function. Hence, a *stationary point* will give the absolute minimum. The stationary point can be obtained by setting the partial derivatives with respect to a and b to zero. Thus, we get the following two equations to solve for a and b:

$$\sum_{t=1}^{n} D_t = na + b \sum_{t=1}^{n} t \tag{2.6}$$

$$\sum_{t=1}^{n} t D_t = a \sum_{t=1}^{n} t + b \sum_{t=1}^{n} t^2 \tag{2.7}$$

Solving the two linear equations, we determine the values of a and b. Then, the forecast for period $t = n + 1$ is given by

$$F_{n+1} = a + b(n + 1)$$

Example 2.4

Consider the monthly demand for a product for the past 6 months given in Table 2.4. The demand data clearly shows an increasing trend. The problem is to determine the forecast for month 7 using the *simple linear trend model*.

Solution

Table 2.5 gives the necessary computations for the linear regression model. Using the last row (sum) in Table 2.5 and substituting them in Equations 2.6 and 2.7, we get the following:

$$1700 = 6a + 21b$$

$$6355 = 21a + 91b$$

Solving the two equations for a and b, we get the following:

$$a = 202.333$$

$$b = 23.143$$

Then, the forecast for month 7 is given by

$$F_7 = 202.33 + (23.14)\ 7 \approx 364$$

The main drawback of the *simple linear trend* model is that it assumes that the level and trend remain constant throughout the forecast horizon.

TABLE 2.4

Data for Example 2.4

t	1	2	3	4	5	6
D_t	220	250	280	295	315	340

TABLE 2.5

Calculations for the Regression Model (Example 2.4)

	t	t^2	D_t	tD_t
	1	1	220	220
	2	4	250	500
	3	9	280	840
	4	16	295	1180
	5	25	315	1575
	6	36	340	2040
Sum	21	91	1700	6355

2.6.2 Holt's Method (Holt, 1957; Hillier and Lieberman, 2001; Chopra and Meindl, 2010)

Regular exponential smoothing model estimates the constant level (L) to forecast future demands. Holt's model improves it by estimating both the *level* (L) and *trend factor* (T). It adjusts both the level and trend factor using exponential smoothing. Holt's method is also known as *double exponential smoothing* or *trend adjusted exponential smoothing* method.

Given the actual values of the random variable $X_t = D_t$ for $t = 1, ..., n$, we are interested in computing the forecast for period $n + 1$, i.e., $F_{n+1} \approx E\,(X_{n+1})$. Given the actual demand D_t, forecast F_t and the estimates of level (L_t) and trend (T_t) for period t, the forecast for period $(t + 1)$ is given by

$$F_{t+1} = L_{t+1} + T_{t+1} \tag{2.8}$$

Under the exponential smoothing method, the estimate of the level for $(t + 1)$ is given by

$$L_{t+1} = \alpha D_t + (1-\alpha)F_t \tag{2.9}$$

The same approach is used to estimate the trend factor for $(t + 1)$ using another smoothing constant β as follows:

$$T_{t+1} = \beta[L_{t+1} - L_t] + (1-\beta)T_t \tag{2.10}$$

where
$(L_{t+1} - L_t)$ is the latest trend based on two recent level estimates
T_t is the most recent estimate of the trend

The term $(L_{t+1} - L_t)$ in Equation 2.10 can be rewritten using Equation 2.9 as follows:

$$L_{t+1} - L_t = \alpha(D_t - D_{t-1}) + (1-\alpha)(F_t - F_{t-1}) \tag{2.11}$$

The term $(D_t - D_{t-1})$ in Equation 2.11 represents the latest trend based on the actual values observed for the last two periods, while $(F_t - F_{t-1})$ is the latest trend based on the last two forecasts. Once L_{t+1} and T_{t+1} are determined, using Equations 2.9 and 2.10 respectively, the forecast for period $(t + 1)$ is given by their sum, as in Equation 2.8.

Example 2.5 (Holt's method)

Consider again the 6-month time series data on demand given in Table 2.4. We will apply Holt's method to determine the forecast for month 7. In order to get started with Holt's method, we need the values of the smoothing constants, α and β and the initial estimates for the level and trend for month 1, namely L_1 and T_1. Let us assume that $\alpha = \beta = 0.3$.

Solution

For illustration, L_1 is assumed to be equal to D_1 and T_1 is the linear slope using D_1 and D_6. In other words,

$$T_1 = \frac{D_6 - D_1}{5} = \frac{340 - 220}{5} = 24$$

Hence, $L_1 = 220$ and $T_1 = 24$.

Then, the forecast for month 1 is given by

$$F_1 = L_1 + T_1 = 220 + 24 = 244$$

For month 2, L_2 and T_2 are computed using Equations 2.9 and 2.10 as follows:

$$L_2 = \alpha D_1 + (1 - \alpha)F_1 = (0.3)\,220 + (0.7)\,244 = 236.8$$

$$T_2 = \beta(L_2 - L_1) + (1 - \beta)T_1 = (0.3)\,(236.8 - 220) + (0.7)\,24 = 5.04 + 16.8 = 21.8$$

The remaining calculations are done on a spreadsheet and are shown in Table 2.6. Note that the cell values in the table have been displayed with one decimal accuracy.

Thus, the forecast for month 7 is given by

$$F_7 = L_7 + T_7 = 338.8 + 20.7 \approx 360$$

TABLE 2.6

Holt's Method (Example 2.5)

Month (t)	Demand (D_t)	Estimate of Level (L_t)	Estimate of Trend (T_t)	Forecast (F_t)
1	220	220.0	24.0	244.0
2	250	236.8	21.8	258.6
3	280	256.0	21.1	277.1
4	295	278.0	21.3	299.3
5	315	298.0	20.9	318.9
6	340	317.8	20.6	338.3
7		338.8	20.7	359.6

$\alpha = \beta = 0.3$; Initial values $L_1 = 220$ and $T_1 = 24$.

2.7 Incorporating Seasonality and Trend in Forecasting

2.7.1 Method Using Static Seasonality Indices (Hillier and Lieberman, 2001)

A straightforward approach to incorporate seasonality and trend is to combine the approaches for seasonality discussed in Section 2.5 and one of the methods for trend discussed in Section 2.6. The basic steps of such an approach are as follows:

Step 1: Compute the *seasonality indices* as discussed in Section 2.5.

Step 2: Using the seasonality indices, compute the *deseasonalized demands*.

Step 3: Apply one of the methods discussed in Section 2.6 to incorporate trend, on the deseasonalized demand data.

Step 4: Determine the deseasonalized forecasts with trend.

Step 5: Compute the actual forecasts using the seasonality indices again.

We will illustrate this method using Example 2.6.

Example 2.6 (Holt's method with seasonality)

Consider the demand data for laptop computer sales for 3 years (2008–2010) given in Table 2.2. Using Holt's method, determine the forecast for the first quarter of 2011 incorporating both seasonality and trend.

Solution

First we compute the seasonality indices and the deseasonalized demands using the approach given in Section 2.5.

Step 1: Compute seasonality indices (*SI*) as described in Example 2.3. The *SI* values are 0.92, 0.907, 0.973, and 1.2 for Quarters 1, 2, 3, and 4, respectively.

Step 2: Compute the deseasonalized demands by dividing the actual quarterly demands by their respective *SI* values as shown in Table 2.7.

In Example 2.3, we then used the exponential smoothing method with $\alpha = 0.2$ to determine the deseasonalized forecasts. Instead, we will apply Holt's method to incorporate trend also.

Step 3: Apply Holt's method on the deseasonalized demand data with $\alpha = 0.2$, $\beta = 0.2$, $L_1 = 600$, $T_1 = 6$.

Step 4: Compute the estimates of level and trend and the deseasonalized forecast. These calculations are done on a spreadsheet and are shown in Table 2.7. Note that the displayed values are shown with one decimal accuracy.

Step 5: Compute the actual forecast by multiplying the deseasonalized forecast by the *SI* values.

Thus, the deseasonalized forecast for Quarter 1 of year 2011 is 634.4 and the actual forecast is $(634.4)(0.92) \approx 584$.

TABLE 2.7

Computations for Example 2.6

Year	Period	Actual Demand	SI	Deseasonalized Demand	Estimate of Level (L)	Estimate of Trend (T)	Deseasonalized Forecast	Actual Forecast
2008	1	540	0.92	587	600.0	6.0	606.0	557.5
2008	2	522	0.907	576	602.2	5.2	607.4	550.9
2008	3	515	0.973	529	601.0	4.0	605.0	588.7
2008	4	674	1.2	562	589.9	0.9	590.8	709.0
2009	5	574	0.92	624	585.0	−0.2	584.7	538.0
2009	6	569	0.907	627	592.6	1.3	593.9	538.7
2009	7	616	0.973	633	600.6	2.7	603.3	587.0
2009	8	712	1.2	593	609.2	3.9	613.1	735.7
2010	9	550	0.92	598	609.1	3.1	612.2	563.2
2010	10	550	0.907	606	609.3	2.5	611.8	554.9
2010	11	629	0.973	646	610.8	2.3	613.0	596.5
2010	12	785	1.2	654	619.7	3.6	623.3	748.0
2011	13		0.92		629.5	4.9	634.4	583.6

2.7.2 Winters' Method (Winters, 1960; Chopra and Meindl, 2010)

A drawback of the previous approach is that the seasonality indices remain static and are not updated during the forecast horizon. Winters (1960) has extended Holt's method by updating seasonality indices also using exponential smoothing.

Given the actual demand D_t, the estimates of level L_t, trend T_t and seasonality indices $SI_1, SI_2, ..., SI_{t+p-1}$, the forecast for period $(t + 1)$ is given by

$$F_{t+1} = (L_{t+1} + T_{t+1})SI_{t+1} \tag{2.12}$$

where

$$L_{t+1} = \alpha\left(\frac{D_t}{SI_t}\right) + (1-\alpha)(L_t + T_t) \tag{2.13}$$

$$T_{t+1} = \beta(L_{t+1} - L_t) + (1-\beta)T_t \tag{2.14}$$

$$SI_{t+p} = \gamma\left(\frac{D_t}{L_t}\right) + (1-\gamma)SI_t \tag{2.15}$$

where α, β, and γ are smoothing constants between 0 and 1 for level, trend and seasonality respectively. Note that Equations 2.13 and 2.14 are very similar to the Holt's model (Equations 2.9 and 2.10). In Equation 2.15, the index p refers to the *periodicity*, periods after which the seasonal cycle repeats itself. Equation 2.15 updates the seasonality index by weighting the observed value and the current estimate of seasonality. We shall illustrate Winters' method using Example 2.7.

Example 2.7 (Winters' method)

Consider the quarterly laptop computer sales data for 3 years (2008–2010) given in Table 2.2. The problem is to determine the forecast for quarter 1 of year 2011 using Winters' method with $\alpha = \beta = 0.2$ and $\gamma = 0.3$.

Solution

To begin Winters' method, we need the initial estimates for level (L_1) and trend (T_1) for Quarter 1 of 2008 and the quarterly seasonality indices (SI) for year 1 (SI_1, SI_2, SI_3, SI_4). Let us assume that $L_1 = 600$, $T_1 = 6$, and the initial SI values are $SI_1 = 0.92$, $SI_2 = 0.907$, $SI_3 = 0.973$, and $SI_4 = 1.2$, as estimated in Example 2.3 (Section 2.5). Note that the *periodicity "p"* is equal to 4 in this example, since the demands are quarterly and the seasonal cycle repeats after every 4 quarters.

Given the actual demand $D_1 = 540$, the SI value for period 5 (SI_5) will be updated. Using $t = 1$ and $p = 4$ in Equation 2.15 and $\gamma = 0.3$, we get the SI value for Quarter 1 (2009) as follows:

$$SI_5 = \gamma\left(\frac{D_1}{L_1}\right) + (1-\gamma)(SI_1) = (0.3)\left(\frac{540}{600}\right) + (0.7)(0.92) = 0.914$$

The forecast for period 2 is given by Equation 2.12 as

$$F_2 = (L_2 + T_2)SI_2$$

where L_2 and T_2 are given by Equations 2.13 and 2.14.

Thus,

$$L_2 = \alpha\left(\frac{D_1}{SI_1}\right) + (1-\alpha)(L_1 + T_1) = (0.2)\left(\frac{540}{0.92}\right) + (0.8)(600+6) = 602.2$$

$$T_2 = \beta(L_2 - L_1) + (1-\beta)T_1 = 0.2\,(602.2 - 600) + (0.8)\,6 = 5.24$$

$$F_2 = (602.2 + 5.24)\,(.907) = 550.9$$

At this time, the seasonality index SI_6 (Quarter 2 of 2009) will be updated using Equation 2.15 as

$$SI_6 = \gamma\left(\frac{D_2}{L_2}\right) + (1-\gamma)(SI_2) = (0.3)\left(\frac{522}{602.2}\right) + (0.7)(0.907) = 0.895$$

TABLE 2.8

Computations for Winters' Method (Example 2.7)

Year	Period	Actual Demand D_t	SI	Estimates		Forecast
				Level	Trend	
2008	1	540	0.920	600.0	6.0	557.4
2008	2	522	0.907	602.2	5.2	551.0
2008	3	515	0.973	601.1	4.0	588.6
2008	4	674	1.200	589.9	0.9	709.0
2009	5	574	0.914	585.0	-0.2	534.4
2009	6	569	0.895	593.4	1.5	532.5
2009	7	616	0.938	603.1	3.1	568.7
2009	8	712	1.183	616.3	5.2	735.1
2010	9	550	0.934	617.5	4.4	580.9
2010	10	550	0.914	615.3	3.0	565.3
2010	11	629	0.963	615.0	2.4	594.6
2010	12	785	1.175	624.5	3.8	738.0
2011	13		0.921	636.3	5.4	591.1

The remaining calculations are shown in Table 2.8 using a spreadsheet. Comparing the forecast of Example 2.6 (Holt's method) and Example 2.7 (Winters' method), we note that the forecasts are the same for the first four quarters (2008). However, they are different from 2009 onward due to the updating of the seasonality indices. The forecast for the first quarter of 2011 is given by

$$F_{13} = (L_{12} + T_{12})SI_{12}$$
$$= (636.3 + 5.4)(0.921) \approx 591$$

2.8 Forecasting for Multiple Periods

So far in our discussion, we were concerned with forecasting for one period only, namely F_{n+1}, given the actual demands $D_1, D_2, ..., D_n$. In practice, one has to forecast for several periods in the future. Methods for forecasting for multiple periods depend on the type of forecasting method selected—constant level, constant level with seasonality, constant level with trend or constant level with seasonality and trend.

Multi-period forecasting problem
Given the actual demands for the last n periods as $D_1, D_2, ..., D_n$, determine the forecasts for the next m periods, $F_{n+1}, F_{n+2}, ..., F_{n+m}$

2.8.1 Multi-Period Forecasting under Constant Level

Under the constant level forecasting approach, the level forecasts are updated based only on the observed demands. Since there is no new demand information beyond period n, the forecast for the next m periods will remain the same.

$$F_{n+i} = F_{n+1} \quad \text{for } i = 1, 2, ... \tag{2.16}$$

In other words, the best forecasts for periods $n + 2, n + 3, ...$ is just F_{n+1}!

To illustrate, consider Example 2.2 where we forecasted the demand for month 7 as 316 based on the last 6 months of demand, using Exponential Smoothing. Forecasts beyond month 7 are given by

$$F_{6+i} = F_7 = 316 \quad \text{for } i = 2, 3, ...$$

2.8.2 Multi-Period Forecasting with Seasonality

Under seasonality, forecasts for multiple periods are given as follows:

$$F_{n+i} = (\overline{F}_{n+1})SI_{n+i} \quad \text{for } i = 1, 2, \ldots \tag{2.17}$$

where
  $\overline{F}_{n+1}$ is the deseasonalized forecast for period $(n + 1)$
  SI_{n+i} is the seasonality index for period $(n + i)$

Since the deseasonalized level forecast does not change, the forecasts for future period are only affected by their respective seasonality indices.

Example 2.8

Consider Example 2.3, where laptop sales by quarter for years 2008, 2009, and 2010 were given and we forecasted the demand for the first quarter of 2011 using exponential smoothing method with seasonality. Suppose the problem now is to forecast the sales for all the four quarters of 2011.

Solution

In Example 2.3 (Table 2.3), the deseasonalized forecast for the first quarter of 2011, namely $\overline{F}_{13}$ was computed as 619. Thus, the actual forecasts for year 2011 are as follows:

$$F_{13} = (619)\,(0.92) \approx 570 \text{ (Quarter 1)}$$

$$F_{14} = (619)\,(0.907) \approx 561 \text{ (Quarter 2)}$$

$$F_{15} = (619)\,(0.93) \approx 576 \text{ (Quarter 3)}$$

$$F_{16} = (619)\,(1.2) \approx 743 \text{ (Quarter 4)}$$

2.8.3 Multi-Period Forecasting with Trend

Using Holt's method for trend (Section 2.6.2), forecasts for multiple periods are given by

$$F_{n+i} = L_{n+1} + (i)T_{n+1} \quad \text{for } i = 1, 2, \ldots, m \tag{2.18}$$

where L_{n+1} and T_{n+1} are the latest estimates of level and trend respectively. Note that we assume a linear trend for the future forecasts.

Example 2.9

Consider Example 2.5, where Holt's method was used to forecast demand for month 7, based on the actual demands for the first 6 months. Determine the forecast of demands for months 7, 8, and 9.

Solution

In the solution to Example 2.5 (Table 2.6), the latest estimates of level and trend for month 7 were computed as

$$L_7 = 338.8; \ T_7 = 20.7$$

Hence, the forecasts for the future months are as follows:

$$F_7 = L_7 + T_7 = 338.8 + 20.7 \approx 360$$

$$F_8 = L_7 + 2T_7 = (338.8) + 2\ (20.7) \approx 380$$

$$F_9 = L_7 + 3T_7 = (338.8) + 3\ (20.7) \approx 401$$

2.8.4 Multi-Period Forecasting with Seasonality and Trend

Using Winters' method to incorporate seasonality and trend (Section 2.7.2), forecasts for multiple periods are given by

$$F_{n+i} = \left[L_{n+1} + (i)T_{n+1}\right]SI_{n+i} \quad \text{for } i = 1, 2, \ldots \tag{2.19}$$

Note that the latest estimates of level, trend, and seasonality indices are used in Equation 2.19.

Example 2.10

Consider Example 2.7, where Winters' method was used to forecast the demand for the first quarter of 2011. Determine the demand forecasts all four quarters of 2011.

Solution

In Example 2.7 (Table 2.8), the latest estimates for level (L_{13}), trend (T_{13}), and seasonality index (SI_{13}) were computed as

$$L_{13} = 636.3, \ T_{13} = 5.4, \ SI_{13} = 0.921$$

and the forecast for the first quarter of 2011 was

$$F_{13} = (L_{13} + T_{13})SI_{13} = (636.3 + 5.4)\ (0.921) \approx 591$$

Since no new data is available on demand, the estimates for level and trend will remain the same as 636.3 and 5.4 respectively for periods 14,

15, and 16 (Quarters 2, 3, and 4 of 2011). However, their seasonality indices can be updated using the data for periods 10, 11, and 12 in Table 2.8. Using Equation 2.15 we get the following:

$$SI_{14} = \gamma\left(\frac{D_{10}}{L_{10}}\right) + (1-\gamma)(SI_{10}) = (0.3)\left(\frac{550}{615.3}\right) + (0.7)(0.914) = 0.908$$

$$SI_{15} = (0.3)\left(\frac{629}{615}\right) + (0.7)(0.963) = 0.981$$

$$SI_{16} = (0.3)\left(\frac{785}{624.5}\right) + (0.7)(1.175) = 1.2$$

Thus, the forecasts for the remaining quarters of 2011 are computed as follows:

Quarter 2: $F_{14} = [636.3 + 2\ (5.4)]\ (0.908) \approx 588$

Quarter 3: $F_{15} = [636.3 + 3\ (5.4)]\ (0.981) \approx 640$

Quarter 4: $F_{16} = [636.3 + 4\ (5.4)]\ (1.2) \approx 790$

2.9 Forecasting Errors

The accuracy of the forecast depends on *forecast errors*, which measure the differences between the forecasted demands and their actual (observed) values. As discussed in Section 2.4.7, the *forecast error*, for period t, denoted by e_t, is given by:

$$e_t = F_t - D_t \quad \text{for } i = 1, 2, \ldots, n$$

where
F_t is the forecast
D_t is the actual demand in period t

Note that e_t can be positive or negative depending on whether the method is over-forecasting or under-forecasting in period t. There are several measures of forecast errors used in practice for determining the accuracy of the chosen forecasting method as given in the following:

- Mean absolute deviation (MAD)
- Mean squared error (MSE)

- Standard deviation of forecast errors (STD)
- Bias
- Mean absolute percentage error (MAPE)

Let us discuss each of the measures in detail, assuming we have n values of forecast errors, $e_1, e_2, \ldots, e_n$.

1. *Mean absolute deviation (MAD)*: MAD measures the dispersion of the forecast errors:

$$\text{MAD} = \frac{1}{n} \sum_{t=1}^{n} |e_t|$$

2. *Mean squared error (MSE)*: MSE also measures the dispersion of the forecast errors, but larger errors get penalized more due to squaring:

$$\text{MSE} = \frac{1}{n} \sum_{t=1}^{n} e_t^2$$

MSE estimates the variance of forecast errors.

3. *Standard deviation of forecast error (STD)*:

$$\text{STD} = \sqrt{\text{MSE}} = \sigma$$

The value of σ can be used to set confidence limits on the mean forecast. For example, if normality is assumed, then we can say that there is 68% probability that the actual forecast will be within one σ of the mean forecast.

4. *Bias*: Bias measures whether the forecast is overestimating or underestimating the actual demand over the forecast horizon:

$$\text{Bias} = \sum_{t=1}^{n} e_t$$

Note that the values of e_t can be positive, negative, or zero. If Bias > 0, the method is overestimating and if Bias < 0, the method is underestimating. Ideally, a good method should have a Bias close to zero. Bias is frequently preferred by managers for measuring forecast accuracy.

5. *Mean absolute percentage error (MAPE)*: MAPE measures the relative dispersion of forecast errors and is given by

$$\text{MAPE} = \frac{1}{n} \sum_{t=1}^{n} \left| \frac{e_t}{D_t} \right| 100$$

MAPE is better than MAD since it takes into account the relative magnitude of the actual demand and is also frequently used in practice.

Let us illustrate all the five measures of forecast errors with an example.

Example 2.11

Consider the actual demand, its forecast, and the errors for five periods given in Table 2.9. The forecasts are obtained using exponential smoothing method with $\alpha = 0.1$. Compute MAD, MSE, STD, Bias, and MAPE for measuring the accuracy of the forecasting method.

Solution

$$\text{MAD} = \frac{20 + 42 + 18 + 76 + 48}{5} = 40.8$$

$$\text{MSE} = \frac{20^2 + 42^2 + 18^2 + 76^2 + 48^2}{5} = 2113.6$$

$$\text{STD} = \sqrt{\text{MSE}} = \sqrt{2113.6} \approx 46$$

$$\text{Bias} = 20 - 42 - 18 - 76 - 48 = -164$$

$$\text{MAPE} = \frac{1}{5}\left(\left(\frac{20}{160}\right) + \left(\frac{42}{220}\right) + \left(\frac{18}{200}\right) + \left(\frac{76}{260}\right) + \left(\frac{48}{240}\right)\right)100 \approx 18\%$$

TABLE 2.9

Forecast Errors (Example 2.11)

Period (t)	Forecast (F_t)	Actual (D_t)	Error ($e_t = F_t - D_t$)
1	180	160	+20
2	178	220	−42
3	182	200	−18
4	184	260	−76
5	192	240	−48

We can make the following observations from the forecast error calculations:

- Large negative value for the Bias indicates that the method is consistently underestimating the actual demands.
- Large values of MAD, STD, MAPE, and MSE indicate that the method is not forecasting well.

Looking at the time series data, it is clear that the demand is generally increasing. Use of a low value of $\alpha = 0.1$, makes the method less reactive to the actual demands. Hence, an increase of α value is definitely warranted.

Uses of forecast errors
There are several uses of the different forecast errors as described in the following:

1. To select a forecasting method by retrospective testing on past demands.
2. To determine the best values of the parameters for a given forecasting method, e.g., selecting the values of α and β in Holt's method.
3. To monitor how well the selected forecasting method is performing based on new data as they become available.

Selecting the best forecasting method
It is recommended that multiple measures of forecast errors be used in selecting the best forecasting method. If a particular method consistently does well in all the measures, then it is clearly the best method to use. However, it is possible that some methods may do well in some measures and poorly in other measures. In such situations, it is quite common to choose the best two or three methods and use the average of their forecasts as the forecast for the future.

2.10 Monitoring Forecast Accuracy

Forecasts are dynamic and they are updated as more information becomes available. Hence, after selecting an appropriate forecasting method and the forecasts based on that method, it is important to continuously monitor the forecast accuracy. For this, one can use one or more of the forecast errors discussed in Section 2.9. In practice, another measure, called *Tracking Signal*, is also commonly used for monitoring forecast accuracy.

Tracking signal: Tracking signal at period k, denoted by TS_k, is the ratio of Bias and MAD up to period k. In other words,

$$TS_k = \frac{\text{Bias}_k}{\text{MAD}_k} = \frac{\sum_{t=1}^{k} e_t}{\frac{1}{k} \sum_{t=1}^{k} |e_t|} \quad \text{for } k = 1, 2, \ldots, n \quad (2.20)$$

NOTES:

1. The numerator of Equation 2.20 can be positive, negative, or zero, while the denominator is always positive. Hence, TS_k can be positive, negative, or zero for any k. Note that TS_1 is always equal to +1 or −1.

2. Unlike the forecast error defined in Section 2.9, tracking signal is not a single number. Instead it is a series of numbers, which can be used to detect changes in the pattern of the forecast.

3. The generally acceptable values of TS_k are ±6. When the tracking signals go outside these limits, the forecaster should be notified to determine the cause of these limit violations.

4. Tracking signals outside the limits do not automatically imply that the forecasting method is not working. Environmental conditions, such as local economy, sales promotions, new competition, etc., can cause sudden fluctuations in tracking signals.

Example 2.12

Consider the data given in Table 2.9 (Example 2.11). Compute the tracking signals and plot.

Solution

$$TS_1 = \frac{20}{20} = +1$$

$$TS_2 = \frac{20 - 42}{(20 + 42)/2} = -0.7$$

$$TS_3 = \frac{20 - 42 - 18}{(20 + 42 + 18)/3} = -1.5$$

$$TS_4 = \frac{20 - 42 - 18 - 76}{(20 + 42 + 18 + 76)/4} = -3$$

$$TS_5 = \frac{-164}{40.8} = -4$$

Since we have data for five time periods, we have five tracking signals. A plot of the tracking signals is shown in Figure 2.5. A downward trend of tracking signals points to forecasting problems ahead.

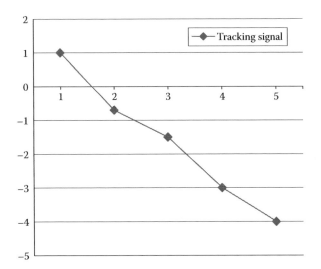

FIGURE 2.5
Plot of tracking signals (Example 2.12).

2.11 Forecasting Software

There are a number of free and commercial forecasting software available in the market. The website www.forecastingprinciples.com has extensive information for both academic and industry users about forecasting. The magazine *OR/MS Today* publishes bi-annual surveys of forecasting software. For the most recent survey, refer to Yurkiewicz (2012). Some of the forecasting software are stand alone dedicated programs, while others are add-ins to Microsoft Excel, SAS, Minitab, etc.

2.11.1 Types of Forecasting Software

Yurkiewicz (2008, 2010, 2012) groups the forecasting software into three categories as follows:

- Automatic (more expensive)
- Semi-automatic (moderate cost)
- Manual (free/inexpensive)

Automatic software: These are designated stand alone programs and may cost thousands of dollars. The user does not have to be proficient in forecasting. The software asks the user to enter the time series data. It selects the appropriate method based on the analysis of the data and recommends a forecasting method. It also computes the optimal values of the parameters (e.g., smoothing constants α, β, and γ for Winters' method) using forecast errors. It then gives the forecasts and provides measures of accuracy, such as confidence interval, MAD, MAPE, Bias, etc. The user can override the recommended forecasting method and investigate other methods. These software programs are quite user friendly. However, tests have shown that they are not very reliable. Note that most automatic software can also operate under semi-automatic and manual mode allowing user intervention.

Semi-automatic software: These are moderately priced but they require the user to have some basic knowledge of forecasting principles and techniques. Here, the user has to select an appropriate forecasting technique based on the analysis of time series data. The software will then compute the optimal parameters for the chosen method using some measure of forecast error. It also gives the forecasts and all the statistics, such as MAD, MAPE, MSE, Bias, etc. The software makes no recommendation as to which forecasting technique is appropriate for the given data.

Manual software: Most of these are free or inexpensive. The user is expected to have very good knowledge of forecasting principles and methods. The user has to select an appropriate method and its parameters. The program will then do the tedious calculations and provide the forecasts and its accuracy in terms of forecast errors. The user has to experiment with different values of the parameters for a particular forecasting method, as well as try different forecasting methods. The user has to decide which method is the best for the given data. The process can be tedious and time-consuming.

A partial list of forecasting software discussed by Yurkiewicz (2012) is given below:

- *Forecast Pro XE* (automatic/semi-automatic/manual)
  A stand alone leading forecasting software for large problems. (Cost: $1300)
  URL: www.forecastpro.com

- *Autobox* (automatic/semi-automatic/manual)
  Another dedicated forecasting software (Cost: $800 to $24,000)
  URL: www.autobox.com

- *PEER forecaster* (automatic/semi-automatic/manual)
  Excel Add-in for Windows (Free)
  URL: www.peerforecaster.com

- *Open forecast* (automatic/semi-automatic/manual)
  A open source *free* forecasting software, part of the Java library.
  URL: openforecast.stevengould.org

- *NCSS 8* (semi-automatic/manual)
  Windows based *EXCEL* Add-in (Cost: $1200)
  URL: www.ncss.com

- *Systat* (manual)
  General purpose statistical software that includes forecasting. (Cost: $1500).
  URL: www.systat.com

- *Minitab statistical software* (semi-automatic/manual)
  Leading statistical software in quality control and education (Cost: $1400)
  URL: www.minitab.com

NOTE: Most vendors have special educational discounts for educational use.

2.11.2 User Experience with Forecasting Software

To evaluate the use and satisfaction with the various forecasting software, Sanders and Manrodt (2003) did a survey of 2400 U.S. Corporations. The survey asked what software they used for forecasting, how satisfied they were with the software, and how well the software did. About 10% (240 companies) responded to the survey. Highlights of the findings by Sanders and Manrodt (2003) are given in the following:

- Only 11% of the respondents used commercial software.
- 48% used Excel spreadsheets.
- 25% used forecasting software developed in-house.
- 6% used outside vendors to develop custom-made software for their companies.
- 10% reported not using any software for forecasting; they primarily used qualitative methods.
- 80% of the firms manually adjusted the forecasts provided by the software.
- 60% of the respondents expressed dissatisfaction with the forecasting software they were using.

2.12 Forecasting in Practice

2.12.1 Real World Applications

There are several *published* results of real world forecasting. We discuss briefly a few of the forecasting applications in practice. For more details and additional applications, the reader is referred to the text by Taylor (Taylor, 2007).

1. *Taco Bell* (Hueter and Swart, 1998): Lunch time sales were forecasted using the moving average method and were used for estimating labor needs at Taco Bell restaurants.

2. *Dell* (Kapuscinski et al., 2004): Monthly forecasts of parts requirements for Dell's suppliers were done using moving average with seasonality. They were used by the suppliers for inventory management.

3. *FedEx* (Xu, 2000): Exponential smoothing with trend and seasonality was used to forecast FedEx's call center loads.

4. *National Car Rental* (Geraghty and Johnson, 1997): Time series forecasting models with seasonality was used by National Car Rental to forecast customer demands for cars and estimate revenues. They were used in the company's revenue management system.

5. *Nabisco* (Amrute, 1998; Barash and Michell, 1998): Sales forecasts of Nabisco products were done using time series forecasting. Moving average method was used for existing products and exponential smoothing with trend was used for new products. The forecasts were then used in production planning decisions.

6. *NBC* (Bollapragada et al., 2008): National Broadcasting Company used a qualitative approach to predict advertising demand for pricing and revenue projections. It used a combination of Delphi method and "grassroots" approach that used customer surveys by sales personnel.

2.12.2 Forecasting in Practice: Survey Results

Sanders and Manrodt (1994) surveyed 500 U.S. companies about their forecasting practices. The objectives of the survey were to ascertain the following:

1. Familiarity with different forecasting methods
2. Choice of forecasting method for short-term, medium-term, and long-term forecasting
3. Reasons for choosing certain forecasting methods
4. Satisfaction with the forecasting methods used

The survey included both manufacturing and service industries with annual sales between $6 million and $15 billion with the median at $1.5 billion. Hence, the results are somewhat skewed toward larger firms. About 20% of the companies (~100) responded to the survey and their responses were used in the analysis. Highlights of the survey results, reported by Sanders and Manrodt (1994), include the following:

- There was no significant difference in the use of the forecasting methods between the small and large companies.
- Not much difference was noted in the responses by manufacturing and service firms. However, manufacturing firms used more quantitative methods (15.5%) compared to service firms (9.6%).
- Overall *moving average method* was the most preferred quantitative method for all firms.

- Most respondents were familiar with the different forecasting methods.
- For long-term (>1 year) and medium-term (6 months to a year) forecasting, most firms preferred qualitative methods, in particular, *executive committee consensus*.
- Among the quantitative methods, *moving average* was the most frequently cited approach for short- and medium-term forecasting. However, for long-term forecasting regression was the most popular choice.
- There was a high degree of satisfaction with *qualitative methods* in general and *executive committee consensus* in particular.
- Among the quantitative methods there was a higher level of satisfaction with *moving average, exponential smoothing, regression,* and *simulation*.
- Difficulty in obtaining data, ease of use, and cost were cited as main reasons for not using sophisticated quantitative methods.

Most managers prefer to *under forecast* (70%) compared to *over forecast* (15%) due to company's reward structure. Company's incentive system generally rewards those who exceed "expectations." Those who over forecast are generally in the manufacturing departments since shortages are expensive and they are held accountable.

2.13 Production Planning Process

As stated at the beginning of the chapter, the success of a manufacturing supply chain depends on its ability to accurately forecast customer demands and plan the production process to meet those demands in time. So far in this chapter, we discussed the forecasting principles and several qualitative and quantitative methods for demand forecasting. Based on the demand forecasts, *aggregate production planning* is done to plan the production.

The general production planning process consists of the following:

1. *Long range planning*: These are strategic planning activities (e.g., location of factories, warehouses, etc.) based on long-term demand forecasts, covering 3–5 years.
2. *Medium range planning*: These are tactical decisions (e.g., labor resources in the near term) covering 6–18 months, based on quarterly or monthly forecasts. They are generally known as *aggregate production planning* decisions or simply *aggregate planning*.

3. *Short range planning*: These are operational decisions, covering 1–2 months, based on daily or weekly demand forecasts. They involve job (order) scheduling, machine loading, overtime planning, and job sequencing and are known as *production scheduling* decisions.

In the remaining sections of this chapter, we will discuss the aggregate planning problem and quantitative methods for its solution.

2.14 Aggregate Planning Problem

Aggregation refers to the idea of focusing on overall capacity, rather than individual products or services. Aggregation is done according to

- Products
- Labor
- Time

Aggregate planning problem: The aggregate planning problem can be stated as follows:

"Given a demand forecast over a planning horizon, develop a plan for production and allocation of resources by making appropriate trade off among capacity, inventory and backlogs."

The aggregate planning problem is basically a medium range tactical supply chain problem. There are primarily three aggregate planning strategies a company can follow.

1. *Chase strategy*: Under this strategy, production rates (workforce levels) are adjusted to match the forecasted demands over the planning horizon. Workforce levels can be changed by hiring or layoff, sub-contracting, use of overtime, use of temporary workers, etc. This is basically a "Follow demand" strategy and maintains very low inventory. Hence, this will be a good strategy when the inventory costs are very high. However, such a policy could create labor unrest.

2. *Level strategy*: Under this strategy, the objective is to maintain a stable workforce, avoiding frequent hiring/firing/layoffs. Here production is not synchronized with demand and inventories are built up during low demand periods for use during high demand periods. This will be a good strategy to follow when the inventory costs are low or labor unions are strong.

3. *Mixed strategy*: Under this strategy, both inventory and workforce levels are allowed to change over the planning horizon. Thus, it is a combination of the "chase" and "level" strategies. This will be a good strategy if the costs of maintaining inventory and changing workforce level are relatively high. Optimization models are generally used to determine an optimal mixed strategy.

2.15 Linear Programming Model for Aggregate Planning

Linear Programming models can be used to determine an optimal strategy between the "chase" and "level" strategies. We shall present the LP model using an Example problem.

Example 2.13

A parts supplier has a contract with an OEM to supply 1550 units of some component C for the next 6 months. Table 2.10 gives the OEM's monthly requirements during the contract period.

The supplier has 20 workers and 50 units of C on hand now. Each worker can produce 10 units of C per month. The company can recruit from the local labor market, but the recruits have to be trained for 1 month by a worker before they can be used in production. Each worker can train at most five recruits during a month. A worker is paid $3000 per month, when used in production or training. A worker can be laid off at a cost of $2000 per month. Firing a worker costs $5000. Each recruit is paid $1500 during training.

Production ahead of schedule incurs an inventory holding cost of $50 per unit per month. Each unit of C not delivered on schedule involves a penalty cost of $75 per month until delivery is completed. However, all deliveries must be completed in 6 months. The supplier requires a final labor force of 30 workers and 50 units of C at the end of the 6th month.

The aggregate planning problem is to decide what hiring, firing, producing, storing, and shortage policy the supplier should follow in order to minimize the total costs during the contract period. Formulate this problem as an LP model and solve. Discuss the production plan.

TABLE 2.10

Demand Data for Example 2.13

Month (t)	Demand (D_t)
1	100
2	200
3	300
4	400
5	300
6	250

Solution

Decision variables: The decision variables of the LP model for month t ($t = 1, 2, \ldots, 6$) are defined as follows:

W_t	Total workers at the beginning of month t, *before firing*
P_t	Workers assigned to production in month t
T_t	Workers assigned to training in month t
L_t	Laid off workers in month t
F_t	Workers fired at the *beginning* of month t
R_t	Total recruits hired at the *beginning* of month t
I_t	Cumulative inventory at the *end* of month t
S_t	Cumulative shortages (backlogs) at the *end* of month t
X_t	Number of units of C produced during month t

Constraints: The model has five constraints to be satisfied each month, as described in the following:

1. Worker balance
   a. Total workforce
   b. Assignment of workforce
   c. Training needs
2. Demand requirements
3. Production capacity

Worker balance constraints (Month t)

1. *Size of the workforce*

$$W_t = W_{t-1} + R_{t-1} - F_{t-1} \quad \text{for } t = 2,3,\ldots,6 \qquad (2.21)$$

Equation 2.21 guarantees that the total number of workers at the beginning of month t will be equal to the number at the beginning of month $t - 1$ plus the number trained in month $t - 1$, minus the number fired at the beginning of month $t - 1$. Since $W_1 = 20$, Equation 2.18 is written for months 2, 3, 4, 5, and 6 only.
   Thus, we write the total workforce constraints as follows:

$$\begin{aligned} W_1 &= 20 & \text{(month 1)} \\ W_2 &= W_1 + R_1 - F_1 & \text{(month 2)} \\ &\vdots \\ W_6 &= W_5 + R_5 - F_5 & \text{(month 6)} \end{aligned}$$

Since we require 30 workers at the end of month 6, we need a constraint:

$$W_6 + R_6 - F_6 = 30$$

2. *Assignment of workforce*

$$W_t = P_t + T_t + L_t + F_t \quad \text{for } t = 1, 2, \ldots, 6 \qquad (2.22)$$

Equation 2.22 guarantees that the total number of workers at the beginning of the month will be assigned to one of the following:

- Production
- Training recruits
- Laid off
- Fired

Thus,

$$W_1 = P_1 + T_1 + L_1 + F_1 \qquad \text{(month 1)}$$
$$\vdots$$
$$W_6 = P_6 + T_6 + L_6 + F_6 \qquad \text{(month 6)}$$

3. *Training*

$$R_t \le 5T_t \quad \text{for } t = 1, 2, \ldots, 6 \qquad (2.23)$$

Equation 2.23 guarantees that each worker can train at most five trainees. Thus,

$$R_1 \le 5T_1 \qquad \text{(month 1)}$$
$$R_2 \le 5T_2 \qquad \text{(month 2)}$$
$$\vdots$$
$$R_6 \le 5T_6 \qquad \text{(month 6)}$$

4. *Demand/inventory balance*

$$X_t + I_{t-1} = D_t + S_{t-1} + I_t - S_t \quad \text{for } t = 1, 2, \ldots, 6 \qquad (2.24)$$

The left hand side of Equation 2.24 is the sum of the current production (X_t) and the inventory carried over (I_{t-1}). Thus, it is the total amount of C available to meet demand in month t. If it exceeds the total requirement, which is the sum of current demand (D_t) and any backlogs carried over (S_{t-1}), then we will have an inventory of I_t at the end of month t. Otherwise, there will be a cumulative backlog of S_t at the end of month t.

Note that the initial inventory (I_0) is 50 and there are no shortages at the beginning ($S_0 = 0$). Hence,

$$X_1 + 50 = 100 + I_1 - S_1 \qquad \text{(month 1)}$$
$$X_2 + I_1 = 200 + S_1 + I_2 - S_2 \qquad \text{(month 2)}$$
$$\vdots$$
$$X_6 + I_5 = 250 + S_5 + I_6 - S_6 \qquad \text{(month 6)}$$

Note that all the requirements of the OEM must be met by the end of the 6th month and the supplier requires 50 units of final inventory. Hence,

$$I_6 = 50$$

$$S_6 = 0$$

5. *Production capacity*

$$X_t \leq 10P_t \quad \text{for } t = 1, 2, \ldots, 6 \qquad (2.25)$$

Equation 2.25 guarantees that each worker can produce *at most* 10 units per month.

Thus,

$$X_1 \leq 10P_1 \qquad \text{(month 1)}$$
$$X_2 \leq 10P_2 \qquad \text{(month 2)}$$
$$\vdots$$
$$X_6 \leq 10P_6 \qquad \text{(month 6)}$$

6. *Non-negativity constraints*

$$P_t, T_t, L_t, F_t, R_t, I_t, S_t, X_t \geq 0 \quad \text{for all } t = 1, 2, 3, \ldots, 6$$

Objective function: The objective function represents the sum of the following costs:

- Wages of production workers
- Wages of laid off workers
- Cost of fired workers
- Cost of trainees hired
- Wages of workers assigned to training
- Inventory holding cost
- Backorder cost

Equation 2.26 represents the objective function to be minimized:

$$\text{Minimize } Z = 3000 \sum_{t=1}^{6} P_t + 2000 \sum_{t=1}^{6} L_t + 5000 \sum_{t=1}^{6} F_t + 1500 \sum_{t=1}^{6} R_t$$

$$+ 3000 \sum_{t=1}^{6} T_t + 50 \sum_{t=1}^{6} I_t + 75 \sum_{t=1}^{6} S_t \qquad (2.26)$$

Solution to the LP model: The LP model resulted in 54 variables and 34 constraints. The LP model was solved using Microsoft Excel's Solver add-in in seconds. The optimal solution is given in Tables 2.11 and 2.12.

A review of Tables 2.11 and 2.12 indicates that the optimal aggregate production plan is much closer to the "level strategy" than the "chase strategy." New workers are hired in months 2 and 3 to meet the surge in demand during months 4, 5, and 6 and the total workforce remains stable otherwise. No worker is fired or laid off and inventory is built in the first 3 months to meet the demand surge later. The main reason for this strategy is that the cost of hiring, firing, and lay off are much higher, compared to the inventory holding cost.

TABLE 2.11

Optimal Production Plan for Example 2.13

Month	Initial Inventory	Production	Demand	Inventory	Shortage
1	50	200	100	150	0
2	150	180	200	130	0
3	130	270	300	100	0
4	100	300	400	0	0
5	0	300	300	0	0
6	0	300	250	50	0

TABLE 2.12

Workforce Analysis for Example 2.13

Month	Total Workforce	Production	Training	Recruits	Fired	Laid Off
1	20	20	0	0	0	0
2	20	18	2	8	0	0
3	28	27	1	2	0	0
4	30	30	0	0	0	0
5	30	30	0	0	0	0
6	30	30	0	0	0	0

2.16 Nonlinear Programming Model for Aggregate Planning

The LP model, illustrated in Section 2.15, can be extended to solve a more general aggregate planning problem. For the development of the general model, we use the following notations:

T	Total number of time periods in the planning horizon
D_t	Forecasted demand in period $t = 1, 2, ..., T$
x_t	Amount produced in period $t = 1, 2, ..., T$
y_t	An *unrestricted variable* (positive or negative) denoting the inventory ($y_t > 0$) or shortages ($y_t < 0$) in period $t = 1, 2, ..., T$

Objective function: We consider three types of cost for the objective function.

1. *Production cost*: Let $C_t(x_t)$ represent the nonlinear cost of producing x_t units in period t. For example, $C_t(x_t)$ could be a convex, piecewise linear, cost function with per unit cost increasing depending on whether the units are produced in regular time, over time, or obtained through outsourcing. Fixed cost can also be included as part of the production cost.

2. *Cost of changing production*: Production quantities can be changed by changing the workforce level (hiring, firing, lay off, etc.). Let $P_t(x_{t-1}, x_t)$ represent the cost of changing production from period $(t-1)$ to period t. Note that $x_t > x_{t-1}$, if the workforce level is increased and $x_t < x_{t-1}$, if there is a reduction in the workforce.

    Some of the functional forms for the production change costs are the following:

    (i) $P_t(x_{t-1}, x_t) = p_t |x_t - x_{t-1}|$

    (ii) $P_t(x_{t-1}, x_t) = \begin{cases} p_t^+(x_t - x_{t-1}), & \text{if } x_t > x_{t-1} \\ p_t^-(x_{t-1} - x_t), & \text{if } x_t < x_{t-1} \end{cases}$

    (iii) $P_t(x_{t-1}, x_t) = p_t(x_t - x_{t-1})^2$

    In case (i), the changeover cost is the same irrespective of whether the production is increased or decreased. In case (ii), the changeover cost is linear, but it depends on whether the production change is positive or negative. In case (iii), the quadratic cost function penalizes larger production changes more.

3. *Inventory/shortage cost*: The unrestricted variable y_t represents inventory, if $y_t > 0$ and backlogs (shortage), when $y_t < 0$. An example of the cost function will be

$$f_t(y_t) = \begin{cases} h_t y_t, & \text{if } y_t > 0 \\ s_t |y_t|, & \text{if } y_t < 0 \end{cases}$$

where
   h_t represents the per unit inventory holding cost
   s_t is the per unit cost of backlogs in period t

The tradeoff between cost incurred in changing production and allowing inventory/shortages is the basic question in most aggregate planning problems. Thus, the general nonlinear programming model for aggregate planning can be stated as follows:

Minimize

$$Z = \sum_{t=1}^{T} C_t(x_t) + \sum_{t=1}^{T} P_t(x_{t-1}, x_t) + \sum_{t=1}^{T} f_t(y_t) \qquad (2.27)$$

Subject to

$$x_t + y_{t-1} - y_t = D_t \quad \text{for } t = 1, 2, \ldots, T \qquad (2.28)$$

$$y_L \le y_t \le y_U \qquad (2.29)$$

$$x_L \le x_t \le x_U \qquad (2.30)$$

$$|x_t - x_{t-1}| \le P_L \qquad (2.31)$$

Equation 2.27 represents the objective function, which is the sum of the production cost, production changeover cost, and inventory shortage costs. Equation 2.28 is the demand/inventory balance constraint. If the unrestricted variable y_t is replaced by the difference of two non-negative variables as

$$y_t = I_t - S_t \qquad (2.32)$$

$$I_t, S_t \ge 0$$

then I_t represents inventory, when $y_t > 0$ and S_t represents cumulative short-age, where $y_t < 0$. Substituting Equation 2.32 in Equation 2.28, we get

$$x_t + I_{t-1} - S_{t-1} - I_t + S_t = D_t \qquad (2.33)$$

Equation 2.33 is the same demand/inventory balance constraint that we developed for the LP model in Section 2.15 (see Equation 2.24).

Equation 2.29 puts limits on the inventory/shortages in any period. If the lower bound y_L is set to zero, then it represents the case when no short-ages are allowed. If y_L is negative, it limits the cumulative shortage to a certain maximum. The upper bound y_U can represent warehouse capacity for storage.

Equation 2.30 limits the production quantity in each period. If the lower limit x_L is positive, then a certain minimum production is guaranteed in each period.

Equation 2.31 can represent maximum changes (increase/decrease) to production due to union contracts or company's labor policy. Note that Equation 2.31 can be written as two linear constraints:

$$x_t - x_{t-1} \le P_t$$

$$-x_t + x_{t-1} \le P_t$$

Thus, the general model, given by Equations 2.27 through 2.33, would represent most aggregate planning situations that may arise in practice. If all the cost functions are linear, then we get the LP model discussed in Section 2.15.

2.17 Aggregate Planning as a Transportation Problem

Transportation problems represent a special class of LP problems that are easier to solve. In this section, we shall discuss the basics of a transportation model, how aggregate planning problems can be formulated as transportation problems, and a "greedy" algorithm to solve special cases of aggregate planning problems by inspection.

2.17.1 Basic Transportation Problem (Ravindran et al., 1987)

Transportation problems are generally concerned with the distribution of a certain product from several sources to numerous facilities at minimum cost. Suppose there are m warehouses where a commodity is stocked, and n markets where it is needed. Let the supply available in the warehouses be

$a_1, a_2, \ldots, a_m$, and the demands at the markets be $b_1, b_2, \ldots, b_n$. The unit cost of shipping from warehouse i to market j is c_{ij}. (If a particular warehouse cannot supply a certain market, we set the appropriate c_{ij} to $+\infty$.) We want to find an optimal shipping schedule that minimizes the total cost of transportation from all warehouses to all the markets.

Linear programming formulation: To formulate the transportation problem as a linear program, we define x_{ij} as the quantity shipped from warehouse i to market j. Since i can assume values from $1, 2, \ldots, m$ and j from $1, 2, \ldots, n$, the number of decision variables is given by the product of m and n. The complete formulation is given in the following:

Minimize

$$Z = \sum_{i=1}^{m} \sum_{j=1}^{n} c_{ij} x_{ij}$$

Subject to

$$\sum_{j=1}^{n} x_{ij} \leq a_i \quad \text{for } i = 1, 2, \ldots, m \text{ (supply restriction at warehouse } i)$$

$$\sum_{i=1}^{m} x_{ij} = b_j \quad \text{for } j = 1, 2, \ldots, n \text{ (demand requirement at market } j)$$

$$x_{ij} \geq 0 \quad \text{for all pairs } (i, j) \text{ (non-negative restrictions)}$$

The supply constraints guarantee that the total amount shipped from any warehouse does not exceed its capacity. The demand constraints guarantee that the total amount shipped to a market meets the minimum demand at that market.

It is obvious that the market demands can be met if and only if the total supply at the warehouse is at least equal to the total demand at the markets. In other words, $\sum_{i=1}^{m} a_i \geq \sum_{j=1}^{n} b_j$. When the total supply equals the total demand $\left(\text{i.e., } \sum_{i=1}^{m} a_i = \sum_{j=1}^{n} b_j\right)$, every available supply at the warehouses will be shipped to meet the minimum demands at the markets. In this case, all the supply and demand constraints would become strict equations, and we call this a *standard transportation problem*.

An *unbalanced* transportation problem, where the total supply *exceeds* total demand, can be converted to a standard transportation problem by creating a dummy market to absorb the excess supply available at the warehouses. The unit cost of shipping from any warehouse to the dummy market is assumed to be zero since in reality the dummy market does not exist and no physical transfer of goods takes place. Thus the unbalanced transportation problem is equivalent to the following standard problem:

Minimize

$$Z = \sum_{i=1}^{m} \sum_{j=1}^{n+1} c_{ij} x_{ij}$$

Subject to

$$\sum_{j=1}^{n+1} x_{ij} = a_i \quad \text{for } i = 1, 2, \ldots, m$$

$$\sum_{i=1}^{m} x_{ij} = b_j \quad \text{for } j = 1, 2, \ldots, n+1$$

$$x_{ij} \geq 0 \quad \text{for all pairs } (i, j)$$

where
$j = n + 1$ is the dummy market with demand $b_{n+1} = \sum_{i=1}^{m} a_i - \sum_{j=1}^{n} b_j$
$c_{i,n+1} = 0$ for all $i = 1, 2, \ldots, m$

Note that the value of $x_{i,n+1}$ denotes the unused supply at warehouse i.

An important feature of the standard transportation problem is that it can be expressed in the form of a table, which displays the values of all the data coefficients (a_i, b_j, c_{ij}) associated with the problem. In fact, the constraints and the objective function of the transportation model can be read directly from the table. A *standard transportation table* for three warehouses and four markets is shown in Table 2.13.

The supply constraints can be obtained by merely equating the sum of all the variables in each row to the warehouse capacities. Similarly the demand

TABLE 2.13

Standard Transportation Table

		Markets				
		M_1	M_2	M_3	M_4	Supplies
Warehouses	W_1	x_{11} $\quad c_{11}$	x_{12} $\quad c_{12}$	x_{13} $\quad c_{13}$	x_{14} $\quad c_{14}$	a_1
	W_2	x_{21} $\quad c_{21}$	x_{22} $\quad c_{22}$	x_{23} $\quad c_{23}$	x_{24} $\quad c_{24}$	a_2
	W_3	x_{31} $\quad c_{31}$	x_{32} $\quad c_{32}$	x_{33} $\quad c_{33}$	x_{34} $\quad c_{34}$	a_3
Demands		b_1	b_2	b_3	b_4	

constraints are obtained by equating the sum of all the variables in each column to the market demands.

Efficient algorithms exist for solving transportation problems. Interested readers are referred to the text by Ravindran et al. (1987).

2.17.2 Aggregate Planning as a Transportation Problem

Before we discuss the general model, we use an example to illustrate how certain aggregate planning problems can be formulated as transportation problems.

Example 2.14 (Ravindran et al., 1987)

Consider the problem of scheduling the weekly production of a certain item for the next 4 weeks. The production cost of the item is $10 for the first 2 weeks, and $15 for the last 2 weeks. The weekly demands are 300, 700, 900, and 800, which must be met. The plant can produce a maximum of 700 units each week. In addition the company can employ overtime during the 2nd and 3rd weeks. This increases the weekly production by an additional 200 units, but the production cost increases by $5 per item. Excess production can be stored at a unit cost of $3 per week. How should the production be scheduled so as to minimize the total costs?

Solution

To formulate this as a transportation problem, we consider the production periods as warehouses, and weekly demands as markets. Since overtime production is possible during the 2nd and 3rd weeks, there are six supply points. The decision variables are as follows:

x_{1j}	Normal production in week 1 for use in week j for $j = 1, 2, 3, 4$
x_{2j}	Normal production in week 2 for use in week j for $j = 2, 3, 4$
x_{3j}	Overtime production in week 2 for use in week j for $j = 2, 3, 4$
x_{4j}	Normal production in week 3 for use in week j for $j = 3, 4$
x_{5j}	Overtime production in week 3 for use in week j for $j = 3, 4$
x_{6j}	Normal production in week 4 for use in week 4

Since the total production (normal and overtime) exceeds the total demand, we create a dummy market to absorb the excess supply. Table 2.14 gives the corresponding transportation model.

Remarks:

1. The cells in the transportation table (Table 2.14) represent the various ways the weekly demands are met. For example, cell (1, 3) represents the amount of week 3's demand met by production in week 1. Hence $c_{13} = 16$, which represents the sum of the production cost ($10) and storage cost for 2 weeks at $3 per week.
2. Some of the cost elements in Table 2.14 are set at M (a very large value) to denote shipments that are not possible. For example,

TABLE 2.14

Transportation Table for Example 2.14

		Demands				
		Week 1	Week 2	Week 3	Week 4	Dummy	
	Week 1						700
		10	13	16	19	0	
	Week 2 (Normal)						700
		M	10	13	16	0	
	Week 2 (Overtime)						200
Supplies		M	15	18	21	0	
	Week 3 (Normal)						700
		M	M	15	18	0	
	Week 3 (Overtime)						200
		M	M	20	23	0	
	Week 4 (Normal)						700
		M	M	M	15	0	
		300	700	900	800	500	

since no shortages are allowed, production during weeks 2, 3, and 4 cannot possibly supply the first week's demand. Hence $c_{i1} = M$ for $i = 2, 3, ..., 6$.

3. Weekly storage cost of $3 per unit is added to the production cost whenever an item is stored to meet future demand.

The problem can now be solved by the transportation algorithm. In fact, Example 2.14 can be solved by inspection using a *greedy algorithm*. That will be discussed in Section 2.17.3.

We shall now discuss a general transportation model that allows for both inventory and shortages in aggregate planning problems.

General transportation model for aggregate planning (Ravindran et al., 1987; Taha, 1987): In order to formulate an aggregate planning problem as a transportation problem, we make the following assumptions to the general aggregate planning model discussed in Section 2.16:

1. Production cost is a piece-wise linear convex cost function.
2. Zero setup cost for production.
3. No production changeover cost.
4. Inventory and shortage costs are linear.

We are given the demand forecasts over the planning horizon T. Regular and overtime productions are available with finite capacities.

Shortages are allowed but all backlogs must be filled by the Tth period. The aggregate planning problem is to determine the optimal production plan that minimizes the total cost of production, inventory, and shortage.

Notations:

D_t	Demand for period t, $t = 1, 2, ..., T$
c_t	Regular time production cost per unit in period t
p_t	Overtime production cost per unit in period t $(p_t > c_t)$
h_t	Inventory holding cost per unit in period t
b_t	Backorder (shortage) cost per unit in period t
R_t	Regular time production capacity in period t
O_t	Overtime production capacity in period t

NOTE: Because all demands have to be satisfied by period T,

$$\sum_{t=1}^{T} (R_t + O_t) \geq \sum_{t=1}^{T} D_t.$$

In order to formulate the aggregate planning problem as a transportation problem, we do the following:

- Each type of production, regular or overtime, in each period will be considered as warehouses. Thus, there will be $2T$ warehouses.
- Each period's demand will be treated as a market. Thus, we will have T markets.
- The cost of shipping from a warehouse to a market will be the sum of the production cost and inventory/backorder cost up to that period.

The transportation table is illustrated in Table 2.15. There are $2T$ warehouses and T markets. The warehouse capacities are the respective regular time and overtime production capacities. The market demands are the forecasted demands for periods $1, 2, ..., T$. The cells in the transportation table (Table 2.15) represent the different ways the market demands can be met. For example, cell (1, 3) represents the amount of period 3's demand met by warehouse 1 (regular time production in period 1). Hence, $c_{13} = c_1 + h_1 + h_2$, which represents the sum of the production cost (c_1) and the storage costs for periods 1 and 2 ($h_1 + h_2$). Similarly, cell (4, 1) represents the amount of period 1's demand met by warehouse 4 (namely overtime production in period 2). This represents the case of backorders. Hence $c_{41} = p_2 + b_1$, the overtime production cost in period 2 plus period 1's backorder cost.

The advantage of the transportation formulation is that the aggregate planning problem can be solved more efficiently by using a transportation algorithm (Ravindran et al., 1987). In fact, for the special case when no shortages are allowed, the transportation problem can be solved by inspection using a *greedy algorithm*. We shall illustrate the greedy algorithm using Example 2.15.

TABLE 2.15

Aggregate Planning as a Transportation Problem

			Markets			
Warehouses	1	2	3	...	T	Supply
1 (RT$_1$)	c_1	$c_1 + h_1$	$c_1 + h_1 + h_2$	...	$c_1 + h_1 + h_2$ $+ \cdots + h_{T-1}$	R_1
2 (OT$_1$)	p_1	$p_1 + h_1$	$p_1 + h_1 + h_2$		$p_1 + h_1 + h_2$ $+ \cdots + h_{T-1}$	O_1
3 (RT$_2$)	$c_2 + b_1$	c_2	$c_2 + h_2$		$c_2 + h_2 + h_3$ $+ \cdots + h_{T-1}$	R_2
4 (OT$_2$)	$p_2 + b_1$	p_2	$p_2 + h_2$		$p_2 + h_2 + h_3$ $+ \cdots + h_{T-1}$	O_2
$\vdots$						
2T-1 (RT$_T$)	$c_T + b_1 + b_2$ $+ \cdots + b_{T-1}$	$c_T + b_2 + b_3$ $+ \cdots + b_{T-1}$	$c_T + b_3 + b_4$ $+ \cdots + b_{T-1}$		c_T	R_T
2T (OT$_T$)	$p_T + b_1 + b_2$ $+ \cdots + b_{T-1}$	$p_T + b_2 + b_3$ $+ \cdots + b_{T-1}$	$p_T + b_3 + b_4$ $+ \cdots + b_{T-1}$		p_T	O_T
Demands	D_1	D_2	D_3		D_T	

2.17.3 Greedy Algorithm for Aggregate Planning

Greedy algorithm (Johnson, 1957; Taha, 1987): The basic principle of the *greedy algorithm* is to begin with week 1 and successively assign the lowest cost cells in column 1 to meet week 1's demand subject to the supply and demand constraints. The warehouse capacities are updated based on week 1's assignment. The algorithm then goes to week 2 and successively assigns the cells in column 2 to meet week 2's demand and so on.

Example 2.15

Consider the aggregate planning problem given in Example 2.14, where no shortages are allowed and weekly demands must be satisfied at all times. Solve the transportation problem using the greedy algorithm.

Solution

The transportation formulation given in Table 2.14 is reproduced in Table 2.16, where the warehouses represent the regular (RT)/overtime (OT) production each week and the markets are the weekly demands.

Applying the greedy algorithm to Example 2.15 (Table 2.16), we get the following steps:

Step 1: Begin with week 1. The lowest cost cell is X_{11}. Assign X_{11} a value as large as possible consistent with the supply at warehouse 1 and demand of market 1. Thus $X_{11} = \max (700, 300) = 300$. This means that week 1's demand is satisfied and the remaining supply in warehouse 1 is now 400. Hence, we set the values of the remaining cells under column 1 to zero, namely $X_{i1} = 0$ for $i = 2, \ldots, 6$.

TABLE 2.16

Solution of Example 2.15 by the Greedy Algorithm

		Weekly Demands				
		1	2	3	4	Capacity
	1 (RT$_1$)	300 · · · 10	0 · · · 13	200 · · · 16	100 · · · 19	~~700, 400,~~ ~~200, 100~~
	2 (RT$_2$)	0 · · · M	700 · · · 10	0 · · · 13	0 · · · 16	~~700,~~ 0
Supplies	3 (OT$_2$)	0 · · · M	0 · · · 15	0 · · · 18	0 · · · 21	200
	4 (RT$_3$)	0 · · · M	0 · · · M	700 · · · 15	0 · · · 18	~~700,~~ 0
	5 (OT$_3$)	0 · · · M	0 · · · M	0 · · · 20	0 · · · 23	200
	6 (RT$_4$)	0 · · · M	0 · · · M	0 · · · M	700 · · · 15	~~700,~~ 0
	Demand	~~300~~ 0	~~700~~ 0	~~900~~ ~~200~~ 0	~~800~~ ~~100~~ 0	

Step 2: Go to week 2. The lowest cost cell in column 2 is X_{22} and assign $X_{22} = \min (700, 700) = 700$. Thus warehouse 2's supply is exhausted and week 2's demand is also met. Hence, the remaining supply at warehouse 2 is set to zero and remaining demand at week 2 is also zero. Thus, set the variables, $X_{2j} = 0$ for $j = 3, 4$ and $X_{i2} = 0$ for $i = 1, 3, 4, 5, 6$.

Step 3: Go to week 3. Even though the lowest cost cell in column 3 is X_{23}, it *has already been assigned* a value zero in step 2 and hence cannot be used to meet week 3's demand. The next lowest cost cell *that has not been assigned* is X_{43}. Set $X_{43} = \min (700, 900) = 700$. Warehouse 3's capacity is exhausted and X_{44} is set to zero. The remaining demand for week 3 is $900 - 700 = 200$. The next lowest cost cell in column 3 is X_{13}. Set $X_{13} = \min (400, 200) = 200$. Week 3's demand is reduced to zero and the capacity of warehouse 1 is reduced to 200.

Step 4: Go to week 4. Using the greedy algorithm, X_{64} is assigned first at 700, followed by X_{14} at 100.

NOTE: Since the total supply exceeds the total demand, there are excess capacities left in warehouse 1 (100), warehouse 3 (200), and warehouse 4 (200). In other words, only 600 of 700 production capacity at regular time in week 1 is used. Overtime capacities in weeks 2 and 3 are never used.

The minimum cost of the production plan is,

$$Z = 300\ (10) + 200\ (16) + 100\ (19) + 700\ (10) + 700\ (15) + 700\ (15) = \$36,100.$$

2.18 Aggregate Planning Strategies: A Comparison

To summarize, there are two distinct strategies a company can follow for aggregate planning. The *chase strategy* adjusts the production quantities to follow the demands, resulting in continuous changes in the production levels and workforce. On the other hand, the *level strategy* maintains a near uniform production level with a constant workforce. Here, inventory/backlogs are used to meet the demand variations.

Table 2.17 gives a comparison of the two aggregate planning strategies with respect to production, workforce, inventory, and other factors.

The managerial issues that affect the choice of the aggregate planning strategy are the following:

1. There is no uniform recommendation that one strategy is better than the other. The optimal strategy has to be tailored to each company and the prevailing environment.

2. Union contracts and/or company policies may constrain the choice of a particular strategy.

3. Even though the optimization models provide an optimal strategy based on all relevant costs, it has to be balanced with managerial judgment and experience before implementation.

4. As we discussed earlier, *aggregate production planning* is a tactical decision in supply chain management. *Production scheduling* is an operational decision. It is important to understand the key differences between the two, with respect to the time horizon and the level of detail in the production plan.

TABLE 2.17

Comparison of Aggregate Planning Strategies

Strategy	Production	Workforce	Inventory	Backlog	Hire/ Layoff	Sub-Contracting
Chase	Large variation	Large variation	Generally low	Generally low	Very high	More
Level	Small variation	Small variation	Generally large	Generally large	Very low	Less

2.19 Summary and Further Readings

Most manufacturing companies follow *make-to-stock* policy, that is, to produce in anticipation of the customer demands. Hence, forecasting customer demands is essential for production planning. Based on the forecast, aggregate production plans are generated to produce on time to meet the customer demands. In this chapter, we discussed commonly used forecasting methods in practice and optimization models for aggregate planning. In this section, we shall summarize the forecasting methods discussed in this chapter and suggest further readings about other forecasting methods and related topics available in the literature.

2.19.1 Demand Forecasting: Summary

We discussed two broad categories of forecasting methods—*Qualitative or judgmental methods* and *Quantitative or statistical methods*. Qualitative methods are usually based on human judgments and do not require past data. They are most useful for new products or services. Under qualitative methods, we discussed *executive committee census, Delphi method, survey of sales force*, and *customer surveys*. These methods are quite frequently used by companies, particularly for making strategic decisions, as well as to supplement quantitative forecasts. Quite often, the companies choose the final forecast as the average of the qualitative and quantitative forecasts. Empirical evidence (Frances, 2011) suggests that the averaging strategy generally produces a much better forecast.

Under quantitative methods, we primarily discussed *time series forecasting* methods. They included *naïve method, averaging method, moving average method, weighted moving average method*, and *exponential smoothing method*. We also discussed modifications to the time series methods to incorporate seasonality and trend (*Holt's method* and *Winter's method*).

2.19.2 ARIMA Method

One of the time series methods that we have not discussed, but very popular in the academic circles, is called *ARIMA (auto regressive integrated moving average)* method developed by Box and Jenkins (1976). In this method, a mathematical model is fitted that is optimal with respect to the historical time series data. In fact, exponential smoothing method becomes a special case of ARIMA model. ARIMA requires an enormous amount of past data. It has been found to be good for short-time (<3 months) and medium-term (3 months to 2 years) forecasting. Because of its mathematical sophistication and huge data requirements, it has not been used widely in practice. In the "Forecasting in practice" survey, (Sanders and Manrodt, 1994), less than 5% of the survey respondents have reported using the ARIMA method, even though nearly half are familiar with the method. Those interested in learning more about ARIMA, can refer to the textbook by Box and Jenkins (1976) and the practical guide to ARIMA by Hoff (1983).

2.19.3 Croston's Method (Croston, 1972)

Another quantitative forecasting method that we did not discuss was Croston's method. This is applicable for intermittent erratic or slow-moving demand. For example, a machine may not fail for several periods and may not need any spare parts. Croston's method considers separately the two random events—the time between (nonzero) demands and the size of the demand when it happens. It then combines the two events, using exponential smoothing, to generate the forecast. Under Croston's method, forecasts change only after a demand happens and is otherwise constant between demands. Forecast increases, when there is a large demand or shorter time between demands. Forecast decreases, when there is small demand or a long time between demands.

Interested readers can refer to the original paper by Croston (1972). Willemain et al. (1994) did a comparative evaluation of Croston's method in forecasting intermittent demand in manufacturing using industrial data. They concluded that Croston's method was superior to the exponential smoothing method and was robust even under situations when Croston's model assumptions were violated.

2.19.4 Further Readings in Forecasting

The forecasting website www.forecastingprinciples.com has extensive information to those interested in learning more about forecasting. There are separate sections for researchers and practitioners. The researchers' section discusses the current state-of-the-art forecasting knowledge and the research needs in forecasting. It also contains a list of current research papers on new forecasting principles. The practitioners' section has information on how to select a forecasting method and essential reading materials for the practitioners. It also has a list of consultants available for forecasting, a discussion group to exchange ideas and a list of forecasting courses offered by universities and companies.

The forecasting website also contains a list of textbooks, trade books, and journals in forecasting. The textbook by Armstrong (2001) is a good handbook on forecasting principles, written for both practitioners and researchers. It is a collection of articles by several authors. A structured approach to demand-driven forecasting is discussed in Chase (2009). In addition to the various time series forecasting methods, the book also discusses *causal forecasting* including regression and Box-Jenkin's ARIMA methods. The textbook by Shim (2000) addresses strategic forecasting issues. Essentials of forecasting methods are presented with examples and real-life cases. In addition to demand forecasting, applications to forecasting cost, earnings, cash flows, interest rates, and foreign exchanges are discussed. The text on business forecasting by Evans (2010) is available as paperback and e-book editions. It is appropriate for a wide range of readers with diverse backgrounds. Case studies are used to discuss both short-term and long-term forecasting approaches.

2.19.5 Production Planning: Further Readings

The total problem of planning production in a supply chain consists of the following decisions:

- Demand forecasting
- Facilities planning
- Facilities layout
- Aggregate planning
- Material requirements planning
- Production scheduling

There are no methods or models that answer all questions simultaneously. There are numerous models and methods that consider each of these decisions as a sub-problem to be optimized.

Demand forecasting forms the basis of production planning. *Facilities planning* are based on long-term demand forecasts. They determine where to locate plants, warehouses, DCs, etc., and what products to make where. Mathematical models and methods for facilities planning will be discussed in Chapter 5. In this chapter, we discussed methods for aggregate planning decisions which are usually made a few months in advance.

Aggregate Production Plans give the optimal allocation of labor and capital resources for the next quarter's operations. Aggregate plans lead to *materials requirement planning* and *production scheduling*. Production scheduling is an operational decision at the shop floor level that occurs a few hours or days in advance of the actual production. Those interested in learning more about *materials requirement planning* and *production scheduling*, should refer to the Industrial Engineering handbooks by Salvendy (2001) and Badiru (2006). Other sources include Foote and Murty (2008, 2009), Nahmias (1993), and Arnold et al. (2012).

2.19.6 Managing Demand

Aggregate planning manages supply (capacity and inventory) to handle demand variability. Hence it is a reactive process given a demand forecast. Typically, manufacturing departments are responsible for aggregate planning.

Another approach to manage demand variability is to influence it through pricing and promotions, which are typically controlled by marketing and sales. The impact of promotions on demand is threefold:

1. *Market growth*: Overall demand increases from existing as well as new customers.
2. *Stealing market share*: Competitor's customers may switch and become new customers.
3. *Forward buying*: Existing customers simply change the timing of the purchase to take advantage of the lower prices.

A good discussion on timing of the promotions and their desired impacts is available in Chopra and Meindl (2010). It is very important that manufacturing and sales coordinate their efforts to successfully match supply and demand.

2.19.7 Bullwhip Effect

In a *decentralized supply chain* the various stages (Suppliers, Manufacturers, Distributors, and Retailers) are owned by different companies. Hence, there is limited communication and information sharing among the trading partners. The retailers are closest to the end-customer demand and can plan their replenishment strategies by forecasting that demand. Since the end-customer demand is generally not shared with the *upstream* trading partners, the manufacturers have to plan production based on *orders* from the downstream partners. It has been found in practice that the demand uncertainty becomes more and more distorted, resulting in increased fluctuations in orders as they move up in the supply chain, from retailers to manufacturers to suppliers. This phenomenon is known as the *bullwhip effect* in a supply chain.

Bullwhip effect was first observed by Procter and Gamble (P&G) in the sale of baby diapers (Lee et al., 1997). Even though diaper sales at the retailers were stable over time, wholesale orders to factories surged up and down, swinging widely over time. Thus, there was very little uncertainty in the end-customer demand, but a very high uncertainty in the orders to factories and suppliers. Similar phenomenon was also observed by HP, apparel manufacturers, and the grocery industry.

The primary causes of the bullwhip effect are the following:

- No knowledge of end-customer demand
- Promotional sales
- Inflated orders by retailers who fear supply shortage
- Long lead times

Bullwhip effect results in poor aggregate production plans that lead to increased safety stocks, reduced customer service due to shortages, increased transportation cost, and inefficient allocation of resources (labor and equipment). By increasing the communication of actual downstream demand and collaboration between trading partners, the bullwhip effect can be minimized. One such collaborative process is known as *collaborative planning, forecasting and replenishment* (CPFR). Refer to the appendix in Chapter 3 of this book for a detailed description of the bullwhip effect in supply chains.

2.19.8 Collaborative Planning, Forecasting and Replenishment (CPFR)

CPFR is a business practice that better aligns supply and demand through the sharing of business data among multiple trading partners. The main focus in CPFR is the creation of a joint forecast of end-customer demand for all supply and demand planning by the trading partners. The earliest reported success of CPFR was the collaboration between Warner Lambert and Wal-Mart. They shared information regarding demand forecasts, production scheduling, and distribution activities. In 1998, the *voluntary inter-industry commerce standards* (VICS) association has published guidelines for CPFR implementation (later updated in 2001). Since then, many companies have successfully implemented CPFR.

For detailed information on CPFR, the reader is referred to the VICS website (www.vics.org). CPFR guidelines, case studies, and other information can be ordered from the website. VICS also provides online CPFR education, workshops, and certification programs. McKaige (2001) gives an excellent Q and A primer on CPFR.

End-to-end communication among trading partners minimizes the bullwhip effect, which results in decreases in inventory at all levels in the supply chain. VICS conducted a survey of retail companies using CPFR pilot programs in 1998–1999. They reported that due to CPFR there was an 80% increase in sales with the CPFR partners, 10% reduction in inventory, improved fill rates, and near 100% service levels. One of the large and best practices of CPFR implementations has been between P&G and Wal-Mart. Several CPFR implementations between other companies are presented as case studies in the VICS website.

Exercises

2.1 Explain the differences between *qualitative* and *quantitative* forecasting methods. Which method is applicable under what conditions?

2.2 What are the advantages and disadvantages of the *Exponential Smoothing* method?

TABLE 2.18

Sales Data for Exercise 2.6

Month	Sales
1	34
2	33
3	42
4	34
5	36
6	43
7	34
8	33
9	43
10	31
11	35
12	41

2.3 Name at least three practical uses of forecast errors.

2.4 State three differences between the *chase strategy* and the *level strategy* in aggregate planning. Which strategy is better under what conditions?

2.5 Under what assumptions an aggregate planning problem can be formulated as a transportation model? What additional assumptions are needed to solve the transportation model by the *greedy algorithm*?

2.6 The monthly sales data for the year 2011 for Datastream, Inc., which produces network routers for small companies, are given in Table 2.18.

The sales manager wants a forecast of sales for the first quarter of year 2012 (months 13–15). Compute the forecasts using the following methods:

(a) Last value method

(b) Averaging method

(c) Three-month moving average method

(d) Exponential smoothing with $\alpha = 0.25$

(e) Holt's method with $\alpha = 0.4$ and $\beta = 0.5$

(f) Given the sales pattern for 2011, do any of these methods seem inappropriate for obtaining the forecasts? Is there another method you would recommend to get a better forecast for months 13, 14, and 15.

2.7 Joe Kool has decided to use *3-month weighted moving average* method for one of his company's products. Last year's monthly sales data are given in Table 2.19.

Joe is trying to decide the appropriate weights to select for his forecasting method that will minimize the sum of the absolute errors

TABLE 2.19

Sales Data for Exercise 2.7

Month	Sales
1	5325
2	5405
3	5200
4	5510
5	5765
6	5210
7	5375
8	5585
9	5460
10	4905
11	5755
12	6320

between the forecasts and the actual values. Show that the determination of the optimal weights can be formulated as an LP problem. Solve the linear program using any optimization software and determine the forecast for month 13.

2.8 Consider the various forecasting methods in Exercise 2.6. Compute their forecasting errors using MAD and BIAS. Based on the forecast errors, which method would you recommend and why?

2.9 ABC company is experimenting with two forecasting methods for its product. Table 2.20 gives the actual sales and the forecasts obtained by the two methods during the past 6 months.

(a) Calculate MAD, MSE, and BIAS for the two methods.

(b) Calculate and plot the tracking signals for the two methods.

(c) Are the methods under-forecasting or over-forecasting? Which method would you recommend for forecasting future sales and why?

TABLE 2.20

Sales and Forecast Data for Exercise 2.9

Month	Actual Sales	Forecast—1	Forecast—2
1	558	532	521
2	490	541	538
3	576	520	546
4	632	550	542
5	515	575	555
6	610	590	575

TABLE 2.21

Demand Data for Exercise 2.12

	June	July	August
Customer A	30	20	15
Customer B	20	20	10

2.10 Consider the production planning problem discussed in Example 2.14 (Section 2.17). Assume now that "backorders" are allowed, that is, if we do not have sufficient stock, we can backorder so that the order can be supplied in a later week. However, this incurs a backorder penalty cost of $4 per unit per week. Reformulate this problem as an LP model assuming that all backorders must be filled by the end of the 4th week. It is sufficient you define all the variables, write out the constraints and the objective function.

2.11 Consider Exercise 2.10 again. Reformulate the production planning problem as a transportation model. Identify the warehouse and markets and set up the transportation table.

2.12 (Ravindran et al., 1987) A company has signed a contract to supply customers *A* and *B* for the next 3 months as given in Table 2.21.

They can produce 40 units per month at a cost of $100 per unit on regular time during June, July, and August. An additional 10 units can be produced in June on overtime at a cost of $120 per unit. No overtime is available during July and August. They can store units at a cost of $10 per unit per month. The contract allows the company to fall short on its supply commitment to customer A during the months of June and July, but this incurs a penalty cost of $5 per unit per month; however, all shipments are to be completed by August. No shortages are allowed for customer B and his demands must be satisfied on the months specified. The company wishes to determine the optimal production schedule that will minimize the total cost of production, storage and shortage. Formulate this as a Transportation Problem by setting up the appropriate Transportation Table. Define what the warehouses and markets are in this case and their respective supplies and demands. Also, compute the appropriate shipping costs for all the cells in the transportation table.

Hint: Your transportation table should have four warehouses and six markets in this case!

2.13 Consider the aggregate planning problem discussed in Example 2.13 (Section 2.15). Discuss how the LP formulation will change if the following constraints are added.

(a) Union regulation requires that no more than 10% of the workforce can be laid off in any month.

(b) The personnel department estimates that no more than 50 trainees can be hired in any month.

(c) The company wants to limit the change in workforce from 1 month to the next to 10 workers.

(d) Company's warehouse limits the monthly inventory to 120 units.

2.14 Sales forecasts for a certain product for the first 6 months of 2012 are given in the following:

January—2,500	February—5,000	March—7,500
April—10,000	May—9,000	June—6,000

Production can be increased from 1 month to the next at a cost of $5 per unit, by hiring new workers. Production can also be decreased at a cost of $3 per unit by laying off existing workers. Inventory is limited to 7000 units due to warehouse capacity. Storage cost is estimated at $3 per unit per month based on the end of the month inventory. December 2011 production is set at 2000 units and the expected inventory at the end of December 2011 is 1000 units. The company does not allow shortages and the demands must be met every month. The company would like to have an inventory of 3000 units at the end of June 2012.

The problem is to determine the production schedule for the first 6 months of 2012 that will minimize the total cost of production change-overs and storage. Formulate this as a LP problem as follows:

(a) Define your variables clearly.

(b) Write out the *linear constraints* that must be satisfied, briefly explaining the significance of each.

(c) Write out the linear objective function that must be minimized.

2.15 (Forecasting case study) Jill Smith, who has joined recently as the forecasting manager for ABC company, is interested in developing quarterly forecasts for one of the company's key products. She has collected data on quarterly sales for this product for the past 5 years and they are given in Table 2.22.

TABLE 2.22

Quarterly Sales Data for 2007–2011 (Exercise 2.15)

Quarter	2007	2008	2009	2010	2011
1	800	1700	2100	2400	3600
2	750	1100	2200	3060	3900
3	600	680	1300	1800	1500
4	1500	2000	3100	4000	3320

Jill decides to use Holt's method with seasonality and trend (discussed in Example 2.6, Section 2.7) with the initial estimates of level and trend as 600 and 50 respectively. She also decides to use the first 4 years of data (2007–2010) for determining the smoothing constants (α and β) and use the 2011 data for validating the forecasting method and the chosen smoothing constants.

(a) Using the data for the years 2007–2010, prepare the initial estimates of the seasonal factors for each quarter.

(b) Using a spreadsheet, develop and select a set of good smoothing constants for both (α and β) for Holt's method. Use the error measures bias and STD to test your parameters. Run your tests using data for years 2007–2010. Use the following combinations of the smoothing constants for the tests:

$(\alpha, \beta) = (0.1, 0.1)$; $(0.1, 0.2)$; $(0.1, 0.3)$; $(0.2, 0.2)$; $(0.2, 0.3)$; $(0.3, 0.3)$

(c) Select the *two best* pairs of values for the smoothing constants (α and β) obtained from part (b) to prepare the quarterly forecasts for 2011. Using the 2011 actual production, validate these forecasts. Use *bias, standard deviation of forecast error, and tracking signals* for the validation.

(d) Suppose Jill wants to try out simple *moving average* forecasting method, by averaging the past four quarter's deseasonalized demands. For example, 2011 Quarter 1 deseasonalized forecast will be the average of the 2010s deseasonalized quarterly demands. Using the moving average method, prepare the quarterly forecasts for 2011. Compare these forecasts with those obtained in part (c) using the three forecast errors. Are they better than the forecasts obtained using Holt's method? Why or why not?

(e) What forecasts, including estimates of their accuracy, should Jill present to her VP for sales for the year 2012?

2.16 (Aggregate planning case study)
XYZ company is preparing its aggregate production plan for 2012. The monthly demand forecasts are given in Table 2.23.

The company is expected to have 100 workers on the payroll and 150 units in inventory at the end of 2011. Each worker can produce 10 units per month on regular time and 2 units per month in overtime.

Additional data:

1. Regular time production cost—$200 per unit
2. Overtime production cost—$300 per unit

TABLE 2.23

2012 Monthly Demand
Forecasts (Exercise 2.16)

Month	Demand
1	500
2	600
3	600
4	800
5	1300
6	2000
7	2500
8	3000
9	2400
10	1800
11	1500
12	1200

3. Hiring cost—$500 per worker
4. Firing cost—$3000 per worker
5. Desired final inventory (December 2012)—100 units
6. Desired final workforce (December 2012)—150 workers
7. Inventory holding cost is charged based on the average inventory at the beginning and end of the month at $50 per unit per month
8. No shortages are allowed
   (a) Determine a production plan using the "chase" strategy.
   (b) Determine a production plan using the "level" strategy.
   (c) Formulate a linear programming (LP) model that will determine the optimal production plan for 2012. You must define your variables clearly, write out the constraints that must be satisfied explaining the significance of each, and write the objective function with explanation.
   (d) Solve the LP model using any optimization software.
   (e) Compare the three production plans (chase strategy, level strategy, and LP solution) with respect to the following:
      (i) Monthly inventory levels
      (ii) Hiring and firing
      (iii) Overtime use
      (iv) Total cost

References

Amrute, S. 1998. Forecasting new products with limited history: Nabisco's experience. *Journal of Business Forecasting*. 17(3): 7–11.

Armstrong, J. S. 2001. *Principles of Forecasting: A Handbook for Researchers and Practitioners*. Springer Series in Operations Research and Management Science, Vol. 30. New York: Springer.

Arnold, J. R. T., S. N. Chapman, and L. M. Clive. 2012. *Introduction to Materials Management*, 7th edn. Upper Saddle River, NJ: Prentice Hall.

Badiru, A. B. 2006. *Handbook of Industrial and Systems Engineering*. Boca Raton, FL: CRC Press.

Barash, M. and D. Mitchell. 1998. Account based forecasting at Nabisco biscuit company. *Journal of Business Forecasting*. 17(2): 3–6.

Bollapragada, S., S. Gupta, B. Hurwitz, P. Miles, and R. Tyagi. 2008. NBC-Universal uses a novel qualitative forecasting technique to predict advertising demand. *Interfaces*. 38(2): 103–111.

Box, G. E. P. and G. M. Jenkins. 1976. *Time Series Analysis, Forecasting and Control*. San Francisco, CA: Holden Day.

Brown, R. G. 1959. *Statistical Forecasting for Inventory Control*. New York: McGraw Hill.

Chase, C. 2009. *Demand—Driven Forecasting: A Structured Approach to Forecasting*. Hoboken, NJ: John Wiley & Sons.

Chopra, S. and P. Meindl. 2010. *Supply Chain Management: Strategy, Planning and Operation*, 4th edn. Eaglewood Cliffs, NJ: Prentice Hall.

Croston, J. D. 1972. Forecasting and stock control for intermittent demand. *Operational Research Quarterly*. 23(3): 1970–1977.

Delbecq, A. L., A. H. VandeVen, and D. Gustafson. 1975. *Group Techniques for Program Planning*. Glenview, IL: Scott Foresman.

Evans, M. K. 2010. *Practical Business Forecasting*. Oxford, U.K.: Blackwell Publishers.

Foote, B. L. and K. G. Murty. 2008. Production systems. In *Operations Research and Management Science Handbook*, ed. A. R. Ravindran, pp. 18-1–18-30. Boca Raton, FL: CRC Press.

Foote, B. L. and K. G. Murty. 2009. Production systems. In *Operations Research Applications*, ed. A. R. Ravindran, Chapter 4. Boca Raton, FL: CRC Press.

Frances, P. H. 2011. Averaging model forecasts and expert forecasts: Why does it work. *Interfaces*. 41(2): 177–181.

Geraghty, M. K. and E. Johnson. 1997. Revenue management saves National Car Rental. *Interfaces*. 27(1): 107–127.

Hillier, F. S. and G. J. Lieberman. 2001. *Introduction to Operations Research*, 7th edn. New York: McGraw Hill.

Hoff, J. C. 1983. *A Practical Guide to Box—Jenkins Forecasting*. Belmont, CA: Lifetime Learning Publications.

Holt, C. C. 1957. Forecasting seasonals and trends by exponentially weighted moving averages. Office of Naval Research, Arlington, VA, Report no. 52.

Hueter, J. and W. Swart. 1998. An integrated labor-management system for Taco Bell. *Interfaces*. 28(1): 75–91.

Johnson, S. M. 1957. Sequential production planning over time at minimum cost. *Management Science*. 3(4): 435–437.

Kapuscinski, R., R. Q. Zhang, P. Carbonneau, R. Moore, and B. Reeves. 2004. Inventory decisions in Dell's supply chain. *Interfaces*. 34(3): 191–205.

Lee, H. L., V. Padmanabhan, and S. Whang. 1997. The bullwhip effect in supply chains. *Sloan Management Review*. 38(3): 93–102.

Lilien, G. L., A. Rangaswamy, and A. D. Bruyn. 2007. *Principles of Marketing Engineering*. Victoria, British Columbia, Canada: Trafford Publishing.

McKaige, W. 2001. Collaborating on the supply chain. *IIE Solutions*. 33(3): 34–37.

Nahmias, S. 1993. *Production and Operations Analysis*, 2nd edn. Boston, MA: Irwin.

Ravindran, A., D. T. Philips, and J. Solberg. 1987. *Operations Research: Principles and Practice*, 2nd edn. New York: John Wiley & Sons, Inc.

Salvendy, G. 2001. *Handbook of Industrial Engineering*, 3rd edn. New York: Wiley-Interscience.

Sanders, N. R. and K. B. Manrodt. 1994. Forecasting practices in US corporations: Survey results. *Interfaces*. 24(2): 92–100.

Sanders, N. R. and K. B. Manrodt. 2003. Forecasting software in practice: Use, satisfaction and performance. *Interfaces*. 33(5): 90–93.

Shim, J. K. 2000. *Strategic Business Forecasting: The Complete Guide to Forecasting Real World Company Performance*, Revised Edition. Boca Raton, FL: CRC Press.

Srinivasan, G. 2010. *Quantitative Models in Operations and Supply Chain Management*. New Delhi, India: Prentice Hall.

Taha, H. A. 1987. *Operations Research*, 4th edn. New York: McMillan.

Taylor, B. 2007. *Introduction to Management Science*, 9th edn. Eaglewood cliffs, NJ: Prentice Hall.

Willemain, T. R., C. N. Smart, J. H. Shockor, and P. A. DeSautels. 1994. Forecasting intermittent demand in manufacturing: A comparative evaluation of Croston's method. *International Journal of Forecasting*. 10(4): 529–538.

Winters, P. R. 1960. Forecasting sales by exponentially weighted moving average. *Management Science*. 6: 324–342.

Xu, W. 2000. Long range planning for call centers at FedEx. *The Journal of Business Forecasting Methods & Systems*. 18(4): 7–11.

Yurkiewicz, J. 2008. Forecasting at steady state? *OR/MS Today*. 35(3): 54–63.

Yurkiewicz, J. 2010. Forecasting: What can you predict for me? *OR/MS Today*. 37(3): 36–43.

Yurkiewicz, J. 2012. Forecasting: An upward trend? *OR/MS Today*. 39(3): 52–61.

3

Inventory Management Methods and Models

3.1 Decision Framework for Inventory Management

Inventory management can be concisely captured in the following policy, of sorts: "Before you run out, get more."* Though indeed concise, the statement captures the essence of the basic problem of inventory management, and immediately begs for an answer to two questions—"How long before you run out (i.e., *when*)?" and "*How much* more do you get?" Thus, managing inventory boils down to answering those two questions, "when?" and "how much?"

Before we dive into tools and techniques for managing inventories, however, let us first step back and make sure that we are clear about the characteristics of the items to be managed. These characteristics, it stands to reason, should have a significant impact on the kinds of tools that are appropriate to the management task. In general, it is a good idea to avoid the temptation to apply a single, elaborate, sophisticated tool to every problem (or buying one—e.g., software—that is intended to solve every problem). Sometimes you really might be better off with the proverbial "hammer" to solve the simple, "nail-like" aspects of your problem. Sometimes, though, an advanced power-tool admittedly might be warranted.

Bozarth (2005) offers an effective framework for placing the inventory management problem in the context of two important aspects of demand for the products being managed. This framework has been reproduced in Figure 3.1. The first of the two context-shaping aspects of demand is whether demand is driven externally by forces outside the firm's direct control or whether demand is driven by internal management decisions. The former situation describes *independent-demand* items, those items whose demand is driven by customer tastes, preferences, and purchasing patterns. Typically, these are finished goods. Inventory management in these cases is influenced greatly by the extent to which the firm can effectively describe the random variations in these customer purchasing patterns.

---

* We owe this concise statement, with only slight alterations applied, to our good friend and colleague Douglas J. Thomas.

	Independent demand	Dependent demand
Stationary demand	Reorder point—order quantity	Kanban/constant work-in-process
Non-stationary demand	Distribution requirements planning (DRP)	Material requirements planning (MRP)

FIGURE 3.1
A framework for inventory management.

The latter situation describes *dependent-demand* items, those items whose demand is driven by the demand of other items.

As an example, consider a finished goods item produced by some firm, let us say a cell phone. To the producing firm, the cell phone itself is an independent-demand item. Its component parts—such as the circuit board, the keypad, the plastic housing, the battery, etc.,—are dependent-demand items. The demand for these items is driven by the production schedule for the cell phones, which is solely under the control of the company that produces the phones. The company's management chooses the production plan in response to projected consumer demand for these phones. Thus, inventory management for dependent-demand items is largely an issue of ensuring that component parts are available in sufficient quantities to execute the production plan.

Given the supply-chain context of this book, we will consider only the management of independent-demand items—i.e., those items that move between firms in the supply chain. Throughout this book, we focus on issues related to node-to-node relationships in the supply chain, consistent with the framework developed in Chapter 1 that defines a supply chain as a *network of nodes*. Dependent demand involves "within-node" effects and is outside the scope of this book, but is discussed extensively in books on production/operations planning and control systems (e.g., Vollmann et al., 2005 or Chapman, 2006, which also contains an excellent discussion on hybrid systems that combine appropriate elements of MRP and kanban control). Note, however, that the classification of an item as an independent-demand item or a dependent-demand item is *not* an absolute characterization. Rather, it only makes sense *in context*. For example, to the company that assembles the cell phones, the keypad is clearly a dependent-demand item, provided that its only demand is derived from the production schedule for cell phones (i.e., not from sales of keypads as stand-alone items). To the firm that produces the keypads and sells them to various cell phone manufacturers, however, the keypad is an *independent-demand* item, with its component parts (e.g., keys, underlying switch membrane, connectors, etc.) being classified as dependent-demand items. This highlights the difference in perspective between a node-to-node view and a within-node view. From a node-to-node standpoint, the independent firms in the supply chain move items that they classify as independent-demand items to their supply chain partners downstream. Thus, every firm in the supply chain has to be concerned with independent-demand inventory management for some part of its activities.

The second important aspect of demand that helps determine which inventory management tools to apply, per Figure 3.1, is the stability of demand over time. Though Bozarth (2005) describes this dichotomy as "stable" versus "variable," we shall characterize demand as *stationary* or *non-stationary*—the former being the case in which the expected value of demand remains constant over time, or at least over the planning horizon, and the latter being the case in which it does not. We consider both situations in the methods presented in this chapter. In either case, though, stationary or non-stationary, demand could be uncertain or known with certainty. Obviously, then, the simplest case would be one in which demand is stationary and known with certainty, and the most challenging would be one in which demand is non-stationary and uncertain.

As indicated by Bozarth (2005), another idea embedded in his framework is that the stationary demand approaches map to *pull systems*, whereas the non-stationary approaches map to *push systems*. Although definitions tend to vary—see Hopp and Spearman (2004) for an excellent discussion—pull systems are those that execute replenishments only in response to actual demand, while push systems drive replenishments from the schedule of projected future demands, which, consistent with Bozarth's framework, will vary over time. Hopp and Spearman (2004) also point out that *kanban*, or "card"—controlled production systems (see, e.g., Vollmann et al., 2005, for descriptions of several examples of this), are only one, specific implementation of what they more generally describe as *constant-WIP* systems (where WIP stands for "work-in-process," or inventory that is not yet fully converted from raw materials into finished goods). Again, since our focus will be on independent-demand inventory management systems, the reader is referred to the Hopp and Spearman article for further details regarding such dependent-demand systems.

Finally, we should point out that the matrix in Figure 3.1 presents what would appear to be *reasonable* approaches to managing demand in these four demand contexts, but not necessarily the *only* approaches. For example, non-stationary reorder point–order quantity systems are certainly possible, but probably not applied very often in practice due to the significant computational analysis that would be required. Indeed, the effort required to manage the system must clearly be considered in deciding which system to apply. While the "optimal" solution (at least from the standpoint of an objective that minimizes solely the costs that are directly impacted by inventory levels, and not necessarily the administrative costs of managing the inventory system) might indeed be a non-stationary reorder point–order quantity system, the data required to systematically update the policy parameters of such a system and the analytical skill required to carry out—and *interpret*— the updates may not be warranted in all cases.

While we will consider only independent-demand inventory systems in this chapter, we will most certainly explore the issue of variability in demand, both from the standpoint of uncertainty as a source of variability, and from the standpoint of stationary versus non-stationary demand environments. One important aspect of inventory management not considered in the framework of Figure 3.1, however, is the issue of the number of items managed, and the number of sites over which those items are managed. Accordingly, later in the chapter, after developing a series of principles related to managing single items at single sites, we turn our attention to the management of multiple items at a single site, and to the management of a single item across multiple sites.

3.2 Some Preliminary Modeling Issues

3.2.1 Two Critical Tasks

The first step in building any inventory management decision model is to *describe demand*. Recalling our simplified statement of inventory management, one quickly realizes that it is impossible to figure out when to "buy more" without some understanding of the nature of demand. At the extremes, demand for an item can be a "spike," a single point in time at which there is demand for the item in question, and after which the item is never again demanded by the marketplace. The other extreme is "perpetual demand," where demand for the item continues at a stationary rate per unit time infinitely into the future.

This last point allows us to reflect back the framework described earlier, contrasting non-stationary (i.e., time-varying) demand from stationary demand (i.e., constant across time). This distinction helps us decide how we should model demand. If demand is stationary, then the expectation of demand in all time periods in our decision horizon is unchanging. All we need to do, then, is to describe demand for some portion of our infinite decision horizon. Moreover, if demand is not known with certainty, then describing demand requires that we specify its probability distribution. If, however, demand is non-stationary, then our demand description task is more complicated. If demand is a "spike," as described earlier, or occurs at only a small number of points across our decision horizon, then our task is similar to the infinite-horizon case. We must define a probability distribution that represents the extent of our knowledge of demand at each of the points in time that we predict that demand will occur. If we have a large number of demand points, we need to employ the types of forecasting techniques discussed in Chapter 2 to complete the "describe demand" task.

The second step in building an inventory decision model is to *assess relevant costs*. Again, if we are to determine the extent of the "buy more" action implied by our two basic inventory management questions, in addition to the timing of the "before we run out" aspect of this, we must have some clear basis for making these decisions. Clearly, the most rational means of establishing this basis is by describing the various costs associated with making the "when" and "how much" decisions. In some cases, the relevant costs are unequivocally clear, specific, and direct, like the cost of purchasing the item being managed. In other cases, the costs are difficult to specify exactly, but real nonetheless, like the cost of running short of stock and failing to fulfill some demand. In some cases, the "costs" we include in the model will actually be prices that we, the decision maker, charge our customers, which, unlike costs are inflows of funds and not outflows, implying *opportunity costs* if, for example, we run short of stock.

3.2.2 ABC Analysis

Most inventory systems contain hundreds of items, and possibly thousands or tens of thousands. Some of the techniques we describe in this chapter are time- and computation-intensive, and the idea of spending a substantial amount of time and computational efforts on every single one of several thousand items sounds like an extremely costly endeavor. Indeed, the most intensive aspect of building any of these models may not be computation of the inventory management parameters themselves, but gathering the data related to describing demand and assessing the relevant costs. The administrative activity required to do this across all items in the system could be a tremendous burden on the firm's management system.

Clearly, a sort of "divide and conquer" approach is warranted. The most popular such approach is typically referred to as "ABC analysis" because it separates inventory items into three categories, "A" items, "B" items, and "C" items. "A" items are those that are determined to be the most important, and therefore the items that require the most management attention and the most careful analysis in terms of specifying the inventory management parameters. "B" items are less important than "A," but still require a moderate level of attention and effort. "C" items are least important and require the least attention and effort. While several means could be used to categorize items into A, B, and C, most often the categorization is made on the basis of *dollar flow*, using either item sales revenue or item "cost of goods sold" (COGS) as the measure of interest.

The underlying idea behind ABC analysis is the *Pareto Principle*—i.e., that among any group of items, 80% of the activity is driven by only 20% of the items, or the familiar 80–20 rule. This "rule" is, of course, only a rule of thumb. The basic idea is that some large portion of the activity is driven by a small group of the items. Thus, the point at which one draws the line between "A" and "B" items does not necessarily have to be at 80% of the sales activity or at 20% of the items, and in fact, in many collections of inventory items, these levels are not likely to match up exactly. Separating "B" from "C" involves similar rules of thumb. The typical one is that roughly 50% of the items are expected to be categorized as "C" items. Thus, the expected breakdown in a given inventory system is about 20% "A," 30% "B," and 50% "C." As with many analytical activities, more can probably be learned merely from gathering and analyzing the data than from computing the results. The following example illustrates the final result of an ABC analysis.

Example 3.1 ABC analysis

Charlie's Bavarian Automotive (CBA) is a distributor that supplies a relatively small set of specialty automotive components to auto repair shops. In total, CBA supplies 287 components. A summary of the ABC categorization of CBA's inventory is shown in Table 3.1 and Figure 3.2. As seen from the table and figure, 80% of the sales volume is captured by slightly less than 20% of the items, and in total, "A" and "B" items represent 45% of the total item count and capture approximately 95% of the sales volume. The top 10 items from CBA's ABC analysis are shown in Table 3.2. Interestingly, the top two items demonstrate a marked contrast, with the top item generating almost $800,000 in sales from only 3,500 units of annual sales volume and the second item generating more than $700,000 in annual sales from almost 169,000 units in volume.

This example illustrates the most common basis for item classification in ABC analysis, based on sales activity (demand). Of course, other bases for categorizing items are possible. Flores and Whybark (1987), for example, propose a categorization on two levels, dollar flow, and the level of importance or criticality assigned to the item. The result is nine categories—AA, AB, ..., CB, CC—as opposed to the traditional three.

TABLE 3.1

ABC Inventory Analysis Results

Class	Number of Items	Percent of Items (%)	Percent of Sales (%)
A	52	18.1	79.8
B	78	27.2	15.1
C	157	54.7	5.1
Grand Total	287	100.0	100.0

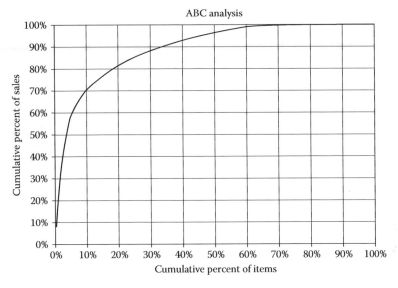

FIGURE 3.2
ABC inventory analysis plot.

TABLE 3.2

Top 10 Items from ABC Inventory Analysis

Stock-Keeping Unit (SKU)	Annual Sales (Units)	Annual Sales ($)	Cumulative % of SKUs	Cumulative % of Sales
TF688	3,500	$795,073	0.35%	8.17%
WN881	168,789	$714,563	0.70%	15.52%
ZG454	4,000	$569,122	1.05%	21.37%
JP909	2,500	$539,064	1.39%	26.91%
MA117	4,757	$486,323	1.74%	31.91%
JN625	4,549	$389,990	2.09%	35.92%
MQ617	1,328	$348,358	2.44%	39.50%
CR841	12,121	$303,303	2.79%	42.62%
RA596	5,495	$282,001	3.14%	45.52%
AY620	490	$256,785	3.48%	48.16%

3.3 Single-Item, Single-Period Problem: The Newsvendor

We will begin our development of the tools and techniques for inventory management with the simplest setting, the management of a single item at a single site. Although we are "starting small," the important issues related to describing demand and assessing the cost tradeoffs in this problem are relevant to *any* inventory management system. Therefore, our discussion of this model sets the stage for the analysis of a wide range of related inventory decision models.

If the entire demand being supported by the inventory decision in question can be considered to occur in a single period, we have the "spike" case discussed earlier. This is the situation of the classic *newsvendor problem*. Clearly, demand for today's newspaper is concentrated solely in the current day. There will be, effectively, no demand for today's newspaper tomorrow (it is "yesterday's news," after all) or at any point in future. Therefore, if we run a newsstand, we will stock papers only to satisfy this 1-day spike in demand. If we do not buy enough papers, we will run out and lose sales—and potentially also lose the "goodwill" of those customers who wanted to buy but could not. If we buy too many papers, then we will have to return them or sell them to a recycler at the end of the day, in either case at pennies on the dollar, if we are lucky.

Now, demand does not have to be concentrated literally into a single day in order for the newsvendor model to be appropriate. Often, this model is applied to seasonal goods (e.g., apparel, calendars, back-to-school items), in which case the operative issue is that there is only a single opportunity to order goods to fulfill demand. Thus, the newsvendor model applies to any situation in which the supply lead time is long relative to the time over which demand occurs— i.e., where the total time for the supplier to respond to our order and for the goods to arrive at the our location exceeds the length of the selling season. In this case, demand is effectively concentrated into a single "point," which is the time covered by the season. Thus, the one key issue in describing the setting for the newsvendor model is fully describing this single-point/single-season demand. The only scenario for which a problem regarding how much to order exists, though, is the one in which demand is uncertain, and therefore, the key issue in describing demand in the newsvendor setting is to specify a probability distribution for the single period (season).

In keeping with the framework we posed earlier, after describing demand we must describe the relevant costs. Typically, some of these costs will be well-specified (e.g., the cost of the item), and some will not (e.g., the cost of failing to fulfill all demand, and perhaps the post-selling-period value of goods left over if the order quantity exceeds the demand). We will address the nature of these costs in the models that follow.

The newsvendor problem, though it seems to be quite basic, actually leads to a rich set of issues in inventory management. In particular, the problem has at its heart the basic tension underlying all inventory decision making— namely, balancing the cost of having too much inventory with the cost of having too little. It does not take too much imagination to see that this too much–too little knife edge can be affected significantly by transportation and network design decisions. The important issue before us at this point, however, is to understand how to evaluate the tradeoff. Note also that in the newsvendor case, one only needs to answer the question of "how much" since the answer to the "when" question, in the case of a demand spike, is simply "sufficiently in advance of the first sale of the selling season."

The newsvendor model, however, is also the basis for many periodic-review inventory models, which we discuss later in this chapter. Unlike the newsvendor model, though, these periodic-review models often involve the case of recurring demand with no cost to reorder, where the answer to "when" is either "at every review of the inventory level," or "when we're close to running out" at the time of review.

Historically, an early version of the newsvendor model was studied by Arrow et al. (1951), but the model appears to have first been stated in the form typically recognized today by Whitin (1953).

Example 3.2 Describing newsvendor demand

The Unlimited, an apparel retailer, foresees a market for hot-pink, faux-leather biker jackets in the Northeast region for the winter 2012–2013 season. The Unlimited's sales of similar products in Northeastern stores in recent winter seasons—e.g., flaming-red, fake-fur, full-length coats in 2011–2012 and baby-blue, faux-suede bomber jackets in 2010–2011—were centered around 10,000 units. The Unlimited's marketing team, however, plans a stronger ad campaign this season and estimates that there is a 90% chance that they can sell as many as 14,000 units. The Unlimited also faces fierce competition in the Northeast from catalog retailer Cliff's Edge, who has recently launched an aggressive direct-mail and Internet ad campaign. Thus, there is concern that sales might come in 3000–4000 units below historical averages for similar products. Jackets must be ordered from an Asian apparel producer in March for delivery in October. Only a single order can be placed and it cannot be altered once it is placed.

It would appear reasonable to assume that the upside and downside information regarding demand implies that the probabilities are well-centered around the mean of 10,000. Therefore, let us assume that demand follows a normal distribution. We can interpret the statement from Marketing to mean that the cumulative probability of demand equaling 14,000 units is 0.90. Thus, we can use a normal look-up table (perhaps the one embedded in the popular Excel spreadsheet software) and use this 90th-percentile estimate from Marketing to compute the standard deviation of our demand distribution. Specifically,

$$z_{0.90} = 1.2816 = \frac{D_{0.90} - \mu}{\sigma} = \frac{14,000 - 10,000}{\sigma}, \text{ meaning that } \sigma = 3{,}121. \text{ Thus,}$$

in this case, we model demand as a normal distribution with mean $\mu = 10{,}000$ and standard deviation $\sigma = 3{,}121$.

As laid out earlier, Example 3.2 gives us all the basics we will need to determine a good order quantity in the single-period setting: an idea of the spread of possible demand values (and, in fact, enough information to infer a probability distribution of demand), a unit sales price, a unit cost, and an end-of-season unit salvage value. In general, if an item sells for p, can be purchased for c, and has end-of-season salvage value of s, then the cost of missing a sale is $p-c$, the profit margin that we would have made on that sale, and the cost of having a unit in excess at the end of the selling season is $c-s$, the cost we paid to procure the item less its salvage value. We denote demand by the random variable D (which we assume, generally, to be defined on $D \in [0, \infty)$), with mean μ and variance σ^2, and let f be its probability distribution function (pdf). The optimal order quantity, therefore, maximizes expected profit, given by

$$\pi(Q) = \int_0^Q \left[(p-c)D - (c-s)(Q-D) \right] f(D)dD + \int_Q^\infty Q(p-c)f(D)dD \qquad (3.1)$$

Note that expression (3.1) indicates that, if the order quantity exceeds (positive-valued) demand, our decision results in receiving the unit profit margin on all D units we sell (leftmost portion of the left side of the sum in expression (3.1)), but we suffer a loss of $c-s$ on the $Q-D$ units in excess of demand (right portion of the left side of the sum). If demand, however, exceeds the order quantity (right side of the sum in expression (3.1)), then we simply obtain the unit profit margin only on the Q units we are able to sell (as long as we assume no goodwill loss from failing to fulfill demand. See Section 3.3.2.)

One approach to finding the best order quantity would be to minimize expression (3.1) directly. Another approach, and one that is better in drawing out the intuition of the model, is to find the best order quantity through a marginal analysis. Assume that we have ordered Q units. What are the benefits and the risks of ordering another unit versus the benefits and the risks of not ordering another unit? If we order another unit—i.e., a total of $Q + 1$— the probability that we will not sell the $(Q + 1)$ st unit is $\Pr \{D \leq Q\} = F(Q)$, where F is the cdf of D. The cost in this case is $c-s$, meaning that the expected cost of ordering another unit beyond Q is $(c - s) F(Q)$. If we do not order another unit, the probability that we could have sold the $(Q + 1)$ st unit is $\Pr\{D > Q\} = 1 - F(Q)$ (provided that D is a continuous random variable). The cost in this case is $p-c$, the opportunity cost of the profit margin foregone by our under-ordering, meaning that the expected cost of not ordering another unit beyond Q is $(p - c)[1 - F(Q)]$. Another way of expressing the costs is to consider $p-c$ to be the *underage cost*, or $c_u = p - c$, and to consider $c-s$ the *overage cost*, or $c_o = c - s$.

A marginal analysis says that we should, starting from $Q = 1$, continue ordering more units until that action is no longer profitable. This occurs where the net difference between the expected cost of ordering another unit—the cost of overstocking—and the expected cost of not ordering another unit— the cost of understocking—is zero, or where $c_o F(Q) = c_u[1 - F(Q)]$. Solving this expression, we obtain the optimal order quantity,

$$Q^* = F^{-1}\left(\frac{c_u}{c_u + c_o} \right) \qquad (3.2)$$

where the ratio in the brackets in expression (3.2) is called the *critical ratio*, which we denote by

$$CR = \frac{c_u}{c_u + c_o} \qquad (3.3)$$

and which specifies the *critical fractile* of the demand distribution amassed by the optimal order quantity.

Example 3.3 Assessing relevant cost tradeoffs

Continuing from the scenario established in Example 3.2, assume that the landed cost of each hot-pink, faux-leather biker jacket is $100 (i.e., the cost to purchase the item, delivered). The jackets sell for $200 each. In March 2013, any remaining jackets will be deeply discounted for sale at $25 each. Thus, we have $c_u = p - c = \$100$ and $c_o = c - s = \$75$, and therefore, $CR = c_u/(c_u + c_o) = 0.571$. The issue now is to translate this critical ratio into a critical fractile of the demand distribution. Recall from the preceding text that $D \sim N(10{,}000, 3{,}121)$. Thus, another table look-up (or Excel computation) gives us $Q^* = 10{,}558$, the value that accumulates a total probability of 0.571 under the normal distribution pdf with mean 10,000 and standard deviation 3,121.

3.3.1 Service Measures in Inventory Models

As it turns out, the optimal order quantity in Example 3.3 is not far from the expected value of demand. This is perhaps not surprising since the cost of overage is only 25% less than the cost of underage. One might reasonably wonder what kind of service outcomes this order quantity would produce. Let us denote α as the *in-stock probability*, or the probability of not stocking out of items in the selling season, and β as the *fill rate*, or the percentage of overall demand that we are able to fulfill. For the newsvendor model, let us further denote $\alpha(Q)$ as the in-stock probability given that Q units are ordered, and let $\beta(Q)$ denote the fill rate if Q units are ordered. It should be immediately obvious that $\alpha(Q^*) = CR$.

Computing the fill rate is a little more involved. If demand is D and unfulfilled demand (i.e., units short) is given by S, then fill rate is given by

$$\beta = \frac{D - S}{D} \qquad (3.4)$$

For a newsvendor setting, both fill rate and units short are functions of the order quantity Q, and since expected demand in the newsvendor model is μ, it follows that the expected fill rate is

$$\beta(Q) = \frac{\mu - S(Q)}{\mu} = 1 - \frac{S(Q)}{\mu} \qquad (3.5)$$

Computing $S(Q)$ requires a *loss function*, which is typically available in look-up table form for the standard normal distribution. Thus, for normally distributed demand

$$\beta(Q) = 1 - \frac{\sigma L(z)}{\mu} \tag{3.6}$$

where
   $L(z)$ is the standard normal loss function*
   $z = (Q - \mu)/\sigma$ is the standardized value of Q

Using the data from Examples 3.1 and 3.2, we can compute $\beta(Q^*) = 0.9016$, meaning that we would expect to fulfill about 90.2% of the demand for biker jackets in the 2012–2013 winter season if we order 10,558 of them.

3.3.2 Service Impact of Shortage Costs

A reasonable question to ask at this point is whether the service levels achieved by the optimal order quantity in our example scenario, a 57% chance of not stocking out and a 90% fill rate are "good" service levels. The best answer to that question is that they are neither "good" nor "bad." They are *optimal* for the overage and underage costs given. Another way of looking at this is to say that, if the decision maker believes that those service levels are too low, then the cost of underage must be understated, for if this cost were larger, the optimal order quantity would increase and the in-stock probability and fill rate would increase commensurately. Therefore, newsvendor models like our example commonly employ another cost factor, what we will call g, the cost of losing customer *goodwill* due to stocking out. This cost inflates the underage cost, making it $c_u = p - c + g$. The problem with goodwill cost, however, is that it is not a cost that one could extract directly from a typical firm's accounting records. Although one can clearly argue that it is "real"—customers do indeed care, in many instances, about the inconvenience of not finding items they came to purchase—it would be hard to evaluate this cost for the "average" customer. The way around this is to allow the service outcome to serve as a proxy for the actual goodwill cost. The decision maker probably has a service target in mind, one that represents the in-stock probability or fill rate that the firm believes is a good representation of their commitment to their customers, but does not "break the bank" in expected overstocking costs.

------

* The standard normal loss function is given by $L(z) = \int_{z}^{\infty}(x - z)\phi(x)dx$, where $x \sim N(0,1)$ and ϕ is the pdf of x (i.e., the standard normal pdf). $L(z)$ can also be computed directly as $L(z) = \phi(z) - z[1 - \Phi(z)]$, where Φ is the standard normal cumulative distribution function (cdf). Modern technology, as it were, has obviated the need to include lengthy statistical tables in textbooks since entities like $L(z)$ can be computed directly for a given value of z. The Excel code for doing this is = NORMDIST(z,0,1,FALSE) – z*(1-NORMSDIST(z)).

TABLE 3.3

Comparative Outcomes of Newsvendor Example for Increasing $g = k(p - c)$

k	$\alpha(Q^*)$	$\beta(Q^*)$	Q^*	Q^*/μ	z^*	Expected Overage[a]
0	0.571	0.901	10,558	1.06	0.179	1543.9
0.2	0.615	0.916	10,913	1.09	0.293	1754.5
0.4	0.651	0.927	11,211	1.12	0.388	1943.2
0.6	0.681	0.935	11,468	1.15	0.470	2114.3
0.8	0.706	0.942	11,691	1.17	0.542	2269.0
1	0.727	0.948	11,884	1.19	0.604	2407.3
2	0.800	0.965	12,627	1.26	0.842	2975.4
4	0.870	0.980	13,515	1.35	1.126	3718.3
6	0.903	0.986	14,054	1.41	1.299	4196.4
8	0.923	0.989	14,449	1.44	1.426	4557.2
10	0.936	0.991	14,750	1.48	1.522	4837.0

[a] Computed via $Q - \mu + \sigma \cdot L(z)$, which translates to "order quantity minus expected demand plus expected lost sales."

Example 3.4 Impact of goodwill cost on optimal service level

Referring back to Examples 3.2 and 3.3, Table 3.3 shows the comparative values of the newsvendor-optimal order quantity, Q, and the values of service measures α and β for goodwill costs $g = k(p - c)$, with k ranging from 0 to 10. Thus, from Table 3.3, as g ranges from 0 to relatively large values, service—measured both by α and β—clearly improves, as does the expected unsold inventory. For comparison, Table 3.3 displays a ratio of the optimal order quantity to the expected demand, Q^*/μ, and also shows the standardized value of the optimal order quantity, $z^* = (Q^* - \mu)/\sigma$. These values are, of course, specific to the parameters specified in our example scenario and would obviously change as the various cost factors and demand distribution parameters change.

Ultimately, it may be more instructive to consider a "parameter-free" view of the service level, which is possible if we focus on the in-stock probability. Figure 3.3 shows a graph similar to Figure 12.2 from Chopra and Meindl (2010, p. 338), which plots $\alpha(Q^*) = CR$, the optimal in-stock probability, versus the ratio* of overage and underage costs, c_o/c_u. This graph shows clearly how service levels increase as c_u increases relative to c_o, in theory increasing to 100% in-stock probability at $c_o = 0$ (obviously an impossible in-stock probability level for a demand distribution that is unbounded in the positive direction). Moreover, for high costs of overstocking, the optimal in-stock probability could be well below 50%.

---

* Note that $CR = c_u/(c_u + c_o)$ can also be expressed as $CR = 1/(1 + c_o/c_u)$.

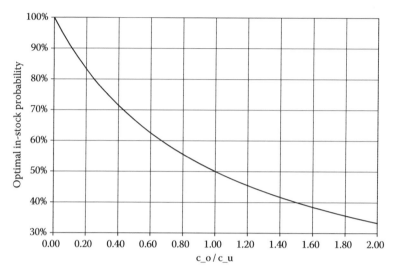

FIGURE 3.3
Newsvendor optimal in-stock probability versus overage-to-underage ratio.

3.3.3 Safety Stock: A First Look

The ideas discussed earlier lead naturally to the idea of *safety stock*. One way to define safety stock is the amount of inventory held in excess of expected demand in order to protect against stockouts. In the case of the newsvendor model, where demand is a spike, the safety stock is simply given by $Q - \mu$. Note also that at $Q = \mu$, the newsvendor safety stock is zero and the expected in-stock probability is 50%.

3.4 Single-Item, Multi-Period Problems

Up to this point, we have carefully considered the key issues that lie behind making a decision regarding how much to order to satisfy demand at a single site in a single period. As we discussed earlier, the essence of making a good decision in this setting boils down to two issues: describing demand and assessing relevant costs. In the newsvendor model, we require a probability distribution of demand—a clear description of the likelihood of "large" or "small" demand values—in order to anchor our order quantity decision. In addition, given that our decision is to maximize expected profit, we need a clear statement of what it costs to acquire materials and of the "monetized" consequences of ordering more than we ultimately needed, or less. These same principles carry over to ordering decisions in systems where we can order at multiple points in time across a horizon. To make the decisions related to "how much" and "when"—the latter issue now more prominent in this multiple-period setting—we must still describe demand and assess the costs relevant to our decision.

Allowing the placement of multiple orders to meet demand in a multi-period horizon, however, introduces a new issue to consider, specifically the cost of placing an order. Since the newsvendor problem involves the placement of only a single order, the consideration of this cost is relevant only if it exceeds the expected cost of ordering nothing, $D \cdot c_u$. Otherwise, the cost of ordering has no impact on the optimal order size. In a multi-period setting, though, we must be cognizant of how a fixed ordering cost may impact the optimal size of an order, and therefore the optimal frequency at which orders are placed across the time horizon. This requires us to formally recognize the "when" aspect of inventory management. In the newsvendor problem, the answer to "when" is simply, "sufficiently in advance of the expected time of the first sale," and the operative question of interest is "how much?" Now, in the multi-period setting, we must provide an answer to both of these formative questions of inventory management.

In deciding when to order, we must also be cognizant of the mechanism by which an order is triggered. In theory, an order would be triggered as soon as inventory falls to a level that indicates that a replenishment is required to maintain sufficient inventory in the system to fulfill demand—i.e., its *reorder point*. This implies a whole host of other issues, namely the length of the replenishment lead time, the accuracy with which this value can be estimated, and the cost of experiencing a shortfall in supply. Moreover, it might not be feasible or desirable, in a system with a large number of items, to manage all of those items so closely that an order is placed precisely when that item, alone, hits its reorder point.

A way of distinguishing how items are managed in the inventory system, therefore, is based on whether their inventory levels are reviewed continuously, implying a *continuous-review* system, or periodically, using a *periodic-review* system. One means of deciding the appropriate system for a given item is through the ABC analysis tool that we discussed earlier. Since "A" items are those that have the greatest impact on business operations, the reasons to manage "A" items via continuous-review are twofold: (i) to ensure availability of these items and (ii) to achieve (i) at the minimum inventory holding cost level for these items. If the fact that a continuous-review approach could achieve the same level of availability at a lower inventory level than a periodic-review approach is not obvious to the reader at face value, this should not be a concern at this point. This fact will become clear later in the chapter, after both the continuous-review and periodic-review approaches are discussed. In addition, we should note here that utilizing continuous review for "A" items is not an "immutable law," but merely a suggestion. In fact, in the discussion of periodic-review approaches in the following, we make the argument that there are clearly situations (namely, in managing multiple items) where "A" items might be better managed that way.

3.4.1 Continuous-Review: Reorder Point–Order Quantity Model

A continuous-review, reorder point–order quantity model is often called a (Q, R) model. In this notation, R represents the reorder point, the inventory value below which an order is placed, in this case for Q units. A sample plot of the inventory levels in a (Q, R) system is shown in Figure 3.4. Note that this graph contains both a solid line and a dashed line plotting inventory levels. The solid line represents the *on-hand inventory* level, which is the actual physical inventory present in the system. The dashed line represents the *inventory position*, which is defined as the sum of on-hand inventory and inventory on-order but not yet received, minus any backlogged demand not met in previous cycles. The inventory position is an important concept in the management of multi-period inventory systems since it is the basis on which ordering decisions are made. For example, let us denote the physical, on-hand inventory by y, and assume that y falls below the reorder point, R. If there is still an order of Q units due, and $Q > R - y$, the inventory position (equal to $y + Q$) still exceeds R, and another order will not be placed. Thus, the ordering rule for a (Q, R) system is: When the inventory position falls below R, place an order for Q units. Note that an order can be placed at any point in time in a (Q, R) system, which is what makes it a continuous-review system.

Following our framework, in order to completely describe the (Q, R) inventory system, we must first describe demand. Let us begin with the case where demand is known with certainty. Let us also consider the case where demand occurs at a constant rate for the foreseeable future, meaning that we consider the extreme multi-period case of perpetual, stationary demand

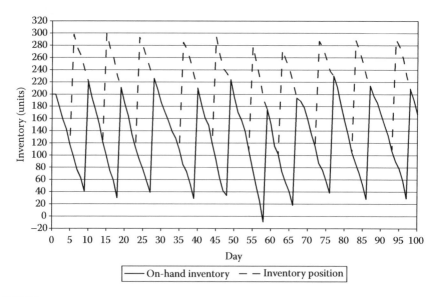

FIGURE 3.4
Sample plot of (Q, R) inventory system levels.

across an infinite horizon. This *infinite-horizon, perpetual demand* setting, in which demand is assumed to continue infinitely far into the future, is clearly the antithesis of the demand spike that characterized the newsvendor setting. Moreover, since we are assuming a constant demand in an infinite horizon, any portion of that horizon we consider should be equivalent. In order to establish a basis for decision making, let us consider 1 year in this infinite decision horizon, and thus, we denote D as the annual demand. In this case, that is all that is required to describe demand.

Now, we turn our attention to assessing the relevant costs. Traditionally, in an infinite- horizon setting, the costs that are factored into the decision model are different from those considered in the single-period, newsvendor model. Specifically, the traditional approach is to treat this problem as a cost-minimization problem, and not a profit-maximization problem. Thus, in this model, we ignore the unit price of the item being managed and focus instead on its cost, c. More precisely, we are assuming that none of our stocking decisions will impact p—consistent with the newsvendor model mentioned earlier—nor that p itself will impact any of our stocking decisions—different from the newsvendor model. Another difference from the newsvendor model is that, since we assume ongoing demand in an infinite horizon, there is no salvage value to consider. Finally, as we indicated earlier, the costs related to placing orders are now relevant to the decision, as are the costs related to holding the inventory that results from these orders.

We denote the *cost of placing an order* by A. This quantity would typically relate to such costs as the administrative activity related to communicating the order to the supplier, and the physical activity related to receiving the goods and placing them into stock. This cost could also stem from the freight charges for transporting the goods from the supplier's location to ours, an issue we address in Chapter 4. Estimating the administrative costs of ordering may not be straightforward, so the issue of the sensitivity of the ordering decision and the cost outcome to the model inputs is a relevant point of discussion that we will undertake in discussing our results.

The *cost of holding inventory* can be roughly dichotomized into (i) the opportunity cost of tying up the firm's capital in goods sitting idle waiting for eventual consumption or sale, and (ii) the various costs that arise as a result of having goods in stock—e.g., owning or leasing warehouse space in which to store the goods, paying warehouse employees to move these goods into and out of stocking locations, insurance to protect against the loss, damage, or theft of these goods, and possibly other risk-related costs such as the potential loss of goods due to spoilage or obsolescence. Again, some of these costs may be difficult to estimate. The cost of tying up capital in inventory should, in theory, be covered by what the finance community calls the *weighted average cost of capital*, a value that reflects the cost to the firm of raising funds, either by selling equity shares in the firm or by securing debt. While the cost of debt is relatively easy to assess, the cost of equity is not so easily established. Whether this cost of capital is easily specified or not, however, it is clear that tying up capital in inventory entails an opportunity cost, reflecting the fact that these funds are not available to the firm to invest in activities that could more immediately and more directly generate a return. The other costs of holding inventory are rather diffuse (i.e., many costs spread across a number of activities) and/or simply not easy to specify (e.g., the cost of obsolescence), so again, we are left to deal with costs that are no doubt real, but not easily specified.

Nevertheless, though the costs related to managing inventory in a perpetual-demand setting are not easy to specify, the best values available must be utilized in a decision framework. The original work on optimal decision making in perpetual-demand inventory systems is credited to Harris (1913). Wilson (1934) offered a more comprehensive treatment of the formulation of the basic optimization problem and implementation of the solution, now most often known simply as the *economic order quantity* model, or EOQ model. In this model, again, ordering cost is given by A. We must, however, specify holding costs on a time scale. Since we have, earlier, chosen an annual time scale for demand, we must develop a per-unit holding cost h on a consistent time scale. Specifically, let

$$h = ic \qquad (3.7)$$

where i denotes the annualized cost of tying up funds in inventory, on a percentage basis—i.e., the percentage of each dollar held in inventory that reflects, in total, the opportunity cost of having that dollar sit in inventory and the additional physical and risk-based costs of holding and managing inventory.

With these pieces in place, we can build a model of the annual cost of managing inventory. First, we consider the *annual cost of ordering goods*. In this constant-demand setting, our sole decision variable is, as it was in the newsvendor model, the order quantity Q. Since each order costs A \$/order, and an average of D/Q orders would be placed annually, the expected annual cost of ordering is $OC = A \cdot D/Q$.

In order to specify the *annual cost of holding inventory*, we first must determine the quantity of inventory held. Returning to Figure 3.4, note that in each *inventory cycle* (i.e., the period of time between successive orders [or successive receipts]) the inventory level rises by Q units and then falls back to some lower level—past the reorder point R—prior to rising again by Q units when the next order arrives. If demand occurs at a known, constant rate, then the cost-minimizing decision is to place each order such that the on-hand inventory level hits zero the instant before the replenishment order arrives. This is possible if the *replenishment lead time*, which we denote by L, is also known with certainty. In that case—known and constant-demand rate, and known and constant lead time—the on-hand inventory level would rise from 0 to Q, decline at a constant rate back to 0, and then rise back to Q again, *ad infinitum*. Clearly, then, the average amount of inventory held would be the average of 0 and Q, or $Q/2$. Another way to look at this is as follows: Assume that you have been given the task of recording the inventory level in the system at random intervals over time. Sometimes, upon observing the inventory level, you would record an inventory level at or near Q, sometimes at or near 0, and sometimes in between these values. Given enough observations at truly random intervals, the average of your observations would necessarily be $Q/2$, which is the average inventory level in this system, also known as the *cycle stock*. Thus, given a cycle stock of $Q/2$, the annualized cost* of carrying this inventory is given by $HC = h \cdot Q/2$.

With these pieces in place, we can express the total annual cost of the constant-demand EOQ model as

$$TAC(Q) = OC + HC = A \cdot \frac{D}{Q} + h \cdot \frac{Q}{2} \tag{3.8}$$

To find the optimal value of Q, we take the derivative of this expression, set it to zero, and solve for Q, obtaining

$$\frac{dTAC(Q)}{dQ} = -\frac{AD}{Q^2} + \frac{h}{2} = 0 \tag{3.9}$$

which we solve for

$$Q^*_{EOQ} = \sqrt{\frac{2AD}{h}} \tag{3.10}$$

Example 3.5 Economic order quantity

J&M Distributors is a supplier of consumer electronics and computer gear to big box retailers. Annual demand for one particular item at J&M, WallShaker surround speakers, is 18,000 units, which occurs at a relatively constant rate throughout the year (comprising 365 sales days—J&M never closes, a matter of pride for its hard-charging CEO). J&M follows a reorder point–order quantity inventory policy. The company estimates that its cost of placing an order is approximately $30, covering both the administrative costs of placing the order and the labor cost of receiving and stocking the item. Given the relatively high costs of obsolescence for retail electronics goods, J&M uses an annual inventory holding cost rate

---

* The authors' experience shows that students sometimes get hung-up on the fact that, while it is rarely the case that any inventory units actually spend an entire year in stock, we still charge ourselves the fully annualized cost of h for each unit of average inventory. The key here is that the inventory charge reflects the fact that the expected *steady-state* level of inventory is $Q/2$, such that *some* units, an amount that in total averages $Q/2$, can be expected to be in inventory at any point in time throughout the year (though, indeed, many of these units will have long passed through the inventory system once a year transpires). Thus, we must charge ourselves the annual cost of $HC = h \cdot Q/2$ for holding this cycle stock.

of 30%. J&M purchases WallShaker speakers from its distributor for $100 per unit (where the "unit" is a set of speakers). Therefore, the economic order quantity is $Q^*_{EOQ} = \sqrt{(2)(30)(18,000)/[(0.3)(100)]} = 189.7\,\text{units}$, or $Q^* = 190$. J&M's distributor, however, requires that WallShaker speaker sets be ordered in multiples of 50 units (5 layers of 10 boxes per layer on a pallet). Thus, J&M should order the nearest multiple of 50 units, or $Q = 200$ units. We will show later that this deviation from the optimal order quantity results in only a slight increase in total annual cost.

We now return to the issue of estimating the various costs involved in the EOQ model. The OC, HC, and TAC curves related to Example 3.5 are shown in Figure 3.5. Note that the TAC curve is relatively flat around $Q^* = 190$. This means that TAC is robust with respect to small misspecifications in the order quantity, or in the parameters used to compute what is assumed to be the optimal order quantity. In fact, the difference between ordering at the optimal value of Q and the value required by the distributor in Example 3.5 is TAC(Q^*) = $5692 versus TAC(200) = $5700. Moreover, from expression (3.10), one can see that a misspecification in either A or h translates into square root effect on the EOQ. In Example 3.5, if A were assumed to be $40, instead of its actual value of $30 (33% larger than estimated), the computed EOQ would be $Q = \sqrt{(2 \cdot 40 \cdot 18000)/(0.3 \cdot 100)} = 219$, which is only $\sqrt{1.3333} = 1.155$ times (or 15.5% larger than) the computed EOQ. Since TAC(Q^*) = $\sqrt{2ADh}$, this result

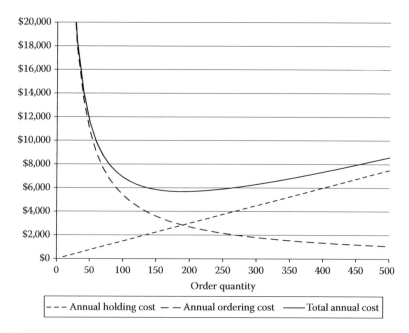

FIGURE 3.5
Cost curves for economic order quantity (EOQ) model.

translates directly to TAC as well. Moreover, in either case, the "nearest big, round number" approach to setting EOQ (as reflected by the final decision value in Example 3.5) would result in an order quantity of $Q = 200$ even with the incorrectly specified value of A.

3.4.2 Continuous-Review under Uncertainty

Recall from Figure 3.4 that the lines reflecting the depletion of inventory in each cycle did not actually reflect certainty in demand. Specifically, they were not exactly straight, indicating some variability in the rate at which demand transpired over time. This is a more realistic situation than the one we used to build our EOQ model earlier. (This does not, however, imply that our EOQ model is not useful.* It is merely a starting point, allowing us to gather useful insights by initially assuming away some of the "clutter.") For example, looking back at Figure 3.4, note that the depletion rate in the cycle, in on-hand inventory, that spans approximately days 28–39, is slower, overall, than the depletion rate in the later cycle that spans days 50–56. In fact, in the latter cycle, the faster depletion causes the inventory system to stock out before the replenishment order arrives, whereas in the earlier cycle, the on-hand inventory level remains comfortably above zero up to the point at which the replenishment arrives.

An alternative pictorial view of a continuous-review, perpetual-demand model with stationary demand is shown in Figure 3.6. Note that the two inventory cycles depicted in this figure essentially "sit on top of" a buffer quantity represented in the figure as SS, which an astute observer would recognize as *safety stock*. Thus, from this pictorial perspective we see that safety stock SS is not only the amount of inventory held in excess of demand—in

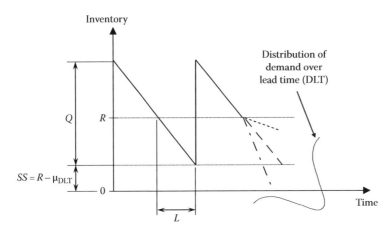

FIGURE 3.6
Uncertainty in demand over lead time in a (Q, R) inventory system.

---

* Quoting the renowned statistician G. E. P. Box, "All models are wrong, but some are useful."

this case demand over the replenishment lead time—in order to protect against stockouts, as we had defined it in our discussion of the newsvendor model, but it can also be seen in this infinite-horizon setting to be the amount of inventory that is expected to be on-hand when a replenishment order arrives. As indicated by the three dashed lines in the second cycle of the figure, if demand over the replenishment lead time is uncertain, the demand rate could track to its expected value over the replenishment lead time (middle line), or it could decrease from the expected value (upper line), or it could increase (lower line), in the latter case leaving the system at risk of a stockout. Setting the reorder point sufficiently high is the means of preventing such stockouts, and therefore, the reorder point, as indicated by the figure is given by

$$R = \mu_{DLT} + SS \qquad (3.11)$$

where μ_{DLT} represents the expected demand over the lead time. More formally, let us define DLT as the random variable representing demand over the lead time, and let its mean and variance be μ_{DLT} and σ^2_{DLT}, respectively.

The problem that we had considered in the constant-demand setting, therefore, has expanded from solving for a single decision variable to finding good values for two decision variables, the reorder point R and the order quantity Q—i.e., both the "when" and "how much" decisions we discussed earlier. Accordingly, the TAC expression that we considered earlier will now expand as well. Clearly, from Figure 3.6, the annual holding cost in our TAC expression must reflect not only the cycle stock of $Q/2$, but also the safety stock, which is now expected to be an ongoing portion of overall inventory. Since $SS = R - \mu_{DLT}$, expression (3.8) expands to become*

$$TAC(Q, R) = \frac{AD}{Q} + h(Q/2 + R - \mu_{DLT}) \qquad (3.12)$$

Some time-honored texts on production and inventory management—e.g., Hax and Candea (1984) and Silver et al. (1998)—include explicit stockout costs in the annual cost expression. Our approach, however, will be to consider a service constraint in finding the (Q, R) solution that minimizes expression (3.12). This is consistent with our earlier discussion of the newsvendor model,

---

* Note that this model implies that expected inventory is $Q/2 + SS$. Technically, however, we should be careful about how demand is handled in the event of a stockout. If demand is backlogged, then expected inventory at any point in time is actually $Q/2 + SS + E[BO]$, where $E[BO]$ is the expected number of backorders. Silver et al. (1998, p. 258), however, indicate that since $E[BO]$ is typically assumed to be small relative to inventory, it is reasonable to use $Q/2 + SS$ for expected inventory. If demand is lost in the event of a stockout, then expected inventory is exactly $Q/2 + SS$.

in which we pointed out that specifying a service level is (conceptually, at least) equivalent to specifying a penalty cost for stockouts, not to mention the fact that it avoids the messy problem of trying to accurately estimate the penalty cost value.

Thus, an important issue in our formulation will be how we specify the service constraint on our optimization problem. From our earlier discussion of in-stock probability and fill rate, recall that the probability distribution of demand figured prominently in measuring the service level. The most general case is the one in which *both* demand per unit time period (where the time period is typically specified as days or weeks) *and* replenishment lead time (correspondingly expressed in days or weeks) are random variables. Let us, therefore, assume that the lead time L follows a normal distribution with mean μ_L and variance σ_L^2. Further, let us assume that the distribution of the demand also follows a normal distribution, such that its mean and variance, μ_D and σ_D^2, are expressed in time units (i.e., demand per unit time) that are consistent with the time units used to express lead time L (e.g., days or weeks).

The normal-distribution assumption on both D and L allows a well-defined convolution of demand over the lead time, which is therefore also normally distributed,[*] with mean

$$\mu_{DLT} = \mu_D \mu_L \tag{3.13}$$

and standard deviation[†]

$$\sigma_{DLT} = \sqrt{\mu_L \sigma_D^2 + \mu_D^2 \sigma_L^2} \tag{3.14}$$

Moreover, given a value of the reorder point R, the expected number of units short in an order cycle under normally distributed lead-time demand can be expressed as

$$S(R) = \sigma_{DLT} L(z) \tag{3.15}$$

where $L(z)$, as presented earlier, is the standard normal loss function evaluated at z, which in this case is given by $z = (R - \mu_{DLT})/\sigma_{DLT}$. If unmet demand in a replenishment cycle is fully backlogged, it should be apparent (see Figures 3.5 and 3.6) that the expected demand met in a replenishment cycle

---

[*] As indicated by Silver et al. (1998), the assumption of the normal distribution for lead-time demand is common for a number of reasons, particularly that it is "convenient from an analytic standpoint" and "the impact of using other distributions is usually quite small" (p. 272).
[†] The general form of this expression, from the variance of a random-sized sum of random variables, can be found in Example 4 in Appendix E of Kulkarni (1995, pp. 577–578).

is Q, meaning that the fill rate in this continuous-review, perpetual demand model is given by[*]

$$\beta(Q, R) = 1 - \frac{S(R)}{Q} = 1 - \frac{\sigma_{DLT}L(z)}{Q} \tag{3.16}$$

Note, therefore, that the problem to minimize expression (3.12) subject to a fill-rate constraint based on expression (3.16) is non-linear in both the objective function and the constraint. The TAC objective, however, is convex[†] in both Q and R, so the solution is fairly straightforward.

A less rigorous approach to finding a (Q, R) solution would be to solve for Q and R separately. Note that $z = (R - \mu_{DLT})/\sigma_{DLT}$ gives a fractile of the distribution of demand over the lead time. Thus, we could set R to achieve a desired in-stock probability, along the lines of the newsvendor problem solution discussed earlier (i.e., to accumulate a given amount of probability under the DLT distribution). In this setting, the in-stock probability is typically referred to as the *cycle service level* (CSL), or the expected in-stock probability in each replenishment cycle.[‡] Specifically, for normally distributed lead-time demand, DLT, we set

$$R = \mu_{DLT} + z_{CSL}\sigma_{DLT} \tag{3.17}$$

where $z_{CSL} = \Phi^{-1}(CSL)$ and where Φ is the standard normal cumulative distribution function (cdf).[§]

With R fixed to achieve a desired CSL, we then use the first-order condition on Q in expression (3.12),

$$\frac{\partial TAC\ (Q, R)}{\partial Q} = -\frac{AD}{Q^2} + \frac{h}{2} = 0 \tag{3.18}$$

---

[*] Again, the backlogging assumption is important. If unmet demand is lost as opposed to backlogged, then the replenishment quantity Q cannot be used to fulfill demand backlogged from previous cycles, and this inflates the expected total demand that is fulfilled in an average cycle from Q to $Q + S(R)$, such that, with normally distributed lead-time demand, we obtain $\beta(R,Q) = 1 - \sigma_{DLT}L(z)/[Q + \sigma_{DLT}L(z)]$.

[†] See Hax and Candea (1984, p. 206) for a proof of the convexity of a closely related model.

[‡] Alternatively, we could determine the CSL target by considering the tradeoff between the cost of underage—i.e., stockout, at some per-unit or per-incidence penalty cost—and the cost of overage—i.e., holding inventory into the next period. This, however, carries with it some challenges related to accurately specifying the time period over which these costs are relevant. Is the penalty cost incurred on a per-period basis, or a one-time basis? Do we annualize the penalty cost? Do we annualize the holding cost? Instead, our choice is simply to specify the desired service level, using the relative knowledge we have of underage and overage costs as a guide to whether this CSL target should be relatively larger or smaller.

[§] Note that if DLT is not normally distributed, then one can still specify the reorder point R as a fractile of the DLT distribution—i.e., $R = F_{DLT}^{-1}$ (CSL).

to solve for

$$Q^* = \sqrt{\frac{2AD}{h}} \qquad (3.19)$$

the familiar EOQ.

Example 3.6 Approximate (Q, R) solution under uncertainty

Let us return to the problem stated in Example 3.5, the WallShaker speakers at J&M Distributors. As we computed in this example, $Q^*_{EOQ} = 189.7$ units, such that, if this value were feasible, we would set $Q = 190$. Now, assume that J&M has collected additional information regarding demand and replenishment lead time. First, note that $\mu_D = 18,000/365 = 49.32$ units/day. J&M has measured the uncertainty in daily demand to be $\sigma_D = 15$ units/day. The company's data indicate, however, that the replenishment lead time is remarkably consistent at $L = 4$ days, such that they are willing to assume that $\sigma_L = 0$. Thus, we can compute $\sigma_{DLT} = \sqrt{\mu_L \sigma_D^2 + \mu_D^2 \sigma_L^2} = \sqrt{(4)(15)^2 + 0} = 15\sqrt{4} = 30$. If J&M desires an in-stock probability, or cycle service level, of CSL = 90%, then since $\mu_{DLT} = \mu_D \mu_L = (49.32)(4) = 197.3$ units, we can compute $R = \mu_{DLT} + z_{CSL}\sigma_{DLT} = 197.3 + (1.282)(30) = 235.7$, with $z_{CSL} = \Phi^{-1}(0.90) = 1.282$. Note that the right side of the sum defining R gives the safety stock level required to achieve the CSL, $SS = z_{CSL}\sigma_{DLT} = R - \mu_{DLT} = 38.46$ units. In order to achieve at least the CSL, we must round the computed value of the reorder point up to $R = \lceil 235.7 \rceil = 236$ units. Thus, the suggested, approximately optimal inventory policy is $(Q, R) = (190, 236)$.

The optimization problem formed by expressions (3.12) and (3.16)—i.e., minimize objective function (3.12) subject to constraint (3.16)—can be solved explicitly with any number of optimization software packages as well. To demonstrate this, let us first express (3.16) in a way that gathers the decision variables on a single side of the expression, specifically,

$$1 - \frac{\sigma_{DLT}L(z)}{Q} \geq \beta_{tgt}$$

which can be expressed as

$$\frac{L\left(\dfrac{R - \mu_{DLT}}{\sigma_{DLT}}\right)}{Q} \leq \frac{1 - \beta_{tgt}}{\sigma_{DLT}} \qquad (3.20)$$

where β_{tgt} is the target fill rate. In addition, we typically wish to achieve the constraint in expression (3.20) by utilizing a positive value of safety stock, and therefore we would also constrain our optimization to require $R - \mu_{DLT} \geq 0 \Leftrightarrow R \geq \mu_{DLT}$. The example given below extends our EOQ-based examples given earlier (Examples 3.5 and 3.6) to demonstrate the process of generating an optimal (Q, R) solution.

Example 3.7 Optimal (Q, R) solution under uncertainty

Figure 3.7 shows an Excel-based implementation of the optimization problem formulated earlier: minimize expression (3.12) subject to expression (3.16) and $R \geq \mu_{DLT}$. To formulate and solve this optimization problem, we must, in contrast to Example 3.6, specify a fill-rate service target. In this case, we specify $\beta_{tgt} = 0.99$, or a fill rate of 99%. In terms of the Excel sheet in Figure 3.7, the optimization problem is set up in the Excel Solver to minimize the value in cell I6, the TAC, which is stated as the sum of OC and HC, given in the cells immediately above this. The computed values of the right-hand and left-hand sides of the fill-rate constraint are also shown in the summary Excel tables. Since this is a non-linear optimization problem consisting of a non-linear objective function and non-linear constraints, the solution process is highly dependent on the initial solution from which the optimization process is started. Using the generalized reduced gradient search (GRG) algorithm embedded in the Excel Solver under various starting solutions, the resulting optimal solution is shown in Figure 3.7, with $(Q^*, R^*) = (210, 230)$. The optimal cost of this solution is $TAC(Q^*, R^*) = \$6704$. In contrast, the cost of the solution specified in Example 3.6 is TAC(190, 236) = \$6854, or about 2.25% over the optimal annual cost. Note, however, that the approximate policy was computed to achieve $CSL \geq 90\%$, and in fact, due to the rounding to an integer value of R in Example 3.6, the $(Q, R) = (190, 236)$ policy actually achieves a cycle service level of $CSL = 90.17\%$ and fill rate of $\beta = 99.27\%$, both of which are higher than the values that result from the optimal policy, as shown in Figure 3.7.

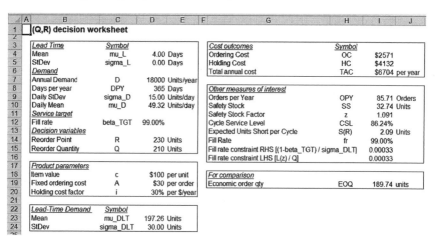

	A	B	C	D	E	F	G	H	I	J
1	(Q,R) decision worksheet									
2										
3	*Lead Time*		*Symbol*				*Cost outcomes*		*Symbol*	
4	Mean		mu_L	4.00	Days		Ordering Cost		OC	$2571
5	StDev		sigma_L	0.00	Days		Holding Cost		HC	$4132
6	*Demand*						Total annual cost		TAC	$6704 per year
7	Annual Demand		D	18000	Units/year					
8	Days per year		DPY	365	Days		*Other measures of interest*			
9	Daily StDev		sigma_D	15.00	Units/day		Orders per Year		OPY	85.71 Orders
10	Daily Mean		mu_D	49.32	Units/day		Safety Stock		SS	32.74 Units
11	*Service target*						Safety Stock Factor		z	1.091
12	Fill rate		beta_TGT	99.00%			Cycle Service Level		CSL	86.24%
13	*Decision variables*						Expected Units Short per Cycle		S(R)	2.09 Units
14	Reorder Point		R	230	Units		Fill Rate		fr	99.00%
15	Reorder Quantity		Q	210	Units		Fill rate constraint RHS [(1-beta_TGT) / sigma_DLT]			0.00033
16							Fill rate constraint LHS [L(z) / Q]			0.00033
17	*Product parameters*									
18	Item value		c	$100	per unit		*For comparison*			
19	Fixed ordering cost		A	$30	per order		Economic order qty		EOQ	189.74 units
20	Holding cost factor		i	30%	per $/year					
21										
22	*Lead-Time Demand*		*Symbol*							
23	Mean		mu_DLT	197.26	Units					
24	StDev		sigma_DLT	30.00	Units					

FIGURE 3.7

Excel set-up for (Q, R) inventory system optimization.

3.4.3 Periodic-Review, Reorder-Point–Order-up-to Models

As we discussed earlier, an alternative to a continuous-review approach to inventory management is to review the inventory level on a periodic interval, specifically every T periods, where T would then be the *review period*. As we will demonstrate in the following, protecting the system against stockouts in this case requires a larger inventory level than the amount required to achieve the same service performance for an inventory system utilizing continuous-review under the same operating conditions. Thus, one might question why a periodic-review system would be used. One reason might be to fulfill the "less intensive management" prescription for "B" items from an ABC analysis. Since the inventory is reviewed and reordered only periodically, the management activity would indeed be less intensive. Or, in cases where reviewing the inventory level of any given item is a manual, labor-intensive process, there would be an obvious desire to limit this activity to being done only periodically. Modern information systems quite likely have dramatically reduced the amount of labor required to monitor inventory levels, though. One last reason for periodic review, however, may be the most important one, and that is to allow groups of items to be reordered together at a common interval. This could allow for significant savings in transportation charges, and possibly savings in other sources of fixed costs from common suppliers.

Among periodic-review systems, the most direct parallel to the continuous-review (Q, R) inventory management system that we discussed earlier is the periodic-review (s, S) system, also called an (s, S, T) system to indicate that it is a reorder point (s)–order-up-to (S) system under periodic review, with period length T. In this case, control of the inventory system proceeds as follows: Every T time units, review the inventory position x (i.e., on-hand plus on-order, as discussed earlier in the chapter), and if $x < s$, order a sufficient amount to bring the inventory level back to the *order-up-to* level S—i.e., order $Q = S - x$. A sample inventory profile from an (s, S, T) system is shown in Figure 3.8, with $(s, S, T) = (170, 350, 5)$. Note that this profile differs from the (Q, R) inventory profile in one important way: The order quantity is not the same every period, as it is in the (Q, R) system, but on the other hand, the inventory position in this system always returns to a constant value, $S = 350$.

Previous researchers (Scarf, 1960; Iglehart, 1963) have shown that an (s, S) inventory policy is optimal for a periodic-review system for which there is a fixed cost of ordering—i.e., where $A > 0$, using the notation we specified earlier. Computing the optimal parameters, s^* and S^*, however, requires rather sophisticated mathematical tools, namely stochastic dynamic programming, and most computational approximations that have been proven to be near-optimal under various conditions are similarly challenging to specify and compute (see Porteus, 1985, for an overview and numerical comparison of a number of such approximations). For purposes of simplicity, however, we propose that a reasonably good solution can be obtained by taking a similar

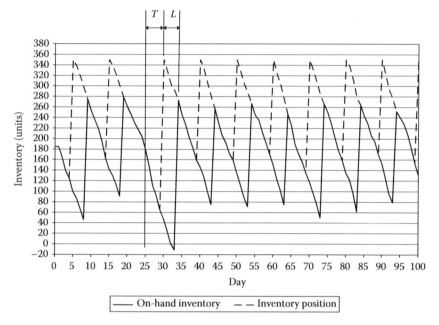

FIGURE 3.8
Sample inventory profile for (s, S, T) system.

approach to the approximate method we specified for the (Q, R) system, namely to compute values of s and S separately.

Thus, let us first consider how we might compute a good value of s, the reorder point. One approach to doing this, as we did in our (Q, R) discussion, is to specify a target cycle service level, CSL. Now, computing the value of s requires us to think carefully about the period of time over which s protects the inventory system against stockouts. Clearly, the system is at risk of stockout over the replenishment lead time, as it was in the (Q, R) system, but in this case, we must also account for the fact that the inventory position will not be reviewed again until T periods have passed. Therefore, s must be set to a level that protects the system against stockouts over a period that covers T + L periods. Visually, this can be seen in Figure 3.8 by considering the cycle, measured by inventory position, that covers days 25–35. Note that this cycle actually spans two order-review periods, and that the order that raised the inventory position back to S = 350 on day 30 was not available to actually protect the system against a stockout until it arrived in stock L = 3 days later. Thus, the review that took place on day 25 had to cover the system up to the next review, plus the lead time for re-supply that results from that review if it generates an order, in this case T + L = 5 + 3 = 8 days later, on day 33. Unfortunately, this review failed to protect the system from a stockout. The inventory position on day 25 was x = 178 units, just slightly above s = 170, and therefore an order was not placed, such that there was no re-supply receipt on day 28. On day 30, however, the inventory position had dropped to a dangerously low $x = 48$ units, resulting in an order of $S - s = 302$ units. When this re-supply arrives on day 33, the inventory position has actually dropped below zero, to −11 units, but that re-supply puts the system into a state where the stock level remains positive throughout the remainder of the cycles shown in Figure 3.8.

Thus, in order to compute a value of s we must specify the distribution of demand over the lead time *plus the review period*, as in Figure 3.8, or what we will call DLTR. Then, if DLTR follows a normal distribution, we can compute the periodic-review reorder point as

$$s = \mu_{\text{DLTR}} + z_{\text{CSL}}\sigma_{\text{DLTR}} \tag{3.21}$$

where μ_{DLTR} and σ_{DLTR} are the mean and standard deviation, respectively, of the DLTR distribution, and z_{CSL} is, as it was defined earlier, the critical fractile of the DLTR distribution that accumulates an in-stock probability equal to CSL. Again, if DLTR does not follow a normal distribution, then provided we know the cdf of DLTR, we can compute $s = F_{\text{DLTR}}^{-1}(CSL)$.

In practice, however, specifying the DLTR distribution may be a bit tricky. Since the review period T is a specified constant, then if the lead time L is constant, DLTR is simply comprised of the sum of $T + L$ independent and identically distributed (iid) demand random variables. In this case, we have

$$\mu_{\text{DLTR}} = (L+T)\mu_{\text{D}} \tag{3.22}$$

and

$$\sigma_{\text{DLTR}} = \sigma_{\text{D}}\sqrt{L+T} \tag{3.23}$$

with the latter expression resulting from the fact that the variance of a sum of independent random variables is equal to the sum of the variances of those variables. If, however, L is a random variable, then DLTR consists of a mixed distribution, with a portion of it comprised of a randomly sized sum of demand random variables (the L portion) and a portion comprised of the sum of T additional demand random variables. For our purposes in this text, we will assume only the constant-L case and use expressions (3.22) and (3.23).

Let us now turn our attention to specifying a value of the order-up-to level S. One straightforward way to do this is to utilize the EOQ expression we discussed earlier in the context of continuous-review models. Accordingly, we could compute

$$\hat{S}_{sST} = s + \text{EOQ} = s + \sqrt{\frac{2AD}{h}} \tag{3.24}$$

Thus, since the inventory position x should, at each review period, be around s, then each order of $\hat{S}_{sST} - x$ units should be around EOQ. This means that each order quantity should, as does EOQ, reasonably balance the costs of ordering and holding inventory.

Example 3.8 Approximate periodic-review, (s, S) solution

Returning to the J&M Distributors scenario we consider in Examples 3.5 through 3.7, let us assume that J&M, at the suggestion of its supplier, moves to semi-weekly reviews of inventory. This allows J&M to limit its order placements to only Monday and Thursday each week, allowing the supplier of WallShaker speaker systems to coordinate its orders with those of other customers located near J&M's distribution center in Nashville, TN. Since this can save freight costs for the supplier, they have agreed to reduce J&M's unit price by 3%, to $c = \$97$.

Part of the challenge here, however, is specifying the review period T. Since J&M operates 7 days per week, the actual value of T differs between the Monday and Thursday orders. If we assume that the inventory position review occurs at the start of the day, then for the Monday orders, $T = 3$ (covering Monday, Tuesday, and Wednesday before the Thursday morning review), and for the Thursday orders, $T = 4$ (covering Thursday–Sunday before the Monday morning review). For purposes of simplicity, we specify only a single inventory policy, and therefore we err on the side of covering the longer of the two review intervals; thus, we use $T = 4$ in computing the value of s. This will result in a higher average inventory in half of the cycles, but it is the price we pay for simplicity. For all orders, however, we continue to assume that the lead time is a fixed value of $L = 4$ days (meaning orders placed on Monday morning arrive on Friday morning, and orders placed on Thursday arrive on Monday morning).

Therefore, with $T + L = 8$ days, we have $\mu_{DLTR} = (T + L) \cdot \mu_D = (8)(49.32) =$ 394.56 units, and $\sigma_{DLTR} = \sigma_D \sqrt{T + L} = 15\sqrt{8} = 42.43$ units. If J&M desires a cycle service level of CSL = 90%, then we can compute $s = \mu_{DLTR} + z_{CSL} \sigma_{DLTR} = 394.6 + (1.282)(42.43) = 449$, with $z_{CSL} = \Phi^{-1}(0.90) = 1.282$. Note that this includes a safety stock of $SS = z_{CSL} \sigma_{DLTR} = s - \mu_{DLTR} = 54.3$ units, which is an increase of about 40% over the computed continuous-review safety stock of 38.46 units from Example 3.5.

Using this computed value of $s = 449$, we can add this to a re-computed EOQ value that accounts for the new value of the item cost, c, specifically $Q_{EOQ} = \sqrt{(2)(30)(18,000)/[(0.3)(97)]} = 192.65$, which we round to $Q = 193$. Thus, we obtain $\hat{S}_{sST} = s + \text{EOQ} = 449 + 193 = 642$. Thus, the recommended order policy for J&M is $(s,S) = (449,642)$, with a review period of $T = 4$ days.*

---

* We will refrain from making cost comparisons back to the continuous-review examples since the evaluation of costs in a periodic-review setting is a little more challenging. For example, the notion of the expected inventory level in a periodic-review system is complicated by the fact that the inventory cycle length—i.e., the time between successive receipts—can vary significantly depending on whether a cycle spans multiple periods (see Figure 3.8). Suffice it to say that a periodic-review system tends to carry more inventory than a continuous-review system, but note from Example 3.8 that the potential savings from moving to periodic-review may be driven by issues beyond inventory, for example improved coordination with other logistical activity.

We note that other approximation schemes are available for computing (s, S, T) inventory system parameters. An excellent reference in this regard is Silver et al. (1998), which provides several methods to account for shortage penalty costs in computing s and for service-related adjustments to EOQ-like formulas for computing S. Porteus (1985) is another valuable reference that reviews and compares a number of computational approaches for computing approximately optimal (s, S, T) policy parameters. In keeping with our broad, rather than deep treatment of various SCM-related computational tools in this text, we will leave our approach to computing (s, S, T) inventory system parameters at the reasonably straightforward method we specify above. In the next section, however, we consider some variations, indeed some further simplifications, to our treatment of periodic-review inventory systems.

3.4.4 Other Periodic-Review Inventory Models

One variation on periodic-review inventory systems considers the case where the cost of placing orders is either zero, negligible, or at least sufficiently small that we can essentially be assured that the optimal behavior is to order at each review period. If the review period is again T, then this results in an *order-up-to system*, wherein every T time units an order of size $Q = S - x$ is placed, where S is the order-up-to level and x is the inventory position. Accordingly, this type of periodic-review inventory system is referred to as an (S, T) system. An inventory profile of such a system is shown in Figure 3.9, with $S = 295$ and $T = 10$ days, where $L = 3$ days.

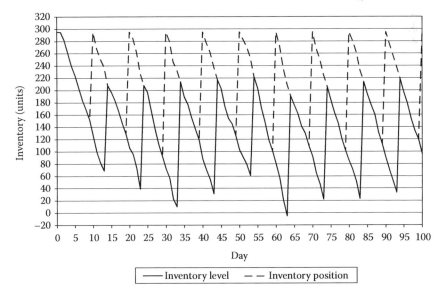

FIGURE 3.9
Sample inventory profile for (S, T) system.

In an (S, T) system, since we know that we will order at every review, the order-up-to level S serves the role of protecting the system from stockouts, more like a reorder point than an order quantity. This can be seen from Figure 3.9. Consider the order placed on day 20 in the figure. Once received, this order must keep the system in positive stock through the next review (on day 30) and receipt of the order that will be placed at that review (on day 33). Thus, a good way to specify the order-up-to level is to consider again the in-stock probability level we would like the system to achieve. This sets our target CSL, and thereby sets the critical fractile of the distribution of demand we must cover. Once again, the demand distribution we must cover, as in the (s, S, T) system discussed earlier, is DLTR, the demand over the lead time plus review period. Thus, we can set $S_{ST} = F_{\text{DLTR}}^{-1}(\text{CSL})$, or if (perhaps by assumption) DLTR is normally distributed,

$$S_{ST} = \mu_{\text{DLTR}} + z_{\text{CSL}}\sigma_{\text{DLTR}} \tag{3.25}$$

where $z_{\text{CSL}} = \Phi^{-1}(\text{CSL})$.

Example 3.9 Order-up-to system

This inventory policy follows immediately from some of the computations made in Example 3.8. In this case, $S_{ST} = \mu_{\text{DLTR}} + z_{\text{CSL}}\,\sigma_{\text{DLTR}} = 449$, which is the same as the computation of s in Example 3.8. For the (S,T) system, we know that an order will be placed every $T = 4$ days, so each order needs to cover uncertain demand until the next order arrives $T + L = 8$ days later. In the (s, S, T) system of Example 3.8, inventory levels were raised higher to allow fewer orders to be placed. Thus, utilizing the (S, T) system, ordering costs will be higher, but inventory levels will be lower.

The final periodic-review system we consider is actually the simplest, the *base-stock* system. In this system, we once again place an order at each review interval, but in this case, the review interval is set equal to the smallest discrete time unit covered by the system. Thus, for this inventory system, the period and the time unit are one-in-the-same. For example, if we manage the system using days as our time units—i.e., if lead time is specified in days—then in a base-stock system, we would review the inventory position x at the start of each day and place an order of size $Q = S - x$, to be received L days later. Thus, for this system, the review interval is $T = 1$ and the *base-stock level* is set to $S_B = F_{L+1}^{-1}(\text{CSL})$, where F is the cdf of the DLTR distribution, which in this case is the distribution of demand over $L + 1$ time units. If the demand rate per time unit, D, is normally distributed, then we can compute

$$S_B = \mu_D(L+1) + z_{\text{CSL}}\sigma_D\sqrt{L+1} \tag{3.26}$$

Sometimes, the base-stock system is referred to as an $(S-1,S)$ system since the reorder point for this system is effectively $s = S - 1$, meaning that, as long as a non-zero demand (i.e., at least one unit of demand) occurs in the reorder interval of $T = 1$, we reorder to bring the inventory position back to S.

Example 3.10 Base-stock system

The computations supporting this inventory policy are similar to those made in Example 3.9. In this case, $S_B = \mu_D(L+1) + z_{CSL}\sigma_D\sqrt{L+1} = (49.32)(5) + (1.282)(15)\sqrt{5} = 290$. For the base-stock system, we know that an order will be placed every day, so each order needs to cover uncertain demand until the order placed at the next review arrives in stock $L + 1 = 5$ days later.

Up to this point, we have discussed a relatively large number of issues and techniques related to the management of a single item at a single site. In all of our analysis, we have assumed either a single point (or season) of demand, or a stable (stationary), repeating series of demand values. Next, we consider situations where many demand values must be estimated over time.

3.4.5 Non-Stationary Demand: Distribution Requirements Planning

We now consider an environment in which we manage a given independent-demand item over multiple time periods, but here we assume that the expected value of demand varies over time. Per the framework established at the outset of the chapter (Figure 3.1), we present distribution requirements planning (DRP) as the appropriate method for this case. DRP is used to specify the time-phased inventory levels that must be present in the distribution system, where finished-goods items (independent-demand items) are collected temporarily in order to fulfill demand downstream. The "time-phasing" aspect of DRP is related to appropriately planning for the production-to-distribution lead time such that inventory is indeed available in time to fulfill projected customer demands. As Vollmann et al. (2005) point out, "DRP's role is to provide the necessary data for matching customer demand with the supply of products being produced by manufacturing" (p. 262).

Note that in any make-to-stock product setting, the linchpin between product manufacturing and customer demand is the distribution system, and indeed, it is distribution system inventory that coordinates customer demand with producer supply. While distribution system inventory levels could be determined using the reorder point–order quantity (ROP-OQ) control methods as described previously, the critical difference between DRP and ROP-OQ is that DRP plans *forward* in time, whereas ROP-OQ systems react to actual demands whenever those demands drive inventory to a level that warrants an order being placed. This is not to say that the two approaches are inconsistent. Indeed, it is quite easy to show that, in a stationary-demand environment with sufficiently small planning "time buckets," DRP and ROP-OQ produce equivalent results. Where

DRP provides a clear benefit, however, is in situations where demand is *not* constant over time, particularly those situations in which surges and drop-offs in demand can be anticipated. Thus, one of the important inputs for DRP is a time-series demand forecast, the subject of the discussion in Chapter 2.

The basic logic of DRP is to compute the time-phased schedule of replenishments for each *stock-keeping unit* (SKU) in the distribution system that keeps the inventory level of that SKU at or above a specified safety stock level (which, of course, could be set to zero). Let the DRP horizon span planning periods $t = 1, \ldots, T$, and assume that the safety stock SS, the order lot size Q, and the lead time L are given for the SKU in question. In addition, assume that we are given a period-by-period forecast of demand for this SKU, D_t, $t = 1, \ldots, T$. Then, we can compute the *planned receipts* for each period t, namely the quantity—as an integer multiple of the lot size—required to keep inventory at or above SS.

Example 3.11 Distribution requirements planning

To illustrate the basic concepts, let us consider an example, altered from Bozarth and Handfield (2006). Assume that MeltoMatic Company manufactures and distributes snow blowers. Sales of MeltoMatic snow blowers are concentrated in the midwestern and northeastern states of the United States. Thus, the company has distribution centers (DCs) located in Minneapolis and Buffalo, both of which are supplied by MeltoMatic's manufacturing plant in Cleveland. Table 3.4 shows the DRP records for the Minneapolis and Buffalo DCs for MeltoMatic's two SKUs, Model SB-15 and Model SBX-25. The first three lines in the table sections devoted to each SKU at each DC provide the forecasted demand (requirements), D_t; the scheduled receipts, SR_t, already expected to arrive in future periods; and the projected ending inventory in each period, I_t. This last quantity, I_t, is computed on a period-by-period basis as follows:

1. For period t, compute *net requirements* $NR_t = \max\{0, D_t - (I_{t-1} + SR_t - SS)\}$—i.e., compute the net demand not covered by available supply (and note that, since we expect inventory to remain at or above SS, this quantity must be subtracted from inventory and scheduled receipts in order to determine the supply available to support current demand).

2. Compute planned receipts $PR_t = \left\lceil \dfrac{NR_t}{Q} \right\rceil \cdot Q$, where $\lceil x \rceil$ gives the smallest integer greater than or equal to x. (Note that this implies that if $NR_t = 0$, then $PR_t = 0$ as well.)

3. Compute $I_t = I_{t-1} + SR_t + PR_t - D_t$—i.e., compute the net supply remaining after accounting for receipts and demand. Set $t \leftarrow t + 1$.

4. If $t \leq T$, go to step 1; else, done.

Then, planned receipts are offset by the lead time for that SKU, resulting in a stream of *planned orders*, PO_t, to be placed so that supply arrives in the DC in time to fulfill the projected demands—i.e., $PO_t = PR_{t+L}$ for $t = 1, \ldots, T - L$. Any planned receipts in periods $t = 1, \ldots, L$ result in orders that are, by definition, *past due*.

TABLE 3.4

DRP Records for MeltoMatic Snow Blowers Example with Actual Orders Equal to Planned Orders

Week		45	46	47	48	49	50	51	52	1	2	3	4
Minneapolis DC													
Forecasted requirements		60	60	70	70	80	80	80	80	90	90	95	95
Scheduled receipts		120											
Model SB-15 — Projected ending inventory (init 80)		140	80	10	60	100	20	60	100	10	40	65	90
LT (weeks): 2 — Net requirements		0	0	0	70	30	0	70	30	0	90	65	40
Safety stock (units): 10 — Planned receipts		0	0	0	120	120	0	120	120	0	120	120	120
Lot size (units): 120 — Planned orders		0	120	120	0	120	120	0	120	120	120	0	0
Actual orders		0	120	120	0	120	120	0	120	120	120	0	0
Forecasted requirements		40	40	50	60	80	90	100	100	110	80	60	40
Scheduled receipts		40											
Model SBX-25 — Projected ending inventory (init 100)		100	60	10	30	30	20	80	60	30	30	50	10
LT (weeks): 2 — Net requirements		0	0	0	60	60	70	90	30	60	60	40	0
Safety stock (units): 10 — Planned receipts		0	0	0	80	80	80	160	80	80	80	80	0
Lot size (units): 80 — Planned orders		0	80	80	80	160	80	0	80	80	80	0	0
Actual orders		0	80	80	80	160	80	0	80	80	80	0	0

(continued)

TABLE 3.4 (continued)

DRP Records for MeltoMatic Snow Blowers Example with Actual Orders Equal to Planned Orders

Week		45	46	47	48	49	50	51	52	1	2	3	4
Buffalo DC													
Forecasted requirements		70	70	80	80	90	90	100	100	120	120	140	140
Scheduled receipts		100											
Model SB-15	(30)												
Projected ending inventory		60	110	30	70	100	10	30	50	50	50	30	10
LT (weeks): 1 — Net requirements		0	20	0	60	30	0	100	80	80	80	100	120
Safety Stock (units): 10 — Planned receipts		0	120	0	120	120	0	120	120	120	120	120	120
Lot size (units): 120 — Planned orders	(0)	120	0	120	120	0	120	120	120	120	120	120	0
Actual orders		120	0	120	120	0	120	120	120	120	120	120	0
Forecasted requirements		50	50	60	70	80	80	100	110	130	130	100	100
Scheduled receipts		60											
Model SBX-25	(30)												
Projected ending inventory		40	70	90	20	20	20	80	50	80	30	90	70
LT (weeks): 1 — Net requirements		0	25	5	0	75	75	95	45	95	65	85	25
Safety Stock (units): 15 — Planned receipts		0	80	80	0	80	80	160	80	160	80	160	80
Lot size (units): 80 — Planned orders	(0)	80	80	0	80	80	160	80	160	80	160	80	0
Actual orders		80	80	0	80	80	160	80	160	80	160	80	0
Cleveland plant													
Master production schedule	SB-15	120	120	240	120	120	240	120	240	240	240		
	SBX-25	80	160	80	160	240	240	160	320	160	240		
	Total	200	280	320	280	360	480	280	560	400	480		

One important aspect of DRP is that it *requires human intervention* in order to turn planned orders into *actual* orders, the last line in Table 3.4 for each SKU. To further illustrate this point, let us extend our example to consider a situation where the human being charged with the task of converting planned orders into actual orders might alter the planned order stream to achieve some other objectives.

Example 3.12 DRP adjustments

Continuing with MeltoMatic from Example 3.11, let us assume that the carrier that ships snow blowers into MeltoMatic's warehouses in Minneapolis and Buffalo provides a discount if the shipment size exceeds a truckload quantity of 160 snow blowers, relevant to shipments of either SKU or to combined shipments containing a mix of both SKUs. Table 3.5 shows an altered actual order stream that generates orders across both SKUs of at least 160 units in all periods $t = 1, ..., T - L$ where $PO_t > 0$ for either SKU. In addition, Table 3.5 also shows the temporary overstock created by that shift in orders, with some units arriving in advance of the original plan. The question would be whether the projected savings in freight transportation exceeds the expense of temporarily carrying excess inventories. Interestingly, we should also note that the shift to larger shipment quantities, somewhat surprisingly, creates a smoother production schedule at the Cleveland plant (after an initial bump in Week 46).

We assume in our examples and algorithm above that the lot size Q was specified as an input to the algorithm. Clearly, if we are given information about the cost per order and the cost of holding inventory, we can compute an economic order quantity (EOQ) to serve as the lot size. This EOQ, however, does not necessarily guarantee an optimal schedule of replenishments—i.e., one that minimizes the costs of ordering and holding inventory. The reason is that this scenario violates one of the key assumptions of the EOQ model, namely that demand occurs at a stationary rate over time. A prime reason for using DRP, however, is that demand is non-stationary.

Several authors have developed methods to address the problem of finding lot sizes that minimize, or nearly minimize, the sum of ordering and holding costs across a planning horizon under non-stationary demand. Wagner and Whitin (1958) developed a widely cited optimal algorithm, based on a dynamic-programming solution to find the minimum-cost lot sizes. Silver and Meal (1973) and DeMatteis and Mendoza (1968) present relatively simple computational approaches that determine the number of periods the next order should cover based on the effect of each successive ordering decision on either the average per-period cost (Silver and Meal) or on the balance between ordering and holding costs accumulated by that decision (DeMatteis and Mendoza). While we do not cover those methods here, the reader is encouraged to seek out these additional references for further reading.

TABLE 3.5

DRP Records for MeltoMatic Snow Blowers Example with Actual Orders Altered for Transportation Discount

Week			45	46	47	48	49	50	51	52	1	2	3	4
Minneapolis DC	Forecasted requirements		60	60	70	70	80	80	80	80	90	90	95	95
	Projected ending inventory	80	140	80	10	60	100	20	60	100	40	10	65	90
Model SB-15	Planned orders	0	0	120	120	0	120	120	0	120	120	120	0	0
	Actual orders		0	120	120	0	120	120	0	120	120	120	0	0
	Forecasted requirements		40	40	50	60	80	90	100	100	110	110	100	100
	Projected ending inventory	100	100	60	10	30	30	20	80	60	80	30	60	40
Model SBX-25	Planned orders	0	0	80	80	80	160	80	80	160	80	80	0	0
	Actual orders		0	80	80	160	80	80	160	80	80	80	0	0
Total for DC	Planned orders		0	200	200	80	280	200	80	280	200	200	0	0
	Actual orders		0	200	200	160	200	200	160	200	200	200	0	0
	Temporary overstock		0	0	0	80	0	0	80	0	0	0	0	0
Buffalo DC	Forecasted requirements		70	70	80	80	90	90	100	100	120	120	140	140
	Projected ending inventory	30	60	110	30	70	100	10	30	50	50	10	30	10
Model SB-15	Planned orders	0	120	0	120	120	0	120	120	120	120	120	120	0
	Actual orders		120	0	120	120	0	120	120	120	120	120	120	0
	Forecasted requirements		50	50	60	70	80	80	100	110	130	100	100	100
	Projected ending inventory	30	40	70	90	20	20	20	80	50	30	90	90	70
Model SBX-25	Planned orders	0	80	80	0	80	80	160	80	160	80	160	80	0
	Actual orders		80	160	80	80	160	80	80	80	80	160	80	0
Total for DC	Planned orders		200	80	120	200	80	280	200	280	200	280	200	0
	Actual orders		200	160	200	200	160	200	200	200	200	200	200	0
	Temporary overstock		0	80	160	160	240	160	160	80	80	160	80	0
Cleveland plant	Master production schedule	SB-15	120	120	240	120	120	240	120	240	240	240	200	
		SBX-25	80	240	160	240	240	160	240	160	160	160	200	
		Total	200	360	400	360	360	400	360	400	400	400	400	

3.5 Multi-Item Inventory Models

Our treatment of inventory management up to this point has, with the exception of Example 3.12 previously given, concerned only independent ordering of a single item at a single location. (Example 3.12 showed how adjustments could be made to independently computed DRP orders for two different items to better coordinate freight transportation.) It would be the rare company, however, that managed only a single item at a single location. As one complication introduced with multiple items, consider a situation in which k items can be ordered jointly at a discount in the ordering cost. Specifically, we consider the case where the cost of ordering k items independently is kA, but ordering these items jointly incurs only an incremental cost, $a_i \geq 0$, for each item i ordered beyond the single base cost of ordering, A, such that $A + \sum_{i=1}^{k} a_i < kA$.

A straightforward way to address this situation is to consider ordering all items jointly on a common replenishment cycle. In essence, this changes the decision from "how much" to order—the individual order quantities, Q_i—to "how often" to order, which is given by n, defined as the number of joint orders placed per year. Then since $n = D_i/Q_i \Rightarrow Q_i = D_i/n$ for each item i, we can build a joint-ordering TAC model with

$$TAC(n) = \left(A + \sum_{i=1}^{k} a_i \right) n + \sum_{i=1}^{k} h_i \left(\frac{D_i}{2n} + SS_i \right) \tag{3.27}$$

where
h_i is the annual cost of holding one unit of item i in inventory
$D_i/(2n) = Q_i/2$ is the cycle stock of item i
SS_i is the safety stock of item i

For fixed values of SS_i (or if demand is certain and $SS_i = 0$, $i = 1,...,k$), the optimal solution to (3.27) is found by setting

$$\frac{\partial TAC(n)}{\partial n} = A + \sum_{i=1}^{k} a_i - \frac{1}{2n^2} \sum_{i=1}^{k} h_i D_i = 0 \tag{3.28}$$

yielding

$$n^* = \sqrt{\frac{\sum_{i=1}^{k} h_i D_i}{2\left(A + \sum_{i=1}^{k} a_i\right)}} \qquad (3.29)$$

Introducing safety stock and reorder points into the decision complicates the idea of joint replenishments. Clearly, it would negate the assumption that all items would appear in each replenishment order since the inventory levels of the various items are unlikely to hit their respective reorder points at exactly the same point in time. Silver et al. (1998) discuss an interesting idea first proposed by Balintfy (1964), a "can-order" system. In this type of inventory control system, two reorder points are specified, a "can-order" point, at which an order *could* be placed, particularly if it would allow a joint-ordering discount, and a "must-order" point, at which an order *must* be placed to guard against a stockout. Although we will not discuss the details here, Silver et al. (1998, p. 435) provide a list of references for computing "must-order" and "can-order" levels in such a system. Thus, one could propose a review period for the joint-ordering system and set reorder points to cover a certain cumulative probability of the distribution of demand over the lead time plus the review period.

Example 3.13 Joint replenishment

Let us consider an example that compares individual orders versus joint orders for two items. Floor-Mart—"We set the floor on prices" (and, apparently, on advertising copy)—is a large, discount retailer. Floor-Mart stocks two models of 34-in., LCD-panel TVs, Toshiba, and LG, that it buys from two different distributors. Annual demand for the Toshiba 34-in. TV is $D_1 = 1600$ units, and the unit cost to Floor-Mart is $400. Annual demand for the LG 34-in. TV is $D_2 = 2800$ units, and the unit cost to Floor-Mart is $350. Assuming 365 sales days per year, this results in an expected daily demand of $\mu_{D,1} = 4.38$ units for the Toshiba TV and $\mu_{D,2} = 7.67$ units for the LG TV. Assume that Floor-Mart also has data on demand uncertainty such that $\sigma_{D,1} = 1.50$ and $\sigma_{D,2} = 2.50$. Annual holding costs at Floor-Mart are estimated to be 20% on each dollar held in inventory, and Floor-Mart's target fill rate on high-margin items like TVs is 99%. Floor-Mart uses the same freight carrier to ship TVs from each of the distributors supplying them. Let us put a slight twist on the problem formulation as it was stated earlier. Assume that Floor-Mart's only fixed cost of placing a replenishment order with its TV distributors is the cost the carrier charges to move the goods to Floor-Mart. The carrier charges $600 for each shipment from the Toshiba distributor, and the mean and standard deviation of the replenishment lead time are $\mu_{L,1} = 5$ and $\sigma_{L,2} = 1.2$ days, respectively. The carrier charges $500 for each shipment from the LG distributor, and the mean and standard deviation of the replenishment lead time are $\mu_{L,2} = 4$ and $\sigma_{L,2} = 1.2$ days, respectively. However, the carrier also offers a discounted "stop-off" charge to pick up TVs from *each* distributor on a *single* truck, resulting in a charge of $700.

With the parameters given earlier, we find that $\mu_{DLT,1} = 21.92$, $\sigma_{DLT,1} = 7.38$, $\mu_{DLT,2} = 30.68$, and $\sigma_{DLT,2} = 10.60$. Solving (3.12) for each TV to find the optimal independent values of Q and R (via Excel Solver), we obtain $(Q_1, R_1) = (166, 25)$ and $(Q_2, R_2) = (207, 36)$. By contrast, independent EOQ solutions would be $Q_{EOQ,1} = 155$ and $Q_{EOQ,2} = 200$. Using (3.29) for the joint solution, we obtain $n^* = \sqrt{(0.2)(400 \cdot 1600 + 350 \cdot 2800)/(2 \cdot 700)} = 15.21$, yielding $Q_1 = 105$ and $Q_2 = 184$. The joint solution—assuming that it does not exceed the truck capacity—saves approximately $5100 in ordering and holding costs over either independent solution. The issue on the table, then, as we discussed earlier, is to set safety stock levels in this joint solution. Since orders would be placed only every 3.4 weeks with $n = 15.21$, a review period in line with the order cycle would probably inflate safety stock too dramatically. Thus, the reader should be able to see the benefit of a "can-order" system in this case, perhaps with a one-week review period, a high CSL—i.e., relatively large value of $z = (R - \mu_{DLT})/\sigma_{DLT}$—on the "can-order" reorder point, and a lower CSL on the "must-order" reorder point. In closing the example, we should point out that we have assumed a constant rate of demand over time, which may not be reasonable for TVs, since it clearly ignores the effects of intermittent advertising and sales promotions. Such effects are considered by distribution requirements planning (DRP), presented earlier, which addresses situations where demand is not stationary over time.

The more general case for joint ordering is to consider what Silver et al. (1998) call *coordinated replenishment*, where each item i is ordered on a cycle that is a multiple, m_i, of a base order cycle τ. For any given item, its order cycle is $\tau_i = Q_i/D_i$, which is obviously expressed in the same time units as D_i (e.g., years, weeks). Thus, $\tau_i = 1/n_i$, where n_i is the number of orders placed in the time span covering demand D_i. In coordinated replenishment, each item is ordered every $m_i\tau$ days. Some subset, possibly a single item, of the k items establishes the base cycle τ and therefore has $m_i = 1$, meaning that this (those) item(s) appear(s) in every replenishment order. Silver et al. (1998, pp. 425–430) offer an algorithmic solution to compute m_i ($i = 1, \ldots, k$) and T for this case. Jackson et al. (1985) consider the restrictive assumption that $m_i \in \{2^0, 2^1, 2^2, 2^3, \ldots\} = \{1, 2, 4, 8, \ldots\}$, which they call a "powers-of-two" policy, and show that this approach results in an easy-to-compute solution whose cost is no more than 6% above the cost of an optimal policy with unrestricted values of m_i. Another situation that could link item order quantities is an aggregate quantity constraint or a budget constraint. Hax and Candea (1984) present this problem and a Lagrangian-relaxation approach to solving it; they refer to Holt et al. (1960) for several methods to solve for the Lagrange multiplier that provides near-optimal order quantities.

3.6 Multi-Echelon Inventory Systems

You will recall from our discussion in Chapter 1 that one of the defining aspects of a supply chain is that it consists of a series of stages that are physically distinct and geographically separated at which inventory is either stored or converted in form and/or value. Some of these supply chain stages may fall under the common ownership of a single firm, and some of them may be distinct from one another in ownership, and therefore portions of different, independent firms. From the standpoint of inventory management, this leads to a number of very interesting problems, all related to the answers to the two questions that, as we have discussed throughout this chapter, comprise all of inventory management: when? and how much? Clearly, though, in a supply chain setting the issue becomes understanding the interdependency of those *when* and *how much* decisions on each other across the stages of the supply chain. Understanding these interdependencies in computing optimal—or approximately optimal—inventory system parameters in a supply chain setting is the crux of studying *multi-echelon inventory systems.*

To begin our study of multi-echelon inventory systems, we must first understand exactly what is meant by an inventory *echelon,* and before we can understand that, we must first characterize supply chain systems according to the nature of the underlying network formed by their various stages. In Figure 3.10, we show a four-site example of the simplest network structure, a serial system, in which each stage by definition has no more than one supplier upstream and no more than one customer downstream. The farthest upstream node in a serial system is assumed to have a supplier of infinite capacity external to the supply chain, and the farthest downstream node is assumed to send its output to a customer that is external to the system. We use the notational convention of Zipkin (2000), wherein supply chain stages are numbered from upstream to downstream, with stage 1 being farthest upstream, supplying stage 2, and so forth, down to the farthest downstream stage, N, which satisfies external demand.

In contrast to the serial system, Figures 3.11 and 3.12 show what are the opposite extremes, in essence. In Figure 3.11 we show a *distribution system,* in which all nodes have no more than one supplier upstream, but may have more than one customer downstream. Again, all nodes at the farthest upstream stages are assumed to have suppliers of infinite capacity, and all nodes at the farthest downstream stages are assumed to send their output to external customers. Figure 3.12 shows an *assembly system,* in which all

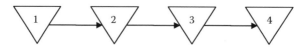

FIGURE 3.10
Four-stage serial system.

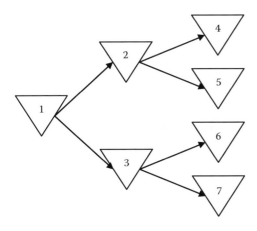

FIGURE 3.11
Example of a distribution network.

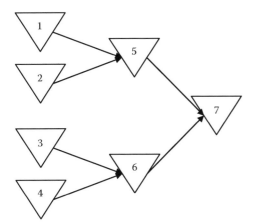

FIGURE 3.12
Example of an assembly network.

nodes have no more than one customer downstream, but may have more than one supplier upstream. Finally, Figure 3.13 shows a general, *arborescent system*, portions of which demonstrate properties of serial, assembly, or distribution systems.

In managing multi-stage inventory systems, the concept of an *echelon* is critically important. The concept was established in the seminal work on multi-echelon inventory systems by Clark and Scarf (1960). In the context of a multi-stage inventory management system, an echelon consists of that stage and all succeeding stages downstream from it. Thus, similar to Figure 5.3.2 in Zipkin (2000), Figure 3.14 pictorially represents the echelons in a four-stage, serial supply chain, where we have denoted each echelon j by the symbol e_j that appears in the rectangle representing that echelon. Each echelon j therefore consists of stage j and all stages downstream from it, up to, and including stage N. We denote the *installation inventory level* at stage j by I_j and define it to cover physical inventory on-hand and backorders, such

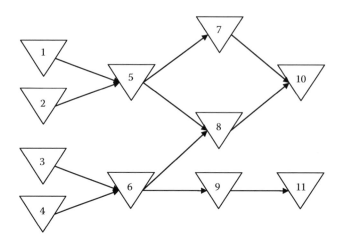

FIGURE 3.13
Example of a general arborescent network.

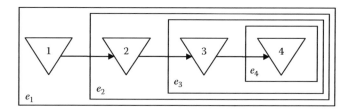

FIGURE 3.14
Echelons in a serial system.

that $-\infty \leq I_j \leq \infty$. We denote *inventory in-transit* to stage j as T_j, and therefore, we define the *echelon inventory* at stage j to be

$$\hat{I}_j = I_j + \sum_{i=j+1}^{N} \left(T_i + I_i \right) \tag{3.30}$$

which means that echelon inventory at stage j includes the installation stock at j, and all installation stock and in-transit stock downstream from j, to the farthest downstream stage in the supply chain. This concept, established by Clark and Scarf (1960), helps significantly in the mathematics related to multi-echelon inventory policies, as we will see in the following.

Given these definitions, we are ready to embark on formally describing the problem of answering the *when* and *how much* questions in a multi-echelon inventory systems. The basic issue in doing this is to determine how to set stocking levels across the supply chain in a way that somehow "coordinates" the stocking decisions and replenishment policies at the various stages. At first blush, this might sound like something that should be relatively straightforward, but the devil is clearly in the details. Much work has been done in studying serial systems, and given the introductory nature of this book, we will contain our discussion to these more easily studied—but still by no means trivially formulated and solved—problems. For a comprehensive treatment of multi-echelon systems, we refer the reader to the foundational text of Zipkin (2000). In addition, interested readers are encouraged to read the widely cited paper by Rosling (1989) on assembly systems, and the seminal work of Schwarz (1973) and Graves and Schwarz (1977) on distribution systems.

3.6.1 Centralized versus Decentralized Control

The two ways to manage such a multi-echelon inventory system are in either a decentralized or a centralized manner. In the decentralized setting, the decision maker that is responsible for the inventory control parameters at each stage determines these independently of the other stages. Still, coordination of this system could be achieved through the terms of a contract established between the firms in any given dyad arrangement in the supply chain. (The interested reader should consult the comprehensive overview of the history of this research on coordination contracts, up to the early 2000s, by Cachon (2003). This continues to be an active area of research in the supply chain management literature.) While such coordination is not guaranteed to result in the "first-best" solution that achieves a system-optimal outcome—i.e., the result of an optimal centralized solution—well-chosen contract terms may generate results that are very close to this, if not frequently equal to the first-best results.

In a centralized solution, a single decision maker is vested with the authority to establish the inventory system parameters for all stages in the supply chain. Coordination of the system is thereby achieved through this common set of system parameters. For centralized multi-echelon inventory systems, Zipkin (2000) defines a *nested* policy as one in which, for all stages j in the chain, whenever stage j orders, so does stage $j + 1$, downstream from stage j. This applies to all multi-echelon systems, continuous-review or periodic-review. In addition, Zipkin defines a *stationary-interval* policy as one in which the time intervals between orders are equal. A stronger definition, that of a *stationary* policy (see Azadivar and Rangarajan, 2008), requires equal order quantities in addition to equal time intervals between orders.

Consistent with our earlier discussion, the inventory system context is important to the nature of the decisions to be made. To illustrate the important issues in multi-echelon inventory management, let us consider two specific settings. First, we will consider the case in which demand is deterministic with a constant rate over time, and where the costs of the system include inventory holding costs and a fixed cost per order. The second scenario we consider is one in which demand is still stationary, but stochastic, and therefore we include a penalty cost for shortages that result from uncertainty in demand. In this setting, however, we assume that fixed ordering costs are negligible. After developing optimal and/or near-optimal inventory management system parameters for these two scenarios, we consider them together, in a setting where neither fixed costs nor demand uncertainty can be ignored or assumed away.

Before moving into developing and discussing our serial supply chain solutions, we point out that in the Appendix to this chapter, we discuss the infamous "bullwhip effect." This phenomenon, whereby variability in the stream of orders increases as one proceeds upstream in the supply chain, has been made known to many students of supply chain management through a game (the "Beer Game") that is played in a four-stage serial supply chain. In the Appendix, we discuss how some of the inventory management methods specified in this chapter can reduce the bullwhip effect, or, in some cases, exacerbate it. Indeed, methods for centralized control like those we discuss below are good means of combatting some of the problems implied by the bullwhip effect. The interested reader is encouraged to visit the Appendix after reading the sections that follow, and to consider how these methods could be employed to improve the management of serial systems, or even more complex supply chains.

3.6.2 Serial Supply Chain with Deterministic Demand and Fixed Ordering Costs

Schwarz (1973) presented the first systematic study of inventory policies for serial systems that optimally balance fixed ordering costs and marginal inventory holding costs to satisfy stationary deterministic demand. Thus, we base our analysis on that of Schwarz, and in a similar vein, begin by studying a system with only two stages. In his model, Schwarz referred to the stages in chain as the warehouse (upstream) and the retailer (downstream). For a two-stage system, with one warehouse and $J \geq 1$ retailer(s), where each installation j experiences a fixed cost of ordering, A_j ($/order), and an installation holding cost, h_j ($/unit/period), under a known and constant rate of demand over an infinite horizon, Schwarz (1973, p. 557) stated the following properties of an optimal ordering policy for the supply chain:

1. A delivery is made to the warehouse only when the warehouse has zero installation stock and at least one retailer also has zero installation stock.

2. A delivery is made to a retailer only when it has zero installation stock.

3. Deliveries to any given retailer made between successive deliveries to the warehouse are of equal size.

The first two of these properties comprise what is known as the "zero-inventory" ordering property, concisely stated by Zipkin (2000) that, under these demand and replenishment conditions, an optimal policy will have "orders planned so as to run out just as each successive order is received" (p. 82). Logically, in a system with a known, constant rate of demand and a constant re-supply lead time at all sites, any policy that results in anything larger than zero units in inventory when a delivery arrives cannot be optimal.

3.6.3 Two-Stage Serial System under Decentralized Control

As a starting point, let us consider the simpler, two-stage serial system, with only an upstream and downstream stage (what Schwarz called the warehouse and retailer, respectively). First, we consider this system under decentralized control, allowing the downstream stage ($j = 2$) to set its own replenishment policy and computing the optimal stage 1 replenishment policy that assumes a "stage-optimal" policy at stage 2. Given the system parameters described earlier, and assuming that the constant rate of demand served by the stage 2 site totals D units annually, then if h_2 is measured in \$/unit/year, and stage 2 sets its optimal order quantity independently, this value will be

$$Q_2^* = \sqrt{\frac{2A_2 D}{h_2}}$$

resulting in an inventory cycle length at stage 2 of

$$\tau_2^* = \frac{Q_2^*}{D} = \sqrt{\frac{2A_2}{Dh_2}} \tag{3.31}$$

Thus, the demand passed upstream from stage 2 to stage 1 in this serial system is an order for Q_2^* units every τ_2^* years (easily converted to weeks or days, as we show in the example that follows).

To maintain a zero-inventory ordering policy at stage 1, the inventory cycle length at stage 1 must be an integer multiple of the inventory cycle for stage 2, and therefore we set $\tau_1 = m\tau_2$, $m \geq 1$ and integer. This becomes apparent upon considering Figure 3.15, which shows the inventory profile over time at stage 2 and stage 1 (with $m = 3$). Note that, due to the lumpy demand passed from stage 2 to stage 1, inventory at stage 1 does not follow the familiar "EOQ sawtooth pattern." However, it is easy to show that the average installation inventory at stage 1 is

$$\bar{I}_1 = \frac{Q_2(m-1)}{2}$$

The reasoning behind this is as follows: Geometrically, from Figure 3.15, take the total area covered by the stage 1 echelon inventory, $(mQ_2 \cdot m\tau_2)/2$,

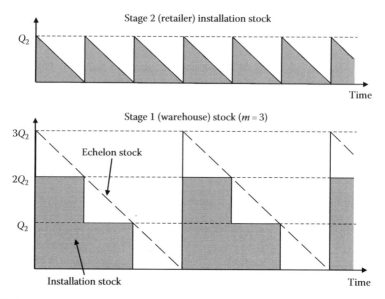

FIGURE 3.15
Installation and echelon stock in a two-echelon system.

subtract the area covered by the stage 2 installation stock portion of this echelon inventory, $m \cdot Q_2 \tau_2 / 2$, and divide this result by the stage 1 cycle length, $\tau_1 = m\tau_2$, to get the time-average inventory. Note also that if $m = 1$, stage 1 carries no installation inventory, which should be apparent upon inspection of Figure 3.15.

Thus, given the stage-2 optimal decision, $\tau_2^* = Q_2^*/D$, the cost at stage 1 can be expressed as a function of m and τ_2^*, specifically

$$TC_1\left(m, \tau_2^*\right) = \frac{A_1}{m\tau_2^*} + h_1 \cdot \frac{D\tau_2^*(m-1)}{2} \tag{3.32}$$

Since expression (3.32) is easily shown to be pointwise convex in m (an integer), $TC\left(m, \tau_2^*\right)$ reaches its minimum where

$$\Delta TC_1\left(m, \tau_2^*\right) = TC_1\left(m+1, \tau_2^*\right) - TC_1\left(m, \tau_2^*\right) \tag{3.33}$$

either equals zero or flips from negative to positive. Thus, setting $\Delta TC_1\left(m, \tau_2^*\right) = 0$ and solving for m gives

$$m(m+1) \geq \frac{2A_1}{h_1 D \cdot \left(\tau_2^*\right)^2} \tag{3.34}$$

Finally, substituting expression (3.31) into expression (3.34), we obtain

$$m(m+1) \geq \frac{A_1}{A_2} \cdot \frac{h_2}{h_1} \tag{3.35}$$

We can find m^* from expression (3.35) by formulating a straightforward optimization problem to minimize the difference between the left- and right-hand sides of this expression, subject to the constraints that this difference must be non-negative and $m \geq 1$ must be an integer. Expression (3.35) clearly shows how the optimal value of m, which dictates the optimal length of the order cycle at stage 1 relative to that at stage 2, is directly proportional to the ratio of the stage 1 and stage 2 ordering costs and inversely proportional to the ratio of the stage 2 to stage 1 holding costs. Thus, as ordering costs go up at stage 1 relative to stage 2, and as holding costs go up at stage 2 relative to stage 1, the optimal solution has a longer inventory cycle (less frequent ordering) at stage 1 relative to stage 2.

3.6.4 Two-Stage Serial System under Centralized Control

To find the optimal centralized control policy for the 2-stage serial system we considered in the previous section, we must find inventory cycle lengths τ_1 and τ_2 that minimize

$$TC(\tau_1, \tau_2) = \frac{A_1}{\tau_1} + \frac{A_2}{\tau_2} + h_1 \bar{I}_1 + h_2 \bar{I}_2 \tag{3.36}$$

Restricting stage 1 to zero-inventory ordering policies, we again have $\tau_1 = m\tau_2$, $m \geq 1$ and integer. Therefore, with $\bar{I}_2 = Q_2/2 = D\tau_2/2$ and $\bar{I}_1 = Q_2(m-1)/2 = D\tau_2(m-1)/2$, expression (3.36) becomes

$$TC(m, \tau_2) = \frac{1}{\tau_2} \left(\frac{A_1}{m} + A_2 \right) + \tau_2 \left[\frac{h_1 D(m-1)}{2} + \frac{h_2 D}{2} \right] \tag{3.37}$$

To simplify this expression, we use *echelon holding costs*, defined as follows: The echelon holding cost at echelon j is $H_j = h_j - h_{j-1}$, where, consistent with the earlier sections of the chapter, h_j is per-unit cost of holding inventory per period. (Since $h_0 \equiv 0$, $H_1 = h_1$.) Thus, echelon stock at echelon j is valued at the incremental installation holding cost for that echelon. Note that in any realistic setting, $H_j \geq 0$ since the installation holding costs h_j are a reflection of the value of the item at that installation, and items will normally increase in value as they move downstream in a supply chain. Thus, $H_2 = h_2 - h_1$ and $H_1 = h_1$, respectively. Now, we define $g_i = H_i D/2$ and, substituting g_i into expression (3.37), we obtain

$$TC(m, T_2) = \frac{1}{\tau_2} \left(\frac{A_1}{m} + A_2 \right) + \tau_2 (mg_1 + g_2) \tag{3.38}$$

The first-order condition on expression (3.38) with respect to τ_2 is

$$\frac{\partial TC}{\partial \tau_2} = \frac{-\left(\dfrac{A_1}{m} + A_2\right)}{\tau_2^2} + mg_1 + g_2 = 0$$

yielding

$$\tau_2^*(m) = \sqrt{\frac{\dfrac{A_1}{m} + A_2}{mg_1 + g_2}} \tag{3.39}$$

Rather than using the first difference on expression (3.38) to find m^*, as in the previous section, we employ the approach suggested by Axsater (2006). First, substitute expression (3.39) into expression (3.38), to obtain

$$TC\left(m, \tau_2^*\right) = 2 \cdot \sqrt{\left(\frac{A_1}{m} + A_2\right)(mg_1 + g_2)}$$

Then, note that

$$\left[TC\left(m, T_2^*\right)\right]^2 = 4\left(\frac{A_1 g_2}{m} + A_2 g_1 m + A_1 g_1 + A_2 g_2\right)$$

has "EOQ form" and is therefore convex in m. Ignoring the fact that m must be integer in the final solution, we use the continuous first-order condition on m,

$$\frac{\partial\left[TC\left(m, T_2^*\right)\right]^2}{\partial m} = \frac{-4A_1 g_2}{m^2} + 4A_2 g_1 = 0$$

to obtain

$$\tilde{m} = \sqrt{\frac{A_1}{A_2} \cdot \frac{g_2}{g_1}} = \sqrt{\frac{A_1}{A_2} \cdot \frac{H_2}{H_1}} \tag{3.40}$$

Axsater (2006) suggests the following approach to finding m^*:

1. Compute $\tilde{m}$ using expression (3.40).
2. If $\tilde{m} \leq 1$, then choose $m^* = 1$ and stop.
3. If $\tilde{m} > 1$, let m' be the largest integer less than or equal to $\tilde{m}$, such that $\tilde{m} \leq m' < \tilde{m} + 1$. Then, if $\tilde{m}/m' \leq (m' + 1)/\tilde{m}$, $m^* = m'$; otherwise, $m^* = m' + 1$.

Note that the conditions stated in step 3 of Axsater's algorithm can be rearranged, resulting in a condition similar to the condition on m^* for independent ordering, namely

$$m(m+1) \geq \frac{A_1}{A_2} \cdot \frac{H_2}{H_1} \qquad (3.41)$$

Expressed using installation holding costs, this becomes

$$m(m+1) \geq \frac{A_1}{A_2} \cdot \left(\frac{h_2}{h_1} - 1 \right) \qquad (3.42)$$

with m^* given by the first integer that satisfies (3.41). This is an interesting parallel to the independent-ordering case (expression (3.35)). Note, however, that in the centralized case the optimal value of m will be no greater than—and could be smaller than—the value of m in the independent-ordering case. Stated another way, the optimal inventory cycle length at stage 1 will tend to be closer to the optimal cycle length at stage 2 in the centralized case than in the independent case.

Continuing the comparison, note that expression (3.39) can be restated using installation-stock holding costs, yielding

$$\tau_2^*(m) = \sqrt{\frac{2\left(\dfrac{A_1}{m} + A_2 \right)}{D\left[(m-1)h_1 + h_2 \right]}} \qquad (3.43)$$

The comparison of the optimal stage 2 inventory cycle value in the independent and centralized cases is less clear here since the additional terms in expression (3.43) as compared to expression (3.31) increase both the numerator and the denominator. We can, however, see the direction of the difference in an example case, as follows.

Example 3.14 Two-echelon order quantities with fixed costs

Consider a two-stage, serial supply chain where the annual demand at stage 2 is $D = 10{,}000$ units. Fixed ordering costs at stages 1 and 2 are $A_1 = \$400/\text{order}$ and $A_2 = \$50/\text{order}$, and inventory holding costs are $h_1 = \$10/\text{unit}/\text{year}$ and $h_2 = \$15/\text{unit}/\text{year}$.

The decentralized solution is computed as follows. First, $\tau_2^* = \sqrt{2A_2/(Dh_2)} = 0.0258$ years (or, at 365 operating days/year, 9.5 days, or stated another way, an average of 38.7 orders per year), meaning that $Q_2^* = D \cdot \tau_2^* = 258$ units. For the upstream order quantity multiplier, m, the condition on its optimal value is $m(m + 1) \geq (A_1/A_2) \cdot (h_2/h_1) = 12$, which we can find by solving

$$\min z$$

subject to

$$m(m+1) - z = 12$$

$$m \geq 1 \text{ and integer}$$

$$z \geq 0$$

Solving this gives $m^* = 3$, and therefore $\tau_1^* = m\tau_2^* = 0.0775$ years (28.1 days, or 12.9 orders per year), meaning that $Q_1^* = D \cdot \tau_1^* = 775$ units. The supply chain cost of this decentralized solution is (from expression (3.37)) TC$(m, \tau_2) = \$11{,}619$, broken down across stage 1 and stage 2 as TC$_1 = \$7{,}746$ and TC$_2 = \$3{,}873$.

For a centralized solution, we start by computing $(A_1/A_2) \cdot (h_2/h_1 - 1) = 4$, which, using the approach above, gives $m^* = 2$, and therefore $\tau_2^* = \sqrt{2\left(A_1/m^* + A_2\right)\Big/\left[D\left((m^*-1)h_1 + h_2\right)\right]} = 0.0447$ years (16.4 days, or 22.4 orders per year), such that $Q_2^* = D \cdot \tau_2^* = 447$ units. Therefore, $\tau_1^* = m\tau_2^* = 0.0894$ years (32.5 days, or 11.2 orders per year), such that $Q_1^* = D \cdot \tau_1^* = 894$ units. The supply chain cost of this centralized solution is TC$(m, \tau_2) = \$11{,}180$, meaning that the gap between the decentralized solution computed earlier and this first-best solution is 3.8%. The centralized solution achieves this by shifting some of the cost burden from stage 1 to stage 2, specifically resulting in TC$_1 = \$6708$ and TC$_2 = \$4472$.

3.6.5 Serial Supply Chain with Stochastic Demand and Negligible Fixed Ordering Costs

Now, let us consider a serial system under stochastic demand with only inventory holding and backorder penalty costs. Thus, in this case, fixed ordering costs are zero, or negligible, and supply shortfalls result in backlogged demand. Shang and Song (2003) consider such a system and prove the existence of easily computed upper and lower bounds for the optimal base-stock levels at each echelon. Shang and Song's results apply to a continuous-review setting, but they point out that these results hold for a periodic-review system with independent and identically distributed demand, and they refer the reader to Chen and Zheng (1994), Gallego and Zipkin (1999), and Zipkin (2000) for details.

The approximation developed by Shang and Song (2003) uses echelon holding costs, as defined earlier, and starts from the optimal echelon N base-stock level, which—by the authors' reference to Chen and Zheng (1994)—is given by computing the critical fractile

$$\theta_N^* = \frac{b + \sum_{i=1}^{N-1} H_i}{b + \sum_{i=1}^{N} H_i} \tag{3.44}$$

and which can be used to specify the echelon N base-stock level, $\hat{S}_N^* = F_{\text{DLT},N}^{-1}\left(\theta_N^*\right)$, where $F_{\text{DLT},N}$ is the cdf of the distribution of demand over the lead time to replenish stage N, and where we use a "hat" to distinguish the echelon base stock from the installation base stock. (Of course, for the stage farthest downstream, $\hat{S}_N = S_N$.) Chen and Zheng (1994), and others, have shown that computing $\hat{S}_N^*$ is the first step in solving a larger set of recursive expressions to obtain the optimal base-stock levels for echelons $N-1$, $N-2$, ..., 1. As Shang and Song cogently point out, however, this process of recursive optimization is not easily communicated to managers and students who lack significant training in higher-level mathematics. Thus, a more accessible, but still reasonably accurate, approximation is desirable.

Shang and Song demonstrate that approximate solutions can be found for each upstream echelon $j \in \{1, 2, ..., N-1\}$ in the serial supply chain from two easily computed newsvendor-type fractiles, θ_j^l and θ_j^u, which they prove to be bounds on θ_j^* $\left(\text{i.e., that } \theta_j^* \in \left[\theta_j^l, \theta_j^u\right]\right)$. These bounds are given by

$$\theta_j^l = \frac{b + \sum_{i=1}^{j-1} H_i}{b + \sum_{i=1}^{N} H_i} \tag{3.45}$$

and

$$\theta_j^u = \frac{b + \sum_{i=1}^{j-1} H_i}{b + \sum_{i=1}^{j} H_i} \tag{3.46}$$

Thus, from expressions (3.45) and (3.46), we can compute $S_j^l = F_{\text{DLT},j}^{-1}\left(\theta_j^l\right)$ and $S_j^u = F_{\text{DLT},j}^{-1}\left(\theta_j^u\right)$, where $F_{\text{DLT},j}$ is the cdf of the distribution of demand over the (nominal) lead time to replenish echelon j, given by $\hat{L}_j = \sum_{i=j}^{N} L_i$ (not accounting for potential stockouts upstream), where L_i is the lead time to replenish stage i from stage $i-1$. From S_j^l and S_j^u, we can compute the approximately optimal echelon base-stock level

$$\hat{S}_j = \left\lceil \frac{S_j^l + S_j^u}{2} \right\rceil \tag{3.47}$$

Computational experiments by Shang and Song (2003) indicate that rounding up from the raw average of S_j^l and S_j^u, as is done in expression (3.47), tends to produce solutions that are closer to the optimal solution, when the penalty cost is relatively high, than those that result from rounding down. As a convenient rule of thumb, we will adopt the convention to simply round up irrespective of the size of the penalty cost.

It is helpful to develop an intuitive interpretation of these bounds on the optimal echelon base-stock levels. Note that θ_j^l can be expressed as

$$\theta_j^l = \frac{\tilde{b}}{\tilde{b} + \tilde{h}_j} \tag{3.48}$$

where $\tilde{b} = b + \sum_{i=1}^{j-1} H_i$ and $\tilde{h}_j = \sum_{i=j}^{N} H_i$ reflect the underage (c_u) and overage (c_o) costs, respectively, of a subsystem of the supply chain that is truncated at echelon j. The intuition behind this as follows: The total value added to a unit in echelon j is $\tilde{h}_j = \sum_{i=j}^{N} H_i$. Thus, any unit that passes through this echelon j-truncated portion of the supply chain, but is not sold, accumulates an overage cost of $c_o = \tilde{h}_j$. Moreover, for any unit ultimately sold to the consumer at the end of the supply chain, the contribution to the overall supply chain gain from echelon j is $h_N - \tilde{h}_j = \sum_{i=1}^{N} H_i - \sum_{i=j}^{N} H_i = \sum_{i=1}^{j-1} H_i$. Thus, failure to fulfill a unit of demand on echelon j incurs the sum of the penalty cost b, plus the gain that will not be realized as a result of the shortfall, such that the underage cost in this case is $c_u = \tilde{b} = b + \sum_{i=1}^{j-1} H_i$. From a total system perspective, however, $\tilde{h}_j$ may overstate the overage cost because stocking decisions at stages $j + 1$, ..., N could cause what was an "overage decision" at echelon j to be "undone" further down the chain. Therefore, since S_j^l is based on an overstated holding cost, it provides a lower bound on the system-optimal base-stock level at echelon j—i.e., because it is based on overage cost that is equal to or larger than the cost that results at echelon j in an optimal system, then $S_j^l \le S_j^*$.

Next, note that θ_j^u can be expressed as

$$\theta_j^u = \frac{\tilde{b}}{\tilde{b} + \underset{\sim}{h}_j} \tag{3.49}$$

where $\underset{\sim}{h}_j = H_j$, which reflects the assumption at the other extreme regarding the overage cost in an "echelon j-truncated" system, namely that the decision made at echelon j that would have resulted in a unit in excess of demand from there through all succeeding echelons is immediately negated by a stocking decision in echelon $j + 1$ so that, in this case, $c_o = \underset{\sim}{h}_j = H_j$.

This potential understatement of the optimal overage cost results in S_j^u being an upper bound on the system-optimal base-stock level at echelon j.

Implementation of these bounds is illustrated in the following computational example.

Example 3.15 Multi-echelon base-stock policies

Assume that your company, Stage Three Incorporated, must make stocking decisions for finished goods that are held at its plant, its regional distribution center, and its field warehouse. In this three-stage, serial supply chain, demand at the field warehouse follows a Poisson distribution (consistent with the demand assumption in the model developed in Shang and Song, 2003) with a mean of $\lambda = 10$ units/day. The stage j replenishment lead times across this serial supply chain are $L_j = 2$ days ($j = 1,2,3$). This results in Poisson lead-time demand at the field warehouse, and therefore at echelon 3, with mean $\lambda_3 = 20$ units/day. Similarly, the lead-time demand for echelon 2 is Poisson with mean $\lambda_2 = 40$ units/day, and lead-time demand for echelon 1 with is Poisson with mean $\lambda_1 = 60$ units/day. We approximate these daily demands using normal distributions, such that for echelons 1, 2, and 3, respectively, $D_1 \sim N(60,\sqrt{60})$, $D_2 \sim N(40,\sqrt{40})$, and $D_3 \sim N(20,\sqrt{20})$. If at any point the field warehouse is short of supply when demand arrives, Stage Three incurs a penalty cost of $b = \$50$ per unit short per period. Installation holding costs are known to be $h_1 = \$1$ per unit per period, $h_2 = \$1.50$ per unit per period, and $h_3 = \$2$ per unit per period.

First, let us compute echelon holding costs $H_1 = h_1 = \$1$, $H_2 = h_2 - h_1 = \$0.50$, and $H_3 = h_3 - h_2 = \$0.50$. Then, we can compute the optimal echelon (and installation) base-stock level for echelon $N = 3$ to be

$$\hat{S}_N^* = \hat{S}_3^* = F_{DLT,3}^{-1}\left(\theta_3^*\right), \text{ where } \theta_N^* = \theta_3^* = \frac{b + \sum_{i=1}^{N-1} H_i}{b + \sum_{i=1}^{N} H_i} = \frac{50 + (1 + 0.5)}{50 + (1 + 0.5 + 0.5)} = $$

0.99038. Since we are using a normal approximation for the underlying Poisson demand, we can compute $\hat{S}_3^* = 20 + \Phi^{-1}(0.99038) \cdot (20) = 30.47$, which we round to $\hat{S}_3^* = 31$ to ensure that we meet or exceed the specified fractile of the demand distribution. For stage $j = 2$ of this

serial system, we compute $\theta_j^l = \dfrac{b + \sum_{i=1}^{j-1} H_i}{b + \sum_{i=1}^{N} H_i} = \dfrac{50 + 1}{50 + (1 + 0.5 + 0.5)} = $

0.98077 and $\theta_j^u = \dfrac{b + \sum_{i=1}^{j-1} H_i}{b + \sum_{i=1}^{j} H_i} = \dfrac{50 + 1}{50 + (1 + 0.5)} = 0.99029$, such that

$S_2^l = \left\lceil 40 + \Phi^{-1}(0.98077) \cdot (40) \right\rceil = 54$ and $S_2^u = \left\lceil 40 + \Phi^{-1}(0.99029) \cdot (40) \right\rceil = 55$.

Therefore, $\hat{S}_2 = \left\lceil \dfrac{S_2^l + S_2^u}{2} \right\rceil = 55$. For stage $j = 1$, the site farthest upstream,

$$\theta_j^l = \frac{b + \sum_{i=1}^{j-1} H_i}{b + \sum_{i=1}^{N} H_i} = \frac{50}{50 + (1 + 0.5 + 0.5)} = 0.96154 \text{ (note that } \sum_{i=1}^{j-1} H_i \text{ is an}$$

empty sum for $j = 1$) and $\theta_j^u = \dfrac{b + \sum_{i=1}^{j-1} H_i}{b + \sum_{i=1}^{j} H_i} = \dfrac{50}{50 + 1} = 0.98039$, such that

$$S_1^l = \left\lceil 60 + \Phi^{-1}(0.96154) \cdot (60) \right\rceil = 74 \text{ and } S_1^u = \left\lceil 60 + \Phi^{-1}(0.99029) \cdot (60) \right\rceil = 76.$$

Therefore, $\hat{S}_1 = \left\lceil \dfrac{S_1^l + S_1^u}{2} \right\rceil = 75$.

Let us interpret the quantities that are computed in Example 3.15. Recall from the initial discussion in this section that Shang and Song's (2003) method applies to continuous-review, base-stock policies, a slight variation on the base-stock policy we discussed earlier in the context of periodic-review systems. This is not dramatically different from our earlier discussion, though. Since the inventory position is reviewed continuously, the ordering rule is that, whenever a demand of size $d > 0$ occurs at stage 3, taking the echelon inventory position to $\hat{x}_3 = \hat{S}_3^* - d$, an order is placed upstream for d units. Since demand simply cascades upstream from stage to stage in the serial supply chain, this results in an order of d units cascading across each stage whenever there is consumer demand for d units at stage 3, farthest downstream in the chain. In the context of the multi-echelon system, then, this centralized policy specifies the echelon inventory positions that are maintained across the system each time an order is placed.

Recall also that, at the farthest downstream site in the serial supply chain, echelon inventory is the same as installation inventory. Thus, Example 3.15 prescribes an installation inventory position at stage 3 of $\hat{S}_3^* = 31$ units. Thus, when the system is first seeded with inventory, we set inventory at the field warehouse to $I_3 = \hat{I}_3 = 31$ units. Similarly, we seed stages 2 and 3—the distribution center and the plant—with inventory levels $I_2 = \hat{S}_3^* - \hat{S}_2 = 24$ units and $I_1 = \hat{S}_2 - \hat{S}_1 = 20$ units, respectively. Therefore, as each demand value d arrives at the field warehouse, the upstream-cascading orders maintain the echelon inventory positions at 31, 55, and 75 for echelons 3, 2, and 1, respectively. The simplicity of the serial system and the continuous-review, base-stock policy allows a similar simplicity in implementation. (Axsater, 2006, is good reference to allow the reader to see how the multi-echelon management of a distribution system, for example (Figure 3.11), is decidedly more complex than that of a serial system.)

3.6.6 Serial Supply Chain with Fixed Costs and Stochastic Demand

Let us combine the two multi-echelon contexts discussed up to this point. Consider a two-stage, serial supply chain that faces stochastic demand and, at each stage, experiences costs of ordering, holding, and penalty from shortfalls in supply. While finding an optimal solution to this system may be a daunting task, the various computational tools we have discussed earlier in this chapter can be employed to generate a reasonable, though not optimal, solution. We show this as an extension to Example 3.14 given earlier.

> **Example 3.16 Multi-echelon setting with fixed costs and stochastic demand**
>
> Extending the setting from Example 3.14, let us now assume that demand remains stationary, but is stochastic. Specifically, assume that daily demand is normally distributed with a mean of $\mu_D = 10,000/365 = 27.4$ units/day and a coefficient of variation of 0.3. Replenishment lead times are assumed to be constant and equal to $L_1 = 7$ days and $L_2 = 3$ days. Thus, lead-time demand at stage 2 is $D_{DLT,2} \sim N(3 \times 27.4, \sqrt{3} \times 0.3 \times 27.4) = N(82.2, 14.2)$. Determining lead-time demand at stage 1 is a little more challenging. If we assume that the underlying consumer demand is the "true" per-period demand on stage 1—ignoring the fact that "internal" demand on stage 1 is actually comprised of batch orders of size Q_2—then we obtain $D_{DLT,1} \sim N(7 \times 27.4, \sqrt{7} \times 0.3 \times 27.4) = N(191.8, 21.75)$.
>
> Again, we note that these are indeed *stage* demands and not *echelon* demands. The reason for taking this approach will be made clear in the following text. Second, we note that the demand distribution at stage 2 is perhaps optimistic since it does not account for potential supply shortfalls at stage 1. There are a few ways to account for this, some more ad hoc than others. One would be to utilize an inflated target service level to set the installation reorder point. Another would be to simulate this multi-echelon system to estimate the effect of supply shortfalls at stage 1 on L_2, resulting in some probability distribution around this lead time value that is otherwise assumed to be constant. (See Exercise 3.11 for an example of this. Other explorations are left as interesting follow-on exercises for the reader.)
>
> Using the information given previously and a CSL target of 95% for each stage, we obtain installation reorder point levels of $R_2 = \lceil \mu_{DLT,2} + z_{CSL} \, \sigma_{DLT,2} \rceil = 82.2 + 1.645 \times 14.2 = 106$ and $R_1 = \lceil \mu_{DLT,1} + z_{CSL} \, \sigma_{DLT,1} \rceil = 191.8 + 1.645 \times 21.75 = 228$. As discussed in Axsater (2006, Chapter 8), any installation-stock reorder point policy can be replaced by an equivalent echelon-stock reorder point policy that has
>
> $$\hat{R}_j = R_j + \sum_{k=j+1}^{N} (R_k + Q_k) \qquad (3.50)$$

This is the reason that we began by computing the installation reorder points. Utilizing this result, and noting that $\hat{R}_2 = R_2 = 106$, we have $\hat{R}_1 = R_1 + R_2 + Q_2 = 228 + 106 + 447 = 781$, using the optimal centralized value from Exercise 3.14, $Q_2^* = 447$.

The resulting (centralized, echelon-stock) policies are $(Q_2, R_2) = (Q_2, \hat{R}_2) = (447, 106)$ and $(Q_1, \hat{R}_1) = (894, 781)$. Therefore, whenever installation inventory at stage 2 (same as echelon inventory at echelon 2) falls below 106 units, we place an order on stage 1 for 447 units, and whenever echelon inventory at echelon 1 falls below 781 units, stage 1 should place an order on its supplier for 894 units.

As a follow-on comment, we note that Graves and Schwarz (1977), in their work on computing order quantities in a general arborescent system, imply that an approach along these lines is a promising one. In the "Concluding Remark" section of their paper, they consider the use of easily computed, approximate values of Q to be "interesting and practical" when coupled with "buffer stocks ... determined independently via the usual newsboy analysis."[*] Indeed, our view is that there is great value in easily computed and easily implemented approximate solutions. The solution obtained in Exercise 3.16 is not guaranteed to be optimal, but it is easily computed and easily implemented. Of course, one could build a simulation-optimization engine to search in the neighborhood of the computed reorder point values for echelon reorder points that improve upon this quickly computed solution.

3.7 Summary and Further Readings

3.7.1 Summary

Once again, therefore, we revisit the enormity of the problem of "optimizing the supply chain," noting the complexity of the aforementioned example for just a small number of items at a single firm, or a single item across a small number of firms. Granted, jointly minimizing inventory cost, transportation cost, and possibly also production-related costs might be a tractable problem at a single firm for a small set of items, but the challenge of formulating and solving such problems clearly grows as the number of components of the objective function increases, as the number of SKUs being planned increases, and as the planning time scale decreases (e.g., from weeks to days). The aforementioned examples, however, give us some insights into how inventory decisions could be affected by transportation considerations, a topic we consider further in Chapter 4.

---

[*] In earlier days when less-inclusive language was used, note that all "newsvendors" were assumed to be boys.

This chapter has covered a lot of ground, but at this point, we can say that the following general principles appear to hold:

- When there is a tradeoff between the fixed cost of placing an order and the marginal holding cost, we should employ an "EOQ-type" solution to exploit this tradeoff.

- When there is a tradeoff between the holding cost and the penalty cost for failing to immediately satisfy uncertain demand, we should employ a "critical fractile-based" solution to exploit this tradeoff.

These principles can serve as a "universal starting point" from which improved solutions can be generated. So, essentially, only two basic tools are required to generate reasonably good solutions to a whole host of inventory management problems. Our wide-ranging discussion, however, implies that there are also many means by which to improve these approximate solutions in order to optimize, or nearly optimize, the objective function of interest.

3.7.2 Further Readings

Given the broad discussion in this chapter, one can see that computing inventory control parameters for even relatively simple multi-echelon systems—e.g., a two-stage, serial system with fixed ordering costs, or a three-stage, serial system without fixed costs—is a complex endeavor. Unfortunately, finding even approximately optimal solutions to more extensive systems is even more challenging, and beyond the scope of this book. While interested readers can consult the end-of-chapter references for extensions to the problems considered in this chapter, we believe that a few additional comments about multi-echelon inventory systems are warranted.

First, note that several authors have studied EOQ-based systems with a single warehouse serving N retailers. As indicated in Simchi-Levi et al. (2008), a solution based on powers-of-two ordering (see Jackson et al., 1985) actually employs an approach that is similar—in essence, at least—to the one we presented earlier. The solution developed in Simchi-Levi et al. (2008), for example, is based on computing a "base" inventory cycle length, and then determining good powers-of-two multiples of that value for each installation in the system.

Note also that the last example in the chapter essentially entails determining the placement of safety inventory in the supply chain (since the reorder point specifies inventory carried over and above the level required to meet expected demand over the replenishment lead time), albeit a simple, serial supply chain. Magnanti et al. (2006) present a general formulation of this problem, specifically a non-linear optimization formulation subject to linear constraints, to solve for the optimal placement of safety stocks in a general supply chain network. Magnanti et al. actually generalize the problem presented in Graves and Willems (2000), who demonstrate that their algorithmic, dynamic programming-based solution may be used to solve networks that have a spanning tree structure. The Magnanti et al. formulation and solution is more general, allowing computation of optimal safety stock levels in any acyclic supply chain network. Still, the approach of Graves and Willems gained notoriety in the success of its implementation, ultimately being incorporated into commercially available software.*

On a final note regarding the multi-echelon inventory problem, note that our analysis has been contained to problems that exhibit stationary demand. In an environment in which demand varies across time periods, we could, of course, re-compute the parameters of the stochastic and/or deterministic models we have developed earlier as we obtain new information about demand. Another option would be to formulate the problem a different way. If we consider the explicit expected values of demand across the planning horizon as deterministic values, we could then formulate a mixed-integer program (MIP) that computes the optimal flows of product across the nodes in the supply chain network, including the optimal levels of inventory to be held each period at each node, across a multi-period planning horizon. Moreover, this problem could be re-solved each period in a rolling planning horizon. This "production scheduling" formulation dates back to the 1960s, first explored in widely cited papers by Zangwill (1966a,b).

We will save the discussion of production planning models, however, for Chapter 5, based on the similarity of that problem to the facility location and location-allocation problems discussed in that chapter. Also in Chapter 5, we introduce the notion of *risk pooling*, which is a means of reducing the safety stock required to support a target service level by aggregating demands across multiple sources, for example across multiple customers or customer regions supported by a single facility.

3.A Appendix: The Bullwhip Effect[†]

Beginning in the 1980s, a number of supply chain management researchers and practitioners began to devote significant attention to the level of variability in the stream of orders placed over time on upstream suppliers by their downstream customers. Empirical evidence (Blinder, 1982, 1986; Lee et al., 1997b; Chen et al., 2000; Callioni and Billington, 2001) shows that replenishment orders demonstrate increasing variability at successive upstream stages of a supply chain. Lee et al. (1997b) report that Procter & Gamble noted this phenomenon, called the *bullwhip effect*, in its supply chain for Pampers diapers.

---

[*] This software was originally sold through a company called Optiant, which was later acquired by Logility.

[†] Portions of this appendix have been adapted with permission from Warsing (2008).

Underlying demand for diapers over a relatively short period of time, say a few months, tends to be quite stable, as one would predict—babies clearly go through them at a pretty steady rate, as any parent knows first-hand! However, P&G noticed that, although underlying consumer demand for Pampers was stable, the orders from retailers to their wholesalers were more volatile than consumer demand, while orders from wholesalers to P&G were more volatile still, and orders from P&G to its suppliers were more volatile still. Indeed, the bullwhip effect is formally defined by this phenomenon, an increasing level of variability in orders at successive upstream stages of the supply chain.

Although the effect has been the subject of much discussion in the supply chain management literature since roughly the mid-1990s, the effect has been studied quite a bit longer than this. As we indicated earlier, economists began to study the effect and its origins in earnest in the mid-1980s (Blinder, 1982, 1986; Caplin, 1985; Khan, 1987). Moreover, the famous systems dynamics studies of Forrester (1961) ultimately have had a tremendous impact on supply chain management education, with scores of students each academic year playing the famous "Beer Game" that resulted from follow-on work to Forrester's by Sterman (1989). In playing the Beer Game, students run a four-stage supply chain for "beer" (typically, only coins that represent cases of beer—to the great relief of many university administrators, and to the great dismay of many students), ultimately trying to fulfill consumer demand at the retailer by placing orders with upstream partners, eventually culminating in the brewing of "beer" by a factory at the farthest upstream stage in the chain. Indeed, the authors' experience in playing the game with students at all levels—undergraduate students, graduate students, and seasoned managers—has never failed to result in bullwhip effects in the chains playing the game, sometimes in quite pronounced fashion.

A natural question, and one that researchers continue to debate, is what causes this effect. If the underlying demand is relatively stable, and orders are meant ultimately to replenish this underlying demand, why is it not the case that orders passed upstream are relatively stable? The astute reader may quickly identify one issue from our discussion of inventory earlier, namely that the economics of the ordering process might dictate that orders should be batched in order to reduce the administrative costs of placing orders or the scale-driven costs of transporting order fulfillment quantities from the supplier to the customer. Indeed, Lee et al. (1997a,b) identify order batching as one of four specific causes of the bullwhip effect, and the economic studies of the 1980s (Blinder, 1982, 1986; Caplin, 1985) clearly note this cause. The other causes as described by Lee et al. are demand forecast updating, price fluctuations, and supply rationing and shortage gaming. The last two effects are fairly straightforward to understand, namely that a customer might respond to volume discounts or short-term "special" prices by ordering out-of-sync with demand, or that a customer might respond to allocations of scarce supply by placing artificially inflated orders to attempt to secure more supply than other customers. In the interest of simplicity, we will focus only on order batching and demand forecast updating, leaving the complexities of pricing and shortage gaming outside the scope of our discussion.

Indeed, the method used to generate the demand forecast at each stage in the supply chain can have a significant impact on the stream of orders placed upstream. Vollmann et al. (2005) provide a simple example to demonstrate the existence of the bullwhip effect. This example is repeated here, with some extensions, and shown in Table 3.6. Specifically, we extend what is a 10-period example from Vollmann et al. to 20 periods that cover two repeating cycles of a demand stream with slight seasonality. In addition, we institute a few policy rules: (1) that orders must be non-negative, (2) that

TABLE 3.6

Bullwhip Example from Vollmann et al. (2005) Altered for Minimum Order Size of 0 and Horizon of 20 Periods

	Consumer	Manufacturer				Supplier			
Period	Sales	Beg Inv	End Inv	Lost Dmd	Order	Beg Inv	End Inv	Lost Dmd	Order
1	50	100	50	0	50	100	50	0	50
2	55	100	45	0	65	100	35	0	95
3	61	110	49	0	73	130	57	0	89
4	67	122	55	0	79	146	67	0	91
5	74	134	60	0	88	158	70	0	106
6	67	148	81	0	53	176	123	0	0
7	60	134	74	0	46	123	77	0	15
8	54	120	66	0	42	92	50	0	34
9	49	108	59	0	39	84	45	0	33
10	44	98	54	0	34	78	44	0	24
11	50	88	38	0	62	68	6	0	118
12	55	100	45	0	65	124	59	0	71
13	61	110	49	0	73	130	57	0	89
14	67	122	55	0	79	146	67	0	91
15	74	134	60	0	88	158	70	0	106
16	67	148	81	0	53	176	123	0	0
17	60	134	74	0	46	123	77	0	15
18	54	120	66	0	42	92	50	0	34
19	49	108	59	0	39	84	45	0	33
20	44	98	54	0	34	78	44	0	24
Min	44		38		34		6		0
Max	74		81		88		123		118
Avg	58.1		58.7		57.5		60.8		55.9
Fill rate				100.0%				100.0%	
Range	30				54				118
Std. dev	9.16				17.81				39.04
				Ratio	1.94			Ratio	4.26

shortages at the manufacturer result in lost consumer demand, (3) that shortages at the supplier are met by the manufacturer with a supplemental source
of perfectly reliable, immediately available supply. The example of Vollman
et al. employs a relatively simple, and essentially *ad hoc*, forecast updating
and ordering policy at each level in the chain: at the end of any period t, place
an order for $2D_t - I_t$, where D_t is actual demand in period t and I_t is the ending inventory in period t, computed after fulfilling as much of D_t as possible
from beginning inventory. Thus, the current demand is used as the forecast of future demand, and since the replenishment lead time is one period
(i.e., an order placed at the end of the current period arrives at the beginning
of the next period—one could obviously debate whether this a lead time of
zero), the policy amounts to carrying a period's worth of safety stock.

We measure the severity of the bullwhip effect by taking the ratio of the
standard deviation of the orders placed at each stage to the standard deviation
of consumer sales, as reported in Table 3.6. As one can see, the example dem-
onstrates a clear bullwhip effect, with manufacturer orders exhibiting about
twice the volatility of consumer sales and supplier orders exhibiting more
than four times the volatility. This is demonstrated pictorially in Figure 3.16,
which plots the demand and orders from the example in Table 3.6. One conclusion that we could draw from this example is that an ad hoc ordering policy can cause a bullwhip effect. Fair enough—but an important aspect of this
conclusion is that many supply chains are, in practice, driven by similarly ad
hoc ordering policies, an observation noted by the author in conversations
with practicing managers and also supported by Chen et al. (2000).

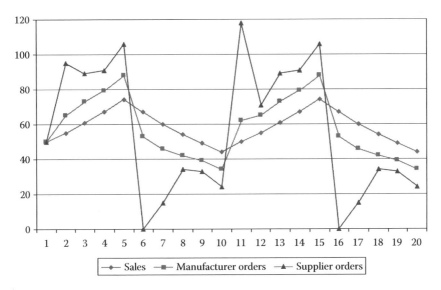

FIGURE 3.16
Bullwhip example from Vollmann et al. (2005) with horizon of 20 periods.

Thus, even without including various complexities in the demand stream such as the serial correlation in demand studied by Khan (1987), Lee et al. (1997a), and Chen et al. (2000), our simple example shows the bullwhip effect in action. Chen et al. (2000), however, demonstrate a key point—that the magnitude of the bullwhip effect can be shown to be a function, indeed in their case a *super-linear* function, of the replenishment lead time. Thus, let us alter the Vollmann et al. (2005) example to consider a situation in which the replenishment lead time is *two* periods—i.e., where an order placed at the end of period t arrives at the beginning of period $t + 2$. Using the same ad hoc ordering policy as mentioned earlier, the results of this revised example are shown in Table 3.7 and Figure 3.17. Clearly, the bullwhip effect gets

TABLE 3.7

Bullwhip Example with Two-Period Lead Time

Consumer		Manufacturer				Supplier			
Period	Sales	Beg Inv	End Inv	Lost Dmd	Order	Beg Inv	End Inv	Lost Dmd	Order
1	50	100	50	0	50	100	100	—	0
2	55	50	0	5	110	100	50	0	170
3	61	50	0	11	122	50	0	60	244
4	67	110	43	0	91	170	48	0	134
5	74	165	91	0	57	292	201	0	0
6	67	182	115	0	19	335	278	0	0
7	60	172	112	0	8	278	259	0	0
8	54	131	77	0	31	259	251	0	0
9	49	85	36	0	62	251	220	0	0
10	44	67	23	0	65	220	158	0	0
11	50	85	35	0	65	158	93	0	37
12	55	100	45	0	65	93	28	0	102
13	61	110	49	0	73	65	0	0	146
14	67	114	47	0	87	102	29	0	145
15	74	120	46	0	102	175	88	0	116
16	67	133	66	0	68	233	131	0	5
17	60	168	108	0	12	247	179	0	0
18	54	176	122	0	0	184	172	0	0
19	49	134	85	0	13	172	172	0	0
20	44	85	41	0	47	172	159	0	0
Min	44		0		0		0		0
Max	74		122		122		278		244
Avg	58.1		59.55		57.4		130.8		55.0
Fill rate				98.6%				94.8%	
Range	30				122				244
Std. dev	9.16				35.23				77.32
				Ratio	3.85			Ratio	8.44

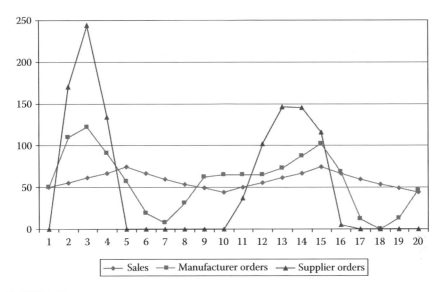

FIGURE 3.17
Bullwhip example with two-period lead time.

dramatically worse in this case, essentially doubling the variability in orders at both stages with respect to consumer demand.

Now, we will try some simple approaches to mitigate the bullwhip effect in this example. First, we will use a base-stock ordering policy at both the manufacturer and the supplier. Since the original example did not specify an ordering cost, we will assume that this cost is zero or negligible. Thus, for *stationary* demand, a stationary base-stock policy would be warranted; however, in our example, demand is not stationary. Let us therefore use an "updated base-stock" policy, where the base-stock level will be updated periodically, in this case every five periods. We will assume that the manufacturer bases its base-stock policy on a forecast of demand, with a forecasted mean and standard deviation of per-period demand of $\mu_1^M = 65$ and $\sigma_1^M = 10$ in periods 1–5 and $\mu_2^M = 55$ and $\sigma_2^M = 10$ in periods 6–10. These parameters are also used in the second demand cycle, with the higher mean being used as the demand rises in periods 11–15 and the lower mean as demand falls in periods 16–20. We will assume that the supplier uses a forecast of the manufacturer's orders to compute its base-stock level in the first five periods, with $\mu_1^S = 60$ and $\sigma_1^S = 20$. In the next three five-period blocks, the supplier uses actual orders from the manufacturer to compute the mean and standard deviation of demand, so that $\left(\mu_i^S, \sigma_i^S\right) = \{(64.8, 20.8), (50.4, 5.3), (68.4, 14.2)\}$, $i = 2, 3, 4$. At both stages, therefore, the base-stock policy is computed as $S_i = \mu_i + z\sigma_i\sqrt{2}$ $(i = 2, 3, 4)$, with $z = 1.282$, yielding a target cycle service level of 90%, and $\sigma_{\text{DLT},i} = \sigma_i\sqrt{2}$. As one can see from Table 3.8 and Figure 3.18, the base-stock policies work quite well in reducing the bullwhip effect, but at

TABLE 3.8

Bullwhip Example with Updated Base-Stock Orders

	Consumer	Manufacturer					Supplier				
Period	Sales	Beg Inv	End Inv	Lost Dmd	B	Order	Beg Inv	End Inv	Lost Dmd	B	Order
1	50	100	50	0	83	33	100	100	—	96	0
2	55	50	0	5	83	83	100	67	0	96	29
3	61	33	0	28	83	83	67	0	16	96	96
4	67	83	16	0	83	67	29	0	54	96	96
5	74	99	25	0	83	58	96	29	0	96	67
6	67	92	25	0	73	48	125	67	0	102	35
7	60	83	23	0	73	50	134	86	0	102	16
8	54	71	17	0	73	56	121	71	0	102	31
9	49	67	18	0	73	55	87	31	0	102	71
10	44	74	30	0	73	43	62	7	0	102	95
11	50	85	35	0	83	48	78	35	0	60	25
12	55	78	23	0	83	60	130	82	0	60	0
13	61	71	10	0	83	73	107	47	0	60	13
14	67	70	3	0	83	80	47	0	26	60	60
15	74	76	2	0	83	81	13	0	67	60	60
16	67	82	15	0	73	58	60	0	21	94	94
17	60	96	36	0	73	37	60	2	0	94	92
18	54	94	40	0	73	33	96	59	0	94	35
19	49	77	28	0	73	45	151	118	0	94	0
20	44	61	17	0	73	56	153	108	0	94	0
Min	44		0			33		0			0
Max	74		50			83		118			96
Avg	58.1		20.65			57.35		45.45			45.8
Fill rate				97.2%					84.0%		
Range	30					50					96
Std. dev	9.16					16.11					36.22
				Ratio		1.76			Ratio		3.95

the expense of lower fill rates at both the manufacturer and the supplier. (Note that our computational experience shows that a higher CSL target does not have much impact on this outcome.)

One key prescription of the literature on the bullwhip effect is to central-ize demand information so that demand forecasts are not inferred from the orders of the stage immediately downstream—the essence of the "demand forecast updating" (Lee et al., 1997b) cause of the bullwhip effect—but instead come directly from consumer sales data. Therefore, let us update our example to account for this. We will assume the same base-stock policy

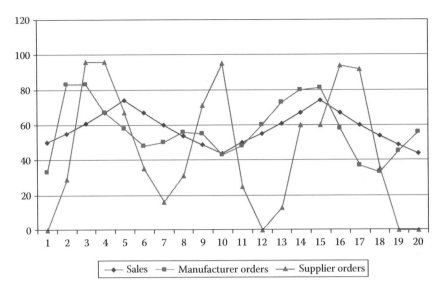

FIGURE 3.18
Bullwhip example with updated base-stock orders.

at the manufacturer as mentioned earlier, based on its forecasts of rising and falling consumer demand. At the supplier, however, we will assume that the initial base-stock level is computed directly from the manufacturer's forecast so that $\mu_1^S = 65$ and $\sigma_1^S = 10$. In the succeeding five-period blocks, the supplier updates its base-stock levels using actual consumer sales data, such that $\left(\mu_i^S, \sigma_i^S\right) = \{(61.4, 9.5), (54.8, 9.0), (61.4, 9.5)\}$, $i = 2, 3, 4$. As proved in the literature (Chen et al., 2000), this will mitigate the bullwhip effect at the supplier, as our results in Table 3.9 and Figure 3.19 demonstrate, but only slightly, and it certainly does not eliminate the effect. Also, the supplier fill rate in this latter case improves from the previous base-stock case, but only slightly.

Finally, let us consider the effect of batch ordering at the manufacturer. Let us assume that the manufacturer has occasion to order in consistent lots, with $Q = 150$. We further assume that the manufacturer follows a (Q, R) policy with R set to the base-stock levels from the prior two examples. In contrast, we assume that the supplier has no need to batch orders and follows a base-stock policy, with $\mu_1^S = 100$ and $\sigma_1^S = 30$, set in anticipation of the large orders from the manufacturer. The results are shown in Table 3.10 and Figure 3.20, in which the supplier's base-stock levels in the remaining five-period blocks are updated using the manufacturer orders, resulting in $\left(\mu_i^S, \sigma_i^S\right) = (60.0, 82.2)$, $i = 2, 3, 4$, since each block contains exactly two orders from the manufacturer. (Note that the assumption of a normal distribution for lead-time demand probably breaks down with such a large standard deviation, due to the high probability of negative values, but we retain this

TABLE 3.9

Bullwhip Example with Updated Base-Stock Orders Using Consumer Demand

Consumer		Manufacturer					Supplier				
Period	Sales	Beg Inv	End Inv	Lost Dmd	B	Order	Beg Inv	End Inv	Lost Dmd	B	Order
1	50	100	50	0	83	33	100	100	—	83	0
2	55	50	0	5	83	83	100	67	0	83	16
3	61	33	0	28	83	83	67	0	16	83	83
4	67	83	16	0	83	67	16	0	67	83	83
5	74	99	25	0	83	58	83	16	0	83	67
6	67	92	25	0	73	48	99	41	0	79	38
7	60	83	23	0	73	50	108	60	0	79	19
8	54	71	17	0	73	56	98	48	0	79	31
9	49	67	18	0	73	55	67	11	0	79	68
10	44	74	30	0	73	43	42	0	13	79	79
11	50	85	35	0	83	48	68	25	0	71	46
12	55	78	23	0	83	60	104	56	0	71	15
13	61	71	10	0	83	73	102	42	0	71	29
14	67	70	3	0	83	80	57	0	16	71	71
15	74	76	2	0	83	81	29	0	51	71	71
16	67	82	15	0	73	58	71	0	10	79	79
17	60	96	36	0	73	37	71	13	0	79	66
18	54	94	40	0	73	33	92	55	0	79	24
19	49	77	28	0	73	45	121	88	0	79	0
20	44	61	17	0	73	56	112	67	0	79	12
Min	44		0			33		0			0
Max	74		50			83		100			83
Avg	58.1		20.65			57.35		34.45			44.85
Fill rate				97.2%					84.9%		
Range	30					50					83
Std. dev	9.16					16.11					29.47
				Ratio		1.76			Ratio		3.22

assumption nonetheless.) From the table and figure, one can see that the variability in the order stream increases dramatically at the manufacturer. Interestingly, however, the significant increase in variability is not further amplified at the supplier when it uses a base-stock policy.

These examples are not intended to prove any general results, per se, but merely to demonstrate the effect of ordering policies on supply chain performance and to emphasize the challenge of identifying policy prescriptions to mitigate these effects. Note that Lee et al. (1997a,b) lay out a more extensive set of prescriptions that may be used in practice to attempt

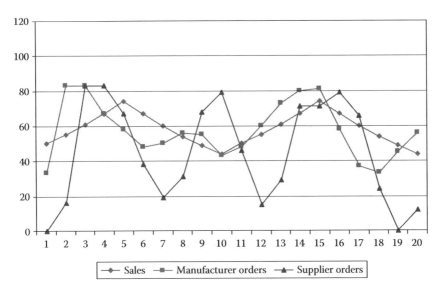

FIGURE 3.19
Bullwhip example with updated base-stock orders using consumer demand.

to address the full set of causes for the bullwhip effect—order batching, forecast updating, price fluctuations, and shortage gaming. From our examples, interestingly, the fill-rate performance is actually best with the ad hoc policy from the original example in Vollmann et al. (2005). This is important, in that lost demand at the supplier in our example serves as a proxy for additional cost at the manufacturer to supplement supply. (A more complex assumption would be to allow backorders at the supplier, further deflating the service performance at the manufacturer and quite likely exacerbating the bullwhip effect at the supplier.) This example really speaks to a need for richer models of the interaction between customer and supplier, especially in a case where there is an underlying pattern to demand, and particularly in the case where the customer has access to more information about consumer demand than the supplier could infer on its own simply from the customer's stream of orders. Indeed, in practice, this has led to burgeoning efforts like collaborative planning, forecasting, and replenishment (CPFR).*

---

* See the website of the Voluntary Interindustry Commerce Standards (VICS) organization at http://www.vics.org/committees/cpfr/ for more details on this set of practices. More information on CPFR is also available in Chapter 2.

TABLE 3.10

Bullwhip Example with (Q, R) Orders at Manufacturer and Base-Stock Orders
at Supplier

Consumer		Manufacturer					Supplier				
Period	Sales	Beg Inv	End Inv	Lost Dmd	R	Order	Beg Inv	End Inv	Lost Dmd	B	Order
1	50	100	50	0	83	150	100	100	—	154	54
2	55	50	0	5	83	150	100	0	50	154	154
3	61	150	89	0	83	0	54	0	96	154	154
4	67	239	172	0	83	0	154	154	0	154	0
5	74	172	98	0	83	0	308	308	0	154	0
6	67	98	31	0	73	150	308	308	0	209	0
7	60	31	0	29	73	150	308	158	0	209	51
8	54	150	96	0	73	0	158	8	0	209	201
9	49	246	197	0	73	0	59	59	0	209	150
10	44	197	153	0	73	0	260	260	0	209	0
11	50	153	103	0	83	0	410	410	0	209	0
12	55	103	48	0	83	150	410	410	0	209	0
13	61	48	0	13	83	150	410	260	0	209	0
14	67	150	83	0	83	0	260	110	0	209	99
15	74	233	159	0	83	0	110	110	0	209	99
16	67	159	92	0	73	0	209	209	0	209	0
17	60	92	32	0	73	150	308	308	0	209	0
18	54	32	0	22	73	150	308	158	0	209	51
19	49	150	101	0	73	0	158	8	0	209	201
20	44	251	207	0	73	0	59	59	0	209	150
Min	44		0			0		0			0
Max	74		207			150		410			201
Avg	58.1		85.55			60.0		169.85			68.2
Fill rate				94.1%					87.8%		
Range	30					150					201
Std. dev	9.16					75.39					75.37
				Ratio		8.23			Ratio		8.23

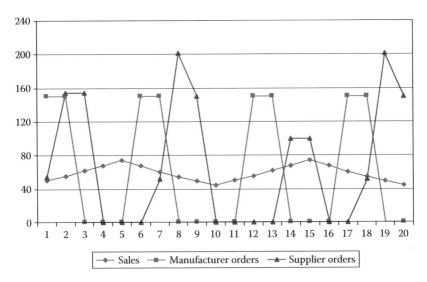

FIGURE 3.20
Bullwhip example with (Q, R) orders at manufacturer and base-stock orders at supplier.

Appendix References

Blinder, A. S. 1982. Inventories and sticky prices: More on the microfoundations of macroeconomics. *American Economic Review.* 72(3): 334–348.

Blinder, A. S. 1986. Can the production smoothing model of inventory behavior be saved? *Quarterly Journal of Economics.* 101(3): 431–453.

Callioni, G. and C. Billington. 2001. Effective collaboration. *OR/MS Today.* October issue, 28(5): 4–39.

Caplin, A. S. 1985. The variability of aggregate demand with (S,s) inventory policies. *Econometrica.* 53(6): 1395–1410.

Chen, F., Z. Drezner, J. K. Ryan, and D. Simchi-Levi. 2000. Quantifying the bullwhip effect in a simple supply chain: The impact of forecasting, lead times, and information. *Management Science.* 46(3): 436–443.

Forrester, J. W. 1961. *Industrial Dynamics.* Cambridge, MA: MIT Press.

Khan, J. A. 1987. Inventories and the volatility of production. *American Economic Review.* 77(4): 667–679.

Lee, H. L., V. Padmanabhan, and S. Whang. 1997a. Information distortion in a supply chain: The bullwhip effect. *Management Science.* 43(4): 546–558.

Lee, H. L., V. Padmanabhan, and S. Whang. 1997b. The bullwhip effect in supply chains. *Sloan Management Review.* 38(3): 93–102.

Sterman, J. D. 1989. Modeling managerial behavior misperceptions of feedback in a dynamic decision making experiment. *Management Science.* 35(3): 321–329.

Vollmann, T., W. Berry, D. C. Whybark, and R. Jacobs. 2005. *Manufacturing Planning and Control for Supply Chain Management*, 5th edn. New York: McGraw-Hill/Irwin.

Warsing, D. P. 2008. Supply chain management, in *Operations Research and Management Science Handbook*, A. Ravi Ravindran, ed., Boca Raton, FL: CRC Press, Chapter 22.

Exercises

3.1 Explain how the same item could be an independent-demand item for one company, but a dependent-demand item for another company.

3.2 Name the two critical tasks necessary for building an inventory decision model and describe briefly how they are executed in general.

3.3 Explain how goodwill cost and the α service measure are related to each other. How could you use this knowledge to suggest a stocking level in the absence of accurate information about goodwill cost?

3.4 Comment on the following statement: In a single-season inventory management scenario where the overage and underage costs are known, as the level of uncertainty in demand increases, all else being the same, the optimal order quantity also increases. Is that statement generally true, generally false, or does it depend on some other information? You should support your answer with facts drawn from your knowledge of the newsvendor model.

3.5 What is the definition of safety stock? Can safety stock be defined in more than one way?

3.6 Explain why a periodic-review system, in general, will carry more inventory than a continuous-review system.

3.7 Describe the primary differences between the following inventory control systems: (a) continuous-review (Q, R), (b) periodic-review (s,S), (c) periodic-review order-up-to S (i.e., an (S,T) system), and (d) periodic-review base stock.

3.8 Given the sales data in Table 3.11, which items would you classify as "A" items, which as "B" items, and which as "C" items? Clearly justify your choices.

3.9 Ephemeral Apparel (EA) has been suffering from stockouts of some items and overstocks of other items. Robert Michaels, vice president of marketing, and Michael Roberts, EA's new vice president of logistics, are battling over how to set inventory targets for several key product lines for the 2013 summer season. Orders must be placed in December 2012 to ensure that products will be available in EA distribution centers by May 2013. Michaels claims that the cost of customer goodwill for stocking out of any item should be set at two times the selling price of that item. Further, he claims that because of this high cost of stocking out, the service target for all items should be a 95% probability of not stocking out in the season. Roberts says that these are overestimates and will cause EA to carry too much inventory. Accordingly, Roberts asked the Finance and Merchandising departments to provide some data to help him develop service level targets more scientifically. One particular item, for example, has a projected gross margin (per-unit profit margin divided by per-unit

TABLE 3.11

Data for Exercise 3.8

Stock-Keeping Unit (SKU)	Annual Sales ($)
J-625	904,366
Z-454	838,481
W-681	757,060
J-909	635,764
T-988	596,075
B-570	482,492
M-117	390,553
H-033	262,363
W-998	212,713
C-841	151,413
R-596	125,234
Y-764	81,392
F-496	48,131
M-154	37,409
M-615	30,011
A-620	29,830
K-388	27,592
B-237	24,633
K-778	22,551
Y-319	19,836
T-670	16,058
S-802	14,996
T-172	10,106
G-676	8,783
M-687	6,624

selling price) of 45% and a projected end-of-season salvage value of 20% of the per-unit cost. (Hint: You should be able to express all relevant costs in terms of the item's price, p.)

a. If Logistics VP Roberts accepts Marketing VP Michaels' estimate of goodwill cost, what is the best service level for this item?

b. If goodwill cost were estimated at *five* times the selling price of the item, what is the best service level for this item?

c. Given these results, what do you believe Mr. Roberts should recommend to Mr. Michaels regarding his proposed 95% service target?

3.10 Suppose you are the director of operations for the southeast region for consumer electronics retailer BigBuy. You have been asked by the chief operating officer of the company to review the inventory management processes and procedures in place at all BigBuy retail outlets in your region.

Accordingly, you have gathered some data from the store managers and regional DC managers.

To begin understanding the stocking processes, you start with a single stock-keeping unit (SKU) at a single retail outlet, a 19-in. LCD computer monitor that is sold at BigBuy's largest store in Raleigh, NC. The data you have collected indicates the following:

- Average demand for this SKU is about 5 units per day, which equates to 1800 units per year on BigBuy's operating schedule of 360 days per year.
- The DC manager indicates that the DC's operating policy specifies next-day fulfillment for all orders placed by retail outlets prior to 6 p.m. on any given day.
- The cost of placing a replenishment order at the retail store is estimated to be $6.25, which covers various administrative functions like entering the day's orders on the Intranet-based ordering system and receiving daily shipments, sorting them, and stocking them in the store room.
- BigBuy's corporate accounting standards specify a 25% inventory holding cost rate for all retail outlets.
- The cost (to BigBuy—i.e., cost on its accounting books) of this 19-in. LCD monitor is $100.
- Since this is a relatively fast-moving item at this store, the current replenishment policy calls for daily (on average) replenishments (i.e., an order quantity of 5 units).

a. Is this a good replenishment policy? What is the annual cost of this replenishment policy?

b. Is the current replenishment policy optimal? If so, why? If not, what needs to change for daily replenishments to be optimal?

3.11 Continuing the scenario described in Question 3.10 earlier, you decide to gather additional data to further explore inventory management processes and procedures at the Raleigh BigBuy store. First, you find that daily demand is, as one would suspect, not constant. The data you gather indicates that mean daily demand is 5 monitors, with a standard deviation of 1.5 monitors, approximately following a normal distribution.

a. Assuming that the DC is able to consistently meet its "next-day" policy, what should the reorder point be in a continuous-review (Q, R) inventory system at the BigBuy store in order to achieve a 90% cycle service level?

b. Relentless in your pursuit of outstanding performance—driven to prove that you deserve to be the next COO—you collect additional data that indicates that the DC is actually not always able to provide next-day fulfillment. The actual performance is an average fulfillment lead time of 1.2 days, with a standard deviation of 0.4 days, approximately following a normal distribution.

c. Given this updated information, what reorder point do you recommend?

d. If BigBuy used the reorder point from (a) above when, in actuality, lead time follows the distribution specified by the data as described earlier, what would be the actual in-stock probability performance (i.e., the cycle service level)?

3.12 Stage Two Incorporated produces a popular item at its Paducah, KY plant and ships this item to its national distribution center (DC) in Columbus, OH. The item has an annual demand of 20,000 units, which transpires at a relatively steady rate throughout the year. The fixed cost of replenishing the finished goods inventory at the Paducah plant is $200, and the fixed replenishment cost at the Columbus DC is $50. Annual holding costs at Paducah are $20 per unit. Annual holding costs at the Columbus DC at $30 per unit. (For the purposes of this problem, you can assume that replenishment of finished goods at the Paducah plant is sufficiently close to instantaneous to allow you to ignore the effects of changes in the average inventory level as production proceeds.)

(a) Use the methods developed in Section 3.6 to compute the optimal order quantity at the plant and at the DC assuming decentralized inventory control.

(b) Use the methods developed in Section 3.6 to compute the optimal order quantity at the plant and at the DC assuming centralized inventory control.

(c) What is the difference in total annual cost of the decentralized and centralized systems? How do your results from (a), (b), and the cost difference compare to the results in Example 3.14?

3.13 Compute the updated echelon base-stock levels for the Stage Three Incorporated example (Example 3.15) given the same distribution of demand at the field warehouse, but with a penalty cost that is 20% of the original value (i.e., $b = \$10$) and a 50% increase in echelon holding costs (i.e., installation holding costs of $h_3 = \$3.00$, $h_2 = \$2.25$, and $h_1 = \$1.50$).

3.14 *Mini-case study:* Milo's Home Improvement was recently named as an authorized distributor of lawn tractors by the Dear John Company (DJC—whose advertising proclaims, "You'll love our tractors so much you won't care if she leaves you for someone who's not so attached to lawn care.") Milo's retail network covers the Southeastern United States. You are the newly hired logistics analyst for Milo's, and you've been asked to evaluate the company's current replenishment policies for optional accessories to support lawn tractors.

While Dear John tractors are built at a plant in central NC, the accessories carried by Milo's are supplied by DJC's distribution center (DC) in Memphis, TN, since these items are built by various manufacturers from around the globe. The current contract terms between DJC and Milo's dictate that Milo's must arrange and pay for freight transportation to its facilities. Milo's currently ships most tractor accessories via truckload (TL) from DJC's Memphis DC to its own DC in Atlanta. Some accessories are shipped via less-than-truckload service, however.

For ease of discussion, let us consider only five items, as specified in Table 3.12. Since sales are concentrated in the relatively temperate Southeastern United States, we can assume that aggregate demand for tractors and accessories is reasonably steady across the year. Daily demand, however, exhibits some uncertainty. Although detailed standard deviation data for demand is currently not available, Milo's has historically used a coefficient of variation (standard deviation divided by mean) of 0.3 to estimate the uncertainty in daily demand in its past inventory planning. The order fulfillment lead time from the manufacturer DC (DJC) to the retailer DC (Milo's) is a fairly consistent 3 days. Both DCs operate 365 days per year.

Other important parameters gathered from company records appear in Table 3.12.

TABLE 3.12

Data for Exercise 3.14

Item	Unit Cost	Volume (cu ft)	Annual Demand (Units)	Current Order Quantity (Units)	Freight Charge (per Shipment)
6.5-bu rear bagger	$250	32.0	6,000	109	$776
3.5-cu ft tow-behind broadcast spreader	$175	27.0	4,500	129	$776
12-V oscillating fan	$18	2.0	9,500	365	$776
Canopy	$28	1.5	5,200	200	$646
Double-bucket holder	$12	0.6	18,500	712	$776
Common Values					
Annual holding cost rate	25%				
Truckload capacity	3500	cu ft			
Memphis-to-Atlanta distance	388	mi			
Memphis-to-Atlanta TL cost	$776				
Cost to receive order at ATL DC	$10				

(a) For each item, compute the economic order quantity. Is the EOQ a feasible order quantity for each item? Why or why not? For any item where the EOQ is not a feasible order quantity, what order quantity do you suggest?

(b) Using your computed or suggested order quantity values from (a), compute the reorder point for each item that achieves an expected target fill rate of 98%. (See expressions (3.16) and (3.20) in the chapter for hints on how to do this.) Your computed reorder points should allow only non-negative safety stock levels.

(c) Given their similar characteristics, you have suggested that the fan, canopy, and double-bucket holder could be ordered jointly in consolidated shipments. Compute the optimal number of orders for this approach. First, do this using the formula developed in the chapter (expression (3.29)), and then as a constrained optimization problem using Excel Solver. Are these solutions feasible?

(d) Since demand is uncertain, using separate (Q, R) ordering policies for the three items considered in (c) is not guaranteed to result in orders that occur at the same points in time. As an alternative, use the optimal order frequency implied by the Solver solution from (c), mentioned earlier, to compute order-up-to levels for those three items for a periodic-review (S, T) system. Your order-up-to levels should be set to achieve a 90% in-stock probability target.

References

Arrow, K. J., T. Harris, and J. Marschak. 1951. Optimal inventory policy. *Econometrica.* 19: 250–272.

Axsater, S. 2006. *Inventory Control*, 2nd edn. New York: Springer.

Azadivar, F. and A. Rangarajan. 2008. Inventory control (Chapter 10). In *Operations Research and Management Science Handbook*, ed. A. R. Ravindran. Boca Raton, FL: CRC Press.

Balintfy, J. 1964. On a basic class of multi-item inventory problems. *Management Science.* 10(2): 287–297.

Bozarth, C. C. 2005. Executive education session notes on Inventory Operations, North Carolina State University, Raleigh, NC.

Bozarth, C. C. and R. B. Handfield. 2006. *Introduction to Operations and Supply Chain Management*. Upper Saddle River, NJ: Pearson Prentice Hall.

Cachon, G. P. 2003. Supply chain coordination with contracts (Chapter 6). In *Handbooks in Operations Research and Management Science: Supply Chain Management*, eds. A. G. de Kok and S. C. Graves. Amsterdam, the Netherlands: Elsevier.

Chapman, S. N. 2006. *Fundamentals of Production Planning and Control*. Upper Saddle River, NJ: Prentice Hall.

Chen, F. and Y. S. Zheng. 1994. Evaluating echelon stock (R,nQ) policies in serial production/inventory systems with stochastic demand. *Management Science.* 40: 1262–1275.

Chopra, S. and P. Meindl. 2010. *Supply Chain Management: Strategy, Planning, and Operation,* 4th edn. Upper Saddle River, NJ: Prentice Hall.

Clark, A. J. and H. Scarf. 1960. Optimal policies for a multi-echelon inventory problem. *Management Science.* 6(4): 475–490.

DeMatteis, J. J. and A. G. Mendoza. 1968. An economic lot sizing technique. *IBM Systems Journal.* 7: 30–46.

Flores, B. E. and D. C. Whybark. 1987. Implementing multiple criteria ABC analysis. *Journal of Operations Management.* 7(1–2): 79–85.

Gallego, G. and P. Zipkin. 1999. Stock positioning and performance estimation in serial production-transportation systems. *Manufacturing & Service Operations Management.* 1(1): 77–88.

Graves, S. C. and L. B. Schwarz. 1977. Single cycle continuous review policies for arborescent production/inventory systems. *Management Science.* 23(5): 529–540.

Graves, S. C. and S. P. Willems. 2000. Optimizing strategic safety stock placement in supply chains. *Manufacturing & Service Operations Management.* 2: 68–83.

Harris, F. W. 1913. How many parts to make at once. *Factory: The Magazine of Management.* 10(2): 135–136, 152.

Hax, A. C. and D. Candea. 1984. *Production and Inventory Management.* Englewood Cliffs, NJ: Prentice-Hall.

Holt, C. C., F. Modigliani, J. F. Muth, and H. A. Simon. 1960. *Planning Production, Inventories, and Work Force.* Englewood Cliffs, NJ: Prentice-Hall.

Hopp, W. J. and M. L. Spearman. 2004. To pull or not to pull: What is the question? *Manufacturing & Service Operations Management.* 6(2): 133–148.

Iglehart, D. 1963. Optimality of (s,S) policies in the infinite horizon dynamic inventory problem. *Management Science.* 9(2): 259–267.

Jackson, P., W. Maxwell, and J. Muckstadt. 1985. The joint replenishment problem with a powers-of-two restriction. *IIE Transactions.* 17(1): 25–32.

Kulkarni, V. G. 1995. *Modeling and Analysis of Stochastic Systems.* Boca Raton, FL: Chapman & Hall/CRC.

Magnanti, T. L., Z. J. M. Shenb, J. Shuc, D. Simchi-Levi, and C. P. Teo. 2006. Inventory placement in acyclic supply chain networks. *Operations Research Letters.* 34: 228–238.

Porteus, E. 1985. Numerical comparisons of inventory policies for periodic review systems. *Operations Research.* 33(1): 134–152.

Rosling, K. 1989. Optimal inventory policies for assembly systems under random demands. *Operations Research.* 37(4): 565–579.

Scarf, H. 1960. The Optimality of (S,s) Policies in the Dynamic Inventory Problem, in *Mathematical Methods in the Social Sciences, 1959. Proceedings of the First Stanford Symposium.* Stanford, CA: Stanford University Press.

Schwarz, L. 1973. A simple continuous review deterministic one-warehouse N-retailer inventory problem. *Management Science.* 19(5): 555–566.

Shang, K. and J. Song. 2003. Newsvendor bounds and heuristic for optimal policies in serial supply chains. *Management Science.* 49(5): 618–638.

Silver, E. A. and H. C. Meal. 1973. A heuristic for selecting lot size requirements for the case of a deterministic time-varying demand rate and discrete opportunities for replenishment. *Production and Inventory Management*. 14: 64–74.

Silver, E. A., D. F. Pyke, and R. Peterson. 1998. *Inventory Management and Production Planning and Scheduling*, 3rd edn. New York: John Wiley & Sons.

Simchi-Levi, D., P. Kaminsky, and E. Simchi-Levi. 2008. *Designing and Managing the Supply Chain: Concepts, Strategies, and Case Studies*, 3rd edn. New York: McGraw-Hill/Irwin.

Wagner, H. M. and T. M. Whitin. 1958. Dynamic version of the economic lot size model. *Management Science*. 5: 89–96.

Whitin, T. M. 1953. *The Theory of Inventory Management*. Princeton, NJ: Princeton University Press.

Wilson, R. H. 1934. A scientific routine for stock control. *Harvard Business Review*. 13(1): 116–128.

Zangwill, W. 1966a. A deterministic multi-period production scheduling model with backlogging. *Management Science*. 13(1): 105–119.

Zangwill, W. 1966b. A deterministic multiproduct, multifacility production and inventory model. *Operations Research*. 14(3): 486–507.

Zipkin, P. H. 2000. *Foundations of Inventory Management*. New York: McGraw-Hill/Irwin.

4

Transportation Decisions in Supply Chain Management

4.1 Introduction

Up to this point in the text, we have discussed methods for generating forecasts of demand for goods in the supply chain, and we have discussed methods for using that demand information, along with our knowledge of other costs, to make decisions regarding stocking levels at inventory locations in the supply chain. Now, we turn our attention to the issue of making decisions regarding the movement of goods through the supply chain system. This is the realm of freight transportation.

An important point to keep in mind as you read this chapter is that, just as "it takes two to tango," so they say, there are two perspectives on the business of freight transportation. One perspective is that of the *carrier*, the service provider to whom goods are consigned for shipment and who executes the shipment. The other perspective is that of the *shipper*, the business or consumer who hires the carrier to move the goods in question. Consistent with the perspective taken throughout this book—that of the firm making decisions regarding the management of various supply chain entities for which is it responsible—this chapter takes the perspective of the shipper.

To put some of these ideas into context, we start by framing an example that will be considered throughout the chapter.

Example 4.1: Basic Scenario

Let us assume that you are the manager of your company's regional distribution center (RDC) for the eastern United States, located in Atlanta, Georgia. Your company regularly receives shipments of goods from a supplier that is headquartered in Asia, but also operates its own U.S. distribution center in Oakland, California, taking advantage of the large amount of ocean-going freight traffic from Asia to the west coast of the United States. Since your company moves a significant amount of freight on the Oakland-Atlanta "lane," your company has, in its price negotiations with the supplier, agreed to assume responsibility for hiring the carrier that will move the goods from Oakland to your RDC in Atlanta.

4.2 Motor Carrier Freight: Truckload Mode

Given the information in Example 4.1, you can clearly imagine that there are a variety of choices available to you in terms of carriers that can be hired to move freight shipments on the Oakland–Atlanta lane. For the sake of simplicity in getting started, let us first consider moving the goods on the Oakland–Atlanta lane via truck, or "motor carrier." From the analysis presented in Kay and Warsing (2009), the average price charged by a carrier to move goods via a dedicated truck—commonly called *truckload* service, or "TL"—in 2004 was approximately \$2.00/mi. To put this estimate in terms that will aid in comparisons we make throughout the chapter, and continuing to use estimation methods suggested by Kay and Warsing (2009), we adjust this estimated TL freight cost to a more current level by multiplying it by the ratio of Producer Price Index values for "General freight trucking, long-distance, TL" (BLS, 2011a) for 2004 and 2010, yielding

$$r_{TL,2010} = \frac{PPI_{TL,2010}}{PPI_{TL,2004}} \cdot r_{TL,2004} = \frac{113.3}{102.7} \times 2.00 \tag{4.1}$$

which results in a rate of approximately \$2.20/mile.

In addition, we must specify the constraints on the size of a truckload shipment. From our company's standpoint as the shipper, will a truckload be too much for us to receive, violating our space constraints in the Atlanta RDC? Is it possible that a truckload could be "too little" to ship, and would we have any recourse if that were the case? These questions are perhaps best addressed in the process of solving a cost-based formulation of the decision, such as expression (3.12) from Chapter 3. From the perspective of the carrier that moves the goods, however, the issue is more rudimentary: What restrictions are posed by the physical constraints on a truck trailer, and possibly also by the regulatory environment prescribed for moving goods on the federal and state highway systems?

Thus, two considerations must be taken into account: the physical volume of the truck trailer and the maximum weight—or *payload*—it is allowed to bear. A standard, full-size trailer measures 53 ft in length, and according to dimensions detailed in Kay (2010), has an interior height (not counting the wheels and external carriage) of between 102 and 110 in. (i.e., between 8 ft 6 in. and 9 ft 2 in.) and an interior width of 98 in. (8 ft 2 in.). Using the average interior height, therefore, a 53-ft trailer has a total physical volume of approximately 3823 ft³. Since the nominal trailer volume can rarely be fully utilized due to the shape and size of the units comprising the truckload, we will assume that "full" truckloads comprise 80% of this total volume, and therefore we state the maximum effective volume—or "cube"—of a standard trailer to be $K_{cu} = 3059$ ft³. Kay (2010) also states the maximum payload of a motor carrier trailer to be 50,000 lb, or 25 ton, based on U.S. Federal

Government regulations of a maximum gross vehicle weight of 80,000 lb for U.S. interstate highway travel (U.S. DOT, 2011). Therefore, we denote the maximum trailer payload as $K_{wt} = 25$ ton. To determine which of these quantities, K_{cu} or K_{wt}, constrains the size of a TL shipment, we must also know the weight and density of the item to be shipped. Let us denote the item weight (in lb) as w, and the item density (in lb/ft^3) as s, and therefore, the maximum size of a shipment (in units shipped) via TL is

$$Q_{max,TL} = \frac{\min\{s \cdot K_{cu}, 2000 \cdot K_{wt}\}}{w} \tag{4.2}$$

Example 4.2: TL Computations

Continuing the scenario described in Example 4.1, let us say that the item we are shipping via TL service from Oakland, CA to our company's RDC in Atlanta, GA has the following characteristics:

- Daily demand follows a normal distribution with mean $\mu_D = 30$ units/day, yielding an expected annual demand of $D = 30 \times 365 = 10{,}950$ units/year, and standard deviation $\sigma_D = 8$ units/day.
- Lead time follows a normal distribution with mean $\mu_L = 5$ days and standard deviation $\sigma_L = 1.5$ days.
- Annual holding cost is $h = \$5$ per unit, based on an item cost of $c = \$20$ and an annual holding cost rate of $i = 25\%$.
- Fixed ordering cost is $A = \$80$ per order, covering various administrative costs related to placing orders with the supplier, coordinating with the TL carrier while goods are en route, and receiving and stocking the item when it arrives at the Atlanta RDC.
- Item weight is $w = 10$ lb/unit.
- Item density is $s = 9.72$ lb/ft^3 (chosen in light of later discussion in the chapter).
- The shipment distance is $d = 2463$ mi.*

Let us assume that, in the interest of "maximum efficiency," we decide to ship only in full truckload quantities, meaning that we fix $Q = Q_{max,TL}$, which by substituting into expression (4.2) given earlier, can be computed to be $Q = 2973$ units. This quantity "cubes out" the trailer at

$$\frac{2973 \text{ units} \times 10 \text{ lb/unit}}{9.72 \text{ lb/cu ft}} = 3059 \text{ cu ft} = K_{cu}$$

but does not "weigh out" the trailer since a 2973-unit load weighs 29,730 lb, or 14.865 ton.

---

* Found via Google Maps, http://maps.google.com/

For a continuous-review inventory system, we solve the optimization problem formulated in Chapter 3, specifically

minimize

$$TAC(R,Q) = \frac{AD}{Q} + h(Q/2 + R - \mu_{DLT}) \tag{4.3}$$

subject to

$$1 - \frac{\sigma_{DLT}L(z)}{Q} \geq \beta_{tgt} \tag{4.4}$$

where β_{tgt} is the target fill rate, $z = (R - \mu_{DLT})/\sigma_{DLT}$ is the standardized value of reorder point R, and $L(z)$ is the standard normal loss function.*

Fixing $Q = Q_{max,TL}$ reduces the optimization problem to simply finding the smallest value of R that satisfies constraint (4.4). With $\sigma_{DLT} = \sqrt{\mu_L \sigma_D^2 + \mu_D^2 \sigma_L^2} = 48.43$, this results in a value of z less than zero (i.e., negative safety stock), specifically $z = -0.375$, which equates to $R = 132$. Let us assume that, as a matter of policy, we will restrict safety stock to only non-negative values, and therefore, we have $R = \mu_{DLT} = \mu_D \mu_L = 150$ units (i.e., zero safety stock).

With $Q = 2973$ units and $R = 150$ units, expression (4.3) mentioned earlier gives inventory-related annual costs of $TAC = \$7727$. In addition, each shipment would result in a *freight charge* of $FC_{TL} = d \cdot r_{TL,2010} = (2463\,\text{mi}) \times (\$2.20/\text{mi}) = \$5418.60$. For this solution, the number of orders per year is $OPY = D/Q = 3.68$, for an average of 3.68 shipments per year, which results in an annual transportation cost of $(3.68)(\$5,418.60) = \$19,958$.

Solving the example above begs a question: Are shipments of $Q = Q_{max,TL} = 2973$ units the best decision for this scenario? First, let us revisit the continuous-review inventory cost model from Chapter 3 that we employed earlier in finding the optimal value of R given our choice of $Q = Q_{max,TL}$. Note that this cost expression includes only the order placement costs and inventory holding costs. More generally, our cost expression should account for transportation cost, to the extent that this cost is relevant to determining the best values of both Q and R. Thus, we should expand the continuous-review cost expression from Chapter 3 more generally to be

$$TAC(R,Q) = OC(Q) + HC(R,Q) + TC(Q) \tag{4.5}$$

---

* Note that we can use standard normal loss function because demand over the lead time (DLT) is the convolution of normal random variables for demand and lead time, and is therefore normally distributed. See the Further Readings section of this chapter for references that address situations where lead-time demand is not normal.

where we have separated the annualized costs into ordering cost, OC, typically a function only of the order quantity Q; holding cost, HC, a function of Q by virtue of cycle stock, and of reorder point R by virtue of safety stock; and transportation cost, TC, which, by its stated form earlier, we imply to be a function of Q, a relationship we establish more clearly both next and later in the chapter.

Note that for TL shipments, the freight charge FC_{TL} effectively serves as a fixed ordering cost. For each order placed, we must pay FC_{TL} to the carrier in order to transport the resulting shipment on a dedicated truck, and therefore the annual transportation cost is

$$TC_{TL}(Q) = \frac{D}{Q} \cdot FC_{TL} \tag{4.6}$$

Thus, a better formulation of the continuous-review inventory problem that accounts for the cost of TL freight transportation would be

minimize

$$TAC_{TL}(R,Q) = \frac{AD}{Q} + h \cdot (Q/2 + R - \mu_{DLT}) + \frac{FC_{TL} \cdot D}{Q}$$

$$= \frac{(A + FC_{TL}) \cdot D}{Q} + h \cdot (Q/2 + R - \mu_{DLT})$$

$$= \frac{(A + r_{TL} \cdot d) \cdot D}{Q} + h \cdot (Q/2 + R - \mu_{DLT}) \tag{4.7}$$

subject to

$$Q \leq Q_{\max,TL} \tag{4.8}$$

$$R \geq \mu_{DLT} \tag{4.9}$$

$$L\left(\frac{R - \mu_{DLT}}{\sigma_{DLT}}\right) - \frac{(1 - \beta_{tgt})Q}{\sigma_{DLT}} \leq 0 \tag{4.10}$$

Regarding this problem formulation, first note that constraint (4.8) restricts the order quantity to a single truckload. Removing this constraint would significantly complicate the behavior of the combined ordering and transportation cost, and therefore, we choose instead to use the simpler formulation here. Also, via constraint (4.9), we restrict our solutions to those with non-negative safety stock. Finally, note that we have stated the fill rate constraint (4.10) both in terms of R (and not z) and in the form of a sum of functions of the decision variables (gathering the terms of this constraint slightly differently than we did in Chapter 3).

	A	B	C	D	E	F	G	H	I	J
1		(Q,R) decision worksheet					Cost outcomes		Symbol		
3		Lead Time	Symbol				Cost outcomes		Symbol		
4		Mean	mu_L	5.00 days			Ordering Cost		OC	$295	
5		StDev	sigma_L	1.50 days			Site Holding Cost		HC	$7,432	
6		Demand					In-transit Holding Cost		ITC	$660	
7		Annual Demand	D	10,950 units/year			Transportation Cost		TC	$19,958	
8		Days per year	DPY	365 days			Total annual cost		TAC	$28,345 per year	
9		Daily StDev	sigma_D	8.00 units/day							
10		Daily Mean	mu_D	30.00 units/day			Constraint terms				
11		Service target					Maximum shipment size		Q_max	2973 units	
12		Fill rate	beta_TGT	99.00%			L((R - mu_DLT)/sigma_DLT)			0.39899	
13		Decision variables					(1-beta_TGT)*Q/sigma_DLT			0.61394	
14		Reorder Point	R	150.00 units							
15		Reorder Quantity	Q	2973.00 units			Other measures of interest				
16							Orders per Year		OPY	3.68 orders	
17		Demand During Lead Time	Symbol				Safety Stock		SS	0.00 units	
18		Mean	mu_DLT	150.00 units			Safety Stock Factor		z	0.000	
19		StDev	sigma_DLT	48.43 units			Cycle Service Level		CSL	50.00%	
20							Expected Units Short per Cycle		S(R)	19.321 units	
21		Product and freight info					Fill Rate		fi	99.35%	
22		Item cost	c	$20.00 per unit							
23		Fixed ordering cost	A	$80 per order			For comparison				
24		Holding cost factor	h	25% per $/year			Economic order qty		EOQ	591.95 units	
25		In-transit holding factor	h_T	22% per $/year							
26		Freight rate	FRPM	$2.20 per mi							
27		Freight cost per shipment	FCPS	$5,418.60 per shipment							

FIGURE 4.1

Excel set-up for truckload (TL)-based optimal inventory decision model.

Example 4.3: TL Optimization

Let us solve the constrained optimization problem posed in expressions (4.7 through 4.10) earlier, using the parameters stated in Example 4.2. The Excel set-up for this is shown in Figure 4.1, and we use Excel Solver to find the optimal solution.* It turns out that our solution validates the decision from Example 4.2 to ship in full truckload quantities, as the optimal solution to the problem is $Q^* = 2973$ and $R^* = 150$.

As a possible alternative solution, note that for fixed value of R, expression (4.7) can be optimized to yield a transportation-inclusive economic order quantity, given by

$$EOQ_{TL} = \sqrt{\frac{2 \cdot (A + FC_{TL}) \cdot D}{h}}$$

which in this case gives $EOQ_{TL} = 4908$. Clearly, this shipment quantity is infeasible since it yields a shipment volume of $5049\,\text{ft}^3$ (exceeding $K_{cu} = 3059\,\text{ft}^3$, but weighing $49,080\,\text{lb}$ and therefore less than $K_{wt} = 50,000\,\text{lb}$).

4.2.1 Accounting for Goods in Transit

In the examples we have considered up to this point, we have assumed that we, the inventory decision makers, bear the cost of freight transportation. An issue we have yet to address, however, is whether any other inventory-related costs are implied by the fact that we have agreed with our supplier that our company hires the carrier and pays the freight bill.

---

* Regarding the in-transit-related parameters shown in Figure 4.1, see the section that follows.

In fact, two questions must be answered regarding the *freight terms of sale* we must negotiate with our supplier, namely, "*Who owns the goods while they are in transit?*" and "*Who pays the freight charges?*" Based on our example, we have clearly answered the second question, but what about the first? Marien (1996) provides a helpful overview of common freight terms between a supplier shipping goods and the customer that ultimately receives them from the carrier hired to move the goods. Note that legally, the carrier that moves the goods never takes ownership of the goods. The question of ownership of goods in-transit is answered by the freight terms, or "F.O.B." terms,* indicating whether the transfer of ownership from supplier (shipper) to customer (receiver—called the *consignee* in logistics terminology) occurs at the point of *origin* or the point of *destination*, hence the terms "F.O.B. origin" and "F.O.B. destination." The next issue is which party, shipper or consignee, pays the freight charges, and here there is a wider array of possibilities. Marien (1996), however, points out that three arrangements are common: (a) F.O.B. origin, freight collect—in which goods in-transit are owned by the consignee, who also pays the freight charges; (b) F.O.B. destination, freight prepaid—in which the shipper retains ownership of goods in-transit and also pays the freight bill; and (c) F.O.B. origin, freight prepaid and charged back, similar to (a), except that the shipper pays the freight bill up-front and then charges it to the consignee upon delivery of the goods. These three sets of terms are described pictorially in Figure 4.2. The distinction of who pays the freight bill, and the interesting twist implied in (c), is important in that the party that

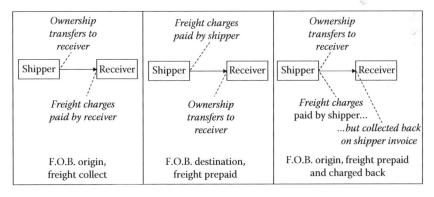

FIGURE 4.2
Common freight transportation terms (based on Marien, 1996).

---

* According to Coyle et al. (2006: p, 473, p. 490), "F.O.B." stands for "free on board," shortened from "free on board ship." Thus, F.O.B. terms differ from the water-borne transportation terms "free alongside ship" (FAS), under which the goods transfer ownership from shipper to consignee when the shipper delivers the goods to the port (i.e., "alongside ship"), leaving the consignee to pay the cost of lifting the goods onto the ship. We consider only the F.O.B. terms in our discussion. Much more detail on freight terms, including the extensive set of terms represented by the international standards known as INCOTERMS, is available in Chapter 13 of Coyle et al. (2006).

pays the freight charges is the one who has the right, therefore, to hire the carrier and choose the transport service and routing.

Note that our TAC expression for inventory management is a decision model for ordering. Thus, it takes the perspective of the *buyer*, the downstream party that will receive the goods being ordered. Therefore, freight transportation terms that involve freight collect or freight prepaid and charged back imply that the decision maker should incorporate a transportation cost term in the TAC expression that is the objective function for the order-placement decision (as we did earlier). If the terms are freight prepaid, then the order-placing decision maker would not include the transportation cost in its objective function. At this point, one might object, pointing out that supply chain management must concern itself with a "supply-chain-wide" objective to minimize costs and maximize profits. The counterargument is that the U.S. Securities and Exchange Commission has not yet required, and likely never will require, any *supply chain* to report its revenues and costs. Moreover, the Wall Street analysts who are so important in determining the market valuation of a firm are concerned only about that particular firm's profits, not those of its supply chain (however one might define that). Thus, decision models that take the perspective of the decision-making firm and incorporate only those costs relevant to the firm are eminently reasonable.

The next question we should address is whether the freight terms have any bearing on inventory holding cost. The answer is—as it often is—"It depends." Specifically, it depends on where the ownership of the goods transfers to the consignee. If the consignee is responsible for the goods in-transit (i.e., "F.O.B. origin" terms), then it stands to reason that this liability should be reflected in annual inventory-related costs. From a conservative standpoint, those goods are "on the books" of the consignee, since, from the shipper's perspective, the goods were "delivered" as soon as the carrier took them under consignment. Although typical invoice terms dictate that money will not change hands between the supplier and the customer for between 10 and 30 days, the goods are technically now part of the payable accounts of the customer. Thus, a conservative perspective would consider them to be "money tied up in inventory," in this case, inventory in-transit. Moreover, under F.O.B. origin terms, if the goods were damaged or lost in-transit, it is the consignee who must file the claim with the carrier to recover the loss, and therefore, the in-transit goods are effectively the consignee's "goods at risk." Recall from Chapter 3 that part of the cost of tying up goods in inventory, h, is driven by the risk of damage or loss.

Using an approach suggested by Coyle et al. (2009, pp. 389–391), let us assume that the re-supply lead time L is composed of two parts, a random transit time T and a constant (and obviously non-negative) order processing time. Under "F.O.B. origin" terms, the consignee assumes ownership of the goods in-transit once the "clock starts" on random variable T. Regarding T, if the expected transit time is μ_T, then every unit shipped spends a fraction μ_T/Y of a year in transit, where Y is the number of days per year. Let h_T be the annual cost of holding inventory in transit. Typically, $h_T < h$ since the latter accounts for storage-related activities (e.g., owning or leasing a warehouse and paying the employees who manage goods in the warehouse). Since an expected total of D units are shipped annually, the annual *in-transit inventory* cost is given by

$$ITC = h_T \cdot D \cdot \frac{\mu_T}{Y} \tag{4.11}$$

The resulting total annual cost expression, under F.O.B. origin terms that also require the receiver to pay the freight charges is

$$TAC(R,Q) = \frac{AD}{Q} + h \cdot \left(\frac{Q}{2} + R - \mu_{DLT} \right) + h_T \cdot D \cdot \frac{\mu_T}{Y} + TC(Q) \tag{4.12}$$

where we have stated the transportation cost generally as a function of order quantity Q (a component of the overall model that we take into consideration later in the chapter). Note that the in-transit inventory cost ITC does not depend on the order quantity Q or the reorder point R. Thus, while it will not affect the optimal values of the decision variables for a given mode, it may affect the ultimate choice *between* modes through the effect of different values of μ_T on in-transit inventory. (We will see this effect in examples presented later in the chapter.) Note also that this approach to computing in-transit inventory costs ignores the time value of money over the transit time; one would assume, however, that such time value effects are insignificant in almost all cases (i.e., for lead times of no more than a few weeks).

Example 4.4: In-transit Holding Cost Computations

Let us assume an in-transit holding cost rate of 22%, less than the on-site rate stated in Example 4.2, reflecting the fact that the in-transit rate covers only the cost of capital tied up in inventory (in this case, tied up in accounts payable and on the books in the "liabilities" column of the firm's balance sheet) and possible other risk-related costs like the risk of loss and/or damage in transit. Thus, with an item cost of $20, the annual in-transit holding cost is $h_T = (0.22)(\$20) = \$4.40/\text{year}$. Let us also assume that the lead time in this case comprises the entire transit time (i.e., not allowing for a separate, constant order processing time), such that $\mu_T = \mu_L = 5$ days. Finally, let us assume that the hard-working people at our Atlanta RDC make sure that it operates to serve our company's retail operations 365 days/year. Therefore, the annual in-transit holding cost is $ITC = h_T \cdot D \cdot \frac{\mu_T}{Y} = (4.40)(10{,}950)(5)/365 = \$660$.

4.3 Stepping Back: Freight Transportation Overview

Up to this point in the chapter, we have considered through our running example only a single mode of freight transportation, via truck (motor carrier), and only one type of carrier for this mode, dedicated truckload (TL). Clearly, there are other options to consider. For example, a different motor freight carrier could be hired to ship our goods in a "less-than-truckload" (LTL) quantity from Oakland to Atlanta, but as you might imagine, our goods are no longer likely to ship directly, point-to-point, from Oakland to Atlanta. Instead, the choice of the specific route that the goods follow between the origin and destination would be up to the LTL carrier, which would do its best to maximize the efficiency of its network of freight-mixing terminals to create various mixed truckloads from the LTL shipments consigned to it by various shippers and move these mixed truckloads, including our goods, across the continental United States. Thus, what truly distinguishes TL service from LTL service is that, with TL, the shipper hires the carrier to execute a direct, point-to-shipment, irrespective of whether the truck is "full" or not. With LTL, the shipper consigns to the carrier a quantity of goods that is clearly less than a truckload, with the knowledge that the carrier may move these goods to several different points between the origin and the destination.

More generally, the problems we consider throughout this chapter revolve around the selection and use of a particular *transport service*, defined by Ballou (2004) as "a set of transportation performance characteristics purchased at a given price" (p. 167). This implies that, in general, we can describe and measure a freight transport mode by the combination of its *cost* and its *service characteristics*. Cost is self-explanatory, but there are various ways we could measure the service characteristics of a particular transport service. Coyle et al. (2009) offer the following list of service characteristics: *accessibility, transit time, reliability,* and *security*. Thus, in choosing a transport service, we (the shipper) wish to answer the following questions:

- How convenient is it to use? (accessibility)
- How fast is it? (transit time)
- How predictable is it? (reliability)
- What is the risk of loss or damage to the goods? (security)
- How much does it cost?

Clearly, this assessment is helpful in understanding the choice of a transportation service, but how do these services and characteristics manifest themselves in the market for freight transportation? Of course, as we have indicated earlier, the market for transportation services includes shippers (the buyers of freight transportation services) and carriers (the sellers of those services). Moreover, carriers may be classified by type along various lines. First, we note that carriers can be classified *legally* (see, e.g., Murphy and Wood, 2011) into *private carriers* and *for-hire carriers*. Private carriers are companies that, in addition to their primary lines of business (e.g., manufacturing, retailing), own a fleet of vehicles and use those vehicles to move their own goods. In that sense, a private carrier is effectively a shipper and a carrier all in one. Recall, however, that our perspective is that of the shipper, and therefore, we consider only contexts regarding for-hire carriers. More specifically, our context deals with either *common carriers*, who have "agreed to serve the general public" (Murphy and Wood, 2011, p. 232) or *contract carriers*, who provide "specialized service to customers on a contractual basis" (Murphy and Wood, 2011, p. 232). (Murphy and Wood also discuss a special class of for-hire carriers, *exempt carriers*, but in the context of freight, this classification applies almost exclusively to the shipment of bulk commodities like grain or gravel.)

The other way in which we distinguish carriers is by the transportation *modes* they utilize. This distinction relates more directly to our modeling efforts in that it very clearly lines up with the various cost and service characteristics discussed earlier. In addition, on any given mode, a carrier can offer various transport services, which we distinguish for some modes in our list given in the following:

- *Motor freight*: Typical services are truckload (TL), which we have discussed in our various examples earlier in the chapter, and LTL, which we mention briefly above and study extensively later in the chapter.

- *Railroads*: Shipment of freight by train (rail). Similar to motor carriers, railroads offer carload (CL) and less-than-carload (LCL) services. As one might suspect CL has a much higher capacity than TL ($6,269\,ft^3$ and over $200,000\,lb$ for a 50-ft, "hi-roof" boxcar—see CSX, 2011). In addition, rail carriers offer *unit train* services for shippers moving large quantities of goods sufficient to cover a single train (e.g., 115–150 cars for a unit coal train—see BNSF, 2011) that moves non-stop from origin to destination.

- *Airfreight*: Since "the great majority of airfreight is carried in the freight compartments of passenger airplanes" (Murphy and Wood, 2011: p. 220), large-quantity shipments or those that are bulky and/or of low density are typically not viable on this mode. For small, high-density shipments, though, this mode offers the advantage of fast transit time, but at a relatively high cost.

- *Water-borne freight*: A huge volume of international freight moves via ocean liner between the various deep-water ports of the world. Of course, the idea of "less-than-liner-load" is clearly nonsensical. Since an enormous volume of freight is required to fill the capacity of an ocean liner, ocean-going freight movements are typically arranged by various third-party consolidators (e.g., *freight forwarders*) that contract directly with the ocean shipping lines. For discrete goods (i.e., not bulk goods), ocean freight shipments are loaded into *shipping containers*, 20- or 40-ft steel boxes of standard dimensions.

This shipping innovation dates to the 1950s (Levinson, 2006) and has led to the international measure of shipping volume of "TEU," or 20-ft-equivalent units (see, e.g., BTS, 2011a). Of course, freight can move domestically via water as well, along various inland water-ways like rivers and canals, or from point-to-point along the coast (e.g., from New York, NY to Charleston, SC in the United States).

- *Pipeline*: Coyle et al. (2009: p. 415) call this mode the "hidden giant of the transportation modes." Obviously, however, goods using this mode must be either in the form of a gas, a liquid, or a relatively fine-grained solid that is capable of being mixed with a liquid such as water to create a "slurry" that can be separated at the pipeline destination.

- *Intermodal*: This entails some combination of the modes mentioned earlier. We include *parcel service* in this category—which applies only to smaller shipments, typically limited to 150 lb or less—since these carriers provide time-definite services that allow them to determine the best mode (typically a mix of air and motor freight) that ensures delivery by the promised time. Parcel carriers in the United Services include the U.S. Postal Service, United Parcel Service (UPS), and FedEx. Intermodal services for larger shipments may be offered by various carriers or logistics service providers, such as rail carriers or the types of freight forwarders mentioned in our discussion of ocean-going freight earlier.

Tables 4.1 and 4.2 show various comparisons, for the domestic United States, of the volume and value of shipments made on the various freight transportation modes.

Many authors build tabular comparisons of the ranked performance of freight transportation modes in terms of cost, speed, reliability, and risk of loss/damage, but we would suggest that there is some danger in trying to do this irrespective of the nature of the cargo or the transportation lane (i.e., origin-destination pair) on which the goods are traveling. Some comparisons are fairly

TABLE 4.1

Ton-Miles of Freight Shipped in the United States

	1980		1990		2000		2007	
Total (millions of ton-miles)	3,403,914		3,621,806		4,328,750		4,608,671	
Air	4,840	0.1%	10,420	0.3%	15,810	0.4%	15,142	0.3%
Truck	629,574	18.5%	848,643	23.4%	1,192,633	27.6%	1,317,061	28.6%
Railroad	932,000	27.4%	1,064,408	29.4%	1,546,319	35.7%	1,819,633	39.5%
Domestic water	921,835	27.1%	833,544	23.0%	645,799	14.9%	553,143	12.0%
Pipeline	915,666	26.9%	864,792	23.9%	928,189	21.4%	903,693	19.6%

Source: National Transportation Survey of the U.S. Bureau of Transportation Statistics, Table 1-50, http://www.bts.gov/publications/national_transportation_statistics/html/table_01_50.html

TABLE 4.2

U.S. Freight Activity by Mode

Mode of Transportation	Value of Goods Shipped (billion $)			% Change (1997–2007)
	1997	2002	2007	
TOTAL all modes	6944	8397	11,685	68.3
Single modes, total	5720	7049	9,539	66.8
Truck	4982	6235	8,336	67.3
For-hire truck	2901	3757	4,956	70.8
Private truck	2037	2445	3,380	66.0
Rail	320	311	436	36.5
Water	76	89	115	51.5
Air (includes truck and air)	229	265	252	10.1
Pipeline	113	149	400	252.1
Multiple modes, total	946	1079	1,867	97.4
Parcel, U.S. postal service or courier	856	988	1,562	82.5
Truck and rail	76	70	187	147.4
Truck and water	8	14	58	608.5
Rail and water	2	3	14	684.4
Other multiple modes	4	4	45	961.6
Other/unknown modes, total	279	269	279	0.2

Source: National Transportation Survey of the U.S. Bureau of Transportation Statistics, Table 1–58, http://www.bts.gov/publications/national_transportation_statistics/html/table_01_58.html

obvious, however: Air is clearly the fastest and most expensive mode, whereas water is the slowest and perhaps the least expensive, although pipeline could probably rival water in terms of cost. In terms of "in-country" surface modes of travel, rail is less expensive than truck, but also much slower—partially because of its lack of point-to-point flexibility—and also more likely to result in damage to, or loss of, the cargo due to the significant number of transfers of goods from train to train as a load makes its way to its destination via rail. In the sections that follow, we incorporate these comparative performance measures—e.g., speed and reliability—into the types of inventory decision-making models discussed in Chapter 3 and earlier in this chapter.

4.4 More General Models of Freight Rates

Up to this point, we have formally specified an annual transportation cost function, *TC*, only for truckload (TL) service. Our general statement of the *TC* function, however, indicated that, for a given item being shipped and stocked in inventory, TC is a function of Q, the quantity shipped and stocked. This was true in the case of TL service, since, as expression (4.6) given earlier indicates, TC(Q) is specified by the product of the freight charge, FC_{TL}, and the number of annual shipments, given by D/Q. Another way of looking at this is that the per-unit rate for goods shipped via TL is

$$r_{TL}(Q) = \frac{FC_{TL}}{Q} \tag{4.13}$$

which is paid for each of the D units shipped annually. Therefore, $D \cdot r_{TL}(Q)$ again gives us the same annual cost of TL transportation as given by expression (4.6).

In building a general model of transportation cost as a function of the shipment quantity, we will continue to focus on truck transit. The reasons for this are as follows: (1) Far and away, a larger value of goods moves via truck transport than any other freight transportation mode in the United States (Table 4.2). (2) Truck-based freight rates are the easiest to estimate (at least in the authors' opinion) from publicly available data, most likely because motor freight carriers can often be hired directly (i.e., not through intermediaries like freight forwarders). (3) The economics of truck transit make the full-load versus less-than-full-load comparison more interesting than it is for other modes.

The basic ideas of our results, however, apply to other transit modes as well. The essential tension for the decision maker is between using smaller shipments more frequently, but at a potentially higher per-unit shipping cost, versus making larger but less costly shipments less frequently. Clearly, this has implications for transportation and inventory costs, and potentially for customer service as well. Thus, as we develop the analysis to follow, we can see some of the connections implied by the discussion in the book up to this point: inventory, transportation, and customer service (and the impact on customer service of uncertainty in describing demand). In Chapter 5, we will incorporate another interconnection, specifically the network that specifies the locations in the supply chain linked by our transportation choices, and at which inventories are stored and/or converted until they move again toward their point of consumption.

From the standpoint of truck transport, this more-frequent/less-frequent decision tradeoff plays out between TL and LTL service. Speigel (2000) points out that it is often difficult to fill truckloads in practice when one considers the many factors driving small shipment sizes: an increasing number of stock-keeping-units driven by customers' demands for greater customization, lean philosophies and the resulting push to more frequent shipments, and customer service considerations that are driving more decentralized distribution networks serving fewer customers per distribution center (see Chapter 5). As we point out later in this chapter, such trends imply that a full and complete analysis would also consider mixed loads as a possible means of building larger, more economical shipments.

Thus, there is clearly a market for carriers that can efficiently move smaller loads of goods more frequently. A question that any carrier that wishes to compete in this segment of the freight transportation must answer is whether there is sufficient revenue to cover its cost. From our perspective in this chapter, that of the shipper, this issue of carrier costs is indeed relevant to understanding the prices that carriers will charge for their services. Along those lines, we observe that a Harvard Business School (1998) note on freight transportation begins with a discussion of the deregulation of the rail and truck freight transport markets in the United States and states that "... an unanticipated impact of regulatory reform was a significant consolidation in the rail and LTL [less-than-truckload] trucking sectors" (p. 3). This stemmed from the carriers in these markets seeking what the HBS note calls "economies of flow," consolidating fixed assets by acquiring weaker carriers to gain more service volume, against which they could then more extensively allocate the costs of managing those fixed assets. This provides interesting insights into the cost structures that define essentially any transport service, *line-haul costs*—the variable costs of moving the goods—versus *terminal/accessorial costs*—the fixed costs of owning or accessing the physical assets and facilities where shipments can be staged, consolidated, and routed.

Even without the benefit of detailed economic analysis, one can infer that modes like pipeline, water, and rail have relatively large terminal/accessorial costs; that air and truck via LTL have less substantial terminal costs; and that truckload (TL), as a point-to-point service, has essentially no terminal costs. Greater levels of fixed costs for the carrier create a stronger incentive for the carrier to offer per-shipment volume discounts on its services, due to its desire to achieve the "economies of flow" discussed earlier. Not only does this have a direct impact on the relative cost of these modes, but it also has implications for the relative lead times of freight transit modes, as shown in Figure 4.3,

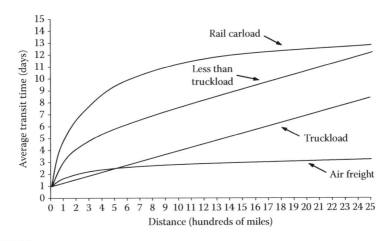

FIGURE 4.3
Comparison of shipping distance versus lead time for various modes. (Reproduced from Piercy, J.E., A performance profile of several transportation freight services, unpublished PhD dissertation, Case Western Reserve University, Cleveland, OH, 1977.)

reproduced (approximately) from Piercy (1977). In this case, more substantial terminal operations—e.g., rail yards to build trains or LTL terminals to build full trailer loads—result in more non-linearity in the distance-versus-transit time relationship. Note from the figure that for TL, this relationship is essentially linear since this mode does not require terminal operations to mix and consolidate loads.

4.5 Building A Rate Model: LTL Service

As we discussed earlier, the cost structure of TL service is exceedingly simple, since the cost of hiring the dedicated truck is given. Therefore, incorporating this cost into our decision model, whose objective function, expression (4.12), is stated in terms of decision variable Q, requires us to simply take the per-shipment freight cost for truckload service, FC_{TL}, and divide by Q. In general, though, for other modes and transport services, as we discussed earlier, it is likely that additional factors will come into play in determining the freight cost charged by the carrier.

Murphy and Wood (2011) note that the structure of transportation service charges actually stems from the earliest days of the freight transportation industry in the late 1800s, when a general and simple rate structure was desirable due to the manual nature of business. The result became the "class-based" rating system that still predominates today and is based on three broad factors: shipment weight, shipment distance, and product characteristics. It is this last factor, or set of factors, actually, that determines the product "class" and that in turn determines the rate structure for various shipment weights and distances. For motor freight in the United States, class-rating guidelines are maintained by the National Motor Freight Transportation Association (NMFTA), through its Commodity Classification Standards Board. NMFTA indicates that motor freight class is determined by the following product and/or shipment characteristics: density, stowability, handling, and liability (NMFTA, 2009). Of these characteristics, density is the most important in determining the rate class: NMFTA's documents state that, "It has been well established through numerous administrative decisions that, absent any unusual or significant stowability, handling or liability characteristics, density is of prime importance in the assignment of classes" (NMFTA, 2009, p. 1).*

---

* As it is with any rule, there are exceptions. In some cases, a carrier may offer a "freight-all-kinds," or FAK rate. Here, only shipment distance and shipment weight determine the freight rate.

Example 4.5: LTL Rate Tariff

Table 4.3 shows a rate tariff for LTL service from Oakland to Atlanta. This data comes from the CzarLite "benchmark rating" for LTL service, which is the basis upon which many LTL providers set their rates (SMC3, 2012). Rates in the table are stated in $/cwt, where cwt stands for "hundred-weight," or 100 lb, the historical basis for LTL service pricing. In the table, one can see that for a given class (row of the tariff table), the rate depends on a given *rate breakpoint* (column of the tariff table). The column headings in the table indicate the shipment weights at which the rate breakpoints occur, such as "L5C," indicating a shipment weight of less than 500 lb, "M5C," a shipment weight of more than 500 lb, "M1M," a shipment weight of more than 1000 lb, and so forth. Note also that Table 4.3 indicates the minimum charge for a shipment on this lane, in this case $203.59.

One important point about the LTL rate tariff in Table 4.3 that is not immediately evident from looking at the numbers is that a shipper could actually benefit from "over-declaring" a shipment, or simply shipping more in an attempt to reduce the overall shipping cost. This stems from the "tapering"

TABLE 4.3

LTL Rate Block for OAK-ATL Lane

Origin ZIP	94601	(CA)
Destination ZIP	30303	(GA)
Minimum Charge	$203.59	

				LTL rate ($/cwt)					
Class	L5C	M5C	M1M	M2M	M5M	M10M	M20M	M30M	M40M
500	676.23	547.80	466.51	392.20	341.01	279.63	245.50	194.22	181.46
400	540.99	438.24	373.21	313.76	272.81	223.70	196.40	155.37	145.17
300	405.74	328.67	279.91	235.32	204.61	167.77	147.30	116.53	108.88
250	338.12	273.89	233.25	196.10	170.50	139.81	122.75	97.11	90.73
200	270.50	219.11	186.60	156.88	136.40	111.85	98.21	77.69	72.59
175	236.68	191.73	163.28	137.27	119.36	97.87	85.92	67.98	63.51
150	202.87	164.34	139.95	117.66	102.30	83.89	73.65	58.27	54.44
125	169.06	136.95	116.63	98.05	85.26	69.91	61.37	48.55	45.37
110	148.77	120.51	102.64	86.28	75.02	61.51	54.01	42.73	39.92
100	135.24	109.56	93.30	78.44	68.20	55.92	49.10	38.84	36.29
92	126.46	102.44	87.24	73.34	63.77	52.29	45.91	36.32	33.94
85	117.32	95.03	80.93	68.04	59.16	48.51	42.59	33.69	31.48
77	107.65	87.20	74.26	62.43	54.29	44.51	39.08	30.92	28.89
70	100.03	81.03	69.01	58.02	50.45	41.37	36.32	28.73	26.84
65	94.09	76.22	64.91	54.57	47.45	38.91	34.15	27.02	25.25
60	89.17	72.23	61.52	51.72	44.97	36.88	32.38	25.62	23.93
55	84.26	68.26	58.13	48.87	42.49	34.84	30.60	24.20	22.62
50	79.27	64.21	54.68	45.97	39.97	32.78	28.78	22.77	21.27

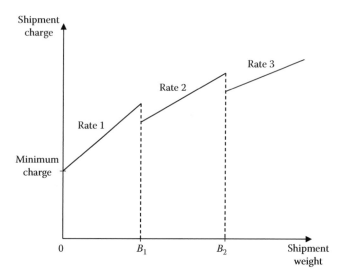

FIGURE 4.4
LTL rate structure as published in tariff.

structure of LTL rates. Figure 4.4 shows a pictorial view of the typical LTL rate structure as explicitly stated in a rate tariff like that of Table 4.3. Note that at the weights that correspond to rate breakpoints, the per-unit freight rate—i.e., the slope of the total shipment charge in Figure 4.4—decreases. Thus, a formal model of the per-unit rates for LTL shipments of arbitrary size requires one to account for *weight breaks,* which are used by carriers to determine the shipment weight at which the customer receives a lower rate, and which are typically *less* than the specific cutoff weight for the next lower rate in order to avoid creating incentives for a shipper to over-declare.*

Example 4.6: LTL Rate Weight Breaks

Using the Class 100 row of the LTL rate data in Table 4.3 for a shipment from Oakland to Atlanta, if one were to ship an 8500-lb (85 cwt) shipment on this lane, the charge without considering a possible weight break would be 85 cwt × $68.20/cwt = $5797. If, however, one were to add 1,500 lb of "dummy weight," this would result in a 10,000-lb shipment, at which point the next rate breakpoint would be reached, and the resulting freight charge would be only 100 cwt × $55.92/cwt = $5592, and therefore, in this case, shipping *more* would cost *less* if the carrier charged according to the rate tariff explicitly. From Figure 4.4, one can see that this situation applies to a range of shipment weights less than B_2. In order to discourage shippers from over-declaring or "over-shipping," however—and therefore leaving space for more goods on the carrier's trailer, thereby providing the carrier

---

* Looking ahead, this type of quantity discount structure also appears in Chapter 5, where we discuss techniques for representing this "all-unit-discount" cost structure in mixed-integer programming models.

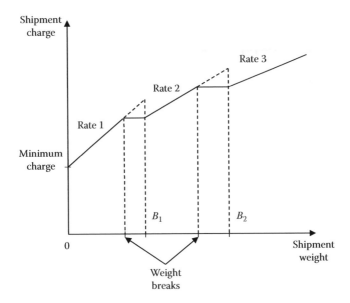

FIGURE 4.5
LTL rate structure as charged.

with an opportunity to generate more revenue—the carrier actually bills any shipment at or above 8199 lb (the *weight break*, or the weight at which the $68.20 rate results in a total charge of $5,592) as if it were 10,000 lb. Therefore, an 8500-lb shipment will actually result in a freight charge of only $5592. Thus, the *effective rate* for this 8500-lb shipment is $5592 ÷ 85 cwt = $65.79/cwt. Figure 4.5 shows a pictorial view of the rates *as actu-ally charged*, with all shipment weights between a given weight break and the next rate breakpoint (i.e., B_1 or B_2) receiving the same freight charge.

Let us formalize the ideas presented in Example 4.6 given earlier. To compute the LTL freight charge, FC_{LTL}, in a way that correctly recognizes the weight-break concept discussed earlier, we utilize a rate lookup structure similar to the one presented in Kay and Warsing (2009). Let i denote the shipment class—as specified by the rows of the LTL tariff table—and let j denote the index of rate breakpoints. Thus, we will map values of $j = \{1, 2, \ldots, 9\}$ to the column headings in the LTL rate tariff (i.e., L5C, M5C, etc.) Let B_j be the weight that corresponds to rate breakpoint j, such that $\{B_1, B_2, B_3, \ldots, B_9\} = \{0, 500, 1,000, \ldots, 40,000\}$.

We will consider the rate tariff table as a matrix $\mathbf{R}$, whose constituent values R_{ij} give the stated rate, in $/cwt, for a shipment of class i corresponding to rate breakpoint j. Let W be the shipment weight (lb), such that if w is the item weight (lb/unit) and Q is the number of units shipped, then $W = Q \cdot w$. Thus, given a shipment weight W, we determine the appropriate rate breakpoint to be

$$j = \arg\left\{B_j : B_j \le W < B_{j+1}\right\}. \tag{4.14}$$

(While the "more than" and "less than" aspects of the rate breakpoint labels in the columns of the tariff table may not match exactly the inequalities expressed in (4.14), one can see from Figure 4.5 that this will not really matter in the final analysis.) Having specified the breakpoint j, the freight charge for this shipment is given by

$$FC_{LTL}(W) = \max\left\{MC, \min\left\{R_{ij} \cdot W/100, R_{i,j+1} \cdot B_{j+1}/100\right\}\right\} \qquad (4.15)$$

where MC is the minimum charge for the rate tariff given by $\mathbf{R}$. This results in an effective rate of

$$r_{LTL}^{cwt}(W) = \frac{FC_{LTL}(W)}{W/100} = 100 \cdot \frac{FC_{LTL}(W)}{W} \qquad (4.16)$$

Example 4.7: Formalizing the LTL Freight Charge

Applying the expressions developed earlier to the data presented in Example 4.6, we see that for an 8,500-lb shipment, the appropriate rate breakpoint to consider in the LTL tariff given in Table 4.3 is $j = 5$, corresponding to the M5M rate, since $B_5 = 5,000$ lb and $B_6 = 10,000$ lb. Thus, $R_{ij} = R_{100,5} = \$68.20$/cwt and $R_{i,j+1} = R_{100,6} = \$55.92$/cwt. Using expression (4.15), we obtain

$$FC_{LTL}(8500) = \max\left\{203.59, \min\left\{68.20 \times 85, \; 55.92 \times 100\right\}\right\}$$

$$= \max\left\{203.59, \; 5592\right\}$$

$$= 5592$$

This results in an effective rate of $r_{LTL}^{cwt}(8500) = 100 \cdot \dfrac{FC_{LTL}(8500)}{8500} = \dfrac{5592}{85} =$ $\$65.79$/cwt, consistent with our computations in Example 4.6.

4.5.1 LTL Mode: Building the Inventory Decision Model

Having formalized the mathematics behind the LTL rate structure, resulting in expressions for r_{LTL}^{cwt} and FC_{LTL}, we need to determine how, and/or if, we can incorporate these mathematical expressions into our overall objective function, $TAC(R,Q)$, as given by expression (4.12) given earlier. Ultimately, we must express the rate function r_{LTL} in terms of the shipment quantity, and not the shipment weight. Before that, though, we need to give some consideration to the relative complexity of the FC_{LTL} function, expression (4.15), that is the basis of the r_{LTL}^{cwt} function. One aspect of this complexity is shown in Figure 4.6, which plots $r_{LTL}^{cwt}(W)$ versus W for the OAK-ATL rate tariff that

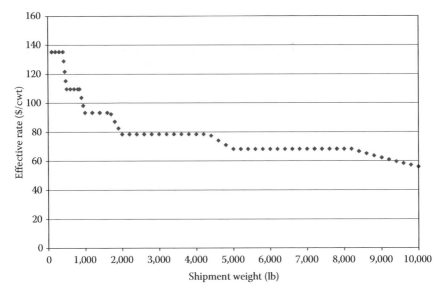

FIGURE 4.6
LTL effective rates on OAK-ATL lane.

is the subject of the previous examples. Note that, using the effective rate concept, which reflects the actual rate charged, results in a continuous function, but clearly not a smooth one that could be easily differentiated if it were included in our *TAC(R,Q)* objective function.

In the academic literature, several authors have studied decision models that incorporate the explicit, discrete rate-discount structures like those of the LTL rate tariff we have explored earlier. In general, this approach involves the use of quantity breaks to represent the shipment weights at which the unit cost changes. Tersine and Barman (1991) provide a review of some earlier versions of this modeling work. More recently, Mutlu and Çetinkaya (2010) extend previous work by Çetinkaya and Lee (2000) and Lee (1986), in which transportation cost is represented as a stepwise function by explicitly reflecting the lower per-unit rates for larger shipments.

In contrast, Swenseth and Godfrey (1996) review a series of models that estimate the freight rate via a smooth, continuous, non-linear function of the shipment weight, building off of earlier work by Ballou (1991), who estimated the accuracy of a linear approximation of trucking rates. A clear advantage of these types of rate models is that they do not require the explicit specification of rate breakpoints for varying shipment sizes, nor do they require any embedded functional complexities (as in expression (4.15) given earlier, for example) to determine if it is economical to increase—or over-declare—the shipping weight on a given route. A second advantage is that a smooth, continuous function is more parsimonious with respect to the specification of the underlying parameters and can therefore be more easily incorporated into optimization models.

Our approach follows that of Tyworth and Ruiz-Torres (2000), who propose a power-function form for the LTL rate, $r(W) = C_w W^b$, where C_w and b are constants found using non-linear regression analysis. A power function structure like this seems reasonable upon inspection of Figure 4.6, where the rate declines at a decreasing rate with the shipment weight—indicating that it is likely that $-1 < b < 0$. With the knowledge of the weight of the item shipped, w, such that $W = wQ$, is it straightforward to express this rate function as $r(Q) = CQ^b$ (in \$/unit), where C is the "quantity-adjusted" equivalent of C_w. Using this form, we can readily incorporate the rate function into an LTL-specific objective function, $TAC_{LTL}(R,Q)$, resulting in a relatively straightforward non-linear optimization problem, as we will demonstrate in the following.

First, however, we note that, although Tyworth and Ruiz-Torres (2000) utilize a non-linear curve-fitting process to build the rate function, we will consider a computationally simpler linear regression approach. Since we wish to build an expression of the form

$$r_{LTL}^{cwt}(W) = C_w W^b \tag{4.17}$$

we can restate this as

$$\ln\left[r_{LTL}^{cwt}(W) \right] = \ln C_w + b \ln W \tag{4.18}$$

and therefore, the parameters of the continuous function that approximates the stepwise nature of LTL rates can be specified by performing a simple linear regression analysis on the logarithms of the effective rate–shipment weight pairs that result from computations like those described earlier to find the effective rate.

Example 4.8: Power Function Estimation for LTL Rate

Figure 4.7 is a restatement of Figure 4.6 in log-log form, based on the effective rate computations for the OAK-ATL rate tariff. Before we discuss the regression estimate, we first note that the original, non-log plot (Figure 4.6) covered values of W up to only 10,000 lb, even though the tariff table gives rates for shipments up to and beyond 40,000 lb. This is not a trivial matter, since our rate estimate can change dramatically depending on the range we consider for W. Our choice of 10,000 lb for the upper limit of the plot (and therefore, of the regression estimate) is not arbitrary, however. It is consistent with the analysis presented in Kay and Warsing (2009), who cite a trucking industry study by Nagarajan et al. (2001) that indicates (p. 132) that the shipment of loads less than 150 lb via LTL is typically not cost-effective as compared to package express (e.g., UPS, FedEx), and that similarly, the shipment of loads exceeding 10,000 lb is typically more cost-effective via TL, leaving 150–10,000 lb as the valid range for LTL shipments.

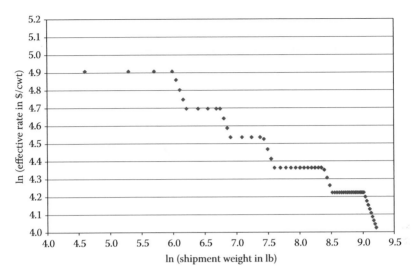

FIGURE 4.7
LTL effective rates on OAK-ATL lane, log-log scale.

Finding the linear regression coefficients displayed on the log-log plot in Figure 4.7 using Excel is straightforward, requiring only the addition of a "trendline" using the Excel "wizard" for this. Alternatively, one could apply the INTERCEPT and SLOPE functions in Excel to the $\ln\left[r_{LTL}^{cwt}(W)\right]$ (i.e., "y") and $\ln W$ (i.e., "x") data, or one could use any number of statistical software packages to perform the linear regression analysis. The resulting linear function is plotted and stated, in x–y terms, on Figure 4.8.

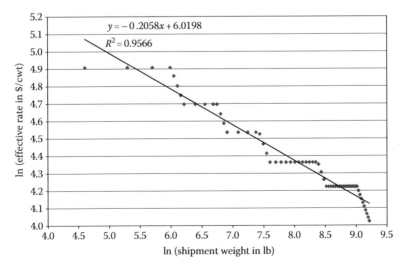

FIGURE 4.8
Linear regression for logarithm of LTL rates on OAK-ATL lane.

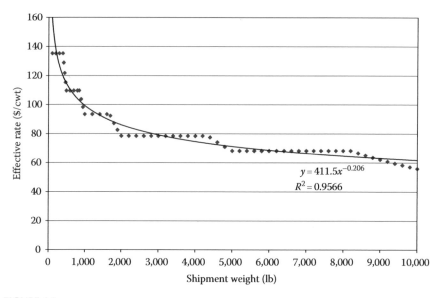

FIGURE 4.9
Resulting power function estimate of LTL rates on OAK-ATL lane.

Converting back to the form of expression (4.17) is also straightforward, requiring only that we compute $C_w = e^{\ln(C_w)}$, which in this case gives $C_w = e^{6.0198} = 411.50$. Thus, the LTL rate function for the OAK-ATL lane is $r_{LTL}^{cwt}(W) = 411.5\, W^{-0.2058}$. A plot of this function applied to the original data is shown in Figure 4.9.

This result gives the LTL rate as a function of shipment weight, and in terms of $/cwt. Our goal, however, is to model the LTL rate as a function of the quantity shipped, stated in $/unit. First, note that

$$r_{LTL}^{cwt}(W) = r_{LTL}^{cwt}(wQ)$$

$$= C_w \cdot (wQ)^b$$

$$= C_w w^b Q^b$$

To express the rate in $/unit, note that each unit equates to $w/100\,\mathrm{cwt}$, and therefore multiplying the rate in $/cwt by a factor of $w/100\,\mathrm{cwt/unit}$ yields a rate in $/unit, specifically,

$$r_{LTL}(wQ) = C_w w^b Q^b \cdot \frac{w}{100}$$

$$= \frac{C_w}{100} \cdot w^{b+1} Q^b \qquad (4.19)$$

Thus, for the OAK-ATL rate tariff, we can estimate the LTL rate by $r_{LTL}(wQ) = 4.115\, w^{0.7942} Q^{-0.2058}$ $/unit.

4.5.2 LTL Mode: Discount from Published Tariff

The challenge inherent in developing models of freight rates from pub-licly available data is that many, if not most, carrier–shipper transactions are governed by privately established contracts, which specify the actual prices charged for carrier services. The value of access to public data like the CzarLite tables available on-line, however, is that they often serve as the basis from which LTL rates are negotiated in carrier–shipper contracts. Commonly, a carrier will quote a shipper a discount from the nominal rates stated in a CzarLite-based rate tariff. Denoting this discount factor by δ_{LTL} ($0 \leq \delta_{LTL} \leq 1$—i.e., a discount of $(1 - \delta_{LTL}) \times 100\%$ off the published rate), the LTL rate estimation function can be stated as

$$r_{LTL}^{\delta}(wQ) = \frac{\delta_{LTL}C_w}{100} \cdot w^{b+1}Q^b \qquad (4.20)$$

where the superscript δ indicates that this is the discounted rate function. Finally, since a total of D units will be shipped across the year-long horizon for which we state our objective function TAC, the estimated annual trans-portation cost is $TC_{LTL} = D \cdot r_{LTL}^{\delta}(wQ) = D \cdot \frac{\delta_{LTL}C_w}{100} \cdot w^{b+1}Q^b$.

Thus, we can state an LTL-specific version of our total annual cost function for inventory decision making, namely

$$TAC_{LTL+ITC}(R,Q) = \frac{AD}{Q} + h \cdot \left(\frac{Q}{2} + R - \mu_{DLT}\right) + h_T \cdot D \cdot \frac{\mu_T}{Y} + D \cdot \frac{\delta_{LTL}C_w}{100} \cdot w^{b+1}Q^b \qquad (4.21)$$

where the subscript $LTL + ITC$ indicates that this cost function applies to LTL shipping where the decision maker also bears responsibility for goods in-transit.

Example 4.9: Comparing TL and LTL for OAK-ATL Lane

Using the problem data we introduced in Example 4.2, let us now con-sider LTL shipments from Oakland to Atlanta as an alternative to TL shipments. We do this merely as an exercise in demonstrating how the computations are structured, since we can immediately presume—from the fact that the $Q \leq Q_{max,TL}$ constraint in Example 4.3 was tight in the optimal solution to the TL optimization—that it is quite unlikely that LTL shipments will be economically justified. One issue to point out here is that, for this problem, we are assuming that the TL and LTL lead times are the same, but this might not always be the case. Since LTL shipments are likely to be shipped to one or more terminals to be mixed with other freight bound for the same destination, the lead time is likely to be a little longer than TL. In this case, however, an OAK-ATL shipment via TL will also incur a delay in order to adhere to government regulations on driver hours, requiring either a stop for the driver to rest, or for a change of drivers.

Additional information for formulating this problem is the LTL rate function $r_{LTL}(wQ) = 4.115\ w^{0.7942}Q^{-0.2058}$ \$/unit, from Example 4.8 given earlier, and $\delta_{LTL} = 1 - 0.462512 = 0.537488$, from Kay and Warsing (2009, described further in the section that follows), so that

$$r_{LTL}^{\delta}(wQ) = 2.2118\ w^{0.7942}Q^{-0.2058}$$

This expression, however, still requires a further adjustment to account for the change in prices for freight transportation services circa 2000, the presumed year of the on-line CzarLite tariff data, to those in 2010, the year upon which all of our examples are based. For LTL service, this process requires two steps. First, we use the producer price index (PPI) tables of the U.S. Bureau of Labor Statistics (BLS, 2011b), to obtain $PPI_{LTL,2010} = 126.8$, which is on a 2003-year basis (series PCU484122484122, with base date 2003-Dec). Then, using a different PPI series, on a 1992-year basis (BLS, 2011c: series PCU4841224841221, with a base date of 1992-June, and for which the 2003-Dec price index is $PPI_{LTL,2003\text{-Dec}}^{1992\text{-Jun}} = 169.3$), we can convert $PPI_{LTL,2010} = 147.5$ to the level of the 2003-Dec base date, yielding $PPI_{LTL,2010}^{2003\text{-Dec}} = 147.5/169.3 = 87.12$. Therefore, to bring 2000-level rates up to 2010, we must multiply the expression given earlier by $126.8/87.12 = 1.4555$.

And, as we discussed earlier, since a total of D units will be shipped across the year-long horizon for which we state our objective function *TAC*, the estimated annual transportation cost will be

$$TC_{LTL} = D \cdot r_{LTL}^{\delta}(wQ) = 3.2193\ D\ w^{0.7942}Q^{-0.2058}$$

Another way of looking at this is that each shipment of Q units will result in a charge of

$$FC_{LTL} = Q \cdot r_{LTL}^{\delta}(wQ)$$

and since D/Q shipments will be made annually, on average, the resulting annual transportation cost would be given by $TC_{LTL} = FC_{LTL} \cdot D/Q$, giving the same expression as given earlier.

Thus, assuming F.O.B. origin, freight collect terms, we solve the following problem for OAK-ATL shipments via LTL, specifically, minimize

$$TAC_{LTL+ITC}(R,Q) = \frac{AD}{Q} + h \cdot \left(\frac{Q}{2} + R - \mu_{DLT} \right)$$

$$+ h_T \cdot D \cdot \frac{\mu_T}{Y} + 3.2193 \ D \ w^{0.7942} Q^{-0.2058}$$

subject to

$$Q \le Q_{max,TL}$$

$$R \ge \mu_{DLT}$$

$$L \left(\frac{R - \mu_{DLT}}{\sigma_{DLT}} \right) - \frac{(1 - \beta_{tgt})Q}{\sigma_{DLT}} \le 0$$

Before we state the solution, note that the upper-bound constraint on Q is based on the size of a truckload, $Q_{max,TL}$. Actually, an LTL carrier would be unlikely to handle a shipment that either "cubes-out" or "weighs-out" the entire truck trailer, mostly because the shipper would clearly get a better rate from a TL carrier for this full-TL-sized shipment. This is, however, the logical upper bound on the size of an LTL shipment.

An Excel implementation of this optimization problem is shown in Figure 4.10. Note from the figure that various cells in the Excel sheet

	A	B	C	D	E	F	G	H	I	J
1	{Q,R} decision worksheet									
3	Lead Time		Symbol				Cost outcomes		Symbol	
4	Mean		mu_L	5.00 days			Ordering Cost		OC	$295
5	StDev		sigma_L	1.50 days			Site Holding Cost		HC	$7,432
6	Demand						In-transit Holding Cost		ITC	$660
7	Annual Demand		D	10950 units/year			Transportation Cost		TC	$42,322
8	Days per year		DPY	365 days			Total annual cost		TAC	$50,709 per year
9	Daily StDev		sigma_D	8.00 units/day						
10	Daily Mean		mu_D	30.00 units/day			Constraint terms			
11	Service target						Maximum shipment size		Q_max	2973 units
12	Fill rate		beta_TGT	99.00%			L((R - mu_DLT)/sigma_DLT)			0.39899
13	Decision variables						(1-beta_TGT) * Q / sigma_DLT			0.61394
14	Reorder Point		R	150.00 units						
15	Reorder Quantity		Q	2973.00 units			Other measures of interest			
16							Orders per Year		OPY	3.68 orders
17	Demand During Lead Time		Symbol				Safety Stock		SS	0.00 units
18	Mean		mu_DLT	150.00 units			Safety Stock Factor		z	0.000
19	StDev		sigma_DLT	48.43 units			Cycle Service Level		CSL	50.00%
20							Expected Units Short per Cycle		S(R)	19.321 units
21	Product and freight info						Fill Rate		fr	99.35%
22	Item cost		c	$20.00 per unit						
23	Fixed ordering cost		A	$80 per order			LTL rate parameters			
24	Holding cost factor		h	25% per $/year			Item weight		w_lb	10 lb / unit
25	In-transit holding factor		h_T	22% per $/year			LTL rate discount		disc	0.537488
26	Freight rate		FRPU	$3.87 per unit			LTL producer price index		PPI_adj	1.4555
27	Freight cost per shipment		FCPS	$11,490.73 per shipment			Rate function coefficient		C_cwt	411.4963
28							Rate function exponent		b	-0.2058

FIGURE 4.10
Excel set-up for LTL-based optimal inventory decision model.

have been set up to capture the appropriate parameters in the LTL rate function $r^\delta_{LTL}(wQ)$ (expressed as "FRPU" in the Excel sheet), stated in \$/unit, as in expression (4.20). Table 4.4 shows the solution to the problem*, as compared to the FTL-constrained solution from Example 4.2 (with $Q = Q_{max,TL}$) and the TL solution from Example 4.3, with Q free to take values $Q \le Q_{max,TL}$. Note that all three problems have the same optimal solution, $Q = Q_{max,TL}$ and $R = \mu_{DLT}$—i.e., ship full truckloads and carry no safety stock. Since the fill rate is a function of both Q and R, the transportation-cost-based incentive to ship infrequently (only 3.68 orders annually, on average) results in a sufficiently large value of Q to force R to its lower bound. For this (Q, R) solution, shipping full truckloads clearly dominates the LTL alternative in terms of the estimated annual transportation cost since the only cost that differs between the TL and LTL solutions is the estimated annual transportation cost.

As a final comment on this example, note that we clearly state that the transportation cost reported in Table 4.4 for this example is *estimated* transportation cost since it is derived from the rate estimating function $r^\delta_{LTL}(wQ)$ (specifically, $r^\delta_{LTL}(wQ) = 2.2118\, w^{0.7942}Q^{-0.2058}$, in this case). Ultimately, this is not problematic because our goal is to make a decision based on the relevant tradeoffs in costs, not to precisely state the costs that we expect to flow back to the accounting ledger.

Example 4.10: F.O.B. Destination, Freight Prepaid

As an extension and point of comparison with the previous example, let us now consider the case where our company has negotiated different freight terms for the shipments from our supplier in Oakland to our DC in Atlanta. Specifically, we assume F.O.B. destination, freight prepaid terms, which means that our company no longer owns the goods in transit (i.e., they are not our liability), nor do we pay the freight charges. Both of these are now the responsibility of our supplier.

For this problem, let us assume that all of the other parameters are the same as those originally given in Example 4.2. Again, we do this merely as an exercise in comparing outcomes because, as any good economist would tell us, there is no such thing as a "free lunch," meaning, in this case, that the costs of bearing the risk of damage or loss to goods in-transit and paying the freight bill would surely be at least partially passed along to us by the supplier in the form of a higher unit price for the item being shipped.

---

* It is instructive to point out the challenge of solving this problem, particularly in Excel Solver. In general, a mathematical program with a non-linear objective and non-linear constraints may be challenging to solve, and its solution is highly dependent on running the solution algorithm from a good starting solution. Indeed, solutions to the LTL version of this problem are no longer likely to be located on the constraint boundaries, and therefore solving this problem requires some experimentation with the "Options" button in Excel Solver in order to obtain what appears to be a reliable solution.

TABLE 4.4

Comparison of Optimization Results on OAK-ATL Lane

Lane/Item Cost	Freight Terms	Mode	Q	R	Orders per Year	Ordering Cost	Holding Cost	In-transit Holding Cost	(Estimated) Transportation Cost	TAC
Oakland-Atlanta	Origin—collect	FTL	2973	150	3.68	$295	$7432	$660	$19,958	$28,345
$c = \$20$	Origin—collect	TL	2973	150	3.68	$295	$7432	$660	$19,958	$28,345
	Origin—collect	LTL	2973	150	3.68	$295	$7432	$660	$42,322	$50,709

A	B	C	D	E	F	G	H	I	J
1	(Q,R) decision worksheet								
2									
3	*Lead Time*	*Symbol*				*Cost outcomes*		*Symbol*	
4	Mean	mu_L	5.00 Days			Ordering Cost		OC	$1411
5	StDev	sigma_L	1.50 Days			Holding Cost		HC	$1738
6	*Demand*					Total annual cost		TAC	$3148 per year
7	Annual Demand	D	10950 Units/year						
8	Days per year	DPY	365 Days			*Other measures of interest*			
9	Daily StDev	sigma_D	8.00 Units/day			Orders per Year		OPY	17.63 Orders
10	Daily Mean	mu_D	30.00 Units/day			Safety Stock		SS	37.00 Units
11	*Service target*					Safety Stock Factor		z	0.764
12	Fill rate	beta_TGT	99.00%			Cycle Service Level		CSL	77.76%
13	*Decision variables*					Expected Units Short per Cycle		S(R)	6.20 Units
14	Reorder Point	R	187 Units			Fill Rate		fr	99.00%
15	Reorder Quantity	Q	621 Units			Fill rate constraint RHS [(1-beta_TGT) / sigma_DLT]		0.00021	
16						Fill rate constraint LHS [L(z) / Q]		0.00021	
17	*Product parameters*								
18	Item value	c	$20 per unit			*For comparison*			
19	Fixed ordering cost	A	$80 per order			Economic order qty		EOQ	591.95 units
20	Holding cost factor	i	25% per $/year						

FIGURE 4.11
Excel set-up for optimal inventory decision model without transportation costs.

Since no transportation-related costs need to be incorporated into the *TAC* function for this formulation, the tabular Excel formulation of this problem is set up in exactly the same way as the optimization examples from Chapter 3. An Excel implementation of this problem is shown in Figure 4.11. The solution, found via Excel Solver, is $(Q,R) = (621,187)$. Let us discuss several aspects of this solution. First, since the only source of fixed ordering costs borne by the decision maker—i.e., our company, the downstream customer—is now the administrative, order-related costs modeled by $A = \$80$, the incentives for infrequent shipping have been reduced dramatically, such that the optimal solution reflects between 17 and 18 shipments per year, on average, as compared to between 3 and 4 when our company paid the per-shipment freight charges. Interestingly, note that this service-constrained optimal order quantity differs from the economic order quantity solution, $EOQ = 592$. The optimization utilizes a higher value of Q to increase the fill rate in a way that adds only $\$h/2$ in cost for each additional unit, as opposed to a full $\$h$ for each additional unit added to the safety stock.

Another important aspect of the solution to Example 4.10 above is what it says about the manner in which the negotiation of freight terms between customer and supplier might dramatically affect logistics-related decisions. Again, it is quite likely that the supplier, if forced to bear the logistics-related costs, would increase the price it charges our company, the customer, for the item. It is not clear, however, that the supplier would simply concede to our decision to ship point-to-point from its facility to ours 17–18 times per year, such that it must be substantially burdened by the full freight charge for each of these shipments. On the contrary, the supplier would probably identify transportation savings by creating a "milk-run" shipping route among several of its customers, including us. In this way, our supplier could create mixed truckloads of goods to be delivered to multiple destinations, such that it would still be economically viable for the supplier to ship our goods upwards of 15 times per year in quantities more in line with our desired $Q = 621$ units.

From the carrier's perspective, the type of multi-stop delivery problem implied by such a "milk-run" is a *vehicle routing problem* (VRP), a widely studied problem in operations research, which is itself a more complex and more constrained version of an even more widely studied problem, the *traveling salesman problem*. Given that we take the perspective in this chapter of a shipper that does not own a fleet of vehicles that must be allocated to shipment routes, the development of multi-stop vehicle routes is beyond the scope of this book. The reader is encouraged to consult some of the references on the VRP listed at the end of this chapter. From the perspective of a firm that is solely a shipper and does not own a fleet of vehicles—the perspective taken in this chapter—consolidating multiple shipments into mixed truckloads for multi-stop deliveries is likely to be a service that is outsourced to a logistics service provider—i.e., a third-party logistics (3PL) company. In that event, we can return to an analytical model more along the lines of what we have presented earlier, wherein we have a carrier pricing schedule for various shipment sizes, which fits into the various model structures presented earlier.

Let us now consider how our models may be used to evaluate shipping alternatives. In this case, we consider an alternative to the Oakland supplier, specifically a supplier that is much closer to our Atlanta DC, with commensurate differences in the item cost, freight charges, and transportation service parameters.

Example 4.11: NYC-ATL Alternative to OAK-ATL Shipments

Assume that we have identified an alternative to the OAK-ATL lane that served as the basis for all of the examples given earlier. Our alternative supplier is located in New York City, a distance of $d = 882$ mi from our DC in Atlanta.* This less distant supplier, however, charges an item price that is 10% higher than our Oakland customer, in this case $22/unit. A benefit of this alternative, however, is that the shorter distance results in a replenishment lead time with $\mu_L = 2$ days and standard deviation $\sigma_L = 0.67$ days. Accordingly, this changes the parameters for the distribution of demand over the replenishment lead time (DLT), and thus the level of safety stock required to meet the stated service target. Specifically, $\mu_{DLT} = \mu_D \mu_L = 60$ and $\sigma_{DLT} = \sqrt{\mu_L \sigma_D^2 + \mu_D^2 \sigma_L^2} = 22.99$, versus 150 and 48.43, respectively, for the OAK-ATL lane. In addition, the LTL carrier on this NYC-ATL lane is willing to provide a larger discount from the published rates than we received on the OAK-ATL lane, specifically a 55% discount, meaning that, in our specification of the LTL rate function next, we should use $\delta_{LTL} = 1 - 0.55 = 0.45$.

---

* Per Google Maps (www.maps.google.com).

For this alternative, we can now compare the same three options we considered for the OAK-ATL lane:

- Full truckload (FTL), with $FC_{TL} = d \cdot r_{TL,2010} = 882\,\text{mi} \times \$2.20/\text{mi} = \$1940.40$ and $Q = Q_{max,TL}$. Since nothing has changed regarding the item characteristics, we still have $Q_{max,TL} = 2973$ units.
- Truckload (TL) with, $FC_{TL} = \$1940.40$, as shown earlier, and $Q \leq Q_{max,TL}$
- Less-than-truckload (LTL), with $Q \leq Q_{max,TL}$ and a rate function $r_{LTL}^{\delta}(wQ) = PPI_{LTL,2010} \cdot \dfrac{\delta_{LTL} C_w}{100} \cdot w^{b+1} Q^b$ that we must specify.

To specify the LTL rate function for the NYC-ATL lane, we start with the CzarLite rate tariff block shown in Table 4.5. As we did earlier, we can use this data to compute and plot-effective rates (in \$/cwt) as a function of the shipment weight (in lb), as shown in Figure 4.12. Performing the log transformation on the data and running the linear regression

TABLE 4.5

LTL Rate Bock for NYC-ATL Lane

Origin ZIP	10001	(NY)
Destination ZIP	30303	(GA)
Minimum Charge	$107.86	

				LTL Rate ($/cwt)					
Class	L5C	M5C	M1M	M2M	M5M	M10M	M20M	M30M	M40M
500	388.20	360.99	295.03	240.69	188.06	133.38	66.97	66.97	66.97
400	311.32	289.51	236.60	193.02	150.81	108.10	54.23	54.23	54.23
300	234.45	218.02	178.17	145.35	113.57	82.30	41.50	41.50	41.50
250	196.33	182.56	149.20	121.72	95.11	68.88	34.61	34.61	34.61
200	157.57	146.52	119.75	97.69	76.33	55.99	28.16	28.16	28.16
175	137.87	128.21	104.78	85.48	66.79	49.02	24.57	24.57	24.57
150	118.81	110.49	90.29	73.66	57.56	42.56	21.42	21.42	21.42
125	99.11	92.17	75.32	61.45	48.02	35.86	18.13	18.13	18.13
110	87.68	81.54	66.63	54.36	42.48	33.80	17.38	17.38	17.38
100	81.96	76.21	62.29	50.81	39.70	33.28	16.93	13.16	12.02
92	76.88	71.49	58.43	47.66	37.24	31.22	16.48	12.81	11.70
85	71.79	66.76	54.57	44.51	34.78	29.15	16.03	12.46	11.38
77	67.34	62.63	51.18	41.76	32.63	27.35	15.58	12.11	11.07
70	63.53	59.08	48.29	39.39	30.78	25.80	14.98	11.65	10.64
65	60.93	56.66	46.31	37.78	29.24	24.74	14.83	11.53	10.53
60	57.75	53.71	43.89	35.81	27.70	23.45	14.68	11.41	10.42
55	55.15	51.29	41.91	34.19	26.47	22.39	14.53	11.30	10.31
50	52.54	48.87	39.93	32.57	25.24	21.34	14.38	11.18	10.21

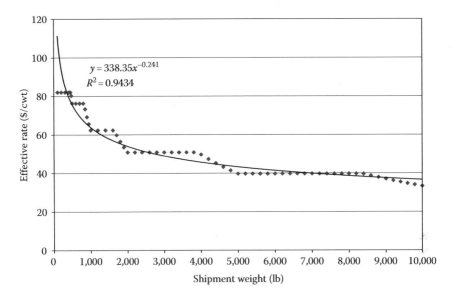

FIGURE 4.12
Power function estimate of LTL rates on NYC-ATL lane.

on this transformed data, we obtain the expression shown in Figure 4.12, $r_{LTL}^{cwt}(W) = 338.3451\ W^{-0.2415}$. Transforming this expression to a \$/unit basis, as in Example 4.9 given above, we obtain

$$r_{LTL}^{\delta}(wQ) = PPI_{LTL,2010} \cdot \frac{\delta_{LTL} C_w}{100} \cdot w^{b+1} Q^b$$

$$= 1.4555 \cdot \frac{(0.45)(338.3451)}{100}\ w^{0.7585} Q^{-0.2415}$$

$$= 2.2161\ w^{0.7585} Q^{-0.2415}$$

and therefore, the objective function for this LTL shipment decision is

$$TAC_{LTL+ITC}(R,Q) = \frac{AD}{Q} + h \cdot \left(\frac{Q}{2} + R - \mu_{DLT}\right) + h_T \cdot D \cdot \frac{\mu_T}{Y}$$

$$+\ 2.2161\ D\ w^{0.7585} Q^{-0.2415}$$

As we did for the OAK-ATL lane, we will assume that $\mu_T = \mu_L$. The constraints for this problem as the same as those for Example 4.3 (i.e., expressions (4.8 through 4.10)). The Excel Solver set-up for the problem is similar to that shown in Figure 4.10.

The solutions to the FTL, TL, and LTL problems on the NYC-ATL lane are given in Table 4.6. Interestingly, the best solution among these three

TABLE 4.6

Comparison of Optimization Results on NYC-ATL Lane

Lane/item cost	Freight Terms	Mode	Q	R	Orders per Year	Ordering Cost	Holding Cost	In-transit holding Cost	(Estimated) Transportation Cost	TAC
NYC-Atlanta	Origin—collect	FTL	2973	60	3.68	$295	$8176	$290	$7,147	$15,908
c = $22	Origin—collect	TL	2869	60	3.82	$305	$7889	$290	$7,407	$15,891
	Origin—collect	LTL	2369	60	4.62	$370	$6516	$290	$25,449	$32,625

is to ship the goods via TL, using a shipment quantity that is slightly less than FTL, 2869 units versus 2973. This inflates the expected annual transportation cost slightly, due to an average of 3.82 shipments per year, as opposed to the 3.68 that would result from full truckloads, but in doing so, cycle stock is reduced, and the average annual inventory holding cost is about $290 lower. In total, though, one can see that the total annual cost outcomes for FTL and TL are nearly the same, just on either side of $15,900, easily within the level of accuracy we could expect to achieve in estimating the various parameters framing the problem (e.g., the fixed ordering cost, A, or the annual holding cost, h). What is clear, however, is that shipping via LTL is not preferred here. Although the optimal LTL solution would generate a 17.4% savings in annual holding costs as compared to TL, it would require estimated annual transportation costs that are more than three times the TL costs.

The net result of comparing all three truck-based motor freight options— FTL, TL, and LTL—for the 10-lb item being shipped in Examples 4.9 and 4.11 is that it would be well worth the 10% premium in unit cost charged by the New York supplier to move our business to that company. Since our best outcome results from shipping this item in (effectively) full truckloads, the substantial savings in mileage by moving our business to the NYC supplier reduces annual transportation costs substantially, with only small increases in annual inventory holding costs—driven by the tradeoff between lower safety stock levels, but a higher cost of inventory per unit—as compared to the OAK-ATL shipments. In the example given below, we consider the effect of a more dramatic difference in item cost on the sourcing decision.

Example 4.12: OAK-ATL versus NYC-ATL Shipments, $200 Item Cost

In this example, all problem parameters remain the same as those specified in the series of previous examples, back to Example 4.2. The only difference is that the item cost in this case is much larger, $c = \$200$/unit. The solutions to all six possibilities—FTL, TL, or LTL on the OAK-ATL lane and those three modes on the NYC-ATL lane—are summarized in Table 4.7.

Note that, with this more expensive item, the cost of tying up money in inventory makes the option of more frequent shipments of smaller quantities the better approach, and among the six alternatives, shipping LTL on the NYC-ATL lane (and accepting the 10% price premium for the product, with $c_{NYC} = \$220$/unit) generates the lowest expected annual cost. Note also that, in this case, minimizing the number of shipments by shipping full truckloads of $Q = 2973$ units results in excessive inventory holding costs. Indeed, in the best alternative, LTL on the NYC-ATL lane, the optimal solution recommends a shipment frequency of more than twice per month (an average of 27.19 shipments annually), but at an annual estimated cost of transportation that best balances with this reduced cost of holding inventory.

TABLE 4.7

Comparison of Optimization Results for $200 Item

Lane/item Cost	Freight Terms	Mode	Q	R	Orders per Year	Ordering Cost	Holding Cost	In-transit Holding Cost	(Estimated) Transportation Cost	TAC
Oakland–Atlanta	Origin—collect	FTL	2973	150	3.68	$295	$74,325	$6600	$19,958	$101,177
$c = \$200$	Origin—collect	TL	1588	157	6.89	$551	$40,075	$6600	$37,354	$84,581
	Origin—collect	LTL	598	188	18.30	$1464	$16,859	$6600	$58,863	$83,786
NYC–Atlanta	Origin—collect	FTL	2973	60	3.68	$295	$81,757	$2904	$7,147	$92,103
$c = \$220$	Origin—collect	TL	915	60	11.96	$957	$25,177	$2904	$23,210	$52,247
	Origin—collect	LTL	397	73	27.60	$2208	$11,653	$2904	$32,804	$49,569

4.6 A More General Rate Model for LTL Service

At this point, it is worthwhile to discuss a modeling alternative for the LTL option that we have introduced in the preceding sections and examples. In the case where the inventory decision maker has access to specific information regarding the LTL rate tariffs being used by the carrier as its basis for pricing, and the discount applied to those rates, the method discussed earlier is a viable approach to building a smooth, continuous functional estimate of LTL rates to use in the extended objective functions we have presented for inventory decision making. If, however, one has not yet hired a specific LTL carrier, and wishes instead to evaluate whether LTL might be a viable alternative to other freight transportation services, the approach mentioned earlier is viable only if one is willing to gather the publicly available CzarLite tariff data and assume a particular discount level from this base rate data.

Moreover, we should be cognizant of the context in which one might apply this transportation cost analysis. For example, if one is attempting to design a distribution network, where the origin points are part of the decision (i.e., facility location decisions, like those considered in Chapter 5, to follow), in addition to specifying the various transportation lanes (i.e., the destination points allocated to each origin—forming various O-D pairs—based on the facility location solution), then the approach mentioned earlier for LTL rate estimation is rather challenging, and is also likely to be extensively data-intensive. In the process of solving the problem, one would need to either generate various rate tariff estimates in the solution process (i.e., selecting the correct tariff table for the given O-D pair, and then performing the estimation process described earlier), or perform all the preliminary work of employing the tariff tables for all possible O-D pairs to estimate the corresponding rate functions in advance (e.g., as in Example 4.8). In either case, this would be a data-intensive and computationally-intensive approach.

To avoid these kinds of modeling challenges, Kay and Warsing (2009) present a more general model of LTL rates, one that is not specific to a transportation lane or to the class of the items being shipped, with the item's density serving as a proxy for its class rating. This generality allows the rate model to be used, as discussed earlier, in the early stages of logistics network design, when location decisions are being made and when the most appropriate shipment size for each lane in the network is being determined.

In their model, Kay and Warsing (2009) use publicly available data—from the CzarLite tables discussed earlier, and from various government and industry sources—to develop an LTL rate model that is scaled to economic conditions, using the PPI for LTL services provided by the U.S. Bureau of Labor Statistics, similar to what was done earlier in the chapter in developing the TL-based decision model at the beginning of this chapter. The Kay–Warsing LTL rate model requires inputs of W_{ton}, the shipment weight (in tons); s, the density (in lb/ft^3) of the item being shipped; and d, the shipment distance (in mi). From a non-linear regression analysis on CzarLite tariff rates for 100 O-D pairs (randomly chosen from the set of all 5-digit ZIP code centroids in the continental United States), with weights ranging from 150 to 10,000 lb (corresponding to the midpoints between successive rate breakpoints in the LTL tariff, and limited by the valid LTL weight range discussed earlier), densities from approximately 0.5 to 50 lb/ft^3 (serving as proxies for class ratings 500–50, respectively), and distances from 37 to 3354 mi (based on reasonable distances for shipping within the continental United States), Kay and Warsing state a generalized LTL rate function, given by

$$r_{LTL}^{\text{ton-mi}}(W_{ton}, s, d) = PPI_{LTL}\left[\frac{\dfrac{s^2}{8}+14}{\left(W_{ton}^{\frac{1}{7}} \cdot d^{\frac{15}{29}} - \dfrac{7}{2}\right)(s^2 + 2s + 14)}\right] \qquad (4.22)$$

in \$/ton-mi,* where, as suggested earlier, PPI_{LTL} is the producer price index for LTL transportation, reported by the U.S. Bureau of Labor Statistics as the index for "General freight trucking, long-distance, LTL."[†] Kay and Warsing (2009) report a weighted average residual error of approximately 11.9% for this functional estimate of LTL rates as compared to actual CzarLite tariff rates. Moreover, since industry-wide revenues are used to scale the functional estimate, the model reflects the average discount provided by LTL carriers to their customers, δ_{LTL}, which Kay and Warsing estimate to be equal to 46.25%.

Thus, for a given lane, with d therefore fixed, the model given by expression (4.22) provides estimates of LTL rates for various shipment weights and item densities. Moreover, the functional estimate can easily be used to evaluate the comparative annual cost of sending truckloads of mixed goods with different weights and densities versus individual LTL shipments of these goods. On the other hand, for a given item, with a given density, expression (4.22) provides a means of analyzing the effect of varying the origin point to serve a given destination, possibly considering different shipment weights as well. These two analytical perspectives are, as we suggested earlier, important in the early stages of supply chain network design, the subject of Chapter 5.

---

* Technically, although this is not pointed out in Kay and Warsing (2009), the proper rate estimate is actually the maximum of the value generated by expression (4.22) and the minimum charge specified in the LTL rate tariff.
† As reported earlier in the chapter, this data is available from http://www.bls.gov/ppi/home. htm. In this case, the series ID is "PCU484122484122."

Example 4.13: Alternative Estimation for LTL Rate

Let us compare the LTL rate functions that result from the approach we built earlier, based on the tariff-specified Class 100 rates for the OAK-ATL lane, and the functional estimate from Kay and Warsing (2009). Based on assumptions from public data, Kay and Warsing (2009) computed the average density for a Class 100 shipment to be $s = 9.72 \, \text{lb/ft}^3$. As we indicated earlier, the distance from Oakland to Atlanta is $d = 2463 \, \text{mi}$. Since s and d are given, we can restate expression (4.22) as

$$r_{LTL}^{\text{ton-mi}}(W_{ton}, s, d) = PPI_{LTL} \left[\frac{G_1}{\left(G_2 \cdot W_{ton}^{\frac{1}{7}} - \frac{7}{2} \right) \cdot G_3} \right]$$

where $G_1 = s^2/8 + 14$, $G_2 = d^{15/29}$, and $G_3 = s^2 + 2s + 14$. For the PPI_{LTL} term, we need to estimate a value circa 2000, since we assume that the CzarLite tariff data is for year 2000 (given the naming convention found on the source website). Chaining together the PPI information from two BLS data series (PCU484122484122, with base date 2003-Dec, as reported earlier, and PCU4841224841221, which contains values back to 2000), we obtain $PPI_{LTL,2000} = 147.5/169.3 = 87.12$—see also Example 4.9. Thus, the resulting LTL rate estimation function is

$$r_{LTL}^{\text{ton-mi}}(W_{ton}|s, d, PPI_{LTL}) = 87.12 \left[\frac{25.8098}{\left(56.7814 \cdot W_{ton}^{\frac{1}{7}} - \frac{7}{2} \right) \cdot 127.9184} \right]$$

$$= \frac{17.5780}{56.7814 \cdot W_{ton}^{0.14286} - 3.5}$$

Recall that this function gives the (circa 2000) LTL rate in units of \$/ton-mi. Thus, we must multiply by 2463 mi, the length of the shipment lane for which we are estimating the LTL rate, to obtain the rate in \$/ton, and then by 1/20 ton/cwt to obtain the rate in \$/cwt, making the result consistent with the rate function we generate from the LTL rate tariff tables. Finally, since $W_{ton} = W/2000$, we obtain

$$r_{LTL}^{cwt}(W|s, d, PPI_{LTL}) = \frac{43294.6167}{19.1703 \cdot W^{0.14286} - 3.5}$$

A plot showing the resulting estimate along with the (circa 2000) tariff-based estimate (discounted by 46%) appears in Figure 4.13.

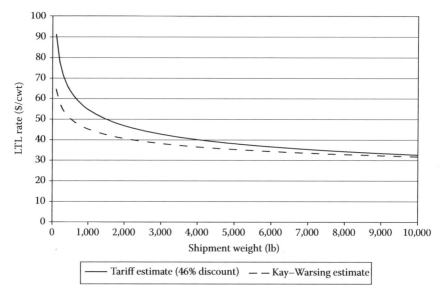

FIGURE 4.13
Comparing tariff-based LTL rate estimate to Kay–Warsing estimation function.

4.7 Beyond Truck Transport: Rail and Air Cargo

Much of the discussion in this chapter has revolved around a number of examples related to truck-based freight transportation. We close the discussion in the chapter, however, by introducing two additional, non-truck-based examples, introducing rail and airfreight as alternatives on the OAK-ATL and NYC-ATL lanes, respectively. In doing so, we extend our earlier discussion about the tradeoffs presented by various freight transportation alternatives.

Example 4.14: Rail Freight Alternative, OAK-ATL Lane

Let us return to the scenario of Example 4.9, where we must determine the best choice among several alternatives for shipping a $20, 10-lb item from our supplier in Oakland, CA to our DC in Atlanta, GA. For truck-based freight, TL was the better option over LTL. We now consider rail as another alternative. Specifically, we consider shipping via the rail equivalent of TL, in this case referred to as *rail carload*, or CL, in which the rail carrier allocates an entire boxcar to our shipment. The possible advantage of this is that the capacity of a rail boxcar is significantly larger than that of a truck trailer, specifically (see CSX, 2012), $K_{wt}^{rail-CL} = 200,000\,lb$ and $K_{cu}^{rail-CL} = 5015\,ft^3$ (80% of the full capacity of a 50-ft, "hi-roof" boxcar listed at CSX, 2012). Using the CL-equivalent of expression (4.2) given earlier, this results in a looser constraint on Q, namely $Q \leq Q_{max,CL} = 4875$ units.

Again, we note that this example is intended to draw out the tradeoffs in choosing a freight transportation alternative. For example, one complication related to assuming rail boxcar shipments, or any rail shipment for that matter, is that we must also address the cost of moving the goods from the dock door of the origin site to the rail yard where the goods are loaded into the railcar that eventually is coupled into a train of many cars. This, of course, is not a complication for truck-based shipments since, in the continental United States, the road network allows trucks to access essentially any point. Rail lines, though, directly link to only a few points (e.g., automotive assembly plants that have rail sidings on site). We can assume for the purposes of this example, however, that the dock-to-rail yard transportation is included in the rail carrier's freight charge. Another possible complication to consider is the additional handling implied by boxcar shipments—i.e., packing the goods into a shuttle truck at the shipper's dock, then unpacking this truck at the rail yard, and repacking the goods into the rail boxcar— which might lead to a decision to instead use a different rail option like trailer-on-flatcar (TOFC) or container-on-flatcar (COFC), whereby the rail carrier does not handle the product but merely lifts the shipper-loaded truck trailer or container onto a railcar built to handle two truck trailers, or four double-stacked containers (see, e.g., Coyle et al., 2006: Fig. 7.1, p. 212).

Let us assume, however, that in spite of these complications, we still choose to use rail CL as our alternative. Moreover, assume that the rail carrier charges $FC_{CL} = \$6500$ per CL shipment. Note that this charge is higher than that for TL on a per-shipment basis, but for FTL versus a full rail carload on the OAK-ATL lane, the comparative rate is \$0.148/ ton-mi for TL versus \$0.108/ton-mi for rail CL. As we discussed earlier in the chapter, however, rail is slower than truck and there is clearly a cost-versus-service tradeoff. In this case, assume that $\mu_L = \mu_T = 8$ days and $\sigma_L = 2.4$ days.

The Excel set-up to generate a Solver-driven solution is similar to the set-up presented in Figure 4.1 from Example 4.3, since both cases (TL and CL), involve a fixed cost per shipment (although in this case it does not stem from a per-mile charge as it did for our TL example). The resulting optimal decision for rail CL is shown in Table 4.8, where the rail option has been appended to the previous truck-based results for the OAK-ATL lane. Note that the rail solution pushes further toward taking advantage of the strong economic incentive to avoid the large per-shipment charge, and thereby results in an optimal solution to ship full boxcars, with $Q_{CL}^{*} = 4{,}875$ units and $R_{CL}^{*} = \mu_{DLT} = 240$ units. Overall, this results in an expected total annual cost of \$28,023, which is slightly less than the *TAC* value that resulted from the FTL option. This reduction in *TAC*, however, is driven by a shift to substantially higher inventory holding costs to balance the high, per-shipment freight charge, yielding only 2.25 shipments per year, on average. In the final analysis, if overall inventory levels are constrained by available space at the decision maker's site, this solution might not be viable, and only a small increase in the required number of annual shipments would clearly push the decision back in favor of FTL.

TABLE 4.8

Comparing Rail to Motor Freight Results on OAK-ATL Lane

Lane/item Cost	Freight Terms	Mode	Q	R	Orders per Year	Ordering Cost	Holding Cost	In-transit Holding Cost	(Estimated) Transportation Cost	TAC
Oakland-Atlanta	Origin—collect	FTL	2973	150	3.68	$295	$7,432	$660	$19,958	$28,345
$c = 20	Origin—collect	TL	2973	150	3.68	$295	$7,432	$660	$19,958	$28,345
	Origin—collect	LTL	2973	150	3.68	$295	$7,432	$660	$42,322	$50,709
	Origin—collect	Rail	4875	240	2.25	$180	$12,187	$1,056	$14,600	$28,023

Example 4.15: Airfreight Alternative, NYC-ATL Lane

As a final example, let us compare the case from Example 4.12 (shipments on the NYC-ATL lane of a 10-lb item with a per-unit cost of $220), where LTL shipments were preferred, with the alternative to ship goods via air-freight. Contrary to the rail example, using airfreight results in a capacity constraint that is much tighter than the truck trailer constraint. Similar to rail, however, airfreight shipments must first be delivered to the air cargo operator's site, which is typically located on-site at an airport. Again, we will assume that the stated freight charge is inclusive of any costs required to move the goods from the shipper's dock to the air cargo operator's site.

The primary constraint on shipping goods via air cargo, most of which, as we stated earlier, is shipped in the cargo hold of commercial flights carrying passengers (and their luggage), is the limited size of this cargo hold. Various sizes and shapes of airfreight containers accommodate this limited space. For this example, we will assume that the carrier uses a relatively large container, specifically an "LD-8" size container (United Cargo, 2012). This container has a weight capacity of $K_{wt}^{air,\,LD-8} = 5400\,lb$ and a volume capacity of only $K_{cu}^{air,\,LD-8} = 195\,ft^3$ (approximately 80% of the full capacity listed at United Cargo, 2012). Given the dramatic reduction in capacity, let us consider a minimum two-container shipment, meaning that we impose the constraint $Q \leq Q_{max,air\text{-}2\times LD8} = 379$ units. We will assume a freight charge for this two-container shipment of $1100, which results in a rate of $0.658/ton-mi for a full 379-unit, two-container load. Note that, as one would suspect, this represents a substantial price premium over the surface modes of truck and rail that we considered in our previous examples. As another point of comparison, we will also consider a three-container shipment with a freight charge of $1500, which equates to $0.581/t-mi for a full, 585-unit shipment (i.e., $Q \leq Q_{max,air\text{-}3\times LD8} = 585$ units).

Finally, we will assume that $\mu_L = \mu_T = 2$ days and $\sigma_L = 0.67$ day for this shipment. These are the same values that we assumed for truck transit between NYC and ATL, but given the transfer time at the origin (shipper-to-air cargo site) and destination (air cargo site-to-consignee), there is no reason to assume that airfreight will be faster than truck transit for this relatively short transportation distance. Effectively, we are claiming that a minimum of 2 days lead time would be the most reasonable expectation for any shipment of freight. Faster door-to-door transit times would require the use of a package express service such as UPS or FedEx, but those modes are typically restricted by the package express carriers to shipments of 150 lb or less.

Again, the Excel set-up to generate a Solver-driven solution is similar to the set-up presented in Figure 4.1 from Example 4.3, since, as for TL and CL, this airfreight-based problem once again involves a fixed cost per shipment. The resulting optimal decisions for the airfreight options are shown in Table 4.9, where these options (two-container [2C] and three-container [3C] shipments) have been appended to the previous truck-based results for the NYC-ATL lane. Interestingly, the best outcome in this case is to utilize the relatively expensive (on a ton-mi rate basis) airfreight option, and in fact, to use the smaller shipment size, the two-container load of 379 units. One might question why this happens—i.e., why it is better to ship less when there is a quantity discount on the airfreight charge. What is

TABLE 4.9

Comparing Air Freight to Motor Freight Results on NYC-ATL Lane

Lane/item Cost	Freight Terms	Mode	Q	R	Orders per Year	Ordering Cost	Holding Cost	In-transit Holding Cost	(Estimated) Transportation Cost	TAC
NYC-Atlanta	Origin—collect	FTL	2973	60	3.68	$295	$81,757	$2904	$7,147	$92,103
$c = \$220$	Origin—collect	TL	915	60	11.96	$957	$25,177	$2904	$23,210	$52,247
	Origin—collect	LTL	397	73	27.60	$2208	$11,653	$2904	$32,804	$49,569
	Origin—collect	Air (2C)	379	74	28.89	$2311	$11,199	$2904	$31,781	$48,195
	Origin—collect	Air (3C)	569	68	19.24	$1540	$16,092	$2904	$28,866	$49,402

happening is that the smaller shipment size for the two-container option significantly reduces annual inventory holding cost at the customer site, by almost \$5000, while increasing annual transportation costs by less than \$3000 over the three-container option. Note that the three-container charge is a fixed charge, not a per-container charge, and therefore it creates an incentive to ship more, which results in higher holding costs.

Moreover, comparing airfreight with the LTL solution in Table 4.9 reveals that the economic incentive to reduce holding cost is strong enough for this example to force the LTL solution to a point where the small shipment size results in a freight rate that is actually higher than that of airfreight. Using the LTL-tariff-based rate expression from Example 4.11, $r_{LTL}^{\delta}(wQ) = 2.2161\, w^{0.7585} Q^{-0.2415}$ \$/unit, a shipment size of $Q_{LTL}^{*} = 397$ units results in a rate of approximately \$3.00/unit, which can be converted to an equivalent rate of \$0.679/ton-mi (converting 10 lb/unit to tons and accounting for the 882-mi distance), and which furthermore explains why airfreight wins out for this shipment lane with the cost parameters given in this example.

4.8 Summary and Further Readings

4.8.1 Summary

This chapter has built on the discussion of the single-site, single-item inventory model from Chapter 3. In this chapter, we have extended the underlying idea of this model, the total annual inventory-related cost (i.e., *TAC*), in an important way. Specifically, we have considered a number of models for annual transportation cost that can be appended to the *TAC* expression in the inventory decision setting. In this way, we have tied together two important aspects of supply chain decision making, inventory and transportation management. We have developed this model in a way that recognizes the important issue of customer-supplier negotiations regarding freight terms, specifically as it regards the issue of which party hires and pays the freight carrier and which party bears the risks related to goods in-transit. Finally, while much of the discussion and the modeling efforts in this chapter revolved around truck-based freight transportation (i.e., motor freight), we have also discussed the characteristics of other modes and presented examples that assess other modes in comparison to motor freight with respect to their impact on *TAC*.

4.8.2 Further Readings

In the interest of brevity, our discussion in this chapter has essentially glossed over the issue of building explicit models of lead time based on the characteristics of the transportation mode. One issue that would surely arise in doing so is the non-normality of actual lead times. Thus, the kinds of assumptions that we made in Chapter 3 regarding the joint distribution of demand and lead time are often not valid in practice. A good resource that lays out the issues regarding non-normal lead time distributions, presents actual data on lead time, and tests the robustness of the normal assumption in a model that also incorporates transportation costs is the paper by Tyworth and O'Neill (1997). These authors also provide a host of references related to the problem of modeling lead-time demand (e.g., the widely cited paper by Bagchi et al., 1984) or avoiding the problem of modeling by utilizing statistical techniques like bootstrapping (Bookbinder and Lordahl, 1989).

Something that we pointed out clearly in this chapter is that our discussion of transportation decision making has taken the shipper's perspective. Doing so meant that we did not discuss the one problem that may well rise first to the mind of anyone trained in management science, operations research, or industrial engineering when they hear the words "transportation" and "optimization" in the same sentence, namely the vehicle routing problem (VRP). Readers who fall into that group and still hunger for more information about this rich and widely studied problem can consult a host of good resources for learning more. Our suggestion is to start with Ballou's (2004) chapter on "Transport Decisions," which offers an excellent overview of VRP, presented using a realistic approach to posing and solving the problem using various heuristics. Among these heuristics are methods for visually generating good starting solutions by grouping delivery points into clusters. Another excellent resource for understanding the basic ideas of vehicle routing in the context of a real problem can be found in Yano et al. (1987). This article documents the formulation and solution of a comprehensive VRP from practice, one that includes both deliveries and pick-ups, delivery time windows, constraints on driver time on the road, and the possibility of using common carriers to supplement private fleet routes. For a more technical treatment of the VRP, the interested reader should consult the book by Simchi-Levi et al. (2005), which contains a comprehensive overview of the problem and various solution techniques. This text also contains helpful background discussion on the more abstract problem underlying the VRP, the traveling salesman problem.

Exercises

4.1 Based on the discussion in this chapter, what are the three shipment characteristics that jointly determine the rate charged for freight transportation?

4.2 What product and/or shipment characteristics would make shipping that product via a truckload service economically more desirable than shipping via less-than-truckload, and vice versa?

4.3 Discuss how you would decide between a freight transit mode that is faster (i.e., lower mean transit time), but less reliable (i.e., higher transit time variance), and a mode that is slower but more reliable? What aspects of the joint transportation-inventory models discussed in this chapter are affected by the mode choice and how could you use that knowledge in building a means of comparing these two hypothetical modes?

4.4 What inventory- and transportation-related aspects of the supplier-customer transaction are determined by the freight terms? What are the most common freight terms, based on the discussion in this chapter? Choose two specific sets of freight terms and explain how they differ in terms of their effects on the supplier and customer.

4.5 Based on the model presented in this chapter, is in-transit inventory holding cost affected by the inventory policy parameters (i.e., order quantity or reorder point)? Why or why not? How does in-transit holding cost affect the transportation decision?

4.6 Given the LTL rate tariff in Table 4.10, compute the effective rate of a 3000-lb shipment and a 4200-lb shipment. Do the effective rates differ? Why or why not?

4.7 Using the Excel linear regression functions, develop a power curve expression ($y = Cx^b$) to estimate the effective rate, in \$/cwt, for a Class 85 shipment from Cleveland, OH (Hopkins International Airport area, ZIP code 44135) to Research Triangle Park, NC (ZIP code 27709). Use the CzarLite web site cited in this chapter (http://smc3apps.smc3.com/Applications/WebCzarLite/Entry.asp) to obtain the rate data and enter it into Excel. Assume that your carrier has agreed to provide a discount of 46% from these rates (including the minimum charge), approximately the same as the estimated national average discount for LTL reported

TABLE 4.10

Rate Tariff for Exercise 4.6

Weight Group	Minimum Weight (lb)	Freight Rate (\$/cwt)
1 (L5C)	1	30.60
2 (M5C)	500	28.15
3 (M1M)	1,000	22.03
4 (M2M)	2,000	18.05
5 (M5M)	5,000	14.21
6 (M10M)	10,000	11.85
7 (M20M)	20,000	6.90
8 (M30M)	30,000	5.31
9 (M40M)	40,000	4.87

Minimum charge = \$83.24.

by Kay and Warsing (2009). What are the actual and estimated effective rates for a 8500-lb shipment on the Cleveland-RTP lane?

4.8 Compute weight breaks for the M5C (500-lb) and M5M (5000-lb) weight groups for the CLE-RTP lane using the rate data from Exercise 4.7.

4.9 According to the Kay and Warsing (2009) paper cited in this chapter, the average density of a Class 85 shipment is $12.72\,\text{lb/ft}^3$. Using this value and the r_{LTL} (s,q,d) expression of Kay and Warsing (2009) presented in the chapter, compute the estimated freight rate in \$/ton-mi, and in \$/cwt, for a 8500-lb shipment on the CLE-RTP lane from Exercise 4.7, given earlier, for the most current PPI for LTL trucking. (See http://www.bls.gov/ppi/data.htm for PPI data, as stated in the chapter. Also, you should use an Internet-based mapping service, such as Google Maps, to find the shipment distance.) How does your answer compare to the answer in Exercise 4.7? Explain the possible sources of the similarities and the differences between the two rate functions.

4.10 Let the power curve expression you developed in Exercise 4.7 given above be denoted by $r_0(x) = C_0 x^{b_0}$, where r_0 is the estimated effective rate in \$/cwt and x is the shipment weight in pounds (lb). Following the steps listed in the chapter, convert this expression to an equivalent (undiscounted) function, $r_1(wQ) = C_1 Q^{b_1}$, where r_1 is the estimated effective rate in \$/unit, Q is the number of units shipped, and w is the per-unit weight in pounds. Explain how these steps ultimately convert \$/cwt to \$/unit.

4.11 *Mini-case study*: As a newly hired supply chain analyst for Milo's Home Improvement, you have been asked to assess inventory and shipping policies at the company's distribution center (DC) in the Atlanta, GA area. Specifically, you have been asked to choose a replenishment option for a particular stock-keeping unit (SKU), a front brush guard, which is an optional accessory for a line of lawn tractors that are also sold by Milo's. Important product- and operations-related data appear in Table 4.11. Your manager, Goober Pyle Jr., the VP of Logistics, is still suspicious of whether all that "fancy college book learnin'" justifies the big salary that he is paying you.

TABLE 4.11

Operations and Product Data for Exercise 4.11

Cost to receive an order in the DC	\$25
Annual on-site holding cost (% of item value)	30%
Annual in-transit holding cost (% of item value)	25%
Item weight	12 lb
Item cost (delivered, TL)	\$35
Projected annual demand	26,000 units
Operating days per year	365

The supplier of the front brush guard has offered two options for the freight terms and price (i.e., cost to Milo's) for this item:

- *Option 1*: F.O.B. destination, freight prepaid terms, provided that the supplier can use its preferred truckload (TL) carrier to make full truckload shipments of 3200 units, on whatever schedule you require—i.e., shipping any day of your 365-day operating year. The order fulfillment lead time is 3 days, 1 day of which covers transit time from the supplier's facility in Memphis, TN. The TL carrier charges the supplier $2.25/mi, or $882 for the Memphis-to-Atlanta shipment.

- *Option 2*: F.O.B. origin, freight collect terms, with a 1.5% discount in the price (i.e., unit cost to Milo's) of the guards. In considering this option, you intend to utilize the same TL carrier that your supplier has been using. The total order fulfillment lead time for this option is the same as that for Option 1, 3 days in total, including a commitment by the TL carrier to the same 1-day transit time. Since you cannot guarantee the level of additional shipment volume that your supplier can, however, the carrier has quoted you a TL rate that is 10% higher than the rate the supplier was paying.

    a. Explain what you believe to be the relevant costs in choosing between these options.

    b. Compute the total relevant annual cost of Option 1, clearly stating each separate component of the total.

    c. For Option 2, what order quantity do you recommend? Why? Compute the total relevant annual cost for this order quantity, clearly stating each separate component of the total.

    d. Which option do you prefer? Why?

    e. Your boss Goober believes that you might be better off to hire the LTL carrier where his cousin Gomer works the night shift. A rate tariff for this carrier's LTL service on this origin-destination lane is given in Table 4.12. For your recommended order quantity from (c), how would the annual transportation cost for LTL service compare to the options given earlier? What other information would you need in order to assess this option completely versus Options 1 and 2?

4.12 *Mini-case study*: You are the new operations manager for the Northeast regional distribution center (RDC) of office supply retailer StapleMax. The distribution center (DC) you mange is located in Carlisle, PA. You have been asked by your new boss, the VP of Logistics, to evaluate inventory replenishment options for a set of stock-keeping units (SKUs)

TABLE 4.12

Rate Tariff for Lane from ZIP 38138 (TN) to ZIP 30060 (GA)

Weight Group	Minimum Weight (lb)	Freight Rate ($/cwt)
1 (L5C)	1	36.59
2 (M5C)	500	33.66
3 (M1M)	1,000	26.34
4 (M2M)	2,000	21.58
5 (M5M)	5,000	16.98
6 (M10M)	10,000	14.16
7 (M20M)	20,000	7.21
8 (M30M)	30,000	5.54
9 (M40M)	40,000	5.09

Minimum charge = $80.33.

supplied by a particular technology products wholesaler. These items are shipped to the StapleMax DC via a series of intermodal transfers. The goods are all manufactured in Asia and arrive in the United States at the port of Long Beach, CA, where they are moved, still container-ized, via rail to the "inland port" of Kansas City, KS, where the whole-saler's break-bulk center is located. The goods are ultimately moved from the wholesaler's facility in Kansas City to StapleMax's DC in Carlisle via truck.

StapleMax's freight terms with the wholesaler are "DDU"*—which means F.O.B. destination (i.e., Carlisle), with StapleMax paying the freight bill for the truck shipments from Kansas City to Carlisle. Currently, the four items supplied by the wholesaler all ship indepen-dently from Kansas City to the Carlisle DC, some via TL and some via LTL. Important logistics-related data and product-related data appear in Tables 4.13 and 4.14.

TABLE 4.13

Logistics Data for Exercise 4.12

Cost to receive an order in the DC	$20/item/shipment
Annual on-site holding cost (% of item value)	35%
Shipment distance	1008 miles
Truckload shipment rate	$2.00/mile
Truckload weight capacity	25 ton
Truckload volume capacity	3200 ft³
Operating days per year	365 days

---

* This is terminology from INCOTERMS. As stated in the chapter, see Coyle et al. (2006) for more information.

TABLE 4.14

Product Data for Exercise 4.12

Item	Annual Demand	Item Cost	Item Weight (lb)	Item Volume (ft³)	Shipment Size	Shipment Mode	Freight Charge
Slimline PC	4000 units	$400/unit	19	1.25	72 units	LTL	$445
Laser printer	2600 units	$160/unit	34	5.19	435 units	TL	$2016
Inkjet printer	6000 units	$60/unit	16	3.13	1022 units	TL	$2016
Wireless mouse (case of 10)	1200 cases	$150/case	10	0.67	48 cases	LTL	$186

a. Given the current replenishment policy information, what is the inventory cycle length, in days, for each of the four items? (See Chapter 3, if necessary, for more information on computing the inventory cycle length.)

b. Given the problem data and your answers to part (a) given earlier, which items do you believe to be good candidates to consider shipping jointly via truckload from Kansas City to Carlisle? Why?

c. For the items you chose in part (b) given earlier, compute a "trial" joint-shipment solution. Your solution should specify the shipment quantity (in units or cases, as appropriate) for each item in the joint solution, and the overall total annual cost of this joint shipment. (See Chapter 3 for more information on setting up a total annual cost function for a two-item, joint shipment solution.)

d. Do you believe that your solution to part (c) given earlier is an optimal joint-shipment solution? If yes, explain what makes it so. If no, explain the steps you would need to take in order to generate the optimal joint-shipment solution for these items.

References

Bagchi, U., J. Hayya, and K. Ord. 1984., Concepts, theory and techniques: Modeling demand during lead time. *Decision Sciences.* 15: 157–176.

Ballou, R. H. 1991. The accuracy in estimating truck class rates for logistical planning, *Transportation Research—Part A.* 25A(6): 327–337.

Ballou, R. H. 2004. *Business Logistics/Supply Chain Management*, 5th edn. Upper Saddle River, NJ: Pearson Prentice Hall.

BLS. 2011a. "General freight trucking, long-distance, TL," Series PCU484121484121, U.S. Bureau of Labor Statistics, http://www.bls.gov/ppi/data.htm

BLS. 2011b. "General freight trucking, long-distance, LTL," Series PCU484122484122, U.S. Bureau of Labor Statistics, http://www.bls.gov/ppi/data.htm

BLS. 2011c. "General freight trucking, long-distance, LTL," Series PCU4841224841221, U.S. Bureau of Labor Statistics, http://www.bls.gov/ppi/data.htm

BNSF. 2011. http://www.bnsf.com/customers/how-can-i-ship/dedicated-train-service/

Bookbinder, J. H. and A. E. Lordahl. 1989. Estimation of inventory re-order levels using the bootstrap statistical procedure. *IIE Transactions*. 21: 302–312.

BTS. 2011a. "Effects of Recent Trends in Container Throughput," http://www.bts. gov/publications/americas_container_ports/2011/html/effects_of_recent_ trends.html

Çetinkaya, S. and C. Y. Lee. 2000. Stock replenishment and shipment scheduling for vendor-managed inventory systems. *Management Science*. 46(2): 217–232.

Coyle, J. J., E. J. Bardi, and C. J. Langley Jr. 2009. *The Management of Business Logistics*, 8th edn. Mason, OH: South-Western.

Coyle, J. J., E. J. Bardi, and R. A. Novack. 2006. *Transportation*, 6th edn. Mason, OH: Thomson – South-Western.

CSX. 2012. http://www.csx.com/index.cfm/customers/equipment/railroad-equipment/

Google Maps, 2012, http://http://maps.google.com/

Harvard Business School. 1998. *Note on the U.S. Freight Transportation Industry*, # 9-688-080. Boston, MA: Harvard Business School Publishing.

Kay, M. 2011. "Freight Transport," from Lecture Notes for Production Systems Design, North Carolina State University Department of Industrial and Systems Engineering, Raleigh, NC, http://courses.ncsu.edu/ise754/common/ Freight%20Transport.pdf

Kay, M. G. and D. P. Warsing. 2009. Estimating LTL Rates Using Publicly Available Empirical Data. *International Journal of Logistics: Research and Applications*. 12(3): 165–193.

Lee, C. Y. 1986. The economic order quantity for freight discount costs. *IIE Transactions*. 18(3): 318–320.

Levinson, M. 2006. *The Box: How the Shipping Container Made the World Smaller and the World Economy Bigger*. Princeton, NJ: Princeton University Press.

Marien, E. J. 1996. Making sense of freight terms of sale. *Transportation & Distribution*. 37(9): 84–86.

Murphy Jr., P. R. and D. Wood. 2011. *Contemporary Logistics*, 10th edn. Upper Saddle River, NJ: Prentice Hall.

Mutlu, F. and S. Çetinkaya. 2010. An integrated model for stock replenishment and shipment scheduling under common carrier dispatch costs. *Transportation Research: Part E*. 46(6): 844–854.

Nagarajan, A., E. Canessa, W. Mitchell, and C. C. White III. 2001. Trucking industry: challenges to keep pace (Chapter 5). In *The Economic Payoff from the Internet Revolution*, eds. R.E. Litan and A.E. Rivlini. Washington, D.C.: Brookings Institution Press.

NMFTA. 2009. "Policies and Directives Pertaining to the National Motor Freight Classification," National Motor Freight Transportation Board, http://www. nmfta.org/Documents/CCSB/CCSB%20Policies%202009.pdf

Piercy, J. E. 1977. A Performance Profile of Several Transportation Freight Services, unpublished PhD. dissertation, Case Western Reserve University, Cleveland, OH.

Simchi-Levi, D., X. Chen, and J. Bramel. 2005. *The Logic of Logistics: Theory, Algorithms, and Applications for Logistics and Supply Chain Management*. New York: Springer.

SMC3. 2012. www.smc3.com/Applications/WebCzarLite/LocalLink.asp

Speigel, R. 2002. Truckload vs. LTL. *Logistics Management*. 47(7): 54–57.

Swenseth, S. R. and M. R. Godfrey. 1996. Estimating freight rates for logistics decisions. *Journal of Business Logistics*. 17(1): 213–231.

Tersine, R. J., and S. Barman. 1991. Lot size optimization with quantity and freight rate discounts. *Logistics and Transportation Review*. 27(4): 319–332.

Tyworth, J. E. and L. O'Neill. 1997. Robustness of the normal approximation of lead-time demand in a distribution setting. *Naval Research Logistics*. 44: 166–186.

Tyworth, J. E. and A. Ruiz-Torres. 2000. Transportation's role in the sole- versus dual-sourcing decision. *International Journal of Physical Distribution and Logistics Management*. 30(2): 128–144.

United Cargo, 2012, http://www.unitedcargo.com/shipping/default.jsp.

U.S. DOT. 2011. http://ops.fhwa.dot.gov/freight/sw/overview/index.htm

Yano, C. A., T. Chan, L. K. Richter, T. Cutler, K. Murty, and D. McGettigan. 1987. Vehicle routing at quality stores. *Interfaces*. 17(2): 52–63.

5

Location and Distribution Decisions in Supply Chains

In our view, a typical manager or engineer's time is concerned with managing demand, inventory, and transportation in an existing supply chain network. Hence, up to this point in the book, we have discussed the issues of forecasting customer demand, determining production and inventory to meet that demand, and managing the transportation process to fulfill the demand in the supply chain. The problems of managing an existing supply chain network are far more pervasive and frequently encountered in practice compared to designing a new network or re-designing an existing one. One might justifiably ask at this point, "How did the decision makers actually decide the location of the facilities in the network and the assignment of those facilities to the customers they serve?" Therefore, in this Chapter, we focus our attention to "location and distribution" strategies or designing and operating the supply chain network. The key questions to answer in this effort are the following:

- How do we design a good (or perhaps the *best*) network?
- What are the key objectives for the network?
- What are the primary decision variables? Are they strategic or tactical decisions?
- What are the key constraints of the network?
- What network alternatives should be considered?

In order to answer these questions, we need to include the perspective of the decision maker, charged with making the design decisions. Typically, the perspective will be that of a single firm. The objectives may be diverse and conflicting. We clearly wish to minimize the cost of operating the network, and since product prices typically will not change as a function of our network design, this objective should be sufficient for most situations. However, customer demand fulfillment and service may also have to be incorporated in the decision making process. That would lead to the use of multiple criteria mathematical programming models for decision making. The key decision variables in the optimization models will be the number and the location of various types of facilities—spanning as far as supplier sites, manufacturing sites, and distribution sites—and the quantities shipped from upstream sites to the downstream sites and ultimately out to customers.

Integer programming (IP) models with binary variables have been successfully used in practice to design supply chain networks. Hence, we begin this chapter with a review of modeling with binary variables. We will then apply the IP models for location and distribution decisions in supply chain management.

Next, we discuss an important aspect of supply chain network design, called *Risk Pooling*. Risk pooling refers to the use of a more *consolidated distribution network* with fewer facilities, each serving a large allocation of customer demand. A consolidated distribution system reduces supply chain costs—inventory holding cost (IHC), order costs, and facilities cost. However, customer service suffers, as time to fulfill customer demand increases. We will study the tradeoff between supply chain cost and customer service under risk pooling.

Next, we present some basic results in *continuous location models* and how they relate to supply chain network design. We conclude the chapter by discussing several real-world applications of IP models used successfully in supply chain network design and other problems.

5.1 Modeling with Binary Variables

This section will be devoted to the use of binary (0-1) variables in modeling real-world problems. Linear programming models with binary decision variables are called *integer programming* (IP) models. Of course, a general IP model may include regular integer variables (non 0-1), as well as, continuous variables. Such IP models are called *mixed integer programming* (MIP) models.

5.1.1 Capital Budgeting Problem

A company is planning its capital spending for the next T periods. There are N projects that compete for the limited capital B_i, available for investment in period i. Each project requires a certain investment in each period once it is selected. Let a_{ij} be the required investment in project j for period i. The value of the project is measured in terms of the associated cash flows in each period discounted for inflation. This is called the net present value (NPV). Let v_j denote the NPV for project j. The problem is to select the proper projects for investment that will maximize the total value (NPV) of all the projects selected.

Formulation: To formulate this as an integer program, we introduce a binary variable for each project to denote whether it is selected or not.

Let

$$x_j = 1, \quad \text{if project } j \text{ is selected}$$
$$x_j = 0, \quad \text{if project } j \text{ is not selected}$$

It is then clear that the following pure integer program will represent the capital budgeting problem:

Maximize

$$Z = \sum_{j=1}^{N} v_j.x_j$$

Subject to

$$\sum_{j=1}^{N} a_{ij}.x_j \le B_i, \qquad\qquad \text{for all } i = 1, 2....., T$$

$$0 \le x_j \le 1, x_j \text{ is a binary variable} \quad \text{for all } j = 1, 2....., N$$

5.1.2 Fixed Charge Problem

Consider a production planning problem with N products such that the jth product requires a fixed production or set-up cost K_j, independent of the amount produced, and a variable cost C_j per unit, proportional to the quantity produced. Assume that every unit of product j requires a_{ij} units of resource i and there are M resources. Given that the product j, whose sales potential is d_j, sells for \$$p_j$ per unit and no more than b_i units of resource i are available $(i = 1, 2, ..., M)$, the problem is to determine the optimal product mix that maximizes the net profit.

Formulation: The total cost of production (fixed plus variable) is a nonlinear function of the quantity produced. But, with the help of binary (0-1) integer variables, the problem can be formulated as an integer linear program.

Let the binary integer variable δ_j denote the decision to produce or not to produce product j.

In other words,

$$\delta_j = \begin{cases} 1, \text{ if product } j \text{ is produced} \\ 0, \text{ otherwise} \end{cases}$$

Let x_j (≥ 0) denote the quantity of product j produced. Then the cost of producing x_j units of product j is $K_j.\delta_j + C_j.x_j$, where $\delta_j = 1$ if $x_j > 0$ and $\delta_j = 0$ if $x_j = 0$. Hence, the objective function is

$$\text{Maximize } Z = \sum_{j=1}^{N} p_j.x_j - \sum_{j=1}^{N} (K_j.\delta_j + C_j.x_j)$$

Subject to
The supply constraint for the ith resource is given by,

$$\sum_{j=1}^{N} a_{ij}.x_j \le b_i \quad \text{for } i = 1, 2...., M$$

The demand constraint for the jth product is given by

$$x_j \le d_j.\delta_j \quad \text{for } j = 1, 2....., N$$

$$x_j \ge 0 \quad \text{and} \quad \delta_j = 0 \text{ or } 1 \quad \text{for all } j$$

Note that x_j can be positive only when $\delta_j = 1$, in which case its production is limited by d_j and the fixed production cost K_j is included in the objective function.

5.1.3 Constraint with Multiple Right-Hand-Side Constants

Consider a problem, where the constraint $a_1x_1 + a_2x_2 + ... + a_nx_n$ must be less than or equal to one of the RHS values b_1, b_2, b_3. In other words, the constraint becomes

$$a_1x_1 + a_2x_2 + ...+ a_nx_n \le b_1, b_2 \text{ or } b_3 \tag{5.1}$$

Constraints similar to Equation 5.1 arise in supply chain network design, where a company has the option to build warehouses of different capacities. In that case, x_i is the quantity of product i stored at that location, a_i is the square footage occupied by one unit of product i and b_1, b_2, b_3 are the three potential warehouse capacities. Using binary variables, one for each b_i value, we can represent Equation 5.1 as a linear constraint. Define $\delta_1, \delta_2, \delta_3$ as the binary variables such that when $\delta_i = 1$, the RHS value is b_i. Then Equation 5.1 can be written as,

$$a_1x_1 + a_2x_2 + ...+ a_nx_n \le \delta_1b_1 + \delta_2b_2 + \delta_3b_3 \tag{5.2}$$

$$\delta_1 + \delta_2 + \delta_3 = 1 \tag{5.3}$$

$$\delta_i \in (0,1), \text{ for } i = 1, 2, 3 \tag{5.4}$$

Equations 5.3 and 5.4 guarantee that only one of the δ_i's will be one and the rest will be zero. Hence, the RHS value of Equation 5.2 can only be b_1, b_2 or b_3. If constraint (5.1) represents the decision to build a warehouse of capacity b_1, b_2 or b_3 only, then the fixed cost of building the warehouse, say K_1, K_2 and K_3, can be easily included in the objective function using the binary variables

δ_1, δ_2, and δ_3 respectively. If we want to include the option of *not building the warehouse* at all at this location, we can write Equation 5.3 as

$$\delta_1 + \delta_2 + \delta_3 \leq 1 \tag{5.5}$$

Equation 5.5 allows the possibility for all the δ_i's to be zero, in which case no warehouse will be built at this location. This would make the RHS value of Equation 5.2 zero, which will force all the storage variables at this location $(x_1, x_2, \ldots, x_n)$ to be equal to zero.

5.1.4 Quantity Discounts

Quantity discounts refer to the practice of offering lower prices for large volume purchases. There are two types of quantity discounts offered by vendors as follows:

1. "All-unit" discount
2. "Graduated" discount

Both types of quantity discount models result in nonlinear cost functions. However, they can be modeled as linear functions with the help of binary variables. We shall discuss the IP formulations of the "quantity discount" models next.

"All-unit" quantity discounts: Under this scenario, the entire purchase will be charged at a lower price based on the order quantity. Figure 5.1 illustrates an example of "all-unit" quantity discount price structure.

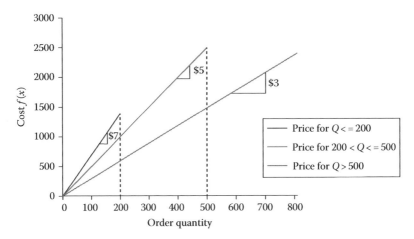

FIGURE 5.1
All-unit quantity discount model.

The mathematical representation of the cost function $f(X)$, where X is the quantity purchased, is as follows:

$$\text{Cost } f(X) = \begin{cases} 7X & 0 \le X \le 200 \\ 5X & 201 \le X \le 500 \\ 3X & X \ge 501 \end{cases}$$

The cost function $f(X)$ is nonlinear in X since the slope changes depending on the value of X. However, we can linearize it using three binary variables, δ_1, δ_2, and δ_3, one for each price range as follows:

$$\delta_i = \begin{cases} 1, & \text{if price range } i \text{ is used} \\ 0, & \text{otherwise} \end{cases}$$

Let X_i = order quantity under price range "i," $i = 1, 2, 3$.

The mixed integer linear programming (MILP) formulation will be as follows:

$$\text{Cost } f(X) = 7.X_1 + 5.X_2 + 3.X_3 \tag{5.6}$$

Subject to:

$$X = X_1 + X_2 + X_3 \tag{5.7}$$

$$0 \le X_1 \le 200.\delta_1 \tag{5.8}$$

$$201.\delta_2 \le X_2 \le 500.\delta_2 \tag{5.9}$$

$$501.\delta_3 \le X_3 \le M.\delta_3 \tag{5.10}$$

$$\delta_1 + \delta_2 + \delta_3 = 1 \tag{5.11}$$

$$\delta_i \in (0,1) \tag{5.12}$$

Equations 5.11 and 5.12 will force exactly one of the δ_i's to be one and the others zero in any solution. Hence, only one of the X_i's will be positive in Equation 5.6. For example, if $\delta_2 = 1$, $(\delta_1 = \delta_3 = 0)$, then X_2 will be positive between 201 and 500 by Equation 5.9. Since δ_1 and δ_3 will be zero, X_1 and X_3 will be forced to zero by Equations 5.8 and 5.10. Note that M is a large positive number in Equation 5.10. In case there is a capacity limit on the maximum order quantity, then M can be replaced by that capacity limit. Note that all the constraints and the objective function are now linear functions.

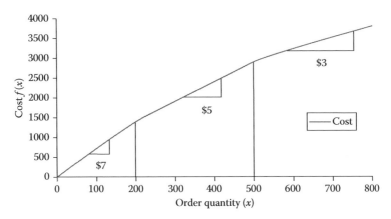

FIGURE 5.2
Graduated quantity discount model.

All-unit quantity discount models are commonly used by trucking companies for the freight rate structure in road transportation.

Graduated quantity discount: Under this, the entire order quantity does not get the lower price; only the additional amount over a price range gets the lower cost. This is illustrated in Figure 5.2.

Under this model, the *first* 200 units are *always* charged at $7/unit, the *next* 300 units are charged at a lower price of $5/unit, and only the amount over 500 is charged at $3/unit. For example, if the order quantity is 300 units, then the cost is given by:

$$f(X) = 200(\$7) + 100(\$5) = \$1900$$

Graduated quantity discount models are frequently used by utility companies for energy usage. Also, the U.S. Postal Service uses graduated rates for its first class mail. The first ounce is always charged at a higher rate and successive ounces carry a lower cost per ounce.

Figure 5.2 also represents a nonlinear cost function. It can be linearized using binary variables, similar to the all-unit quantity discount model. Let δ_1, δ_2, and δ_3 be the binary variables for each price range and X_1, X_2 and X_3 are the quantity purchased under price ranges 1, 2, and 3. Then, the linear IP formulation becomes as follows:

$$\text{Cost } f(X) = 7.X_1 + 5.X_2 + 3.X_3 \tag{5.13}$$

Subject to:

$$X = X_1 + X_2 + X_3 \tag{5.14}$$

$$0 \leq X_1 \leq 200.\delta_1 \tag{5.15}$$

$$X_1 \geq 200.\delta_2 \tag{5.16}$$

$$0 \leq X_2 \leq (500 - 200).\delta_2 \tag{5.17}$$

$$X_2 \geq (500 - 200).\delta_3 \tag{5.18}$$

$$0 \leq X_3 \leq M.\delta_3 \tag{5.19}$$

$$\delta_1, \delta_2, \delta_3 \in (0, 1) \tag{5.20}$$

Unlike the all-unit discount model, one or more of the δ_i's can be one in a solution. Hence, one or more of the X_i's can be positive and their sum equals to the total order quantity X, as given by Equation 5.14. Note that X_3 can be positive only if $\delta_3 = 1$ (Equation 5.19). When $\delta_3 = 1$, $X_2 \geq 300$ by Equation 5.18. Since X_2 cannot be positive unless $\delta_2 = 1$, it will force $\delta_2 = 1$ by Equation 5.17. Now both Equations 5.17 and 5.18 reduce to:

$$X_2 \leq 300$$

$$X_2 \geq 300$$

Hence, the only solution is $X_2 = 300$, its maximum value under the second price range. Similarly when $\delta_2 = 1$, it will force $\delta_1 = 1$ and $X_1 = 200$. Thus for X_3 to be positive (to buy at the lowest price), we have to reach the maximum under the previous two price ranges, namely $X_1 = 200$ and $X_2 = 300$.

Application: A case study illustrating the IP formulations of quantity discount models in supplier selection and order allocation is given in Chapter 6, Example 6.3.

5.1.5 Handling Nonlinear Integer Programs

Nonlinear IP problems can be converted to linear integer programs using binary variables. We shall illustrate this with a numerical example.

Example 5.1 (Ravindran et al., 1987)

Consider a nonlinear (binary) IP problem.

Maximize

$$Z = x_1^2 + x_2 x_3 - x_3^3$$

Subject to

$$-2x_1 + 3x_2 + x_3 \leq 3$$

$$x_1, x_2, x_3 \in (0, 1)$$

This nonlinear integer problem can be converted to a linear IP problem for solution. Observe the fact that for any positive k and a binary variable x_j, $x_j^k = x_j$. Hence, the objective function immediately reduces to $Z = x_1 + x_2x_3 - x_3$. Now consider the product term x_2x_3. For binary values of x_2 and x_3, the product x_2x_3 is always 0 or 1. Now introduce a binary variable y_1 such that $y_1 = x_2x_3$. When $x_2 = x_3 = 1$, we want the value of y_1 to be 1, while all other combinations of y_1 should be zero. This can be achieved by introducing the following two constraints:

$$x_2 + x_3 - y_1 \leq 1$$

$$-x_2 - x_3 + 2y_1 \leq 0$$

Note that when $x_2 = x_3 = 1$, this constraint reduces to $y_1 \geq 1$ and $y_1 \leq 1$ implying $y_1 = 1$. When $x_2 = 0$ or $x_3 = 0$ or both are zero, the second constraint, $y_1 \leq \left(\dfrac{x_2 + x_3}{2} \right)$ forces y_1 to be zero.

Thus the equivalent linear (binary) integer program becomes

Maximize

$$Z = x_1 + y_1 - x_3$$

Subject to

$$-2x_1 + 3x_2 + x_3 \leq 3$$

$$x_2 + x_3 - y_1 \leq 1$$

$$-x_2 - x_3 + 2y_1 \leq 0$$

$$x_1, x_2, x_3, y_1 \in (0,1)$$

A drawback of the aforementioned procedure for handling product terms is that an integer variable is introduced for each product term. It has been observed in practice that the solution time for IP problems increases with the number of integer variables. An alternate procedure has been suggested by Glover and Woolsey (1974) that introduces a continuous variable rather than an integer variable. This procedure replaces x_2x_3 by a continuous variable x_{23} and introduces three new constraints as follows:

$$x_2 + x_3 - x_{23} \leq 1$$

$$x_{23} \leq x_2$$

$$x_{23} \leq x_3$$

$$x_{23} \leq 0$$

where, x_{23} replaces the product term x_2x_3.

Whenever x_2, x_3 or both are zero, the last two constraints force x_{23} to be zero. When $x_2 = x_3 = 1$, all the three constraints together force x_{23} to be 1. The primary disadvantage of this procedure is that it adds more constraints than the previous method.

Remarks

The procedure for handling the product of two binary variables can be easily extended to the product of any number of variables. For example, consider the product terms $x_1 x_2 \ldots x_k$. Under the first procedure, a binary variable y_1 will replace $x_1 x_2 \ldots x_k$ and the following two constraints will be added:

$$\sum_{j=1}^{k} x_j - y_1 \leq k - 1$$

$$-\sum_{j=1}^{k} x_j + k y_1 \leq 0$$

$$y_1 \in (0,1)$$

$$x_j \in (0,1)$$

Under the second procedure, the product terms $x_1 x_2 \ldots x_k$ will be replaced by a nonnegative variable x_0 and the following $(k + 1)$ constraints will be added:

$$\sum_{j=1}^{k} x_j - x_0 \leq k - 1$$

$$x_0 \leq x_j \quad \text{for all } j = 1, 2, \ldots, k$$

5.1.6 Set Covering and Set Partitioning Models

Set covering and set partitioning models are linear integer programs with binary variables. They have been successfully applied in warehouse location decisions, airline crew scheduling, aircraft scheduling, supply chain network design, and package delivery problems. They are discussed in Section 5.5. We shall first look at the formal mathematical statement of the set covering problem.

5.1.6.1 Set Covering Problem

Consider an $(m \times n)$ matrix A, called the *set covering matrix*, whose elements a_{ij}'s are either 0 or 1. If $a_{ij} = 1$, we say that column j "covers" row i. If not, $a_{ij} = 0$. The set covering problem is to select the minimum number of columns such that *every row* is covered by *at least one column*.

To formulate the set covering problem as an integer programming (IP) model, we define a binary variable for each column such that

$$X_j = \begin{cases} 1, \text{ if column } j \text{ is selected} \\ 0, \text{ otherwise} \end{cases}, \quad \text{for } j = 1, 2, \ldots n$$

The IP model is given by

$$\text{Minimize } Z = \sum_{j=1}^{n} X_j \tag{5.21}$$

Subject to

$$\sum_{j=1}^{n} a_{ij} X_j \geq 1 \quad \text{for } i = 1, 2, \ldots, m \tag{5.22}$$

$$X_j \in (0,1)$$

Constraints denoted by Equation 5.22 guarantee that every row is "covered" by at least one column. In other words, for row i, when at least one $a_{ij} = 1$, the corresponding X_j must be one. The objective function, given by Equation 5.21 guarantees that the minimum number of columns is selected to cover all the rows.

Example 5.2

Consider the set covering matrix given by

$$A = \begin{bmatrix} 1 & 1 & 0 & 1 & 1 & 1 & 0 \\ 0 & 1 & 0 & 0 & 1 & 1 & 1 \\ 1 & 0 & 1 & 1 & 0 & 1 & 0 \\ 0 & 0 & 1 & 0 & 1 & 0 & 1 \end{bmatrix}$$

The set covering problem becomes

$$\text{Minimize } Z = \sum_{j=1}^{7} X_j$$

Subject to

$$X_1 + X_2 + X_4 + X_5 + X_6 \geq 1 \tag{5.23}$$

$$X_2 + X_5 + X_6 + X_7 \geq 1 \tag{5.24}$$

$$X_1 + X_3 + X_4 + X_6 \geq 1 \qquad (5.25)$$

$$X_3 + X_5 + X_7 \geq 1 \qquad (5.26)$$

$$X_j \in (0,1) \quad \text{for } i = 1,2,\dots,7$$

Equation 5.23 guarantees that row 1 will be covered by at least one column. Similarly, Equations 5.24, 5.25, and 5.26 guarantee that rows 2, 3, and 4 will be covered respectively by at least one column.

5.1.6.2 Set Partitioning Problem

In the set covering problem every row has to be covered by *at least* one column. In the set partitioning problem, every row has to be covered by *exactly* one column. Otherwise the two problems are the same. Thus, the only change in the IP model is that constraints given by Equation 5.22 will now become equalities:

$$\sum_{j=1}^{n} a_{ij} X_j = 1 \quad \text{for } i = 1,2,\dots,m$$

The objective function given by Equation 5.21 remains the same.

5.1.6.3 Application to Warehouse Location

In applying the set covering model to the warehouse location problem in supply chain, we treat the potential warehouse locations as "columns" and the customer regions as "rows" of the set covering matrix A. We construct the matrix A, by setting its elements a_{ij} as follows:

$$a_{ij} = \begin{cases} 1, & \text{if customer region } i \text{ can be supplied by warehouse location } j, \\ & \text{considering service level and distance criteria} \\ 0, & \text{otherwise} \end{cases}$$

By including the cost of building a warehouse at location j as K_j, we will minimize the total cost of building warehouses such that every customer region can be supplied by at least one warehouse. We will illustrate this with Example 5.3 in the next section. In addition to the warehouse location problem, Section 5.2 will also include other examples in supply chain network design and distribution problems using binary variables for modeling.

5.2 Supply Chain Network Optimization

In this section, we will apply the IP models, discussed in Section 5.1, to different supply chain network optimization problems, including warehouse location, network design, and distribution problems.

5.2.1 Warehouse Location

Given a set of potential warehouse sites, the problem is to choose the best site that will serve all the customers at minimum cost.

Example 5.3 (Ravindran et al., 1987)

A firm has four possible sites for locating its warehouses. The cost of locating a warehouse at site i is $\$K_i$. There are nine retail outlets, each of which must be supplied by at least one warehouse. It is not possible for any one site to supply all the retail outlets as shown in Figure 5.3.

   The problem is to determine the location of the warehouse such that the total cost is minimized.

Solution

The warehouse location problem is basically a set covering problem. The first step is to define the *set covering matrix* (A) based on the network configuration shown in Figure 5.3. The rows of the matrix will be the nine retail outlets and the columns will be the four potential warehouse locations. The elements of matrix A, a_{ij}, will be set to 1 if retailer i (R_i) can be supplied by warehouse location j (W_j), that is, there is a direct link between R_i and W_j. Otherwise, we set $a_{ij} = 0$.

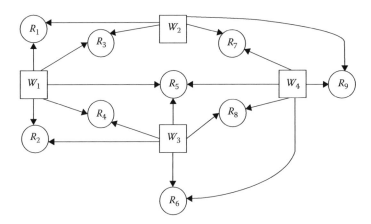

FIGURE 5.3
Supply chain network (Example 5.3).

For example, W_1 can supply R_1, R_2, R_3, R_4 and R_5. Hence, we set $a_{11} = a_{21} = a_{31} = a_{41} = a_{51} = 1$ and $a_{61} = a_{71} = a_{81} = a_{91} = 0$. The complete set covering matrix A is given as follows:

Set covering matrix

$$
A_{(9 \times 4)} = \begin{array}{c} \\ R_1 \\ R_2 \\ R_3 \\ R_4 \\ R_5 \\ R_6 \\ R_7 \\ R_8 \\ R_9 \end{array}
\begin{bmatrix}
W_1 & W_2 & W_3 & W_4 \\
1 & 1 & 0 & 0 \\
1 & 0 & 1 & 0 \\
1 & 1 & 0 & 0 \\
1 & 0 & 1 & 0 \\
1 & 0 & 1 & 1 \\
0 & 0 & 1 & 1 \\
0 & 1 & 0 & 1 \\
0 & 0 & 1 & 1 \\
0 & 1 & 0 & 1
\end{bmatrix}
$$

Model formulation

Define binary variables X_j for $j = 1, 2, 3, 4$ such that

$$X_j = \begin{cases} 1, \text{ if site } j \text{ is selected for a warehouse location} \\ 0, \text{ otherwise} \end{cases}$$

The IP formulation of the warehouse selection problem becomes

$$\text{Minimize } Z = K_1X_1 + K_2X_2 + K_3X_3 + K_4X_4$$

Subject to

$$
\begin{array}{lll}
X_1 + X_2 \geq 1 & (R_1) & \\
X_1 + X_3 \geq 1 & (R_2) & \\
X_1 + X_2 \geq 1 & (R_3) & \text{(Redundant)} \\
X_1 + X_3 \geq 1 & (R_4) & \text{(Redundant)} \\
X_1 + X_3 + X_4 \geq 1 & (R_5) & \\
X_3 + X_4 \geq 1 & (R_6) & \\
X_2 + X_4 \geq 1 & (R_7) & \\
X_3 + X_4 \geq 1 & (R_8) & \text{(Redundant)} \\
X_2 + X_4 \geq 1 & (R_9) & \text{(Redundant)}
\end{array}
$$

Note that several of the set covering constraints are redundant and can be omitted before solving the integer program.

5.2.2 Distribution Planning

The selection of optimal sites for warehouse location is a *strategic decision*. We shall now consider the *tactical decision* of distributing products to retail outlets from a given set of warehouses. The distribution problem is basically a *transportation problem* that we discussed in Chapter 2 (Section 2.17.1). We shall illustrate this with a numerical example.

Example 5.4

Consider a distribution planning problem with 3 warehouses and 12 retailers. Tables 5.1 and 5.2 give the supply available at the warehouses and the retailer demands, respectively.

The unit cost of shipping the product from a given warehouse to each retailer is given in Table 5.3. The problem is to determine the optimal distribution plan that will minimize the total cost.

TABLE 5.1

Warehouse Supplies
(Example 5.4)

	Warehouses		
	W_1	W_2	W_3
Supply	5000	5000	3000

TABLE 5.2

Retailer Demands (Example 5.4)

	Retailers											
	R_1	R_2	R_3	R_4	R_5	R_6	R_7	R_8	R_9	R_{10}	R_{11}	R_{12}
Demand	650	400	850	1900	3100	250	350	400	500	400	1350	450

TABLE 5.3

Unit Shipping Cost in Dollars
(Example 5.4)

Retailer	Warehouse		
	W_1	W_2	W_3
R_1	3.50	4.30	3.00
R_2	3.60	4.90	3.90
R_3	4.30	3.80	3.30
R_4	4.80	3.70	4.20
R_5	3.10	4.40	3.40
R_6	3.70	3.90	3.10
R_7	3.80	4.10	3.50
R_8	3.80	4.00	2.80
R_9	3.70	4.30	3.10
R_{10}	3.80	4.75	3.50
R_{11}	4.50	3.00	3.80
R_{12}	3.90	4.90	4.10

Solution

The total supply available at the warehouses is 13,000 units, while the total retailer demand is 10,600. Since any warehouse can supply any retailer, it is feasible to meet all the retailer demands with the available supply. To determine the least cost distribution plan, we formulate the following transportation problem:

Decision variables: X_{ij} = Amount shipped to Retailer i (R_i) from warehouse j (W_j); i = 1, 2,…,12 and j = 1, 2, 3. Thus we have 36 decision variables.

Supply constraints at each warehouse: Each warehouse cannot supply more than its capacity. Thus, we get,

$$\sum_{i=1}^{12} X_{i1} \leq 5000 \quad \text{(Warehouse 1)}$$

$$\sum_{i=1}^{12} X_{i2} \leq 5000 \quad \text{(Warehouse 2)}$$

$$\sum_{i=1}^{12} X_{i3} \leq 3000 \quad \text{(Warehouse 3)}$$

Demand constraints for each retailer: The total amount shipped to a retailer from the three warehouses should be equal to the retailer's demand.

$$
\begin{aligned}
X_{11} + X_{12} + X_{13} &= 650 \quad &&(R_1 \text{ - demand}) \\
X_{21} + X_{22} + X_{23} &= 400 \quad &&(R_2 \text{ - demand}) \\
X_{31} + X_{32} + X_{33} &= 850 \quad &&(R_3 \text{ - demand}) \\
X_{41} + X_{42} + X_{43} &= 1900 \quad &&(R_4 \text{ - demand}) \\
X_{51} + X_{52} + X_{53} &= 3100 \quad &&(R_5 \text{ - demand}) \\
X_{61} + X_{62} + X_{63} &= 250 \quad &&(R_6 \text{ - demand}) \\
X_{71} + X_{72} + X_{73} &= 350 \quad &&(R_7 \text{ - demand}) \\
X_{81} + X_{82} + X_{83} &= 400 \quad &&(R_8 \text{ - demand}) \\
X_{91} + X_{92} + X_{93} &= 500 \quad &&(R_9 \text{ - demand}) \\
X_{10,1} + X_{10,2} + X_{10,3} &= 400 \quad &&(R_{10} \text{ - demand}) \\
X_{11,1} + X_{11,2} + X_{11,3} &= 1350 \quad &&(R_{11} \text{ - demand}) \\
X_{12,1} + X_{12,2} + X_{12,3} &= 450 \quad &&(R_{12} \text{ - demand})
\end{aligned}
$$

Objective function: Minimize total cost of shipping given by

$$Z = \left(3.5X_{11} + 4.3X_{12} + 3X_{13}\right) + \left(3.6X_{21} + 4.9X_{22} + 3.9X_{23}\right) + \ldots\ldots$$

$$+ \left(4.5X_{11,1} + 3X_{11,2} + 3.8X_{11,3}\right) + \left(3.9X_{12,1} + 4.9X_{12,2} + 4.1X_{12,3}\right)$$

Thus the transportation problem has 36 variables and 15 constraints. Solving it in Microsoft's Excel Solver add in, we get the optimal solution as shown in Table 5.4. The minimum shipping cost is $34,830.

5.2.3 Location-Distribution Problem

Let us now consider an integrated example, where both the location decisions and distribution decisions have to be made simultaneously.

TABLE 5.4

Optimal Distribution Plan
(Example 5.4)

Retailer	Warehouse		
	W_1	W_2	W_3
R_1	0	0	650
R_2	400	0	0
R_3	0	0	850
R_4	0	1900	0
R_5	3100	0	0
R_6	0	0	250
R_7	0	0	350
R_8	0	0	400
R_9	0	0	500
R_{10}	400	0	0
R_{11}	0	1350	0
R_{12}	450	0	0

Example 5.5 (Ravindran et al., 1987)

A retail firm is planning to expand its activities in an area by opening
two new warehouses. Three possible sites are under consideration as
shown in Figure 5.4. Four customers have to be supplied whose annual
demands are $D_1, D_2, D_3,$ and D_4.

Assume that any two sites can supply all the demands but site 1 can
supply customers 1, 2, and 4 only; site 3 can supply customers 2, 3, and 4;
while site 2 can supply all the customers. The unit transportation cost
from site i to customer j is C_{ij}. For each warehouse, Table 5.5 gives the
data on capacity, annual investment, and operating costs.

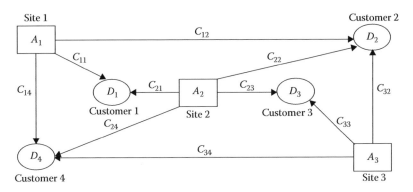

FIGURE 5.4
Supply chain network for Example 5.5.

TABLE 5.5

Warehouse Data (Example 5.5)

Site	Capacity	Initial Capital Investment ($)	Unit Operating Cost ($)
1	A_1	K_1	P_1
2	A_2	K_2	P_2
3	A_3	K_3	P_3

The optimization problem is to select the proper sites for the two warehouses which will minimize the total costs of investment, operation, and transportation.

Solution

Each warehouse site has a fixed capital cost independent of the quantity stored, and a variable cost proportional to the quantity shipped. Thus the total cost of opening and operating a warehouse is a nonlinear function of the quantity stored. Through the use of binary integer variables, the warehouse location-distribution problem can be formulated as an integer program.

Let the binary integer variable δ_i denote the decision to select site i. In other words,

$$\delta_i = \begin{cases} 1, \text{ if site } i \text{ is selected} \\ 0, \text{ otherwise} \end{cases}$$

Let X_{ij} denote the quantity shipped from site i to customer j. The supply constraint for site 1 is given by

$$X_{11} + X_{12} + X_{14} \leq A_1\delta_1 \quad \text{(Site 1)}$$

When $\delta_1 = 1$, site 1 is selected with capacity A_1 and the quantity shipped from site 1 cannot exceed A_1. When $\delta_1 = 0$, the nonnegative variables X_{11}, X_{12}, and X_{14} will automatically become zero, implying no possible shipment from site 1.

Similarly for sites 2 and 3, we obtain

$$X_{21} + X_{22} + X_{23} + X_{24} \leq A_2\delta_2 \quad \text{(Site 2)}$$
$$X_{32} + X_{33} + X_{34} \leq A_3\delta_3 \quad \text{(Site 3)}$$

To select exactly two sites, we need the following constraint:

$$\delta_1 + \delta_2 + \delta_3 = 2$$

Since the δ_i's can assume value 0 or 1 only, the new constraint will force two of the δ_i's to be one.

The demand constraints can be written as follows:

$$X_{11} + X_{21} = D_1 \quad \text{(Customer 1)}$$
$$X_{12} + X_{22} + X_{32} = D_2 \quad \text{(Customer 2)}$$
$$X_{23} + X_{33} = D_3 \quad \text{(Customer 3)}$$
$$X_{14} + X_{24} + X_{34} = D_4 \quad \text{(Customer 4)}$$

To write the objective functions, we note that the total cost of investment, operation, and transportation for site 1 is

$$K_1\delta_1 + P_1(X_{11} + X_{12} + X_{14}) + C_{11}X_{11} + C_{12}X_{12} + C_{14}X_{14}$$

When site 1 is not selected, δ_1 will be zero. This will force X_{11}, X_{12}, and X_{14} to become zero. Similarly, the cost functions for sites 2 and 3 can be written. Thus the complete formulation of the warehouse location problem reduces to the following mixed integer program:

Minimize

$$Z = K_1\delta_1 + P_1(X_{11} + X_{12} + X_{14}) + C_{11}X_{11} + C_{12}X_{12} + C_{14}X_{14}$$
$$+ K_2\delta_2 + P_2(X_{21} + X_{22} + X_{23} + X_{24}) + C_{21}X_{21} + C_{22}X_{22} + C_{23}X_{23} + C_{24}X_{24}$$
$$+ K_3\delta_3 + P_3(X_{32} + X_{33} + X_{34}) + C_{32}X_{32} + C_{33}X_{33} + C_{34}X_{34}$$

Subject to

$$X_{11} + X_{12} + X_{14} \leq A_1\delta_1$$

$$X_{21} + X_{22} + X_{23} + X_{24} \leq A_2\delta_2$$

$$X_{32} + X_{33} + X_{34} \leq A_3\delta_3$$

$$\delta_1 + \delta_2 + \delta_3 = 2$$

$$X_{11} + X_{21} = D_1$$

$$X_{12} + X_{22} + X_{32} = D_2$$

$$X_{23} + X_{33} = D_3$$

$$X_{14} + X_{24} + X_{34} = D_4$$

$$\delta_i \in (0,1) \quad \text{for } i = 1,2,3$$

$$X_{ij} \geq 0 \quad \text{for all } (i, j)$$

5.2.4 Location-Distribution with Dedicated Warehouses (Srinivasan, 2010)

In Example 5.5, we allowed multiple deliveries to a customer. In other words, a customer can receive his demand from more than one warehouse. Suppose the customers demand single deliveries. In this case, each customer has to be supplied by one warehouse only, even though a warehouse may supply more than one customer. We call these as "dedicated warehouse" problems. This condition can be easily modeled with a minor change in the definition of variables X_{ij}. In Example 5.5, X_{ij} was a continuous variable denoting the quantity shipped from site i to customer j. To incorporate dedicated warehouses, we define X_{ij} as a binary variable.

$$X_{ij} = \begin{cases} 1, \text{ if site } i \text{ supplies customer } j;\ i = 1,2,3 \quad j = 1,2,3,4 \\ 0, \text{ otherwise} \end{cases}$$

To guarantee that customer 1 receives supply from one of the sites only, we write the constraint:

$$X_{11} + X_{21} = 1$$

The supply constraint at site 1 will become

$$D_1 X_{11} + D_2 X_{12} + D_4 X_{14} \le A_1 \delta_1$$

The complete formulation is given in the following:

Minimize $Z = K_1 \delta_1 + P_1 \left(D_1 X_{11} + D_2 X_{12} + D_4 X_{14} \right)$

$\qquad + C_{11} D_1 X_{11} + C_{12} D_2 X_{12} + C_{14} D_4 X_{14} + K_2 \delta_2$

$\qquad + P_2 \left(D_1 X_{21} + D_2 X_{22} + D_3 X_{23} + D_4 X_{24} \right) + C_{21} D_1 X_{21} + C_{22} D_2 X_{22}$

$\qquad + C_{23} D_3 X_{23} + C_{24} D_4 X_{24} + K_3 \delta_3 + P_3 \left(D_2 X_{32} + D_3 X_{33} + D_4 X_{34} \right)$

$\qquad + C_{32} D_2 X_{32} + C_{33} D_3 X_{33} + C_{34} D_4 X_{34}$

Subject to

$$D_1 X_{11} + D_2 X_{12} + D_4 X_{14} \le A_1 \delta_1$$

$$D_1 X_{21} + D_2 X_{22} + D_3 X_{23} + X_4 X_{24} \le A_2 \delta_2$$

$$D_2 X_{32} + D_3 X_{33} + D_4 X_{34} \le A_3 \delta_3$$

$$\delta_1 + \delta_2 + \delta_3 = 2$$

$$X_{11} + X_{21} = 1$$

$$X_{12} + X_{22} + X_{32} = 1$$

$$X_{23} + X_{33} = 1$$

$$X_{14} + X_{24} + X_{34} = 1$$

$$\delta_i \in (0,1) \quad \text{for } i = 1,2,3$$

$$X_{ij} \in (0,1) \quad \text{for all } (i,j)$$

5.2.5 Supply Chain Network Design

Since warehouses can be built of different capacities and cost, we shall consider a multi-state supply chain network design problem, which determines not only the locations of the warehouses but their right capacities to meet the customer demand.

Example 5.6

XYZ Company is looking at improving its distribution system. XYZ's 3-stage supply chain—a factory, two potential warehouse sites, and four retailers, is shown in Figure 5.5.

At each of the potential sites, the company can build a warehouse in three different sizes—small, medium, or large. The investment cost and capacities at the two warehouse sites are given in Table 5.6. The company can build a warehouse at either locations or both. However, it cannot build more than one warehouse at each location. In other words, the company cannot build both a large and a medium size warehouse at the same location. The customer demands must be met and are given in Table 5.7. The unit costs of shipping XYZ's product from the factory to the warehouse sites and from each warehouse site to the customer are given in Table 5.8.

Note that the construction costs are lower at site 2, but its distribution costs are higher. The company has a total investment limit of $30,000 for construction. The objective is to minimize the total cost of investment and transportation such that all customers' demands are met.

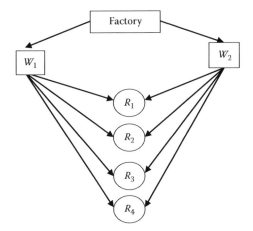

FIGURE 5.5
XYZ's 3-stage supply chain (Example 5.6).

TABLE 5.6

Warehouse Capacities and Investment Cost
(Example 5.6)

| Size | Capacity | Investment Cost (Annualized) | |
		Site 1 (W_1)	Site 2 (W_2)
Small	3,000 units	$10,000	$8,000
Medium	5,000 units	$15,000	$12,000
Large	10,000 units	$20,000	$17,000

TABLE 5.7

Annual Demand (Example 5.6)

Retailer Demands			
R_1	R_2	R_3	R_4
2000	2500	1000	3500

TABLE 5.8

Unit Shipping Cost (Example 5.6)

(i) Factory to warehouse

	W_1	W_2
Factory	$10	$15

(ii) Warehouse to retailers

| Warehouse | Retailer | | | |
	R_1	R_2	R_3	R_4
W_1	$2	$3	$5	$4
W_2	$5	$4	$6	$7

Solution

Binary variables

$$\delta_{iS} = \begin{cases} 1, \text{ if small warehouse is built at site } i \\ 0, \text{ otherwise} \end{cases}, \text{ for } i = 1, 2$$

Similarly we define additional binary variables for δ_{iM} and δ_{iL} for building medium and large warehouses at site i.

Continuous variables

X_{Fi} = Amount shipped from the factory to warehouse i ($i = 1, 2$)
X_{ij} = Amount shipped from warehouse i to retailer j ($i = 1, 2$ and $j = 1, 2, 3, 4$)

Thus, we have 16 decision variables (6 binary and 10 continuous)

Constraints

1. At each site, no more than one warehouse can be built.

$$\text{Site 1: } \delta_{1S} + \delta_{1M} + \delta_{1L} \leq 1 \tag{5.27}$$

$$\text{Site 2: } \delta_{2S} + \delta_{2M} + \delta_{2L} \leq 1 \tag{5.28}$$

2. Amount shipped from the factory to the warehouse cannot exceed that warehouse's capacity.

$$\text{Site 1: } X_{F1} \leq 3,000\delta_{1S} + 5,000\delta_{1M} + 10,000\delta_{1L} \tag{5.29}$$

$$\text{Site 2: } X_{F2} \leq 3,000\delta_{2S} + 5,000\delta_{2M} + 10,000\delta_{2L} \tag{5.30}$$

(Recall the formulation of constraints with multiple right-hand side constants discussed in Section 5.1.3.)

3. A warehouse cannot ship more than what it receives from the factory.

$$\text{Site 1: } X_{11} + X_{12} + X_{13} + X_{14} \leq X_{F1} \tag{5.31}$$

$$\text{Site 1: } X_{21} + X_{22} + X_{23} + X_{24} \leq X_{F2} \tag{5.32}$$

4. Retailer demands must be met.

$$\text{Retailer 1: } X_{11} + X_{21} = 2000 \tag{5.33}$$

$$\text{Retailer 2: } X_{12} + X_{22} = 2500 \tag{5.34}$$

$$\text{Retailer 3: } X_{13} + X_{23} = 1000 \tag{5.35}$$

$$\text{Retailer 4: } X_{14} + X_{24} = 3500 \tag{5.36}$$

5. Total investment for construction cannot exceed \$30,000.

$$10,000\delta_{1S} + 15,000\delta_{1M} + 20,000\delta_{1L} + 8,000\delta_{2S} + 12,000\delta_{2M}$$

$$+ 17,000\delta_{2L} \leq 30,000 \tag{5.37}$$

Thus we have 11 constraints in the model.

Objective function: The total cost of investment and transportation is to be minimized.

1. Investment Cost (IC):

$$IC = 10,000\delta_{1S} + 15,000\delta_{1M} + 20,000\delta_{1L} + 8,000\delta_{2S}$$
$$+ 12,000\delta_{2M} + 17,000\delta_{2L}$$

2. Shipping Cost (SC):

$$SC = 10X_{F1} + 15X_{F2} + 2X_{11} + 3X_{12} + 5X_{13} + 4X_{14} + 5X_{21} + 4X_{22}$$
$$+ 6X_{23} + 7X_{24}$$

$$\text{Minimize Total Cost} = TC = IC + SC$$

NOTES

1. In Equations 5.29, the right-hand side values can only be equal to 0, 3,000, 5,000, or 10,000, since no more than one binary variable (δ_{1S}, δ_{1M}, δ_{1L}) can be one for site 1 due to Equation 5.27.
2. If no warehouse is built at site 1, then all the three binary variables δ_{1S}, δ_{1M}, and δ_{1L} will be zero. In such a case, the right-hand side value of Equation 5.29 will zero and it will force $X_{F1} = 0$. When $X_{F1} = 0$, all the shipping variables from site 1 to the 4 retailers (X_{11}, X_{12}, X_{13}, X_{14}) will be forced to zero due to Equation 5.31. Then, Equations 5.33 through 5.36 will force all the shipping to the retailers from site 2 only.

Optimal solution: The optimal solution to this integer program was obtained by using Microsoft's Excel Solver software. Table 5.9 gives the optimal solution.

From Table 5.9, it can be inferred that one large warehouse is built at site 1. No other warehouses are built. All shipments to and from the second site are zero which again ensures that all retailer demands are met from the large warehouse at site 1 alone.

TABLE 5.9

Optimal Solution for Example 5.6

(i) Warehouse locations and capacities

Binary Variable	Value
δ_{1S}	0
δ_{1M}	0
δ_{1L}	1
δ_{2S}	0
δ_{2M}	0
δ_{2L}	0

(ii) Factory to warehouse shipments

	Warehouses	
	W_1	W_2
Factory	9000	0

(iii) Warehouse to retailer shipments

	Retailers			
	R_1	R_2	R_3	R_4
W_1	2000	2500	1000	3500
W_2	0	0	0	0

5.3 Risk Pooling or Inventory Consolidation

Chopra and Meindl (2010) suggest that supply chain network design encompasses four phases: Phase 1—Supply Chain strategy, Phase 2—Regional facility strategy, Phase 3—Desirable sites, Phase 4—Location choices. They also suggest a number of factors that enter into these decisions. We discussed models and methods for Phases 3 and 4 until now. We shall now address the first two phases.

Two primary strategic choices in the treatment of supply chain network design will be how *consolidated* or *deconsolidated* the network design should be. A *consolidated distribution network* will have fewer facilities each serving a large allocation of customer demand. On the other hand, a *deconsolidated system* will have more facilities, each serving a smaller region. For example, Barnes & Noble is an example of a deconsolidated supply chain network. Individual bookstores are used to satisfy customer demand for books locally. Amazon.com is an example of a consolidated network. It only uses two regional warehouses to supply all the customer demands for the entire United States.

Pooling refers to consolidation. *Risk pooling* is the practice of consolidating facilities in fewer locations, by using a consolidated distribution strategy that will reduce supply and demand risk. We will show that risk pooling or the use of consolidated distribution strategy, reduces supply chain costs—IHC (inventory holding cost), order cost, and facilities cost. However, customer service will suffer as time to fulfill customer demand increases. We will study the tradeoff between supply chain cost and customer service under risk pooling. Given the complexity of the network design process, risk pooling will be an iterative process, by evaluating several network options that range from highly consolidated to highly deconsolidated. Ultimately, a choice on this consolidated–deconsolidated spectrum will lead to how many facilities there should be at each stage of the supply chain network. Once the decision maker has an idea of the number of facilities and regional configuration, he or she can answer the question of where the good candidate locations would be and the optimal location choices using the models described in Section 5.2.

5.3.1 Principles of Risk Pooling

The basic principles of risk pooling are as follows:

- Use of fewer warehouses or DCs to supply customers by consolidating customer regions
- Aggregating customer demands reduces demand risk
- Reduction in demand risk reduces total inventory in the supply chain

We shall illustrate the principles of risk pooling with a simple example. Consider a 3-stage supply chain network shown in Figure 5.6. The factory

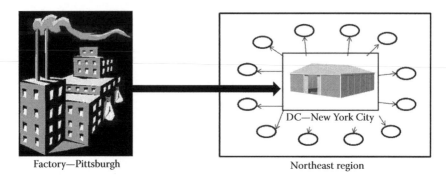

Factory—Pittsburgh   Northeast region

FIGURE 5.6
Option 1: One central warehouse supplies northeast region demand.

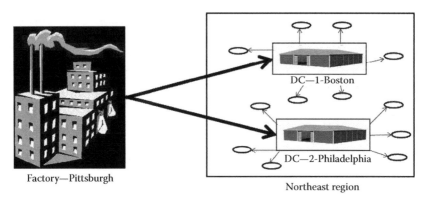

Factory—Pittsburgh

Northeast region

FIGURE 5.7
Option 2: Two warehouses each serving one half of the northeast region demand.

located in Pittsburgh, Pennsylvania, uses one central DC, located in New York City, to meet its product demands in the northeast United States. Assume that the annual demand for the northeast region is D. Now consider an alternate supply chain network design (option 2), as shown in Figure 5.7. Under option 2, two regional DCs, located in Boston and Philadelphia, are used to meet the northeast region's demand. For simplicity, let us assume that the annual demand D is split evenly between the two DCs as $D/2$.

Note that option 1 is the risk pooled alternative. Let us now look at the impact on the following supply chain metrics under each option:

1. Order quantity
2. Average inventory
3. Time between orders
4. Number of orders per year

5. Annual inventory cost
6. Customer responsiveness
7. Logistics cost
   a. Inbound to DC from the factory
   b. Outbound from DC to the customers
8. Supply risk
9. Warehouse facility cost

First consider the *average inventory* in the supply chain under both options. We will use the economic order quantity (EOQ) formula to compute the average inventory.

Option 1 (one central DC)

$$\text{Optimal Order Quantity} = Q_1^* = \sqrt{\frac{2AD}{h}}$$

where A is the fixed cost per order, D is the annual demand, and h is the annual IHC per unit. (See Equation 3.10.)

$$\text{Average Inventory} = \bar{I}_1 = \frac{Q_1^*}{2}$$

Total Inventory in the Supply Chain $= \bar{I}_1(SC) = \bar{I}_1$, since there is only one DC.

Option 2 (Two Regional DCs)
Since each DC supplies an annual demand of $D/2$, the optimal order quantity *at each DC* will be equal to:

$$Q_2^* = \sqrt{\frac{2A(D/2)}{h}} = \frac{Q_1^*}{\sqrt{2}}$$

$$\text{Average Inventory at each DC} = \left(\frac{1}{2}\right)\left(\frac{Q_1^*}{\sqrt{2}}\right) = \frac{\bar{I}_1}{\sqrt{2}}$$

where $\bar{I}_1$ is the average inventory in the central DC under option 1. Thus the average inventory at each DC decreases by $\frac{1}{\sqrt{2}}$. However, the total inventory in the supply chain (in both DCs) becomes

$$I_2(SC) = 2\left(\frac{\bar{I}_1}{\sqrt{2}}\right) = \sqrt{2}\bar{I}_1(SC)$$

Key Result: Total inventory in the supply chain, under option 2, increases $\sqrt{2}$ times of the amount under risk pooling (option 1), *i.e., a 40% increase in inventory carried in the supply chain for the northeast region!*

Impacts on other inventory measures under option 2

1. *Time between orders*

   Let T_1 be the time between orders at the central DC (option 1) and T_2 is the time between orders for each regional DC (option 2). Then,

$$T_1 = \frac{Q_1^*}{D}$$

$$T_2 = \frac{Q_2^*}{\left(D/2\right)} = \frac{2Q_2^*}{D}$$

$$\text{Since, } Q_2^* = \frac{Q_1^*}{\sqrt{2}}$$

$$\text{we get, } T_2 = \frac{2Q_1^*}{\sqrt{2}D} = \sqrt{2}T_1$$

   Thus, the time between orders increases by $\sqrt{2}$.

2. *Number of orders per year*

   Option 1:

$$N_1 = \frac{1}{T_1} = N_1(SC)$$

   Option 2:

$$N_2(SC) = 2N_2 = \sqrt{2}N_1 = \sqrt{2}N_1(SC)$$

Thus, the *total number of replenishment orders per year increases by 40% under option 2, which will increase the order cost by 40%.* These results can be easily extended from two DCs to any number of DCs. We will discuss those results in the next section.

5.3.2 General Risk Pooling Model

Let us now generalize the results of Section 5.3.1 to n identical DCs, such that the annual demand D is split evenly across all the DCs, each with a demand of $\dfrac{D}{n}$. Since the derivations of results are very similar to those in Section 5.3.1, we will just present the final results.

- Order quantity at each DC:

$$Q_n^* = \sqrt{\frac{2A\left(D/n\right)}{h}} = \frac{Q_1^*}{\sqrt{n}}$$

- Average inventory at each DC:

$$\overline{I}_n = \frac{1}{2}Q_n^* = \frac{Q_1^*}{2\sqrt{n}}$$

- Total inventory in the supply chain:

$$\overline{I}_n(SC) = n\overline{I}_n = n\left(\frac{Q_1^*}{2\sqrt{n}}\right) = \sqrt{n}\,\frac{Q_1^*}{2}$$

Since $\dfrac{Q_1^*}{2}$ is the average inventory in the supply chain with one central DC, denoted by $\overline{I}_1(SC)$, we get the key result:

$$\overline{I}_n(SC) = \sqrt{n}\,\overline{I}_1(SC) \tag{5.38}$$

Thus the inventory in the supply chain increases by $\sqrt{n}$.

- Total number of orders in the supply chain:

$$N_n(SC) = \sqrt{n}\,N_1(SC) \tag{5.39}$$

where $N_1(SC)$ and $N_n(SC)$ are the total number of orders in the supply chain with one central DC and n regional DCs respectively. Thus, the annual number of orders increases by $\sqrt{n}$.

- Annual inventory holding cost (IHC)

$$IHC = (h)(\text{average inventory})$$

Hence,

$$IHC_1 = (h)\overline{I}_1(SC)$$

$$IHC_n = (h)\overline{I}_n(SC)$$

Using Equation 5.38:

$$IHC_n = \left(\sqrt{n}\right)IHC_1 \tag{5.40}$$

Thus, the IHC increases by $\sqrt{n}$.

- Annual order cost (AOC)

  The annual order cost (AOC) is given by

$$AOC = (A)(\text{Number of orders per year})$$

Where A is the fixed cost per order.

Thus,

$$AOC_n = AN_n(SC)$$

$$= A\sqrt{n}N_1(SC)$$

$$= \sqrt{n}AOC_1 \tag{5.41}$$

Thus, the AOC increases by $\sqrt{n}$ due to deconsolidation.

Example 5.7

A company's distribution network has 10 regional warehouses for the U.S. market. The customer zones are assigned to the warehouses in such a way that their demands are equal. The company decides to close four warehouses and assign their customer demand equally to the remaining six.

(a) What will be the impact of this risk pooling strategy on the total inventory carried in the supply chain?
(b) Suppose the company wants to reduce the inventory by 50%, how many warehouses should it close?

Solution

(a) Under the current system with 10 warehouses, the inventory carried in the supply chain, $\overline{I_{10}}(SC)$ equals:

$$\overline{I_{10}}(SC) = \sqrt{10}\,\overline{I_1}(SC) \tag{5.42}$$

where, $\overline{I_1}(SC)$ is the inventory in the supply chain with just one warehouse.

After closing four warehouses, the supply chain inventory becomes

$$\overline{I_6}(SC) = \sqrt{6}\, \overline{I_1}(SC) \qquad (5.43)$$

Using Equations 5.42 and 5.43, we get

$$\frac{\overline{I_6}(SC)}{\overline{I_{10}}(SC)} = \sqrt{\frac{6}{10}} \cong 0.775$$

Hence, the supply chain inventory is reduced by 22.5%.

(b) To reduce the inventory by 50%, we want,

$$\frac{\overline{I_n}(SC)}{\overline{I_{10}}(SC)} = 0.5 \qquad (5.44)$$

where n is the number of warehouses in the risk pooled system. Since $\overline{I_n}(SC) = \sqrt{n}\, I_1$ and $\overline{I_{10}}(SC) = \sqrt{10}\, I_1$, Equation 5.44 reduces to

$$\frac{\sqrt{n}}{\sqrt{10}} = 0.5$$

$$n = 10(0.5)^2 = 2.5 \text{ warehouses}$$

Thus, by closing eight warehouses and keeping just two central warehouses, the supply chain inventory will be reduced by nearly 55%. By keeping three warehouses, it will reduce the supply chain inventory by nearly 45%.

5.3.3 Pros and Cons of Risk Pooling

In Sections 5.3.1 and 5.3.2, we derived the benefits of risk pooling, assuming equal distribution of demand among the warehouses. This was done primarily for ease of calculations and to estimate quickly the exact reduction in inventory and number of orders. However, the demands do not have to be distributed evenly among the DCs. When customer demands are divided unequally, the average inventory for each DC has to be calculated based on its share of the demand, in order to calculate the total inventory in the supply chain. Still, the final results, namely the reduction in supply chain inventory and annual orders, will be true.

Let us now summarize the advantages and drawbacks of the risk pooling strategy, which consolidates customer demands by using fewer warehouses for distribution.

Pros and cons of risk pooling

- Supply chain inventory is reduced, resulting in lower IHC.
- Total replenishment orders at the warehouses are reduced, resulting in lower annual procurement cost.
- Since the order quantities at the consolidated DCs are higher, inbound transportation cost from the factory to the DCs will be reduced due to economies of scale and volume discounts in shipping.
- With fewer warehouses, the outbound transportation cost will increase.
- With fewer DCs, the facility cost of operating the warehouses will reduce.
- With fewer warehouses, the risk of supply chain disruption will increase.

So far in our discussion, we assumed that the demand is deterministic and used the EOQ formulas to derive the results to show the advantages of risk pooling. Naturally, the question arises, is risk pooling as effective when the demand is a random variable? In fact, the benefits of risk pooling increase significantly when there is high demand uncertainty. As discussed in Chapter 3, companies maintain *safety stock* to manage uncertain demand and unreliable supply. We will derive results in the next section to show how the safety stock is reduced under risk pooling.

5.3.4 Risk Pooling under Demand Uncertainty

Consider a 3-stage supply chain—a plant, warehouses, and several demand regions. Let $X_1, X_2, ..., X_N$ are the random variables denoting the *weekly demand* in regions 1, 2, ..., N. Assume that the X_i's are distributed Normal with mean μ_i and variance σ_i^2. Also, let ρ_{ij} be the correlation coefficient between demands in regions i and j. Note that ρ_{ij}'s are between -1 and $+1$. Let L represent the replenishment lead time in weeks between the plant and the warehouses, irrespective of their locations. To illustrate the impact on safety inventory due to risk pooling, we will consider the following two *extreme network configurations*:

Case I: N warehouses, one for each demand region (deconsolidated)

Case II: One central warehouse supplying all the N demand regions (consolidated/risk pooled)

1. *Completely deconsolidated distribution network*: Here each retail region is supplied by its own dedicated warehouse. To determine the safety stock, using *service level* (SL) criterion, we need the distribution of the

lead time demand (LTD) for each warehouse. Let LTD_i represent the lead time demand for warehouse i. Then,

$$E(LTD_i) = L\mu_i, \quad i = 1, 2, \ldots, N \tag{5.45}$$

$$V(LTD_i) = L\sigma_i^2, \quad i = 1, 2, \ldots, N \tag{5.46}$$

The *safety stock* to be maintained in warehouse i, denoted by SS_i, is given by,

$$SS_i = K_{SL}\sqrt{V(LTD_i)} = K_{SL}\sqrt{L}\sigma_i, \quad i = 1, 2, \ldots, N \tag{5.47}$$

where K_{SL} = safety factor for a given service level SL, for a normal distribution. For example, for 90% service level, $K_{SL} \approx 1.29$.

Thus, the total safety stock carried in the supply chain denoted by SS (SC), is given by,

$$SS(SC) = \sum_{i=1}^{N} SS_i = K_{SL}\sqrt{L}\left[\sum_{i=1}^{N} \sigma_i\right] \tag{5.48}$$

2. *Risk Pooled/Consolidated network*: Here the regional demands are pooled and one central warehouse supplies them all. Let X_p denote the random variable representing the "pooled demand" at the central warehouse. Hence,

$$X_p = X_1 + X_2 + \cdots + X_N$$

$$E_p = E(X_p) = \sum_{i=1}^{N} E(X_i) = \sum_i \mu_i$$

$$V_p = V(X_p) = V\left(X_1 + X_2 + \cdots + X_N\right) = \sum_{i=1}^{N} V(X_i) + 2\sum_{i=1}^{N-1}\sum_{j=i+1}^{N} Cov(X_i, X_j)$$

$$= \sum_{i=1}^{N} \sigma_i^2 + 2\sum_{i=1}^{N-1}\sum_{j=i+1}^{N} \rho_{ij}\sigma_i\sigma_j \tag{5.49}$$

Note that the term $Cov(X_i, X_j)$ in the aforementioned equation represents the covariance of demands between regions i and j and $\rho_{ij} = \dfrac{Cov(X_i, X_j)}{\sigma_i\sigma_j}$.

Assuming the same replenishment time L, from the plant to the central warehouse, the lead time demand at the warehouse is given by

$$E(LTD) = LE_p$$

$$V(LTD) = LV_p$$

$$\sigma_{LTD} = \sqrt{L}\sqrt{V_p} = \sqrt{L}\sigma_p$$

where σ_p is the standard deviation of the pooled demand.

Since there is only one warehouse, the safety stock at the warehouse is the same as the supply chain safety stock. Hence,

$$SS(SC) = K_{SL}\sqrt{L}[\sigma_p] \qquad (5.50)$$

We can now compare the supply chain safe stocks under the two cases, by comparing the terms in the double brackets in Equations 5.48 and 5.50. To prove that the safety stock in the pooled system is lower than that of the decentralized system, we have to show:

$$\sigma_p \leq \sum_i \sigma_i \qquad (5.51)$$

Squaring both sides of Equation 5.51, and substituting Equation 5.49 for V_p, we get

$$V_p \leq \left(\sum_i \sigma_i\right)^2$$

$$\sum_{i=1}^{N} \sigma_i^2 + 2\sum_{i=1}^{N-1}\sum_{j=i+1}^{N} \rho_{ij}\sigma_i\sigma_j \leq \sum_{i=1}^{N} \sigma_i^2 + 2\sum_{i=1}^{N-1}\sum_{j=i+1}^{N} \sigma_i\sigma_j \qquad (5.52)$$

Since ρ_{ij}'s are between -1 and $+1$, Equation 5.52 is always true! Hence, the safety stock in the risk pooled system is *never greater* than the safety stock in the deconsolidated system. Let us now study some special cases under the risk pooled system.

Case II-A: When the regional demands are independent, i.e., $\rho_{ij} = 0$ for all i and j.
In this case, Equation 5.52 becomes a strict inequality:

$$\sum_{i=1}^{N} \sigma_i^2 < \sum_{i=1}^{N} \sigma_i^2 + 2\sum_{i=1}^{N-1}\sum_{j=i+1}^{N} \sigma_i\sigma_j$$

Hence, the safety stock for the risk pooled system will be *significantly reduced*.

Case II-B: When there exist some negative correlations between the regional demands, namely, when some of the ρ_{ij}'s are negative.

In this case, the left-hand-side of inequality (5.52) will be lower than the right-hand-side. Thus, the safety stock in the risk pooled system will be *reduced*.

Case II-C: When the regional demands are perfectly correlated such that $\rho_{ij} = 1$ for all i and j.

In such a case, inequality (5.52) will become a strict equation and risk pooling does *not* provide any advantage in reducing the safety stock. However, in practice, such perfect correlations do not exist. Hence, we can conclude that for all practical situations, *risk pooling reduces the safety stock inventory* in the supply chain for the same service level. However, if we maintain the same amount of safety stock, then risk pooling will *increase the service level to the customers!*

5.3.5 Risk Pooling Example

We shall illustrate the reduction in safety stock due to risk pooling with a numerical example.

Example 5.8

ABC Tools has divided its northeast U.S. market into three customer regions. The weekly demands at the regions are normally distributed with mean and standard deviation as shown in Table 5.10. The demand correlations are given in Table 5.11. The company wants

TABLE 5.10

Weekly Demands (Example 5.8)

Region	Mean	Standard Deviation
1	2000	100
2	3000	50
3	1500	50

TABLE 5.11

Demand Correlation Matrix (Example 5.8)

Region	1	2	3
1	1	0.3	0.1
2	0.3	1	0
3	0.1	0	1

to provide 90% service level to its customers and is considering three distribution options:

1. Three dedicated DCs, one for each demand region
2. Two regional DCs, one covering region 1 and another covering regions 2 and 3 and
3. One central DC covering all three demand regions

The replenishment lead time from the plant to the DC is 1 week under all the options. Compare the total safety stocks ABC Tools has to carry under each option.

Solution

Option 1 (Three Regional DCs): Since the lead time is 1 week, the weekly demand and the lead time demands are the same. For 90% service level, the safety factor $K_{SL} = 1.29$. Using Equation 5.47, the safety stocks to be carried at the three DCs are as follows:

$$DC - 1 : SS_1 = (1.29)\,100 = 129$$

$$DC - 2 : SS_2 = (1.29)\,50 \approx 65$$

$$DC - 3 : SS_3 = (1.29)\,50 \approx 65$$

Total safety stock in the supply chain, $SS\ (SC) = 129 + 65 + 65 = 259$.

Option 2 (Two DCs): Assume that DC – 1 supplies demand region 1 and DC – 2 supplies regions 2 and 3. The safety stock for region 1 will be the same as in option 1, i.e., $SS_1 = 129$.

To compute the safety stock for DC – 2, we need the covariance of the pooled demand for DC – 2. Using Equation 5.49 we get,

$$V_p = \sigma_2^2 + \sigma_3^2 + 2\rho_{23}\sigma_2\sigma_3 = 50^2 + 50^2 + 2(0)(50)(50) = 5000$$

The safety stock to be carried at DC – 2 is given by

$$SS_2 = (1.29)\sqrt{5000} \cong 91$$

Total safety stock in the supply chain,

$$SS_{SC} = SS_1 + SS_2 = 129 + 91 = 220$$

Note that the supply chain safety stock has reduced from 259 in option 1 to 220 in option 2, a 15% decrease!

Option 3 – (One central DC): To compute the safety stock at the central DC, we need the variance of the pooled demand faced by one DC.

$$V_p = \sigma_1^2 + \sigma_2^2 + \sigma_3^2 + 2\rho_{12}\sigma_1\sigma_2 + 2\rho_{13}\sigma_1\sigma_3 + 2\rho_{23}\sigma_2\sigma_3$$

$$= 100^2 + 50^2 + 50^2 + 2(0.3)(100)(50) + 2(0.1)(100)(50) + 2(0)(50)(50)$$

$$= 10,000 + 2,500 + 2,500 + 3,000 + 1,000 = 19,000$$

The supply chain safety stock is given by,

$$SS(SC) = 1.29\sqrt{19000} \cong 178$$

Note that the supply chain safety stock is further reduced in option 3 to just 178. *This amounts to a decrease by nearly 31% from option 1 and 19% from option 2!*

Determining the *best level* of risk pooling or inventory aggregation is a difficult problem and needs extensive analysis of all possible scenarios. Minimizing costs in the supply chain is not always the only priority of companies. Maintaining a high level of customer responsiveness is also considered important. Since risk pooling worsens customer responsiveness, the conflicting nature of these two criteria—cost and responsiveness, makes the solution to the risk pooling problem difficult.

Example 5.8 is a smaller version of a larger case study on risk pooling reported by Gaur and Ravindran (2006). In their paper, the authors develop a stochastic bi-criteria nonlinear IP model to determine the best supply chain distribution network to meet customer demands, where minimizing supply chain costs while maintaining high levels of responsiveness is important. They develop a two-stage optimization algorithm to solve the problem.

5.3.6 Practical Uses of Risk Pooling

In Section 5.3.3, we discussed the pros and cons of risk pooling. Here we will discuss certain environmental factors and product considerations that lead to potential opportunities for risk pooling. Listed in the following are the situations where risk pooling or inventory consolidation can be used to get maximum benefits:

1. *Product with high demand uncertainty or forecast errors*: During the growth phase of a new product, the demand variability is very high. Since risk pooling reduces the variance of the demand, it can reduce the safety stock for the same level of service or increase the service level for the same amount of safety stock.

2. *Products with high value-to-weight ratio*: Electronic goods (laptops, smart phones, cameras) and jewelry are examples of such products. Their IHC will be much higher compared to the distribution cost. Risk pooling will reduce the total inventory carried in the supply chain and hence, the total IHCs. Since they weigh less, air freight can be used inexpensively to reduce their delivery time.

3. *Large warehousing cost*: When facility costs are high for the warehouses, they form a larger share of the total supply chain costs.

Since risk pooling reduces the number of facilities, it provides significant opportunities to reduce the supply chain cost.

4. *Large customer orders*: One of the drawbacks of risk pooling is that it would increase the outbound transportation cost of distributing the products to the customers due to fewer warehouses. However, if there are several large customer order, economies of scale can be used to lower the outbound transportation costs also.

5.4 Continuous Location Models

In the previous section, we discussed *risk pooling*, a strategy to consolidate or centralize the distribution system by using fewer warehouse facilities. Given the complexity, the network design process is an iterative one that evaluates several network options that range from highly consolidated to highly deconsolidated. Ultimately, a choice on the risk pooling spectrum will lead to how many facilities there should be at each echelon of the network. Once the decision maker has an idea of the number of facilities and a rough regional configuration, he or she can answer the question where the good candidate locations would be using a *continuous location model*. This "ballpark" set of locations can then be subjected to qualitative analysis that considers *service-oriented issues* (roadway access for trucks, nearness to customers) and *local factors* (low taxes, tax exemptions, tax credits, cost of labor, union profile, expansion opportunity, and community disposition to industry). Of course, property and labor costs and low taxes could be directly incorporated into the IP models discussed in Section 5.2. *Continuous location models* examine all possible locations along a *space continuum* and select the best one based on a *weighted distance metric*. In contrast, the IP models we studied in Section 5.2 can be called *discrete location models* since they select the best sites for the facilities given a list of potential sites. In fact, in designing a supply chain network with multiple facilities for plants and DCs, *discrete locations* are used extensively in practice.

5.4.1 Continuous Location Model: Single Facility

We will begin with the discussion of determining the best location for a single facility. We will present the *gravity model* and an iterative method for its solution.

5.4.1.1 Gravity Model

The *gravity model* considers several existing facilities (plants and demand regions), with known locations on a plane. Each facility has an associated weight. These weights may represent volume or quantity handled at these facilities, if the freight rates (\$/mile) are constants or the distribution cost for each facility if the freight rates vary depending on the location of the existing facility. The objective of the *gravity model* is to determine the location of the new facility (warehouse/DC) that will receive supplies from the existing plants and delivers them to the given customer regions. The model assumes that there is no capacity constraint and the fixed cost for the new facility is the same everywhere. The *criterion* used for the optimization is to minimize the sum of the weighted Euclidean distance between the new and existing facilities.

Assume that there are N existing facilities. The weight of facility n ($n = 1$, 2, ..., N) is w_n and its coordinate location on a plane is (x_n, y_n). If the new facility's location is (x, y), then the distance d_n for each facility is given by

$$d_n = \sqrt{(x - x_n)^2 + (y - y_n)^2}, \quad n = 1, 2, \ldots, N \tag{5.53}$$

Then the optimization problem is to

$$\text{Minimize } Z = \sum_{n=1}^{N} w_n d_n \tag{5.54}$$

Note that the weights w_n for location n can be computed in two ways:

1. $w_n = Q_n$, where Q_n is the quantity handled (supply/demand) at location n, assuming the same freight rate (e.g., \$ per ton-mile) at all locations.
2. $w_n = Q_n R_n$, where R_n is the freight rate (\$ per ton-mile) at location n.

5.4.1.2 Iterative Method

The objective function of the gravity model is given by

$$\text{Minimize } Z = \sum_{n=1}^{N} w_n \left[\sqrt{(x - x_n)^2 + (y - y_n)^2} \right] \tag{5.55}$$

We can show that Equation 5.55 is a convex function. Hence, setting its partial derivatives with respect to x and y equal to zero, will provide the optimal location, denoted by (x^*, y^*), as given in the following:

$$x^* = \frac{\sum_{n=1}^{N} w_n x_n / d_n}{\sum_{n=1}^{N} w_n / d_n} \tag{5.56}$$

$$y^* = \frac{\displaystyle\sum_{n=1}^{N} w_n y_n / d_n}{\displaystyle\sum_{n=1}^{N} w_n / d_n} \qquad (5.57)$$

Unfortunately, a direct solution to Equations 5.56 and 5.57 is not possible since

$$d_n = \sqrt{(x^* - x_n)^2 + (y^* - y_n)^2}$$

Hence, an iterative algorithm is used to find the optimal solution. Assuming an initial trial value for x^* and y^*, Equations 5.56 and 5.57 are used to update the initial values. These values, in turn, are used to update d_n, and get the next set of values for x^* and y^*. The process is repeated until two successive values of x^* and y^* are very close, within a user-specified accuracy of $\varepsilon\%$. The most common initial values are those obtained by setting $d_n = 1$ for all $n = 1, 2, \ldots, N$ (assuming that the initial location is at the centroid). The origin $(0,0)$ can also be used as the initial value. We shall now illustrate the iterative method with an example.

5.4.1.3 Illustrative Example: Gravity Model

In the risk pooling Example 5.8, ABC Tools examined three network options, from completely deconsolidated to fully consolidated, based on inventory in the supply chain. Suppose ABC Tools decides to go with the option of one central warehouse supplying all three customer regions. Then, the next question is where the central location should be such that it minimizes the distribution cost.

> **Example 5.9 (Gravity model)**
>
> Consider the 3-stage supply chain given in Figure 5.8.
> The supply at the plant, the demands at the retailers, their coordinate locations in miles and their freight rates are given in Table 5.12.
> The problem is to determine the location of the single DC that will minimize the total distribution cost. Use an accuracy of 5% for termination of the algorithm.
>
> **Solution**
>
> First we calculate the weights for each existing location as follows:
>
> $$w_1 = (200)(2) = 400$$
>
> $$w_2 = (300)(2) = 600$$
>
> $$w_3 = (100)(2) = 200$$
>
> $$w_4 = (600)(1) = 600$$

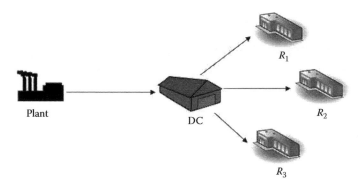

FIGURE 5.8
Three-stage supply chain (Example 5.9).

TABLE 5.12

Data for Example 5.9

Location	Facility	Quantity	Freight Rate ($ per unit per mile)	Coordinates (Miles)	
				x	y
1	R_1	200	$2	20	50
2	R_2	300	$2	30	80
3	R_3	100	$2	70	90
4	Plant	600	$1	80	20

Assuming $d_n = 1$ for $n = 1, 2, 3, 4$ in Equations 5.56 and 5.57, we get the initial values of x^* and y^* for iteration 1 as follows:

$$x^*(1) = \frac{\sum_{n=1}^{4} w_n x_n}{\sum_{n=1}^{4} w_n} = \frac{88,000}{1,800} \cong 48.89$$

$$y^*(1) = \frac{\sum_{n=1}^{4} w_n y_n}{\sum_{n=1}^{4} w_n} = \frac{98,000}{1,800} \cong 54.44$$

Using these values, we calculate the distances for the existing facilities using Equation 5.53 as $d_1 = 29.23$, $d_2 = 31.78$, $d_3 = 41.35$ and $d_4 = 46.41$. The total distribution cost is given by

$$Z_1 = \sum_{n=1}^{4} w_n d_n = \$66,878.50$$

TABLE 5.13

Iteration 1 Solution (Example 5.9)

Location n	x_n	y_n	w_n	$w_n x_n$	$w_n y_n$	d_n	$\dfrac{w_n}{d_n}$	$\dfrac{w_n x_n}{d_n}$	$\dfrac{w_n y_n}{d_n}$
1	20	50	400	8,000	20,000	29.23	13.68	273.70	684.25
2	30	80	600	18,000	48,000	31.78	18.88	566.34	1510.25
3	70	90	200	14,000	18,000	41.35	4.84	338.54	435.27
4	80	20	600	48,000	12,000	46.41	12.93	1034.25	258.56
						SUM	50.33	2212.83	2888.33

Using the aforementioned distances in Equations 5.56 and 5.57, we update the values of x^* and y^*. These calculations can be easily done on a spreadsheet as shown in Table 5.13.

Thus,

$$x^*(2) = \frac{2212.83}{50.33} \cong 43.97$$

$$y^*(2) = \frac{2888.33}{50.33} \cong 57.39$$

Using these values, the new distances are calculated for iteration 2. The results for the first four iterations of the gravity model are shown in Table 5.14.

The algorithm is terminated at iteration 4, since we have achieved 5% accuracy level on the values of x^* and y^*. Thus, the approximate location for the DC is (40.35, 58.43) with a minimum cost of $65,648.30. Figure 5.9 shows the locations of the existing facilities and the new DC on a planar graph.

TABLE 5.14

Results for 4 Iterations (Example 5.9)

Iteration	x	y	d_1	d_2	d_3	d_4	Total Cost
1	48.89	54.44	29.23	31.78	41.35	46.41	66,878.5
2	43.97	57.39	24.38	29.13	44.07	49.84	65,947.5
3	41.56	58.18	22.01	28.05	45.53	51.61	65,710.2
4	40.35	58.43	20.83	27.58	46.3	52.52	65,648.3

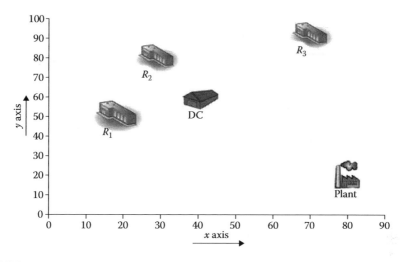

FIGURE 5.9
Facility location (Example 5.9).

5.4.1.4 Limitations of Gravity Model

Even though the gravity model is easy to understand and the iterative method is easy to implement on a spread sheet, there are limitations in the model's applicability. Listed in the following are few of its limitations:

1. Fixed facility costs are ignored or assumed to be the same irrespective of the location.
2. Shipping costs are assumed to be linear with the distance. As discussed in Chapter 4, freight rates are nonlinear with fixed and variable cost that depend on the volume shipped.
3. There is no capacity constraint at any location.
4. Euclidean distance metric (straight line route) is an approximation, since the road network is rectilinear. However, conversion factors are available for U.S. intercity transport network.
5. The entire demand for a region is assumed to occur at a particular coordinate. Unless the region is very small, this is not accurate for estimating the distribution cost to the region.

5.4.2 Multiple Facility Location

It is not easy to extend the gravity model to multiple facilities. The main problem is that one has to decide not only the location of the multiple warehouses, but also the quantity shipped from each warehouse to the various demand regions. Thus, the optimization problem becomes very complex. There are two ways to overcome this complexity:

1. *Assume a set of potential sites for the multiple facilities*: This is more typical in practice and the IP models we discussed in Section 5.2 can then be applied directly to choose the best sites and the optimal distribution plan. Here we can take into account fixed facility cost, warehouse capacities, etc.

2. *Pre-assign the customer regions to the warehouses*: Once the warehouses have been assigned their customer regions, the continuous location problem for multiple facilities becomes a set of independent single facility problems. The gravity model and the iterative algorithm can then be applied to each facility. One way to select the optimal number of facilities and their assignments to the customer regions is to use the *risk pooling* algorithm we discussed in Section 5.3. To illustrate this approach consider Example 5.7 for risk pooling. Three warehouse options are analyzed under risk pooling. Suppose we decide to use option 2 based on safety stock and customer service. Option 2 uses 2 DCs: DC–1 serving demand region 1 and DC–2 serving demand regions 2 and 3. Since there are no interactions (each region can receive all its demand from one dedicated DC only), we solve two gravity models—one for DC–1, using the locations for Region 1 and the plant, and another for DC–2, using the locations for Regions 2 and 3 and the plant.

There are other approaches that cluster the demand regions and pre-assign them to the warehouses. The clustering can be done by grouping the demand regions based on how close they are with respect to one another. The process can be repeated by regrouping the regions and forming new clusters and evaluating them using the total distribution cost.

5.5 Real-World Applications

There are several published results of real-world applications using IP models for supply chain network design. We discuss briefly a few of the applications in practice. For interested readers, the cited references will provide more details on the case studies.

5.5.1 Multi-National Consumer Products Company

Here we discuss two case studies illustrating the applications of integer programming (IP) for location and distribution problems. Both are based on the dissertations of the author's doctoral students (Portillo, 2009; Cintron, 2010). Both case studies are related to a leading global health and hygiene company listed in the Fortune 500, selling paper, and personal care products. The company's sales are close to $20 billion a year. Its global brands are sold in 150 countries holding first or second market positions.

5.5.1.1 Case 1: Supply Chain Network Design

The first case study, described in Portillo (2009), was related to the company's largest international division; selling products in 22 countries, across one continent with 21 manufacturing plants, 45 DCs, 100 customer zones, and 22 brands. The company's supply chain was formed by different types of customers, from multi-national chains, and large distributors to thousands of small "mom and pop" stores. A mixed integer linear programming (MILP) model was developed for a complete strategic and tactical optimization of the manufacturing and distribution network based on customer demand projections for a 5 year horizon. The MILP model, with the objective of maximizing profits, consisted of three stages:

Stage 1: Determine the ability of the company to fulfill present and projected sales based on the current supply chain design.

Stage 2: Evaluate how the company's ability would be improved, considering the potential expansions of plants and DCs already in the horizon.

Stage 3: Optimize the global supply chain design that would deliver the best results for the entire 5 year time horizon.

The MILP model had 7500 variables, of which 300 were binary and 7000 constraints. The mathematical model was coded in ILOG and solved using the CPLEX solver (www.ILOG.com). The solver reached optimality in 2 min.

The stage 1 analysis showed a very close to full capacity utilization of the existing facilities under current demand levels. When considering future demand, the overall results showed that the existing supply chain could fulfill only 75% of the total projected demand, primarily restricted by production and distribution capacity.

The stage 2 analysis included capacity expansions in production and distribution facilities already considered by the management. Close to a dozen new production lines were planned within a 2 year horizon. Although management had already decided on their locations, multiple options were allowed in the model to confirm their choices. The results showed that the majority of the chosen locations were optimal. Although the locations were the best to maximize profit, the supply chain network was capable of meeting only 85% of the projected demand.

During the stage 3 analysis, additional levels of facility expansions were considered. The results provided production, distribution, and sales levels to maximize profits. However, the optimal demand fulfillment ratio was only 96%, highlighting specific product-market combinations that were not profitable.

Additional details on the MILP model are available in Chapter 8 as an illustrative case study for global supply chain network design.

5.5.1.2 Case 2: Distribution Planning

This case study (Cintron, 2010) was applied to determine the best distribution network in one of the countries for the same consumer product company discussed in Case 1. The country under study had 4 manufacturing plants, 66 retailers, 5 independent distributors, and 2 DCs (one company-owned and one leased). Multiple products were made at the plants to meet customer demand. No product was made in more than one plant. The company was the global competitor in that country with sales over $86M annually. The company was looking to reduce their distribution costs by improving the region's distribution network design. Specifically, the company was interested in designing the flow of products from the manufacturing plants to the customers.

The case study considered four distribution options for the customers to receive their products. Products can be supplied from the following:

1. The manufacturing plant directly
2. The company's DC
3. An independent distributor, who is supplied directly by the plant
4. An independent distributor, who is supplied by the company's DC

A mixed integer linear programming (MILP) model was developed to select the best option for each customer to maximize the profitability and customer responsiveness among other things. The MILP model had 2790 variables, of which 2500 were binary and 900 constraints. The model was solved using the GAMS software in less than 30 s. The optimal network recommended by the MILP model, eliminated the need for the leased DC for the company, and increased the direct shipment from plants to customers from 33% of all demands (existing policy) to 83%. This resulted in the reduction of supply chain distribution cost from 12% to 3% of net sales, a saving of $7M annually. The complete details of the model and the results are available in Cintron et al. (2010).

5.5.2 Procter and Gamble (P&G)

P&G sells over 300 brands of consumer products around the world. In 1993, P&G's Operations Research (OR) team undertook a major study, called *Strengthening Global Effectiveness*, to restructure P&G's global supply chain. A major part of the study was to examine the North American supply chain, which had 60 plants, 15 DCs, and more than 1000 customer zones. To be globally competitive, P&G decided to consolidate manufacturing plants to reduce cost and improve speed to market. The OR team decomposed the overall supply chain problem into two sub-problems: one dealing with the *location of the DCs*, and the other, dealing with *product sourcing*, one for each product category. Thus the DC locations were chosen independent of the plant locations. This was justified based on the facts that only 10%–20% of product volume goes through DC, the manufacturing costs are much larger than the distribution cost and the need to locate DCs closer to the customer zones to provide good customer service.

The *DC location model* was formulated as an uncapacitated facility location model, as discussed in Section 5.2.4. The solution to this integer program determined the location of the DCs and the customer zones assigned to each DC. The *product sourcing model* determined the location of the manufacturing plants, the products each plant makes and the distribution of the products to either directly the customer zone or through the DCs. Instead of formulating this as a MIP (mixed integer programming) model, the OR team developed several scenarios for plant locations and their products. Thus the product sourcing model was essentially reduced to solving a series of transportation problems, similar to the one we discussed in Example 5.4 (Section 5.2.2).

The OR study was completed in 1994 and was implemented in mid-1996. It resulted in closing 20% of the manufacturing plants at 12 sites and a savings of over $200 million annually. For more details on the study, readers are referred to Camm et al. (1997).

5.5.3 Ford Motor Company

Before introducing new model cars into the market, automobile companies go through a time-consuming and expensive process of developing prototypes of the model and subjecting them to multiple tests to check for any design flaws. Since the prototypes are generally one-of-a-kind, they are very expensive to make and may cost more than $250,000 per prototype. Complex vehicle design programs may require between 100 and 200 prototypes for product development. Naturally it is imperative to keep this cost down, but at the same time, have enough prototype vehicles to conduct all the required tests. Ford, with the help of the Engineering Management students at Wayne State University in Detroit developed a prototype optimization model (POM) to solve this problem.

POM was an IP model based on the set covering model formulation we discussed in Section 5.1.6. The rows of the set covering matrix are the various tests that need to be conducted, while the columns represent the different buildable vehicle configurations (prototypes). The elements of the set covering matrix, called *buildable combination matrix* (BCM), indicated whether a particular test can be done on a certain vehicle configuration. The objective function is to determine the minimum number of prototypes needed to complete all the required tests.

The set covering model was applied to the European Transit Vehicle program. Even though the BCM matrix initially contained 38,800 columns, the POM's optimal solution identified just 27 prototypes needed to cover all the required vehicle tests in the design-verification plan, resulting in a cost savings of $12M. Based on that success, Ford used POM to manage other complex vehicle design programs for Taurus, Windstar, Explorer, and Ranger. The total cost savings achieved for prototype vehicles were $250M from 1995 to 2000.

Interested readers should refer to the paper by Chelst et al. (2001) for more details on this case study.

5.5.4 Hewlett-Packard (HP)

The supply chain team at HP, called SPaM (Strategic Planning and Modeling), has been using quantitative methods to solve its global supply chain design problems since 1994 (Lee and Billington, 1995). The SPaM team's approach is to generate alternate scenarios based on intuition and expert knowledge for HP's global supply chain network design problems and use Operations Research for analyzing the scenarios. SPaM's initial success was solving the problems of spiraling inventory and declining customer satisfaction in the early 1990s. Their initial project dealt with the personal computer and desk jet printer divisions. The development of the *worldwide inventory network optimizer* (WINO) became the building block of the complete Supply Chain Management models at HP. Customer fill rate and finished goods inventory were the two conflicting objectives of WINO. WINO achieved inventory reduction of 10%–30% and increased customer satisfaction simultaneously. From inventory modeling, SPaM moved to supply chain design strategies in manufacturing and distribution.

In a recently published case study (Laval et al., 2005), the SPaM team reported on how scenario analysis and optimization were combined to design the global supply chain for HP's Imaging and Printing Group. The main objective of the group was to reduce the number of its contract manufacturing partners worldwide. Using expert knowledge, the SPaM team selected scenarios that included closing several existing sites and opening some new ones centrally. The scenario-based modeling approach helped to reduce the number of binary variables and constraints of the IP model. Each scenario defined a set of contract manufacturing locations, the products handled at those locations and the demand areas they served. Uncertainties in demands and lead times were also included in the scenarios. The optimization model was a MIP model. The objective function was to minimize the total supply chain cost that included outbound and inbound transportation cost, and manufacturing cost (fixed production cost plus variable labor cost). The model's recommendations resulted in cost savings of over $10M in supply chain cost, without sacrificing the existing service levels.

5.5.5 BMW

BMW produces cars in eight plants located in Germany, United Kingdom, the United States, and South Africa. Its engines are manufactured at four other sites. It also has six assembly plants. This case study reported on the development of a strategic planning model to design BMW's global supply chain for a 12 year time horizon (Fleischmann et al.2006). The optimization model addressed the allocation of car models to its global production facilities, supply of materials, and the distribution of cars to the global markets. The global market was aggregated into 10 sales regions.

The basic supply chain model was a multi-period MILP. The binary variables identified whether a certain product was produced at a given plant. The continuous variables represented the annual volume in supply, production, and distribution. The objective function included fixed capital expenditure as well as variable costs for supply, production, and distribution. An example model in the case study had 6 plants, 36 products, and 8 sales regions for a 12 year planning horizon. The MILP model had 60,000 variables (2000 binary) and 145,000 constraints. The model was solved using the ILOG/CPLEX solver on a 1.6 GHz processor in just 4 min!

The use of the MILP model improved BMW's long-term strategic planning process. It made the planning process more transparent and reduced the planning effort. It also resulted in the reduction of 5.7% in investment and the cost of materials, production, and distribution.

5.5.6 AT&T

AT&T provides telecommunication equipment and long-distance services to the telemarketing industry. To help its telemarketing customers for their site location decisions, AT&T developed a decision support system (DSS) (Spencer et al., 1990). When the "800-service" was introduced in 1967, the cost of 800-service determined their locations, since the labor cost was minimal. The Midwest, particularly Omaha (Nebraska), became the "800-captial" of the world. However, in the 1980s, labor cost started becoming very significant. AT&T provided the *decision support system* (DSS) as a value-added service to its customers.

The DSS was basically a MILP model, similar to the capacitated facility location model we discussed in Section 5.2.3 (Example 5.5). The MILP model determined the best locations for the telemarketing centers from a set of candidate locations, and the volume of customer traffic from different regions handled by each center. The objective was to minimize the total cost of labor, fixed facility, and communication. For example, in 1988, the model helped 46 AT&T telemarketing customers with their site location decisions. AT&T got business worth $375M in annual communication revenues and $31M in equipment sales.

5.5.7 United Parcel Service (UPS)

Began as the American Messenger Service in Seattle in 1907, the United Parcel Service (UPS) was established in 1919 to deliver packages, first in

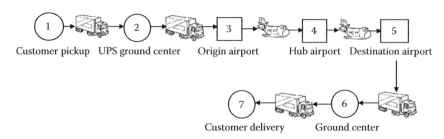

FIGURE 5.10
UPS's next-day-air service.

the United States and then worldwide. In 2004, UPS delivered more than 13 million packages daily to over 200 countries. UPS competes with FedEx and others to provide next-day-air express delivery service to its customers. The next-day-air service produced over $5B in revenue in 2002 and delivered more than 1 million packages overnight every day. Figure 5.10 illustrates UPS's next day air service.

UPS has more than 300 aircrafts of nine different types. It has more than 100 airports in the United States and 7 hub airports (six regional hubs and one all-points hub). With the tight pickup and delivery window for the next-day-air service, the planning of the aircraft routes is a very complex problem.

For example, each type of aircraft has different cargo capacity, maximum flying range, and speed and can only fly to certain airports. The maximum number of airports, a plane can serve is two (excluding the hub). Until 1993, UPS used manual planners, who used to take 9 months to produce a single plan! In 1994, UPS's Operations Research team, working jointly with the MIT faculty, developed an optimization model, called *VOLCANO–Volume, Location and Aircraft Network Optimizer* (Armacost et al., 2004).

VOLCANO was basically an integer program, based on the set covering and partitioning model discussed in Section 5.1.6. The "rows" of the set covering matrix were the "flight-legs." Each flight leg was a pair of take-off and destination airports for a plane. To serve the more than 100 airports in the United States (Figure 5.10), UPS had more than 2000 flight legs. The "columns" of the set covering matrix were the feasible routes for the different aircrafts. There could be more than one feasible route for each aircraft based on its cargo capacity, speed, flying range, permissible airports, etc. The elements of the set covering matrix, a_{ij}'s, were set to one, if flight leg i could be served by route j and zero otherwise. The cost of each feasible route was also determined. The objective was to cover all the flight legs such that the total cost of the aircraft routes was minimized. The set covering problem was solved using the commercial optimization software, ILOG-CPLEX solver.

Use of VOLCANO saved UPS over $87M between 2000 and 2002. UPS estimated additional savings of over $189M for the next 10 years.

5.6 Summary and Further Readings

5.6.1 Summary

In this Chapter, we answered the question of selecting the best location for the facilities (plants and DCs) in the supply chain network and the assignment of those facilities to the customers they serve. Since IP models with binary variables have been successfully used in practice to design supply chain networks, we began the chapter with a review of modeling with binary variables. We then applied the IP models for location and distribution decisions. The key decision variables were the number and location of various types of facilities—spanning as far as supplier sites, manufacturing locations and distribution centers, and the quantities shipped from upstream sites to downstream sites and ultimately out to the customers. The objective function was to minimize the cost of operating the supply chain network or maximize its profitability.

Next we discussed an important aspect of supply chain network design, called risk pooling or the use of consolidated distribution network to serve the customers. We showed risk pooling reduced inventory and facility costs. Even though the inbound shipping cost could be reduced under risk pooling, the outbound shipping cost would be increased. Moreover, risk pooling reduced supply chain responsiveness and increased disruption risk in the supply chain.

We then presented the basics of the "continuous location" models. We presented the "gravity model" for single facility location and the iterative algorithm for its solution. Extensions to the multiple facility location models were also discussed.

Finally we presented several published real-world applications using IP models for network design. Case studies from a major consumer products company, Ford Motor Company, Procter and Gamble, Hewlett-Packard, BMW, AT&T, and UPS were discussed.

We shall now discuss further readings on these topics for those interested.

5.6.2 Further Readings

5.6.2.1 Multiple Criteria Models for Network Design

The optimization models discussed in this chapter had a single objective— either to minimize supply chain costs or to maximize supply chain profitability, in case the product prices vary by location or customer. However, customer demand fulfillment and service are also important in designing a supply chain network. More recently, supply chain risk is emerging to be another important criterion (Supply chain risk is discussed in detail in Chapter 7). Hence, recent applications of optimization models have used multiple criteria optimization models for decision making.

The first case study on designing the supply chain footprint for a consumer products company (Section 5.5.1.1) was extended to a multiple criteria model by incorporating gross profit, customer service, and sourcing risk as three conflicting criteria (Portillo, 2009). Customer service was measured in terms of demand fulfillment (fill rates) and lead time (delivery time). Since this case study involved a global supply chain network, sourcing risk considered both facility-specific and country-specific risks. Facility risk was a risk score based on expert opinion on qualitative assessment of risk factors such as transportation infrastructure, existence of unions, zoning risk, and occurrence of natural disasters. The risk score was computed using *analytic hierarchy process* (AHP) and the multiple criteria optimization model was solved using *goal programming*. Both AHP and goal programming techniques are discussed in detail in Chapter 6. In Chapter 8, we present, as a case study, a multiple criteria model for global supply chain network design and discuss its results.

The second case study for the consumer products company (Section 5.5.1.2) to determine the best distribution network for the company's supply chain was also a multiple criteria model (Cintron et al., 2010). It considered five conflicting criteria-gross profit, delivery time, power, credit performance, and distributor's reputation. Except for delivery time, all other objectives were maximized. Power ratings were assigned to the customers based on their potential to grow their relationship with the company. Credit ratings were assigned to customers based on their past credit history. The company wanted to move the customers with lower power and credit ratings to the third party distributor to handle. This model was also solved using goal programming techniques. The basic model was then extended to a multi-period strategic tactical model (Cintron, 2010).

5.6.2.2 Risk Pooling

Eppen (1979) demonstrated the reduction in total cost due to risk pooling, when the demands follow a normal distribution. Chen and Lin (1989) extended Eppen's results for non-normal distributions also. Evers (2001) discussed the benefits of statistical risk pooling, the square root law, and its impact on physical aggregation. Collier (1982) studied the effects of component commonality and physical aggregation of inventory. Zeng (1998) developed a spreadsheet model to determine the optimal number of sources and the lot sizes. She also used sensitivity analysis to show that transportation had a significant impact on the decisions.

Since risk pooling has both positive and negative impacts on supply chain performance, multiple criteria models have been used to analyze the tradeoff between inventory reduction and customer service. Glasserman and Wang (1998) quantified the tradeoff between lead times and inventory levels to achieve required fill rates. They developed a linear slope relationship between lead time and inventory level for a given fill rate. Lodree et al. (2004)

considered responsiveness to determine order policies. Jang (2006) discussed an improved demand forecasting and then optimizing excess inventory and backorder. Kulkarni et al. (2005) evaluated the tradeoff between the reduction in logistics cost and loss of risk pooling benefits in plant networks, which spread component manufacturing over several plants as compared to those that consolidate component manufacturing in a single plant.

Finally, Gaur and Ravindran (2006) developed a stochastic bi-criteria nonlinear IP model to determine the best supply chain network that minimizes cost and delivery time.

5.6.2.3 Facility Location Decisions

Decisions regarding facility locations have significant *qualitative* aspects to them. As with customer service, these factors may be challenging to incorporate into a quantitative decision model. Consider the "10 most powerful factors in location decisions," from a 1999 survey of the readers of *Transportation and Distribution* (T&D) magazine (Scwartz, 1999a), listed, not necessarily in order of importance, in the following:

- Reasonable cost for property
- Roadway access for trucks
- Nearness to customers
- Cost of labor
- Low taxes
- Tax exemption
- Tax credits
- Low union profile
- Ample room for expansions
- Community disposition to industry

In fact, only two or perhaps three of these ten—property cost, labor cost, and low taxes—could be directly incorporated into one of the quantitative facility location models. The other could be grouped into what another T&D article (Schwartz, 1999b) referred to as *service-oriented issues* (roadway access, nearness to customers) and *local factors* (tax exemptions, tax credits, union profile, expansion opportunity, and community disposition). The T&D article goes on to state, mostly from consulting sources, that service-oriented issues are becoming more dominant. In a different trade publication, Atkinson (2002) states, "While everyone can identify critical elements, not all of them agree on the order of priority." In Chapter 6, Section 6.3, we discuss methods for assessing the importance of criteria and their ratings using a single decision maker as well as a group of decision makers. These criteria can then be incorporated explicitly as objective functions in a multiple criteria model for facility locations.

5.6.2.4 Case Studies

In addition to the case studies we discussed in Section 5.5, a number of real-world applications in supply chain management are available in the *Interfaces* journal. *Interfaces* is published bi-monthly, by the *institute of operations research and management science* (INFORMS). Each issue contains real applications of Operations Research models in practice. Supply chain management applications appear frequently in this journal.

Exercises

5.1 Explain the differences between *set covering* and *set partitioning* models. Give two applications of each.

5.2 What is *risk pooling*? How does it affect safety stock and transportation costs? For what type of products risk pooling is beneficial?

5.3 Discuss the differences between the two types of quantity discounts. Is one better than the other? If so, under what conditions?

5.4 What are the pros and cons of discrete and continuous location models in facility location decisions? Discuss their practical applications.

5.5 Consider a supplier order allocation problem under multiple sourcing, where it is required to buy 2000 units of a certain product from three different suppliers. The fixed set-up cost (independent of the order quantity), variable cost (unit price), and the maximum capacity of each supplier are given in Table 5.15 (two suppliers offer quantity discounts).

The objective is to minimize the total cost of purchasing (fixed plus variable cost). Formulate this as a linear integer programming problem. You must define all your variables clearly, write out the constraints to be satisfied with a brief explanation of each and develop the objective function.

TABLE 5.15

Supplier Data for Exercise 5.5

Supplier	Fixed cost	Capacity	Unit Price
1	$100	600 units	$10/unit for the first 300 units
			$7/unit for the remaining 300 units
2	$500	800 units	$2/unit for all 800 units
3	$300	1200 units	$6/unit for the first 500 units
			$4/unit for the remaining 700 units

5.6 Reformulate the problem in Exercise 5.5 under the assumption that both suppliers 1 and 3 offer "all units" discount, as described in the following:

- Supplier 1 charges $10/unit for orders up to 300 units and for orders more than 300 units, the *entire order* will be priced at $7/unit.
- Supplier 3 charges $6/unit for orders up to 500 units and for orders more than 500 units, the *entire order* is priced at $4/unit.

5.7 (Ravindran et al., 1987) A company manufacturers three products A, B, and C. Each unit of product A requires 1h of engineering service, 10h of direct labor, and 3lb of material. To produce one unit of product B requires 2h of engineering, 4h of direct labor, and 2lb of material. Each unit of product C requires 1h of engineering, 5h of direct labor, and 1lb of material. There are 100h of engineering, 700h of direct labor, and 400lb of materials available. The cost of production is a nonlinear function of the quantity produced as shown in Table 5.16.

Given the unit selling prices of products A, B, and C as $12, 9, and 7 respectively, formulate a linear mixed integer program to determine the optimal production schedule that will maximize the total profit.

TABLE 5.16

Data for Exercise 5.7

Product A		Product B		Product C	
Production (Units)	Unit Cost ($)	Production (Units)	Unit Cost ($)	Production (Units)	Unit Cost ($)
0–40	10	0–50	6	0–100	5
41–100	9	51–100	4	over 100	4
101–150	8	over 100	3		
over 150	7				

Note: If 60 units of A are made, the first 40 units cost $10/unit and the remaining 20 units cost $9/unit.

5.8 Explain how the following conditions can be represented as linear constraints using binary variables.

(a) Either $x_1 + x_2 \le 3$ or $3x_1 + 4x_2 \ge 10$

(b) Variable x_2 can assume values 0, 4, 7, 10, and 12 only

(c) If $x_2 \le 3$, then $x_3 \ge 6$; Otherwise $x_3 \le 4$ (assume x_2 and x_3 are integers)

(d) At least 2 out of the following 5 constraints must be satisfied:

$$x_1 + x_2 \le 7$$

$$x_1 - x_2 \ge 3$$

$$2x_1 + 3x_2 \le 20$$

$$4x_1 - 3x_2 \geq 10$$

$$x_2 \leq 6$$

$$x_1, x_2 \geq 0$$

5.9 (Ravindran et al., 1987) Convert the following nonlinear integer program to a linear integer program.

Minimize: $Z = x_1^2 - x_1 x_2 + x_2$

Subject to: $x_1^2 + x_1 x_2 \leq 8$

$x_1 \leq 2$

$x_2 \leq 7$

$x_1, x_2 \geq 0$ and integer

Hint: Replace x_1 by $2^0 \delta_1 + 2^1 \delta_1$ and x_2 by $2^0 \delta_3 + 2^1 \delta_4 + 2^2 \delta_3$, where $\delta_i \in (0, 1)$ for $i = 1, 2, ..., 5$.

5.10 (Adapted from Srinivasan, 2010) Consider a supply chain network with three potential sites for warehouses and eight retailer regions. The fixed costs of locating warehouses at the three sites are given as follows:

Site 1: $100,000

Site 2: $80,000

Site 3: $110,000

The capacities of the three sites are 100,000, 80,000, and 125,000 respectively. The retailer demands are 20,000 for the first four retailers and 25,000 for the remaining.

The unit transportation costs ($) are given in Table 5.17:

(a) Formulate a mixed integer linear program to determine the optimal location and distribution plan that will minimize the total cost. You must define your variables clearly, write out the constraints, explaining briefly the significance of each and write the objective function. Assume that the retailers can receive supply from multiple sites. Solve using any optimization software. Write down the optimal solution.

TABLE 5.17

Data for Exercise 5.10

	R_1	R_2	R_3	R_4	R_5	R_6	R_7	R_8
Site 1	4	5	5	4	4	4.2	3.3	5
Site 2	2.5	3.5	4.5	3	2.2	4	2.6	5
Site 3	2	4	5	2.5	2.6	3.8	2.9	3.5

(b) Reformulate the optimization problem as a linear integer program, assuming dedicated warehouses, that is, each retailer has to be supplied by *exactly* one warehouse. Solve the integer programming model. What is the new optimal solution?

(c) Compare the two optimal solutions and comment on their distribution plans.

5.11 A company's distribution network has eight regional warehouses for the U.S. market. The total demand is equally distributed among the eight warehouses. Suppose the company decides to close three warehouses and assign their demands equally to the remaining five warehouses.

(a) What will be impact of this risk pooling strategy on the total inventory carried in the supply chain?

(b) What are the negative impacts of this risk pooling strategy?

5.12 DataStream currently uses one central DC to supply the entire northeast region of the United States. In order to improve customer service, the company decides to open two more warehouses and distribute the total NE demand as follows:

(i) 50% to the central DC

(ii) 25% each to the two new DCs

To examine the impact of this strategy on DataStream, derive the expressions for the following inventory measures:

(a) Total inventory in the supply chain

(b) Total number of orders in the supply chain

(c) Frequency of orders from the DCs

5.13 ABC Tools has divided its northeast U.S. market into three customer regions. The weekly demands at the regions are normally distributed with mean and standard deviation as shown in Table 5.18. Assume that the regional demands are independent. Currently, the company has three dedicated DCs, one for each demand region and maintains an 80% service level to its customers. The company is studying a new option to have just one central DC to serve all three demand regions.

TABLE 5.18

Weekly Demands (Exercise 5.13)

Region	Mean	Standard Deviation
1	2000	100
2	3000	50
3	1500	50

Given the replenishment lead time from the plant to any DC is 1 week, answer the following questions:

(a) Determine the current level of safety stock in the supply chain under the present (three DC) system.

(b) Compute the new level of safety stock in the supply chain under the new option (one central DC), if the company continues to maintain 80% service level to its customers.

(c) Suppose the company wants to maintain the current level of safety stock found in part (a) for the new (one DC) option also. What will be the new Service Level to its customers in that case?

5.14 Consider the following design proposals and financial performance measures for Mighty Manufacturing given in Table 5.19. Both sets of numbers are for 2012 with all possible markets open and plants open in both Denver and Covington. Scenario A has DCs in Denver, Chicago, Pittsburgh, and Atlanta. Scenario B has DCs in Denver, Los Angeles, Portland, Dallas, Chicago, Boston, Pittsburgh, and Atlanta.

(a) Which scenario offers better customer service? Why?

(b) Which scenario would you suspect has higher transportation costs? Why?

(c) Under which scenario is Mighty Manufacturing in a better position if forecasts for new markets are too high? Why?

(d) Under which scenario is Mighty Manufacturing in a better position if forecasts for new markets are too low? Why?

5.15 (*Supplier selection and order allocation: Case study*)

Introduction:

One of the most strategic decisions facing a company in supply chain management is the purchasing strategy. In most industries, cost of raw materials and procured components consume a significant portion of the company's budget. For example, in technology firms, purchased materials and services account for up to 80% of the total product cost.

TABLE 5.19

Data for Mighty Manufacturing (Exercise 5.14)

	Scenario A	Scenario B
Total sales ($000)	$76,537	$75,092
Return on assets	62.03%	57.93%
Average time through supply chain (days)	42.68	40.03
Order cycle time to customer (days)	4.38	3.42
Achieved fill rate	98.95%	97.08%
Average production capacity utilization	89.22%	82.47%

The identification of right sources ensures that firms receive proper quality, quantity, time and price from suppliers; and hence, is vital to a firm's survival.

Problem description:

In this case study, you will solve a supplier selection and order allocation problem with two products, two buyers,* two suppliers and each supplier offering "incremental" price discounts (not "all unit discount"). The objective is to minimize total cost, which consists of fixed cost and the variable cost. Fixed cost is a one-time cost that is incurred if a supplier is used for any product, irrespective of the number of units bought from that supplier.

The constraints in the model include the following:

1. Capacity constraints of the suppliers
2. Buyer's demand constraints
3. Buyer's quality constraints (use weighted average quality)
4. Buyer's lead time constraints (use weighted average lead time)
5. Price break constraints for products

The data regarding the various model parameters are given in Tables 5.20 through 5.26.

TABLE 5.20

Product Demand (Exercise 5.15)

Product	Buyer	Demand
1	1	150
1	2	175
2	1	200
2	2	180

TABLE 5.21

Supplier Capacities (Exercise 5.15)

Product	Supplier	Capacity
1	1	400
1	2	450
2	1	480
2	2	460

---

* Multiple buyers represent situations when different divisions of a company buy through one central purchasing department.

TABLE 5.22

Fixed Cost of Suppliers
(Exercise 5.15)

Supplier	Fixed Cost
1	$3500
2	$3600

TABLE 5.23

Lead Time of Products in Days
(Exercise 5.15)

Product	Buyer	Supplier	Lead Time
1	1	1	8
1	1	2	17
1	2	1	14
1	2	2	24
2	1	1	28
2	1	2	8
2	2	1	16
2	2	2	12

TABLE 5.24

Quality of Product (Measured by
Percentage of Rejects) (Exercise 5.15)

Product	Supplier	Quality (%)
1	1	3
1	2	9
2	1	6
2	2	2

The average quality levels that each buyer requires from all the suppliers are given as follows:

Buyer 1: 6%

Buyer 2: 4%

The price break points of the suppliers are given in Tables 5.25 and 5.27. Level 1 break points represent the quantity at which price discounts apply. Level 2 break points represent the maximum quantity of a particular product a supplier can provide to that buyer. For example, for product 1, Buyer 1 and Supplier1, the first 85 units will cost $180/unit and the next 65 units (i.e., 150–85) will cost $165/unit; no more than 150 units of product 1 can be purchased from Supplier 1 by Buyer 1.

TABLE 5.25

Product Prices (Including Shipping Cost)
of Suppliers with Quantity Discounts
(Exercise 5.15)

Product	Buyer	Supplier	Level	Unit Price
1	1	1	1	180
1	1	1	2	165
1	1	2	1	178
1	1	2	2	166
1	2	1	1	180
1	2	1	2	165
1	2	2	1	178
1	2	2	2	166
2	1	1	1	80
2	1	1	2	70
2	1	2	1	83
2	1	2	2	69
2	2	1	1	80
2	2	1	2	70
2	2	2	1	83
2	2	2	2	69

TABLE 5.26

Average Lead Time Requirements
of the Buyers (Exercise 5.15)

Product	Buyer	Lead Time
1	1	12.5
1	2	19
2	1	18
2	2	14

You are asked to determine the company's procurement strategy.

(a) Formulate the supplier selection and order allocation problem as a mixed integer linear programming (MILP) problem. You must define your variables clearly, write out the constraints explaining their significance, and write down the objective functions. Specify the problem size: the number of variables (binary and continuous) and the constraints.

(b) Solve the MILP model using any optimization software.

(c) Write out the optimal solution and discuss the optimal procurement plan.

TABLE 5.27

Price Break Points by Suppliers (Exercise 5.15)

Product	Buyer	Supplier	Level	Break Point
1	1	1	1	85
1	1	1	2	150
1	1	2	1	95
1	1	2	2	175
1	2	1	1	85
1	2	1	2	150
1	2	2	1	95
1	2	2	2	175
2	1	1	1	120
2	1	1	2	170
2	1	2	1	125
2	1	2	2	160
2	2	1	1	120
2	2	1	2	170
2	2	2	1	125
2	2	2	2	160

5.16 (*Risk pooling*: *Case study*)

Pip Boys (motto: "Cars tolerate us, people are getting used to us") currently holds inventories in distribution centers (DCs) in five regions of the continental United States—Northeast, Southeast, North Central, South Central, and West—to supply its many retail stores across the country. As the rising star in her organization, you have been asked by Sandy McNeill, the V.P. of Logistics at Pip Boys, to determine whether it would be beneficial to centralize the stocking of some items and hold them only in the North Central DC in Omaha. Ms. McNeill has suggested that you first consider two specific SKUs, the Zippy Spark Plug, and the Control-Tac Leather-wrapped Steering Wheel.

TABLE 5.28

Data for Pip Boys (Exercise 5.16)

Common Parameters	Control-Tac Steering Wheel	Zippy Spark Plug
Unit cost	$45.00/unit	$1.00/unit
Annual demand	6,000 units	230,000 units
Coefficient of variation of demand	0.90	0.10
Holding cost factor (annual, as % of item value)	20.00%	20.00%
Cycle service level	97.50%	97.50%
Safety stock factor (z)	1.960	1.960

TABLE 5.29

Data for Pip Boys (Exercise 5.16)

Regional Parameters	Control-Tac Steering Wheel	Zippy Spark Plug
Mean demand over LT—Northeast	200 units	5200 units
Mean demand over LT—Southeast	130 units	6000 units
Mean demand over LT—North Central	90 units	3400 units
Mean demand over LT—South Central	110 units	4400 units
Mean demand over LT—West	180 units	3800 units
Std. deviation of demand over LT—Northeast	180 units	520 units
Std. deviation of demand over LT—Southeast	117 units	600 units
Std. deviation of demand over LT—North Central	81 units	340 units
Std. deviation of demand over LT—South Central	99 units	440 units
Std. deviation of demand over LT—West	162 units	380 units

The necessary data is given in Tables 5.28 and 5.29. Assuming that the regional demands are independent and the company continues to maintain the same 97.5% service level, answer the following:

(a) Determine the current levels of safety stock for the two components at each of the five regional DCs.

(b) Determine the safety stock that would result from centralization, i.e., hold them only in North Central Omaha DC.

(c) Compare the annual savings due to centralization for each component using the following measures.

    (i) Total safety stock

    (ii) Safety stock holding cost

    (iii) Cost savings as a percent of the annual material cost

(d) Is there a significant difference in savings as a percent of the annual material cost? Explain the reason.

References

Armacost, A. P., C. Barnhart, K. A. Ware, and A. M. Wilson. 2004. UPS optimizes its air network. *Interfaces*. 34(1): 15–25.

Atkinson, W. 2002. DC siting – What makes most sense? *Logistics Management and Distribution Report*. 41(5): 563–565.

Camm, J. D., T. E. Chorman, F. A. Dill, J. R. Evans, D. J. Sweeney, and G. W. Wegryn. 1997. Blending OR/MS, judgment and GIS: Restructuring P&G's supply chain. *Interfaces*. 27: 128–142.

Chelst, K., J. Sidelko, A. Przebienda, J. Lockedge, and D. Mihailidis. 2001. Rightsizing and management of prototype vehicle testing at Ford motor company. *Interfaces*. 31: 91–107.

Chen, M.-S. and C.-T. Lin. 1989. Effects of centralization in a multi-location newsboy problem. *Journal of the Operational Research Society.* 40(6): 597–602.

Chopra, S. and P. Meindl. 2010. *Supply Chain Management: Strategy, Planning and Operation*, 4th edn. Upper Saddle River, NJ: Prentice Hall.

Cintron, A. 2010. Optimizing an integrated supply chain. PhD dissertation. The Pennsylvania State University, University Park, PA.

Cintron, A., A. Ravindran, and J. A. Ventura. 2010. Multi-criteria mathematical models for designing the distribution network of a consumer products company. *Computers and Industrial Engineering.* 58: 584–593.

Collier, D. A. 1982. Aggregate safety stock levels and component commonality. *Management Science.* 28(11): 1296–1303.

Eppen, G. D. 1979. Effects of centralization on expected costs in a multi-location newsboy problem. *Management Science.* 25(5): 498–501.

Evers, P. T. 2001. Managing supply chain inventories. *National Institute of Standards and Technology.* 4: 1–15.

Fleischmann, B., S. Ferber, and P. Henrich. 2006. Strategic planning of BMW's global production network. *Interfaces.* 36(3): 194–208.

Gaur, S. and A. Ravindran. 2006. A bi-criteria model for the inventory aggregation problem under risk pooling. *Computers and Industrial Engineering.* 51: 482–501.

Glasserman, P. and Y. Wang. 1998. Leadtime-inventory trade-offs in assemble to order systems. *Operations Research.* 46(6): 858–871.

Glover, F. and E. Woolsey. 1974. Converting 0–1 polynomial programming problem to a 0 -1 linear program. *Operations Research.* 22: 180–182.

Jang, W. 2006. Production and allocation policies in a two-class inventory system with time and quantity dependent waiting costs. *Computers and Operations Research.* 33(8): 2301–2321.

Kulkarni, S., M. Magazine, and A. Raturi. 2005. On the trade-offs between risk pooling and logistics costs in a multi-plant network with commonality. *IIE Transactions.* 37(3): 247–265.

Laval, C., M. Feyhland, and S. Kakaouros. 2005. Hewlett-Packard combined OR and expert knowledge to design its supply chains. *Interfaces.* 35(3): 238–247.

Lee, H. L. and C. Billington. 1995. The evolution of supply chain management models and practice at Hewlett-Packard company. *Interfaces.* 25(5): 42–46.

Lodree, E. Jr., W. Jang, and C. Klein. 2004. Minimizing response time in a two-state supply chain system with variable lead time and stochastic demand. *International Journal of Production Research.* 42(11): 2263–2278.

Portillo, R. C. 2009. Resilient global supply chain network design optimization. PhD dissertation. The Pennsylvania State University, University Park, PA.

Ravindran, A., D. T. Philips, and J. Solberg. 1987. *Operations Research: Principles and Practice*, 2nd edn. New York: John Wiley & Sons, Inc.

Schwartz, B. M. 1999a. Map out a site route. *Transportation & Distribution.* 40(11): 67–69.

Schwartz, B. M. 1999b. New rules for hot spots. *Transportation & Distribution.* 40(11): 43–45.

Spencer, T., A. J. Brigani, D. R. Dargon, and M. J. Sheehan. 1990. AT&T's telemarketing site selection system offers customer support. *Interfaces.* 20(1): 83–96.

Srinivasan, G. 2010. *Quantitative Models in Operations and Supply Chain Management*, New Delhi, India: Prentice Hall.

Zeng, A. Z. 1998. Single or multiple sourcing: an integrated optimization framework for sustaining time-based competitiveness. *Journal of Marketing Theory and Practice.* 6(4): 10–25.

6

Supplier Selection Models and Methods*

6.1 Supplier Selection Problem

6.1.1 Introduction

The contribution of the purchasing function to the profitability of the supply chain has assumed greater proportions in recent years; one of the most critical functions of purchasing is selection of suppliers. For most manufacturing firms, the purchasing of raw material and component-parts from suppliers constitutes a major expense. Raw material cost accounts for 40%–60% of production costs for most U.S. manufacturers. In fact, for the automotive industry, the cost of components and parts from outside suppliers may exceed 50% of sales (Wadhwa and Ravindran, 2007). For technology firms, purchased materials, and services account for 80% of the total production cost. It is vital to the competitiveness of most firms to be able to keep the purchasing cost to a minimum. In today's competitive operating environment, it is impossible to successfully produce low-cost, high-quality products without good suppliers. A study carried out by the Aberdeen group (2004) found that more than 83% of the organizations engaged in outsourcing achieved significant reduction in purchasing cost, more than 73% achieved reduction in transaction cost and over 60% were able to shrink sourcing cycles.

Supplier selection process is difficult because the criteria for selecting suppliers could be conflicting. Figure 6.1 illustrates the various factors which could impact the supplier selection process (Sonmez, 2006). Supplier selection is a multiple criteria optimization problem that requires trade-off among different qualitative and quantitative factors to find the best set of suppliers. For example, the supplier with the lowest unit price may also have the lowest quality. The problem is also complicated by the fact that several conflicting criteria must be considered in the decision making process.

---

* Portions of this chapter have been adapted with permission from Ravindran and Wadhwa (2009).

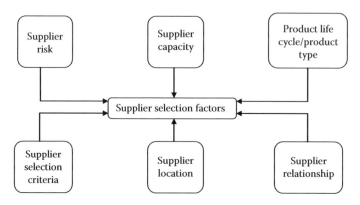

FIGURE 6.1
Supplier selection factors.

In most instances, buyers have to choose among a set of suppliers by using some predetermined criteria such as, quality, reliability, technical capability, lead-times, etc., even before building long-term relationships. To accomplish these goals, two basic and interrelated decisions must be made by a firm. The firm must decide which suppliers to do business with and how much to order from each supplier. Weber et al. (1991) refer to this pair of decisions as the supplier selection problem.

6.1.2 Supplier Selection Process

Figure 6.2 illustrates the steps in the supplier selection process. The first step is to determine whether to *make or buy* the item. Most organizations buy those parts which are not core to the business or not cost effective if produced in-house. The next step is to define the various criteria for selecting the suppliers. The criteria for selecting a supplier of critical product may not be the same as a supplier of maintenance, repair, and operating items. Once a decision to buy the item is taken, the most critical step is selecting the right supplier. Once the suppliers are chosen, the organization has to

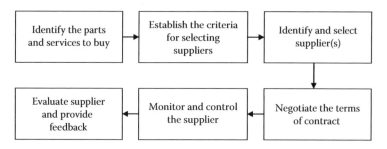

FIGURE 6.2
Supplier selection steps.

negotiate terms of contract and monitor their performance. Finally, the suppliers have to be constantly evaluated and the feedback should be provided to the suppliers.

6.1.3 In-House or Outsource

As illustrated in Figure 6.2, the important sourcing decision is to decide whether to make the part or perform the service in-house or to outsource the activity. In the past, companies used to be vertically integrated in such a way that all the necessary parts were produced internally and support services, such as transportation, procurement, payroll, etc. were provided by the companies themselves. However, in recent years, companies have started outsourcing their non-core functions. For example, many manufacturing companies outsource their shipping to Third Party Logistics (3PL) providers, such as, UPS, FedEx, Penske, etc. This generally reduces cost and provides better service by taking advantage of the 3PL's expertise in transportation. However, the company loses control and flexibility. When manufacturing of critical parts is outsourced, the company may also lose intellectual property. Thus, Step 1 in Figure 6.2, whether to do it in-house or outsource, is an important strategic decision for critical parts and services.

In addition to the qualitative factors, a company can also use a cost criterion to decide whether to make or buy a certain part. This process is illustrated in Example 6.1.

Example 6.1: Make or Buy Problem

A manufacturing company is currently producing a part required in its final product internally. It uses about 3000 parts annually. The set-up of a production run costs $600 and the part costs $2. The company's inventory-carrying cost is 20% per year. The company is considering an option to purchase this part from a local supplier since the fixed cost of ordering will only be $25. However, the unit cost of each part will increase from $2 to $2.50. The problem is to determine whether the company should buy the parts or produce them internally.

Solution

(a) In-House Production Option
The optimal number of parts per production run P can be calculated using the EOQ formula given in Chapter 3:

$$P = \sqrt{\frac{(2)(600)(3000)}{(0.2)(2)}} = 3000 \text{ units}$$

$$\text{Total yearly cost} = \sqrt{(2)(600)(3000)(0.4)} + (2)(3000)$$

$$= \$1200 + \$6000 = \$7200$$

(b) OutSourcing Option

Optimal order quantity $= \sqrt{\dfrac{(2)(25)(3000)}{(0.2)(2.5)}} = 548$ units

Total yearly cost $= \sqrt{(2)(25)(3000)(0.5)} + (2.5)(3000) = \$274 + \$7500 = \$7774$

Hence, based on minimum cost, the company should continue to make the parts internally.

6.1.4 Chapter Overview

In Section 6.2, we review the literature and discuss some basic methods that are available for supplier selection. In Section 6.3, we present multiple criteria optimization models for supplier selection and discuss methods for ranking the suppliers. When there are a large number of suppliers, these methods can be used for pre-qualification or pre-screening of the supplier base. The methods are illustrated with examples. In Section 6.4, we present multiple criteria models for supplier order allocation when there are several pre-qualified suppliers and multiple products to purchase. These models will determine the optimal order quantities from each supplier for the various products under conflicting criteria. In Section 6.4, we also discuss different goal programming models for solving the multiple criteria supplier selection and order allocation problem. A case study is used to illustrate the various goal programming approaches. The chapter ends with concluding remarks and topics for further reading in Section 6.5.

6.2 Supplier Selection Methods

As mentioned earlier, the supplier selection activity plays a key role in cost reduction and is one of the most important functions of the purchasing department. Different mathematical, statistical, and game theoretical models have been proposed to solve the problem. References Weber et al. (1991), Aissaoui et al. (2007), De Boer et al. (2001), and Dickson (1966), provide an overview of supplier selection methods. De Boer et al. (2001) stated that that supplier selection is made up of several decision making steps as shown in Figure 6.3.

6.2.1 Sourcing Strategy

It is important for a company to recognize the importance of the sourcing strategy. It can be either strategic or tactical depending on the type of item being purchased. For example, if the item being purchased is a critical

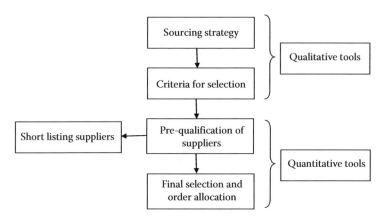

FIGURE 6.3
Supplier selection problem.

component or the most expensive part of the final product, then the supplier selection becomes a strategic decision. The company would most likely want to develop a long-term relationship with the supplier. In this case, subjective criteria such as trust and keeping abreast of emerging technologies would be more important than lead-time or even base price. On the other hand, when an item is a commodity which can be easily found in the market from several suppliers then the supplier selection is a tactical decision. Quantitative criteria such as cost, lead-time, delivery, and geographic location play a key role.

6.2.2 Criteria for Selection

Depending on the buying situation, different sets of criteria may have to be employed. Criteria for supplier selection have been studied extensively since the 1960s. Dickson's (1966) study was the earliest to review the supplier selection criteria. He identified 23 selection criteria with varying degrees of importance for the supplier selection process. Dickson's study was based on a survey of purchasing agents and managers. A follow-up study was done by Weber et al. (1991). They reviewed 74 articles published in the 1970s and 1980s and categorized them based on Dickson's 23 selection criteria. In their studies, net price, quality, delivery, and service were identified as the key selection criteria. They also found that under Just-In-Time manufacturing, quality and delivery became more important than net price. Zhang et al. (2003) conducted a similar study and found that price, quality, and delivery are the top three key criteria for supplier selection. Table 6.1 compares the relative importance of the various supplier selection criteria identified in the three studies. Supplier selection criteria may also change over time. Wilson (1994) examined studies conducted in 1974, 1982, and 1993 on the relative importance of the selection criteria. She found

TABLE 6.1

Importance of Supplier Selection Criteria

Criterion	Dickson (1966) Ranking	Rating[b]	Weber et al. (1991) Ranking[a]	(%)	Zhang et al. (2003) Ranking[a]	(%)
Net price	6	A	1	80	1	87
Quality	1	A +	3	53	2	82
Delivery	2	A	2	58	3	73
Production facilities and capacity	5	A	4	30	4	44
Technical capacity	7	A	6	20	5	33
Financial position	8	A	9	9	6	29
Geographical location	20	B	5	21	7	11
Management and organization	13	B	7	13	7	11
Performance history	3	A	9	9	7	11
Operating controls	14	B	13	4	7	11
Communication systems	10	B	15	3	7	11
Reputation and position in industry	11	B	8	11	12	7
Repair service	15	B	9	9	13	4
Packaging ability	18	B	13	4	13	4
Training aids	22	B	15	3	13	4
Procedural compliance	9	B	15	3	13	4
Labor relations record	19	B	15	3	13	4
Warranties and claims policies	4	A	—	0	13	4
Attitude	16	B	12	8	19	2
Reciprocal arrangements	23	C	15	3	19	2
Impression	17	B	15	3	21	0
Desire for business	12	B	21	1	21	0
Amount of past business	21	B	21	1	21	0

[a] Ranking is based on the frequency that the criterion is discussed in reviewed papers.
[b] A+: Extremely important; A: considerably important; B: averagely important; C: slightly important.

that quality and service criteria began to dominate price and delivery. She concluded that globalization of the market place and an increasingly competitive atmosphere had contributed to the shift. Table 6.2 shows the top 15 supplier selection criteria as given in Weber et al. (1991). Mandal and Deshmukh (1994) proposed interpretive structural modeling (ISM) as a technique based on group judgment to identify and summarize relationships between supplier choice criteria through a graphical model. Vokurka et al. (1996) developed an expert system for the formulation of the supplier selection criteria. Nominal group technique involving all the stakeholders

TABLE 6.2

Key Criteria for Supplier Selection

Rank	Criteria
1	Net price
2	Delivery
3	Quality
4	Production facilities and capabilities
5	Geographical location
6	Technical capability
7	Management and organization
8	Reputation and position in industry
9	Financial position
10	Performance history
11	Repair service
12	Attitude
13	Packaging ability
14	Operational control
15	Training aids

Source: Weber, C.A. et al., 1991. *Eur. J. Oper. Res.*, 50: 2–18.

of the supplier selection decision can also be used to identify the important supplier selection criteria (Goicoechea, 1982).

6.2.3 Pre-Qualification of Suppliers

Pre-qualification (Figure 6.3) is defined as the process of reducing a large set of potential suppliers to a smaller manageable number by ranking the suppliers under a pre-defined set of criteria. The primary benefits of pre-qualification of suppliers are (Holt, 1998) as follows:

1. The possibility of rejecting good suppliers at an early stage is reduced.
2. Resource commitment of the buyer toward purchasing process is optimized.
3. With the application of pre-selected criteria, the pre-qualification process is rationalized.

Pre-qualification is an orderly process of ranking various suppliers under conflicting criteria. It is a multiple-criteria ranking problem that requires the buyer to make trade-off among the conflicting criteria, some of which may be qualitative. There are a number of multiple criteria ranking methods available for pre-qualification of suppliers. Section 6.3 discusses in detail the various ranking methods available for supplier selection.

6.2.4 Final Selection

Most of the publications in the area of supplier selection have focused on final selection. In the final selection step (Figure 6.3), the buyer identifies the suppliers to do business with and allocates order quantities among the chosen supplier(s). In reviewing the literature, there are basically two kinds of supplier selection problem, as stated by Ghodsypour and O'Brien (2001):

- *Single sourcing*, which implies that any one of the suppliers can satisfy the buyer's requirements of demand, quality, delivery, etc.
- *Multiple sourcing*, which implies that there are some limitations in suppliers' capacity, quality, etc. and multiple suppliers have to be used.

The primary objective of a single sourcing strategy is to minimize the procurement costs. A secondary objective is to establish a long-term relationship with a single supplier. However, the single sourcing strategy exposes a company to a greater risk of supply chain disruption. Toyota's 1977 brake valve crisis is an example of such a strategy (Mendoza, 2007). Toyota was using a single supplier (Aisim Seiki) for brake valves used in all its cars (Nishiguchi and Beaudet, 1998). The supplier's plant was shut down due to a fire, which resulted in the closure of Toyota's assembly plants for several days. It impacted the production of 70,000 cars, at an estimated cost of $195 million. Thereafter, Toyota decided to use at least two suppliers for every part (Treece, 1997).

A multiple sourcing strategy results in obtaining raw materials and parts from more than one supplier and provides greater flexibility to a company. It introduces competition among the suppliers, which helps to drive the procurement costs down and improve quality of parts (Jayaraman et al., 1999). It also protects the company from major supply disruptions. However, the overhead costs will increase as the company has to deal with multiple suppliers for each part. Bilsel and Ravindran (2011) discuss a hybrid sourcing strategy, wherein the company uses one main supplier but has several "back-up" suppliers that can be used in case of supply disruption at the main supplier.

6.2.4.1 Single Sourcing Methods

Single sourcing is a possibility when a relatively small number of parts are procured. In this section, we will discuss two most commonly used methods in single sourcing.

1. *Linear weighted point (LWP) method*: The linear weighted point (LWP) method is the most widely used approach for single sourcing. The approach uses a simple scoring method that heavily depends on human judgment. Some of the references that discuss this approach include Wind and Robinson (1968) and Zenz (1981). The multiple

TABLE 6.3

Supplier Selection by LWP method

Criteria	Weights	Supplier A	Supplier B	Ideal Value
Quality	15	10	9	10
Service	20	8	7	8
Capacity	10	10	8	10
Price	45	7	10	10
Risk	10	6	9	9
Total	100%	785	895	

criteria ranking methods discussed in Section 6.3 can also be used for single sourcing by selecting the top-ranked supplier.

To illustrate the LWP method, consider a simple supplier selection problem with two suppliers and five criteria (Quality, service, capacity, price, and risk) as shown in Table 6.3. Each criterion is measured over a scale of 1–10, with higher numbers preferred over lower ones. Note that Supplier A has higher quality, service, and capacity but costs more and has higher risk of supply disruptions. The *ideal values* in Table 6.3 represent the best values for each criterion. However, the ideal values are not achievable simultaneously for all the criteria, because the criteria conflict with one another.

In the LWP method, the purchasing manager assigns a weight to each criterion in a scale of 1–100. In Table 6.3, price has the highest weight (45%) followed by service and quality. Capacity and risk are tied for the last place. A total weighted score is then calculated by summing the product of each criterion weight and its value (See Table 6.3). Based on the total score, Supplier B would be selected even though Supplier A was better in three out of the five criteria.

The LWP method has several drawbacks. It is generally difficult to come up with exact criteria weights. The method also requires all the criteria values to be scaled properly if the measurement units are very different. For example, quality may be measured in "percent defectives," service may be measured by lead-time in days, capacity in hundreds of units, and price in dollars. The use of linear additive value function violates the normal economic principle of "Diminishing Marginal Utility." In Section 6.3, we will discuss several methods for ranking suppliers that overcome these drawbacks.

2. *Total cost of ownership (TCO)*: Total cost of ownership (TCO) is another supplier selection method used in practice for single sourcing. TCO looks beyond the unit price of the item to include costs of other factors, such as quality, delivery, supply disruption, safety stock, etc. Unlike the LWP method, TCO assigns a cost to each criterion and computes the TCO with respect to each supplier.

Finally, the business is awarded to the supplier with the lowest total cost. General Electric Wiring Devices have developed a total cost supplier selection method that takes into account risk factors, business desirable factors, and measurable cost factors (Smytka and Clemens, 1993). TCO approach has also been used by Ellram (1995), Degreave and Roodhoft (1999, 2000, 2004). Example 6.2 illustrates the TCO approach for a simple supplier selection problem.

Example 6.2

A company is considering two potential suppliers, one domestic and one foreign, for one of its parts needed in production. The relevant supplier data are given in Table 6.4.

The daily usage of this part is 20 units. The company follows an inventory policy of maintaining a safety stock of 50% of the lead-time demand and uses an inventory-carrying charge of 25% per year (i.e., $0.25/ dollar/year). Defective parts encountered in manufacturing cost the company $5/unit. The company wishes to choose one of the suppliers using the TCO approach.

The TOC includes the following costs:

- Procurement cost
- Inventory holding costs (Cycle inventory + Safety stock)
- Cost of quality

Solution

We will compute the annual cost of ownership for each supplier.

Supplier A (Domestic):

Annual procurement cost = $(10) (20) (365) = $73,000
Lead-time demand = (10) (20) = 200 units
Safety stock = (50%) (200) = 100 units
Average cycle inventory = $\frac{500}{2}$ = 250 units
Annual inventory holding cost = (250 + 100) (0.25) ($10) = $875
Annual cost of quality = ($5) (7300) (5%) = $1825
TCO for supplier A = $73,000 + $875 + $1825 = $75,700/year

TABLE 6.4

Supplier Data for Example 6.2

	Supplier A (Domestic)	Supplier B (Foreign)
Item cost/unit	$10	$9
Minimum order	500	3000
Lead-time	10 days	90 days
Quality (% defective)	5%	20%

Supplier B (Foreign)

Annual procurement cost = $ (9) (20) (365) = $65,700
Lead-time demand = (90) (20) = 1800 units
Safety stock = (50%) (1800) = 900 units
Average cycle inventory = $\frac{3000}{2}$ = 1500 units
Annual inventory holding cost = (1500 + 900) (0.25) ($9) = $5400
Annual cost of quality = ($5) (7300) (20%) = $7300
TCO for supplier B = $65,700 + $5,400 + $7,300 = $78,400/year

Even though the per unit cost of the item is lower with the foreign supplier, the TCO is less with the domestic supplier when we take into account the cost of inventory and cost of quality.

6.2.4.2 Multiple Sourcing Methods

In multiple sourcing, a buyer purchases the same item from more than one supplier. Multiple sourcing can offset the risk of supply disruptions. Mathematical programming models are the most appropriate methods for multiple sourcing decisions. It allows the buyer to consider different constraints including capacity, delivery time, quality, etc., while choosing the suppliers and their order allocations. Two types of mathematical models are found in the literature, single objective and multiple objective models. The single objective models invariably use cost, as the main objective function. We shall illustrate the single objective model in this section using Example 6.3. Multiple objective models will be discussed in detail in Section 6.4.

Example 6.3 (Supplier Selection and Order Allocation)

Consider a supplier selection and order allocation problem with two products, two buyers (multiple buyers represent situations when different divisions of a company buy through one central purchasing department), two suppliers and each supplier offering "incremental" price discounts to each buyer (not "all unit discount"). The objective is to minimize *Total cost*, which consists of the fixed cost and the variable cost. Fixed cost is a one-time cost that is incurred if a supplier is used for any product, irrespective of the number of units bought from that supplier.

The constraints in the model include the following:

1. Capacity constraints of the suppliers
2. Buyer's demand constraints
3. Buyer's quality constraints
4. Buyer's lead-time constraints
5. Price break constraints for products

The customer demand data is given in Table 6.5. Data regarding supplier capacities, lead-time, and product quality are given in Tables 6.6 through 6.8.

TABLE 6.5

Product Demand

Product	Buyer	Demand
1	1	150
1	2	175
2	1	200
2	2	180

TABLE 6.6

Supplier Capacities

Product	Supplier	Capacity
1	1	400
1	2	450
2	1	480
2	2	460

TABLE 6.7

Lead-time of Products in Days

Product	Buyer	Supplier	Lead-Time
1	1	1	8
1	1	2	17
1	2	1	14
1	2	2	24
2	1	1	28
2	1	2	8
2	2	1	16
2	2	2	12

TABLE 6.8

Quality of Product (Measured by Percentage of Rejects)

Product	Supplier	Quality (%)
1	1	3
1	2	9
2	1	6
2	2	2

TABLE 6.9

Product Prices (Including Shipping Cost) of
Suppliers with Quantity Discounts and Their
Corresponding Price Break Points

Product	Buyer	Supplier	Level	Unit Price	Break Point
1	1	1	1	180	85
1	1	1	2	165	150
1	1	2	1	178	95
1	1	2	2	166	175
1	2	1	1	180	85
1	2	1	2	165	150
1	2	2	1	178	95
1	2	2	2	166	175
2	1	1	1	80	120
2	1	1	2	70	170
2	1	2	1	83	125
2	1	2	2	69	160
2	2	1	1	80	120
2	2	1	2	70	170
2	2	2	1	83	125
2	2	2	2	69	160

The quantity discounts and price breaks offered by the suppliers are given in Table 6.9. The Level 1 break points represent the quantity at which price discounts apply. Level 2 break points represent the maximum quantity of a particular product a supplier can provide to that buyer. For example, for product 1, Buyer 1 and Supplier 1, the first 85 units will cost \$180/unit and the next 65 units (i.e., 150–85) will cost \$165/unit; no more than 150 units of product 1 can be purchased from Supplier 1 by Buyer 1.

The lead-time requirements of the buyers for the two products are given in Table 6.10.

Buyers 1 and 2 limit the average reject levels to 6% and 4% respectively from all suppliers. The fixed costs of using suppliers 1 and 2 are \$3500 and \$3600, respectively.

TABLE 6.10

Average Lead-Time
Requirements of the Buyers

Product	Buyer	Lead
1	1	12.5
1	2	19
2	1	18
2	2	14

Solution

Decision Variables

x_{ijkm} — Quantity of product i, that buyer j buys from supplier k, at price level m

δ_k — $\begin{cases} 1, \text{ if supplier } k \text{ is used} \\ 0, \text{ Otherwise} \end{cases}$

β_{ijkm} — $\begin{cases} 1, \text{ if buyer } j \text{ buys product } i \text{ from supplier } k \text{ at price level } m \\ 0, \text{ Otherwise} \end{cases}$

Product, i   =   1, 2
Buyer, j   =   1, 2
Supplier, k =   1, 2
Price level, m = 1, 2

Thus, there are 18 binary variables and 16 continuous variables.

Objective function

Minimize cost = Fixed cost + Variable cost

$$\begin{aligned} = \ & 3500\,\delta_1 + 3600\,\delta_2 + 180\,x_{1111} + 165\,x_{1112} + 178\,x_{1121} + 166\,x_{1122} \\ & + 180\,x_{1211} + 165\,x_{1212} + 178\,x_{1221} + 166\,x_{1222} + 80\,x_{2111} \\ & + 70\,x_{2112} + 83\,x_{2121} + 69\,x_{2122} + 80\,x_{2211} + 70\,x_{2212} + 83\,x_{2221} \\ & + 69\,x_{2222} \end{aligned}$$

The objective function is to minimize the sum of fixed and variable costs.

Constraints

1. Buyers' demand constraints

$$\begin{array}{ll} x_{1111} + x_{1112} + x_{1121} + x_{1122} = 150 & \text{(Product 1, Buyer 1)} \\ x_{1211} + x_{1212} + x_{1221} + x_{1222} = 175 & \text{(Product 1, Buyer 2)} \\ x_{2111} + x_{2112} + x_{2121} + x_{2122} = 200 & \text{(Product 2, Buyer 1)} \\ x_{2211} + x_{2212} + x_{2221} + x_{2222} = 180 & \text{(Product 2, Buyer 2)} \end{array}$$

2. Suppliers' capacity constraints

$$\begin{array}{ll} x_{1111} + x_{1112} + x_{1211} + x_{1212} \le 400\,\delta_1 & \text{(Product 1, Supplier 1)} \\ x_{1121} + x_{1122} + x_{1221} + x_{1222} \le 450\,\delta_2 & \text{(Product 1, Supplier 2)} \\ x_{2111} + x_{2112} + x_{2211} + x_{2212} \le 480\,\delta_1 & \text{(Product 2, Supplier 1)} \\ x_{2121} + x_{2122} + x_{2221} + x_{2222} \le 460\,\delta_2 & \text{(Product 2, Supplier 2)} \end{array}$$

3. Buyers' quality constraints
   Buyer 1: The weighted average rejects of both products received from the two suppliers (measured in percent defectives) cannot exceed 6% for Buyer 1.

$$\frac{0.03(x_{1111} + x_{1112}) + 0.09(x_{1121} + x_{1122}) + 0.06(x_{2111} + x_{2112}) + 0.02(x_{2121} + x_{2122})}{150 + 200} \le 0.06$$

Buyer 2: Similarly for Buyer 2, it is limited to 4%

$$\frac{0.03(x_{1211}+x_{1212})+0.09(x_{1221}+x_{1222})+0.06(x_{2211}+x_{2212})+0.02(x_{2221}+x_{2222})}{175+180} \leq 0.04$$

4. Buyer's lead-time constraints

Buyer 1: The weighted average lead-time is 12.5 days for product 1 and 18 days for product 2 as follows:

$$\frac{8(x_{1111}+x_{1112})+17(x_{1121}+x_{1122})}{150} \leq 12.5$$

$$\frac{28(x_{2111}+x_{2112})+8(x_{2121}+x_{2122})}{200} \leq 18$$

Buyer 2: Similarly, the lead-time constraints for Buyer 2 are as follows:

$$\frac{14(x_{1211}+x_{1212})+24(x_{1221}+x_{1222})}{175} \leq 19$$

$$\frac{16(x_{2211}+x_{2212})+12(x_{2221}+x_{2222})}{180} \leq 14$$

5. Price Break Constraints for Products

   (i) Product 1, Buyer 1, Supplier 1

   $0 \leq x_{1111} \leq 85\, \beta_{1111}$

   $x_{1111} \geq 85\, \beta_{1112}$

   $0 \leq x_{1112} \leq 65\, \beta_{1112}$

   Note that if β_{1112} is one, then it would force β_{1111} to be one and $x_{1111} = 85$ units. In other words, Buyer 1 cannot buy product 1 from Supplier 1 at the level 2 (lower) price unless he has ordered at least 85 units at the level 1 (higher) price. Similarly we get the other price break constraints for the other supplier, product and buyer as given in the following:

   (ii) Product 1, Buyer 1, Supplier 2

   $0 \leq x_{1121} \leq 95\, \beta_{1121}$

   $x_{1121} \geq 95\, \beta_{1122}$

   $0 \leq x_{1122} \leq 80\, \beta_{1122}$

   (iii) Product 1, Buyer 2, Supplier 1

   $0 \leq x_{1211} \leq 85\, \beta_{1211}$

   $x_{1211} \geq 85\, \beta_{1212}$

   $0 \leq x_{1212} \leq 65\, \beta_{1212}$

   (iv) Product 1, Buyer 2, Supplier 2

   $0 \leq x_{1221} \leq 95\, \beta_{1221}$

   $x_{1221} \geq 95\, \beta_{1222}$

   $0 \leq x_{1222} \leq 80\, \beta_{1222}$

TABLE 6.11

Optimal Order Allocation (Units)

		Supplier 1		Supplier 2	
		Level 1	Level 2	Level 1	Level 2
Buyer 1	Product 1	85	65	0	0
	Product 2	40	0	125	35
Buyer 2	Product 1	85	65	25	0
	Product 2	20	0	125	35

(v) Product 2, Buyer 1, Supplier 1
$$0 \leq x_{2111} \leq 120\,\beta_{2111}$$
$$x_{2111} \geq 120\,\beta_{2112}$$
$$0 \leq x_{2112} \leq 50\,\beta_{2112}$$

(vi) Product 2, Buyer 1, Supplier 2
$$0 \leq x_{2121} \leq 125\,\beta_{2121}$$
$$x_{2121} \geq 125\,\beta_{2122}$$
$$0 \leq x_{2122} \leq 35\,\beta_{2122}$$

(vii) Product 2, Buyer 2, Supplier 1
$$0 \leq x_{2211} \leq 120\,\beta_{2211}$$
$$x_{2211} \geq 120\,\beta_{2212}$$
$$0 \leq x_{2212} \leq 50\,\beta_{2212}$$

(viii) Product 2, Buyer 2, Supplier 2
$$0 \leq x_{2221} \leq 125\,\beta_{2221}$$
$$x_{2221} \geq 125\,\beta_{2222}$$
$$0 \leq x_{2222} \leq 35\,\beta_{2222}$$

(ix) Binary and Non-negativity Constraints
$$\delta_k \in (0,1) \;\; \forall k = 1,2;$$
$$\beta_{ijkm} \in (0,1) \;\; \forall i, j, k, m$$
$$x_{ijkm} \geq 0$$

This mixed integer program with 18 binary variables, 16 continuous variables, and 38 constraints was solved using Excel Solver. The optimal order allocation is given in Table 6.11. Both suppliers are used for purchases. The policy results in a total cost of $93,980.

6.3 Multi-Criteria Ranking Methods for Supplier Selection

Many organizations have a large pool of suppliers to select from. The supplier selection problem can be solved in two phases. The first phase reduces the large number of candidate suppliers to a manageable size (pre-qualification of suppliers). In Phase II, a multiple criteria optimization

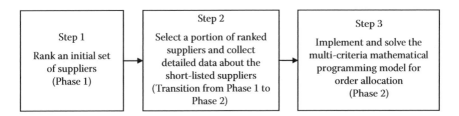

| Step 1
Rank an initial set of suppliers
(Phase 1) | Step 2
Select a portion of ranked suppliers and collect detailed data about the short-listed suppliers
(Transition from Phase 1 to Phase 2) | Step 3
Implement and solve the multi-criteria mathematical programming model for order allocation
(Phase 2) |

FIGURE 6.4
Two phase supplier selection model.

model is used to allocate order quantities among the shortlisted suppliers (order allocation). Figure 6.4 shows the steps involved in the two phase supplier selection model.

6.3.1 Ranking of Suppliers

The problem of ranking suppliers (pre-qualification) represents a class of multiple criteria optimization problems that deal with the ranking of a finite number of alternatives, where each alternative is measured by several conflicting criteria. Several ranking methods have been developed in the multiple criteria optimization literature to solve the problem (Masud and Ravindran, 2008, 2009). In this section, we present several multiple criteria ranking approaches for the supplier ranking problem, namely, the pre-qualification of suppliers.

In the pre-qualification process (Phase I), readily available qualitative and quantitative data are collected for the various suppliers. This data can be obtained from trade journals, Internet, and past transactions to name a few sources. Once this data is gathered, these suppliers are evaluated using multiple criteria ranking methods. The decision maker (DM) then selects a portion of the suppliers for extensive evaluation in Phase II.

6.3.1.1 Case Study 1: Ranking of Suppliers

The first step in pre-qualification is defining the selection criteria. For this case study, we have used the following 14 pre-qualification criteria as an illustration. The pre-qualification criteria have been split into various categories, such as organizational criteria, experience criteria, etc. The various pre-qualification criteria are described in the following:

1. Organizational criteria:
    a. Size of company (C1): Size of the company can be either its number of employees, or its market capitalization.
    b. Age of company (C2): Age of the company is the number of years that the company has been in business.
    c. R&D activities (C3): Investment in research and development.

2. Experience criteria:
   a. Project type (C4): Specific types of projects completed in the past.
   b. Project size (C5): Specific sizes of projects completed in the past.
3. Performance criteria:
   a. Cost overruns (C6): Cost overruns in the past.
   b. Capacity (C7): Capacity of the supplier to fulfill orders.
   c. Lead-time (C8): Meeting promised delivery time.
4. Quality criteria:
   a. Responsiveness (C9): If there is an issue concerning quality, how fast the supplier reacts to correct the problem.
   b. Acceptance rate (C10): Perfect orders received within acceptable quality.
5. Cost criteria:
   a. Order change and cancellation charges (C11): Fee associated with modifying or changing orders after they have been placed.
   b. Cost savings (C12): Overall *reduction* in procurement cost.
6. Miscellaneous criteria:
   a. Labor relations (C13): Number of strikes or any other labor problems encountered in the past.
   b. Procedural compliances (C14): Conformance to national/international standards (e.g., ISO 9000).

In this case study, we assume there are 20 suppliers during pre-qualification. The 14 supplier criteria values for the initial set of 20 suppliers are given in Table 6.12. Smaller values are preferred for criteria C6, C11, and C13; larger values are preferred for other criteria. Next, we discuss several multiple criteria ranking methods for short listing the suppliers and illustrate them using the case study data. Each method has advantages and limitations. The methods that we discuss are as follows:

1. L_p metric method
2. Rating method
3. Borda count
4. Analytic hierarchy process (AHP)
5. Cluster analysis

For a more detailed discussion of multi-criteria ranking methods, the reader is referred to Masud and Ravindran (2008, 2009).

TABLE 6.12

Supplier Criteria Values for Case Study 1

	C1	C2	C3	C4	C5	C6	C7	C8	C9	C10	C11	C12	C13	C14
S1	0.75	1	0.46	1	0.92	0.9	1	0	0.13	0.18	0.18	0.01	0.26	0.79
S2	0.22	0	0.33	1	0.94	0.35	0.9	0.13	0.02	0	0.38	0.95	0.88	0.72
S3	0.53	0	0.74	0	0.03	0.89	0.1	0.12	0	0.3	0.66	0.08	0.86	0.22
S4	0.28	1	0.8	0	0.54	0.75	0.85	1	1	0.87	0.33	0.5	0.78	0.12
S5	0.3	0	0.79	1	0.6	0.49	0.8	0.15	0.97	0.79	0.83	0.13	0.46	0.15
S6	0.5	1	0.27	0	0.43	0.52	0.12	0	0	0.25	0.9	0.07	0.26	0
S7	0.25	1	0.6	1	0.1	0.18	0	0.13	1	0.85	0.51	0.59	0.12	1
S8	0.76	1	0.68	1	0.55	0.87	0	0.14	0	1	0.98	0.19	0.86	0.99
S9	0.25	1	0.5	1	0.26	0.92	0.94	0.03	0.15	1	0.7	0.41	0.95	1
S10	0.16	1	0.7	0	0.46	0.62	0.9	0	0.03	0	0.3	0.68	0.61	1
S11	0.31	0	0.3	0	0.09	0.73	1	1	1	0	0.87	0.3	0.98	0
S12	0.34	1	0.39	1	0.75	0.94	0.78	0.3	0	0.85	0.94	0.61	0.46	0.3
S13	0.08	0	0.27	0	0.14	0.42	1	0.91	0	0.82	0.45	0.42	0.81	1
S14	0.62	1	0.02	1	0.15	0.97	0.15	0.01	0.18	0.92	0.55	0.23	0.12	0.97
S15	0.49	0	0.98	0	0.52	0.68	0	0.24	0.06	0	0.52	0.84	0.05	0.76
S16	0.1	1	0.32	1	0.67	0.21	1	0.85	0.16	0.29	0.49	0.41	0.29	0.27
S17	0.08	0	0.19	1	0.24	0.87	0	0.72	0.26	1	0.84	0.99	0.64	0.04
S18	0.86	0	0.28	1	0.95	0.08	1	0.12	0.2	0	0.4	0.76	0.66	1
S19	0.72	0	0.88	0	0.15	0.93	0.97	1	1	1	0.75	0.64	0.26	1
S20	0.15	1	0.92	1	0.77	0.63	0	0	0.3	0.22	0.22	0.94	0.93	0.26

6.3.2 Use of L_p Metric for Ranking Suppliers

Mathematically, the L_p metric represents the distance between two vectors $\mathbf{x}$ and $\mathbf{y}$, where $x, y \in R^n$, and is given by:

$$\|\mathbf{x} - \mathbf{y}\|_p = \left[\sum_{j=1}^{n} |x_j - y_j|^p \right]^{1/p} \tag{6.1}$$

One of the most commonly used L_p metrics is the L_2 metric ($p = 2$), which measures the Euclidean distance between two vectors. The ranking of suppliers is done by calculating the L_p metric between the Ideal solution (H) and each vector representing the supplier's ratings for the criteria. The Ideal solution represents the best values possible for each criterion from the initial list of suppliers. Since no supplier will have the best values for all criteria (e.g., a supplier with minimum cost may have poor quality and delivery time), the ideal solution is an artificial target and cannot be achieved. The L_p metric approach computes the distance of each supplier's attributes from the ideal solution and ranks the supplier's based on that distance (the smaller

TABLE 6.13

Ideal Values (*H*) for Case Study 1

Criteria	Ideal Value	Criteria	Ideal Value
C1	0.86	C8	1
C2	1	C9	1
C3	0.98	C10	1
C4	1	C11	0.18
C5	0.95	C12	0.99
C6	0.08	C13	0.05
C7	1	C14	1

the better). We illustrate the steps of the L_2 metric method using the supplier data in Table 6.12.

6.3.2.1 Steps of the L_2 Metric Method

Step 1: Determine the Ideal solution. The ideal values (H) for the 14 criteria of Table 6.12 are given in Table 6.13.

Step 2: Use the L_2 metric to measure the closeness of supplier to the ideal values. The L_2 metric for supplier k is given by

$$L_2(k) = \sqrt{\sum_{j=1}^{n} (H_j - Y_{jk})^2} \qquad (6.2)$$

where, H_j is the ideal value for criterion j and Y_{jk} is the jth criterion value for supplier k.

Step 3: Rank the suppliers using the L_2 metric. The supplier with the smallest L_2 value is ranked first, followed by the next smallest L_2 value, etc. Table 6.14 gives the L_2 distance from the ideal value for each supplier and the resulting supplier rankings.

6.3.3 Rating (Scoring) Method

Rating is one of the simplest and most widely used ranking methods under conflicting criteria. This is similar to the linear weighted point (LWP) method discussed in Section 6.2.4. First, an appropriate rating scale is agreed to (e.g., from 1 to 10, where 10 is the most important and 1 is the least important selection criteria). The scale should be clearly understood by the decision maker (DM) to be used properly. Next, using the selected scale, the DM

TABLE 6.14

Supplier Ranking Using L_2 Metric (Case Study 1)

Supplier	L_2 Value	Rank	Supplier	L_2 Value	Rank
Supplier 1	2.105	7	Supplier 11	2.782	18
Supplier 2	2.332	11	Supplier 12	2.083	5
Supplier 3	3.011	20	Supplier 13	2.429	15
Supplier 4	1.896	3	Supplier 14	2.347	13
Supplier 5	2.121	8	Supplier 15	2.517	16
Supplier 6	2.800	19	Supplier 16	1.834	2
Supplier 7	1.817	1	Supplier 17	2.586	17
Supplier 8	2.357	4	Supplier 18	2.092	6
Supplier 9	2.206	9	Supplier 19	1.970	4
Supplier 10	2.339	12	Supplier 20	2.295	10

provides a rating r_j for each criterion, C_j. The same rating can be given to more than one criterion. The ratings are then normalized to determine the weights of the criteria j. Assuming n criteria:

$$W_j = \frac{r_j}{\sum_{j=1}^{j=n} r_j} \quad \text{for } j = 1,2,\ldots,n$$

(6.3)

$$\text{Note: } \sum_{j=1}^{n} W_j = 1$$

Next, a weighted score of the attributes is calculated for each supplier as follows:

$$S_i = \sum_{j=1}^{n} W_j f_{ij}, \quad \forall \, i = 1\ldots K$$

where, f_{ij}'s are the criteria values for supplier i. The suppliers are then ranked based on their scores. The supplier with the highest score is ranked first. Rating method requires relatively little cognitive burden on the DM.

Table 6.15 illustrates the ratings and the corresponding weights for the 14 criteria. Since criteria C6, C11, and C13 are to minimize, their respective weights have been set as negative values.

Table 6.16 shows the final scores for different suppliers and their rankings using the rating method. Appendix A (Section A.2) has another numerical example illustrating the rating method for ranking.

TABLE 6.15

Criteria Weights Using Rating Method (Case Study 1)

Criterion	Rating	Weight	Criterion	Rating	Weight
C1	6	0.073	C8	1	0.012
C2	7	0.085	C9	8	0.098
C3	5	0.061	C10	7	0.085
C4	9	0.110	C11	(−) 6	−0.073
C5	10	0.122	C12	7	0.085
C6	(−) 2	−0.024	C13	(−) 4	−0.049
C7	3	0.037	C14	7	0.085

TABLE 6.16

Supplier Ranking Using Rating Method (Case Study 1)

Supplier	Total Score	Rank	Supplier	Total Score	Rank
Supplier 1	0.475	2	Supplier 11	0.095	19
Supplier 2	0.360	13	Supplier 12	0.404	9
Supplier 3	0.032	20	Supplier 13	0.196	17
Supplier 4	0.408	7	Supplier 14	0.394	10
Supplier 5	0.375	12	Supplier 15	0.247	16
Supplier 6	0.131	18	Supplier 16	0.394	11
Supplier 7	0.522	1	Supplier 17	0.250	15
Supplier 8	0.412	5	Supplier 18	0.450	3
Supplier 9	0.411	6	Supplier 19	0.405	8
Supplier 10	0.308	14	Supplier 20	0.430	4

6.3.4 Borda Count

This method is named after Jean Charles de Borda, eighteenth century French physicist. The method is as follows:

1. The n criteria are ranked 1(most important) to n (least important)
   a. Criterion ranked 1 gets n points, 2nd rank gets n-1 points, and the last place criterion gets 1 point.
2. Weights for the criteria are calculated as follows:

$$\text{Criterion ranked } 1 = \frac{n}{s}$$

$$\text{Criterion ranked } 2 = \frac{n-1}{s}$$

$$\text{last criterion} = \frac{1}{s}$$

where, s is the sum of all the points $= \dfrac{n(n+1)}{2}$.

Table 6.17 illustrates the calculations of criteria weights using the Borda count method. For example, criterion 3 is ranked first among the 14 criteria and gets 14 points. Criterion 11 is ranked last and gets 1 point. Thus, the weight for criterion 3 $= \dfrac{14}{105} = 0.133$. Using these criteria weights, the supplier scores are calculated as before for ranking as shown in Table 6.18. Appendix A (Section A.2) has another numerical example illustrating the Borda count method for ranking.

TABLE 6.17

Criteria Weights Using Borda Count (Case Study 1)

Criterion	Ranking Points	Weight	Criterion	Ranking Points	Weight
C1	9	0.086	C8	13	0.124
C2	7	0.067	C9	2	0.019
C3	14	0.133	C10	8	0.076
C4	6	0.057	C11	(–) 1	−0.010
C5	5	0.048	C12	12	0.114
C6	(–) 10	−0.095	C13	(–) 4	−0.038
C7	11	0.105	C14	3	0.029

TABLE 6.18

Supplier Ranking Using Borda Count (Case Study 1)

Supplier	Total Score	Rank	Supplier	Total Score	Rank
Supplier 1	0.342	9	Supplier 11	0.238	17
Supplier 2	0.335	11	Supplier 12	0.387	6
Supplier 3	0.085	20	Supplier 13	0.331	12
Supplier 4	0.478	2	Supplier 14	0.230	18
Supplier 5	0.345	8	Supplier 15	0.274	15
Supplier 6	0.138	19	Supplier 16	0.462	3
Supplier 7	0.399	5	Supplier 17	0.270	16
Supplier 8	0.325	13	Supplier 18	0.416	4
Supplier 9	0.351	7	Supplier 19	0.503	1
Supplier 10	0.312	14	Supplier 20	0.336	10

6.3.5 Pair-Wise Comparison of Criteria

When there are many criteria, it would be difficult for a DM to rank order them precisely. In practice, pair-wise comparison of criteria is used to facilitate the criteria ranking required by the Borda count. Here, the DM is asked to give the relative importance between two criteria C_i and C_j, whether C_i is preferred to C_j, C_j is preferred to C_i or both are equally important. When there are n criteria, the DM has to respond to $\dfrac{n(n-1)}{2}$ pair-wise comparisons. Based on the DM's response, the criteria rankings and their weights can be computed, following the steps given in the following:

Step 1: Based on the DM's response, a pair-wise comparison matrix, $P_{(n \times n)}$, is constructed, whose elements p_{ij} are as given in the following:

$p_{ii} = 1$ for all $i = 1, 2, \ldots, n$
$p_{ij} = 1, p_{ji} = 0$, if C_i is preferred to C_j $(C_i > C_j)$
$p_{ij} = 0, p_{ji} = 1$, if C_j is preferred to C_i $(C_i < C_j)$
$p_{ij} = p_{ji} = 1$ if C_i and C_j are equally important.

Step 2: Compute the row sums of the matrix P as, $t_i = \displaystyle\sum_j p_{ij}$, for $i = 1, 2, \ldots, n$

Step 3: Rank the criteria based on the t_i values and compute their weights,

$$W_j = \frac{t_j}{\displaystyle\sum_i t_i}, \ \forall j = 1, 2, \ldots n$$

Example 6.4

Five criteria A, B, C, D, and E have to be ranked based on 10 pair-wise comparisons given in the following:

- $A > B, A > C, A > D, A > E$
- $B < C, B > D, B < E$
- $C > D, C < E$
- $D < E$

Solution:

Step 1: Construct the pair-wise comparison matrix P:

$P_{(5 \times 5)} =$	A	B	C	D	E
A	1	1	1	1	1
B	0	1	0	1	0
C	0	1	1	1	0
D	0	0	0	1	0
E	0	1	1	1	1

Step 2: Compute the row sums as $t_A = 5, t_B = 2, t_C = 3, t_D = 1,$ and $t_E = 4$.

Step 3: The ranking of the five criteria is $A > E > C > B > D$ and their weights are $W_A = \dfrac{5}{15}, W_B = \dfrac{2}{15}, W_C = \dfrac{3}{15}, W_D = \dfrac{1}{15}, W_E = \dfrac{4}{15}$.

6.3.6 Scaling Criteria Values

The major drawback of the ranking methods discussed so far (L_p metric, rating, Borda count, and pair-wise comparison) is that they use criteria weights that require the criteria values to be scaled properly. For example, in Table 6.12, all the supplier criteria values ranged from 0 to 1. In other words, they have been already scaled. In practice, supplier criteria are measured in different units. Some criteria values may be very large (e.g., cost) while others may be very small (e.g., quality, delivery time). If the criteria values are not scaled properly, the criteria with large magnitudes would simply dominate the final rankings, independent of the assigned weights. In this section, we shall discuss some common approaches to scaling criteria values.

Consider a supplier selection problem with m suppliers and n criteria, where f_{ij} denotes the value of criterion j for supplier i. Let F denote the supplier criteria matrix.

$$F_{(m \times n)} = [f_{ij}]$$

Determine $H_j = \max_i f_{ij}$ and $L_j = \min_i f_{ij}$

H_j will be the ideal value if criterion j is maximizing and L_j is its ideal value if it is minimizing.

6.3.6.1 Simple Scaling

In simple scaling, the criteria values are multiplied by 10^K where "K" is a positive or negative integer including zero. This is the most common scaling method used in practice. If a criterion is to be minimized, its values should be multiplied by (-1) before computing the weighted score.

6.3.6.2 Ideal Value Method

In this method, criteria values are scaled using their ideal values as given in the following:

$$\text{For "max" criterion: } r_{ij} = \frac{f_{ij}}{H_j} \tag{6.4}$$

$$\text{For "min" criterion: } r_{ij} = \frac{L_j}{f_{ij}} \tag{6.5}$$

Note that the scaled criteria values will always be ≤ 1, and all the criteria have been changed to maximization. The best value of each criterion is 1, but the worst value need not necessarily be zero. In the next approach, all criteria values will be scaled between 0 and 1, with 1 for the best value and 0 for the worst.

6.3.6.3 Simple Linearization (Linear Normalization)

Here the criteria values are scaled as given in the following:

$$\text{For "max" criterion: } r_{ij} = \frac{f_{ij} - L_j}{H_j - L_j} \tag{6.6}$$

$$\text{For "min" criterion: } r_{ij} = \frac{H_j - f_{ij}}{H_j - L_j} \tag{6.7}$$

Here all the scaled criteria value will be between 0 and 1 and all the criteria are to be maximized after scaling.

6.3.6.4 Use of L_p Norm (Vector Scaling)

The L_p norm of a vector $X \in R^n$ is given by L_p norm $= \left[\sum_{j=1}^{n} |X_j|^p \right]^{\frac{1}{p}}$, for $p = 1, 2, \ldots \infty$.

The most common values of p are, $p = 1, 2$, and ∞.

$$\text{For } p = 1, L_1 \text{ norm} = \sum_{j=1}^{n} |X_j| \tag{6.8}$$

$$\text{For } p = 2, L_2 \text{ norm} = \left[\sum_{j=1}^{n} |X_j|^2 \right]^{\frac{1}{2}} \quad \text{(Length of vector } X) \tag{6.9}$$

$$\text{For } p = \infty, L_\infty \text{ norm} = \max\left[|X_j| \right] \quad \text{(Tchebycheff's norm)} \tag{6.10}$$

In this method, scaling is done by dividing the criteria values by their respective L_p norms. After scaling, the L_p norm of each criterion will be one.

We shall illustrate the different scaling methods with an example.

6.3.6.5 Illustrative Example of Scaling Criteria Values

Example 6.5: Scaling Criteria Values

Consider a supplier selection problem with 3 suppliers A, B, C and three selection criteria—Total Ownership Cost (TCO), Service, and Experience. The criteria values are given in Table 6.19. TCO has to be minimized, while the Service and Experience criteria have to be maximized.

TABLE 6.19

Supplier Criteria Values for Example 6.5

	TCO (Min)	Service (Max)	Experience (Max)
Supplier A	$125,000	10	9
Supplier B	$95,000	5	6
Supplier C	$65,000	3	3

Solution

Note that the high cost supplier (A) gives the best service and has the most experience, while supplier C has the lowest cost and experience and gives poor service. The criteria values are not scaled properly, particularly cost measured in dollars. If the values are not scaled TCO criterion will dominate the selection process irrespective of its assigned weight. If we assume that the criteria weights are equal $\left(\dfrac{1}{3}\right)$, then the weighted score for each supplier would be:

$$S_A = \frac{(-125,000+10+9)}{3} = -41,660$$

$$S_B = \frac{(-95,000+5+6)}{3} = -31,663$$

$$S_C = \frac{(-65,000+3+3)}{3} = -21,665$$

Note that the cost criterion (TCO) has been multiplied by (−1) to convert it to a maximization criterion, before computing the weighted score. Supplier C has the maximum weighted score and the rankings will be Supplier C > Supplier B > Supplier A in that order. In fact, even if the weight for TCO is reduced, cost will continue to dominate, as long as it is not scaled properly. Let us look at the rankings after scaling the criteria values by the methods given in this Section.

6.3.6.6 Simple Scaling Illustration

Dividing TCO values by 10,000, we get the scaled values as 12.5, 9.5, and 6.5, which are comparable in magnitude with the criteria values for service and experience. Assuming equal weights again for the criteria, the new weighted scores of the suppliers are as follows:

$$S_A = \frac{(-12.5+10+9)}{3} = 2.17$$

$$S_B = \frac{(-9.5+5+6)}{3} = 0.5$$

$$S_C = \frac{(-6.5+3+3)}{3} = -0.17$$

Now, Supplier A is the best followed by Suppliers B and C.

TABLE 6.20

Scaled Criteria Values by the Ideal
Value Method (Example 6.5)

	TCO	Service	Experience
Supplier A	0.52	1	1
Supplier B	0.68	0.5	0.67
Supplier C	1	0.3	0.33

6.3.6.7 Scaling by Ideal Value Illustration

For Example 6.5, the maximum and minimum criteria values are as given in
the following:

$$C_1 - \text{TCO: } H_1 = 125{,}000, L_1 = 65{,}000$$

$$C_2 - \text{Service: } H_2 = 10, L_2 = 3$$

$$C_3 - \text{Experience: } H_3 = 9, L_3 = 3$$

The ideal values for the three criteria are 65,000, 10, and 9 respectively.
Of course, the ideal solution is not achievable.

The scaled criteria values, using the Ideal Value method are computed
using Equation 6.5 for TCO and Equation 6.4 for Service and Experience.
They are given in Table 6.20.

Note that the scaled values are such that all criteria (including TCO) have
to be maximized. Thus, the new weighted scores are as follows:

$$S_A = \frac{(0.52 + 1 + 1)}{3} = 0.84$$

$$S_B = \frac{(0.68 + 0.5 + 0.67)}{3} = 0.62$$

$$S_C = \frac{(1 + 0.3 + 0.33)}{3} = 0.54$$

The final rankings are suppliers A, B, and C; the same ranking obtained
using the simple scaling.

6.3.6.8 Simple Linearization (Linear Normalization) Illustration

Under this method, the scaled values are computed using Equations 6.6
and 6.7 and are given in Table 6.21.

TABLE 6.21

Scaled Criteria Values by Simple
Linearization (Example 6.5)

	TCO	Service	Experience
Supplier A	0	1	1
Supplier B	0.5	0.29	0.5
Supplier C	1	0	0

Note that the best and worst values of each criterion are 1 and 0, respectively and all the criteria values are now to be maximized. The revised weighted sums are as follows:

$$S_A = \frac{(0+1+1)}{3} = 0.67$$

$$S_B = \frac{(0.5+0.29+0.5)}{3} = 0.43$$

$$S_C = \frac{(1+0+0)}{3} = 0.33$$

The rankings are unchanged with Supplier A as the best, followed by B and C.

6.3.6.9 Scaling by L_p Norm Illustration

We shall illustrate using L_∞ norm for scaling. The L_∞ norms for the three criteria are computed using Equation 6.10.

L_∞ norm for TCO = Max (125000, 95000, 65000) = 125,000
L_∞ norm for Service = Max (10, 5, 3) = 10
L_∞ norm for Experience = Max (9, 6, 3) = 9

The criteria values are then scaled by dividing them by their respective L_∞ norms and are given in Table 6.22.

TABLE 6.22

Scaled Criteria Values Using L_∞ Norm
(Example 6.5)

	TCO	Service	Experience
Supplier A	1	1	1
Supplier B	0.76	0.5	0.67
Supplier C	0.52	0.3	0.33

Note that the scaling by L_p norm did not convert the minimization criterion (TCO) to maximization as the previous two methods (Ideal value and Simple Linearization) did. Hence, the TCO values have to be multiplied by (−1) before computing the weighted score. Note also, that the L_∞ norm of each criterion (column) in Table 6.22 is always 1.

The new weighted scores are as follows:

$$S_A = \frac{(-1+1+1)}{3} = 0.33$$

$$S_B = \frac{(-0.76+0.5+0.67)}{3} = 0.14$$

$$S_C = \frac{(-0.52+0.3+0.33)}{3} = 0.04$$

Once again, the rankings are the same, namely Supplier A, followed by Suppliers B and C.

It should be noted that even though the scaled values, using different scaling methods, were different, the final rankings were always the same. Occasionally, it is possible for rank reversals to occur.

6.3.7 Analytic Hierarchy Process

The analytic hierarchy process (AHP), developed by Saaty (1980), is a multi-criteria decision making method for ranking alternatives. Using AHP, the DM can assess not only quantitative but also various qualitative factors, such as financial stability, feeling of trust, etc. in the supplier selection process. The buyer establishes a set of evaluation criteria and AHP uses these criteria to rank the different suppliers. AHP can enable the DM to represent the interaction of multiple factors in complex and unstructured situations. AHP does not require the scaling of criteria values.

6.3.7.1 Basic Principles of AHP

- Design a hierarchy: Top vertex is the main objective and bottom vertices are the alternatives. Intermediate vertices are criteria/sub-criteria (which are more and more aggregated as you go up in the hierarchy).
- At each level of the hierarchy, a paired comparison of the vertices criteria/sub-criteria is performed from the point of view of their "contribution (weights)" to each of the higher-level vertices to which they are linked.
- Uses both rating method and comparison method. A numerical scale 1–9 (1-equal importance; 9-most important).

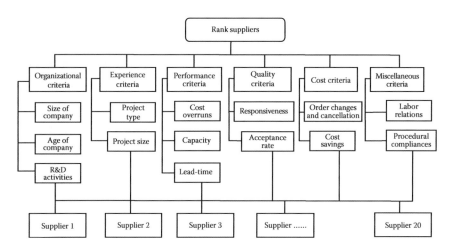

FIGURE 6.5
AHP hierarchy for supplier criteria (Case study 1).

- Uses pair-wise comparison of alternatives with respect to each criterion (sub-criterion) and gets a numerical score for each alternative on every criterion (sub-criterion).
- Computes total weighted score for each alternative and ranks the alternatives accordingly.

To design the hierarchy for Case Study 1 given in Section 6.3.1, the 14 supplier criteria are grouped into six major criteria and several sub-criteria as shown in Figure 6.5.

6.3.7.2 Steps of the AHP Model

Step 1: In the first step, carry out a pair-wise comparison of criteria using the 1–9 degree of importance scale shown in Table 6.23.

TABLE 6.23

Degree of Importance Scale in AHP

Degree of Importance	Definition
1	Equal importance
3	Weak importance of one over other
5	Essential or strong importance
7	Demonstrated importance
9	Absolute importance
2,4,6,8	Intermediate values between two adjacent judgments

TABLE 6.24

Pair-wise Comparison of Criteria (Case Study 1)

	Organizational	Experience	Performance	Quality	Cost	Miscellaneous
Organizational	1	0.2	0.143	0.33	0.33	1
Experience	5	1	0.5	2	2	5
Performance	7	2	1	5	4	7
Quality	3	0.5	0.2	1	1	3
Cost	3	0.5	0.25	1	1	3
Miscellaneous	1	0.2	0.143	0.33	0.33	1

If there are n criteria to evaluate, then the pair-wise comparison matrix for the criteria is given by, $A_{(n \times n)} = [a_{ij}]$, where a_{ij} represents the relative importance of criterion i with respect to criterion j. Set $a_{ii} = 1$ and $a_{ji} = \dfrac{1}{a_{ij}}$. The pair-wise comparisons, with the degree of importance, for the six major criteria in Case Study 1, are shown in Table 6.24.

Step 2: Compute the normalized weights for the main criteria. We obtain the weights using L_1 norm. The two step process for calculating the weights is as follows:

- Normalize each column of A matrix using L_1 norm:

$$r_{ij} = \frac{a_{ij}}{\sum_{i=1}^{n} a_{ij}}$$

- Average the normalized values across each row to get the criteria weights:

$$w_i = \frac{\sum_{j=1}^{n} r_{ij}}{n}$$

Table 6.25 shows the criteria weights for Case Study 1 obtained as a result of Step 2.

Step 3: In this step we check for consistency of the pair-wise comparison matrix using Eigen value theory as follows (Saaty, 1980):

1. Using the pair-wise comparison matrix A (Table 6.24) and the weights W (Table 6.25), compute the vector AW. Let the vector $X = (X_1, X_2, X_3 \dots X_n)$ denote the values of AW.

TABLE 6.25

Final Criteria Weights Using
AHP (Case Study 1)

Criteria	Weight
Organizational	0.047
Experience	0.231
Performance	0.430
Quality	0.120
Cost	0.124
Miscellaneous	0.047

2. Compute

$$\lambda_{max} = Average \left[\frac{X_1}{W_1}, \frac{X_2}{W_2}, \frac{X_3}{W_3} \cdots \frac{X_n}{W_n} \right]$$

3. Consistency index (CI) is given by

$$CI = \frac{\lambda_{max} - n}{n - 1}$$

Saaty (1980) generated a number of random positive reciprocal matrices with $a_{ij} \in (1, 9)$ for different sizes and computed their average CI values, denoted by RI, as given in the following:

N	1	2	3	4	5	6	7	8	9	10
RI	0	0	0.52	0.89	1.11	1.25	1.35	1.4	1.45	1.49

He defines the consistency ratio (CR) as $CR = \dfrac{CI}{RI}$. If CR < 0.15, then accept the pair-wise comparison matrix as consistent.

Using these steps, CR is found to be 0.009 for our example problem. Since the CR is less than 0.15, the response can be assumed to be consistent.

Step 4: In the next step, we compute the relative importance of the sub-criteria in the same way as done for the main criteria. Steps 2 and 3 are carried out for every pair of sub-criteria with respect to their main criterion. The final weights of the sub-criteria are the product of the weights along the corresponding branch. Table 6.26 illustrates the final weights of the various criteria and sub-criteria for Case Study 1.

TABLE 6.26

AHP Sub-Criteria Weights for Case Study 1

Criteria (Criteria Weight)	Sub-Criteria	Sub-Criteria Weight	Global Weight (Criteria Weight × Sub-Criteria Weight)
Organizational (0.047)	Size of company	0.143	0.006
	Age of company	0.429	0.020
	R&D activities	0.429	0.020
Experience (0.231)	Project type	0.875	0.202
	Project size	0.125	0.028
Performance (0.430)	Cost overruns	0.714	0.307
	Capacity	0.143	0.061
	Lead-time	0.143	0.061
Quality (0.120)	Responsiveness	0.833	0.099
	Acceptance rate	0.167	0.020
Cost (0.124)	Order change	0.833	0.103
	Cost savings	0.167	0.020
Miscellaneous (0.047)	Labor relations	0.125	0.005
	Procedural compliances	0.875	0.041

Step 5: Repeat steps 1, 2, and 3 and obtain,

(a) Pair-wise comparison of alternatives with respect to each criterion using the ratio scale (1–9).

(b) Normalized scores of all alternatives with respect to each criterion. Here, an $(m \times m)$ matrix S is obtained, where S_{ij} = normalized score for alternative i with respect to criterion j and m is the number of alternatives.

Step 6: Compute the total score (TS) for each alternative as follows $TS_{(m \times 1)} = S_{(m \times n)} W_{(n \times 1)}$, where W is the weight vector obtained after step 4. Using the total scores, the alternatives are ranked. The total scores for all the suppliers obtained by AHP for Case Study 1 are given in Table 6.27.

NOTE: There is commercially available software for AHP called Expert Choice. Interested readers can refer to www.expertchoice.com for additional information.

6.3.8 Cluster Analysis

Clustering analysis (CA) is a statistical technique particularly suited to grouping of data. It is gaining wide acceptance in many different fields of research such as data mining, marketing, operations research, and bioinformatics. CA is used when it is believed that the sample units come from an unknown population. Clustering is the classification of similar objects into

TABLE 6.27

Supplier Ranking Using AHP
(Case Study 1)

Supplier	Total Score	Rank
Supplier 1	0.119	20
Supplier 2	0.247	14
Supplier 3	0.325	12
Supplier 4	0.191	18
Supplier 5	0.210	16
Supplier 6	0.120	19
Supplier 7	0.249	13
Supplier 8	0.328	11
Supplier 9	0.192	17
Supplier 10	0.212	15
Supplier 11	0.427	7
Supplier 12	0.661	1
Supplier 13	0.431	6
Supplier 14	0.524	5
Supplier 15	0.422	8
Supplier 16	0.539	4
Supplier 17	0.637	2
Supplier 18	0.421	9
Supplier 19	0.412	10
Supplier 20	0.543	3

different groups, or more precisely, the partitioning of a data set into subsets (clusters), so that the data in each sub-set share some common trait. CA develops sub-sets of the raw data such that each sub-set contains member of like nature (similar supplier characteristics) and that difference between different sub-sets is as pronounced as possible.

There are two types of clustering algorithms (Khattree and Dayanand, 2000), namely

- *Hierarchical*: Algorithms that employ hierarchical clustering find successive clusters using previously established clusters. Hierarchical algorithms can be further classified as *agglomerative* or *divisive*. Agglomerative algorithms begin with each member as a separate cluster and merge them into successively larger clusters. On the other hand, divisive algorithms begin with the whole set as one cluster and proceed to divide it into successively smaller clusters. Agglomerative method is the most common hierarchical method.
- *Partitional*: In partitional clustering, the algorithm finds all the clusters at once. An example of partitional methods is *K*-means clustering.

6.3.8.1 Procedure for Cluster Analysis

Clustering process begins with formulating the problem and concludes with carrying out analysis to verify the accuracy and appropriateness of the method. The clustering process has the following steps:

1. Formulate the problem and identify the selection criteria.
2. Decide on the number of clusters.
3. Select a clustering procedure.
4. Plot the dendrogram (A dendrogram is a tree diagram used to illustrate the output of clustering analysis) and carry out analysis to compare the means across various clusters.

Let us illustrate the cluster analysis for supplier selection using Case Study 1.

Step 1: In the first step every supplier is rated on a scale of 0–1 for each attribute as shown in Table 6.12.

Step 2: In this step we need to decide on the number of clusters. We want the initial list of suppliers to be split into two categories, good suppliers (shortlisted) and bad suppliers (rejected); hence the number of clusters is two.

Step 3: Next we apply both hierarchical and partitional clustering methods to supplier data. We choose the method which has the highest R^2 value pooled over all the 14 attributes. A summary table showing the pooled R^2 values is shown in Table 6.28.

The R^2 value for K-means is the highest among different methods; hence K-means is chosen for clustering. There are several other methods available for determining the goodness of fit (Khattree and Dayanand, 2000).

TABLE 6.28

Pooled R^2 Values by Different Cluster
Methods (Case Study 1)

Method	Pooled R^2 over All the Variables
Single linkage	0.05
Wards linkage	0.1301
Centroid linkage	0.06
Complete linkage	0.1307
Average linkage	0.1309
K-means	0.135

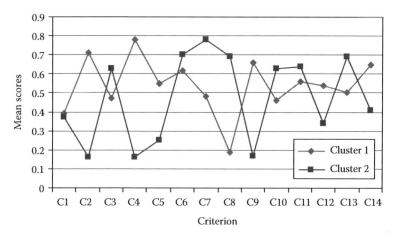

FIGURE 6.6
Comparison of mean across clusters (Case study 1).

Step 4: In this step, we apply the *K*-means clustering to the given data. Graphical comparison of the two clusters is shown in Figure 6.6. Figure 6.6 helps to identify the cluster which has suppliers of higher mean scores. Cluster 1 has 6 members and cluster 2 has 14 members. From Figure 6.6 it can be seen that suppliers of cluster 1 have better mean scores than suppliers in cluster 2 (on most criteria); hence 6 suppliers of cluster 1 are chosen as shortlisted suppliers.

6.3.9 Group Decision Making

Most purchasing decisions, including the ranking and selection of suppliers, involve the participation of multiple DMs and the ultimate decision is based on the aggregation of DMs' individual judgments to arrive at a group decision. The rating method, Borda count and AHP discussed in this section can be extended to group decision making as described in the following:

1. *Rating method*: Ratings of each DM for every criterion are averaged. The average ratings are then normalized to obtain the group criteria weights.

2. *Borda count*: Points are assigned based on the number of DMs that assign a particular rank for a criterion. These points are then totaled for each criterion and normalized to get criteria weights (This is similar to how the college polls are done to get the top 25 football or basketball teams).

3. *AHP*: There are two methods to get the group rankings using AHP.

    a. *Method 1*: Strength of preference scores assigned by individual DMs are aggregated using geometric means and then used in the AHP calculations.

    b. *Method 2*: First, all the alternatives are ranked by each DM using AHP. The individual rankings are then aggregated to a group ranking using Borda count.

6.3.10 Comparison of Ranking Methods

Different ranking methods can provide different solutions resulting in rank reversals. In extensive empirical studies with human subjects, it has been found (Powdrell, 2003; Patel, 2007) that Borda count (with pair-wise comparison of criteria) rankings are generally in line with AHP rankings. Given the increased cognitive burden and expensive calculations required for AHP, Borda count might be selected as an appropriate method for supplier rankings. Even though the Rating method is easy to use, it could lead to several ties in the final rankings, thereby making the results less useful.

Velazquez et al. (2010) studied the best combination of weighting and scaling methods for single and multiple decision makers. The scaling methods considered were ideal value, linear normalization, and vector normalization using L_p norm for $p = 1, 2, 3$, and ∞. The weighting methods included in the study were rating, ranking (Borda count), and AHP. The L_p metric method for $p = 1, 2, 3$, and ∞ were also included for ranking. Experiments were done with real DMs. They found that the best scaling method was influenced by the chosen weighting method. The best combination was scaling by L_∞ norm and ranking by Borda count. The worst combination was scaling by L_∞ norm and ranking by L_∞ metric. The conclusions were the same for both single and multiple DMs.

6.4 Multi-Objective Supplier Allocation Model

As a result of pre-qualification, the large number of initial suppliers is reduced to a manageable size. In the second phase of the supplier selection, detailed quantitative data such as price, capacity, quality, etc. are collected on the shortlisted suppliers and are used in a multi-objective framework for the actual order allocation. We consider multiple buyers, multiple products, and multiple suppliers with volume discounts. The scenario of multiple buyers is possible in case of a central purchasing department, where different divisions of an organization buy through one purchasing department. Here the number of buyers will be equal to the number of divisions buying through the central purchasing. In all other cases, the number of buyers is equal to one.

We consider the least restrictive case where any of the buyers can acquire one or more products from any suppliers, namely, a multiple sourcing model.

In this phase of the supplier selection process an organization will make the following decisions:

- To choose the most favorable suppliers who would meet its supplier selection criteria for the various components.

- To order optimal quantities from the chosen most favorable suppliers to meet its production plan or demand.

The mathematical model for the order allocation problem is discussed next.

6.4.1 Notations Used in the Model

Model indices

i Set of products to be purchased, $i = 1,2...I$

j Set of buyers who procure multiple units in order to fulfill some demand, $j = 1,2...J$

k Potential set of suppliers, $k = 1,2...K$

m Set of incremental price breaks for volume discounts, $m = 1,2...M$

Model parameters

p_{ikm} Cost of acquiring one unit of product i from supplier k at price level m

b_{ikm} Quantity at which incremental price breaks occurs for product i by supplier k

F_k Fixed ordering cost associated with supplier k

d_{ij} Demand of product i for buyer j

l_{ijk} Lead-time of supplier k to produce and supply product i to buyer j. The lead-time of different buyers could be different because of geographical distances

q_{ik} Quality that supplier k maintains for product i, which is measured as percent of defects

CAP_{ik} Production capacity for supplier k for product i

N Maximum number of suppliers that can be selected

Decision variables in the model

X_{ijkm} Number of units of product i supplied by supplier k to buyer j at price level m

Z_k $\begin{cases} 1, & \text{if supplier } k \text{ is chosen to supply any product} \\ 0, & \text{Otherwise} \end{cases}$

Y_{ijkm} $\begin{cases} 1, & \text{if price level } m \text{ is used} \\ 0, & \text{Otherwise} \end{cases}$

6.4.2 Mathematical Formulation of the Order Allocation Problem

The conflicting objectives used in the model are simultaneous minimization of price, lead-time, and rejects. It is relatively easy to include other objectives also. The mathematical form for these objectives is as follows:

1. Price (z_1): Total cost of purchasing has two components; fixed and the variable cost.

   Total variable cost: The total variable cost is the cost of buying every additional unit from the suppliers and is given by:

$$\sum_i \sum_j \sum_k \sum_m p_{ikm} \cdot X_{ijkm} \qquad (6.11)$$

   Fixed cost: If a supplier k is used then there is a fixed cost associated with it, which is given by:

$$\sum_k F_k \cdot Z_k \qquad (6.12)$$

   Hence the total purchasing cost is

$$\sum_i \sum_j \sum_k \sum_m p_{ikm} \cdot X_{ijkm} + \sum_k F_k \cdot Z_k \qquad (6.13)$$

2. Lead-time (z_2):

$$\sum_i \sum_j \sum_k \sum_m l_{ijk} \cdot X_{ijkm} \qquad (6.14)$$

   The product of lead-time of each product and quantity supplied is summed over all the products, buyers, and suppliers and should be minimized. This represents weighted average lead-time.

3. Quality (z_3):

$$\sum_i \sum_j \sum_k \sum_m q_{ik} \cdot X_{ijkm} \qquad (6.15)$$

   The product of rejects and quantity supplied is summed over all the products, buyers, and suppliers and should be minimized. This represents weighted average quality. Quality in our case is measured in terms of percentage of rejects.

The constraints in the model are as follows:

1. *Capacity constraint*: Each supplier k has a maximum capacity for product i, given by CAP_{ik}. Total order placed with this supplier must be less than or equal to the maximum capacity. Hence the capacity constraint is given by:

$$\sum_j \sum_m X_{ijkm} \leq (CAP_{ik})Z_k \quad \forall i \text{ and } k \quad (6.16)$$

The binary variable on the right-hand side of the constraint implies that a supplier cannot supply any products if not chosen, i.e., if Z_k is 0.

2. *Demand constraint*: The demand of buyer j for product i has to be satisfied using a combination of the suppliers. The demand constraint is given by:

$$\sum_k \sum_m X_{ijkm} = d_{ij} \quad \forall i,j \quad (6.17)$$

3. *Maximum number of suppliers*: The maximum number of suppliers chosen must be less than or equal to the specified number. Hence this constraint takes the following form:

$$\sum_k Z_k \leq N \quad (6.18)$$

4. *Linearizing constraints*: In the presence of incremental price discounts, objective function is nonlinear. The following set of constraints are used to linearize it:

$$X_{ijkm} \leq (b_{ikm} - b_{ikm-1}) * Y_{ijkm} \qquad \forall i,j,k, 1 \leq m \leq m_k \quad (6.19)$$

$$X_{ijkm} \geq (b_{ikm} - b_{ikm-1}) * Y_{ijkm+1} \qquad \forall i,j,k, 1 \leq m \leq m_k - 1 \quad (6.20)$$

$0 = b_{i,k,0} < b_{i,k,1} < \cdots < b_{i,k,m_k}$ is the sequence of quantities at which price break occurs. p_{ikm} is the unit price of ordering X_{ijkm} units from supplier k at level m, if $b_{i,k,m-1} < X_{ijkm} \leq b_{i,k,m}$ ($1 \leq m \leq m_k$).

Constraints (6.19) and (6.20) force quantities in the discount range for a supplier to be incremental. Because the "quantity" is incremental, if the order quantity lies in discount interval m, namely, $Y_{ijkm} = 1$, then the quantities in interval 1 to $m-1$, should be at the maximum of those ranges. Constraint (6.19) also assures that a quantity in any range is no greater than the width of the range.

5. Non-negativity and binary constraint: $X_{ijkm} \geq 0$; Z_k, $Y_{ijkm} \in (0,1)$.

6.4.3 Goal Programming Methodology
(Masud and Ravindran, 2008, 2009)

The mathematical model formulated in Section 6.4.2 is a multiple objective integer linear programming problem. One way to treat multiple criteria is to select one criterion as primary and the other criteria as secondary. The primary criterion is then used as the optimization objective function, while the secondary criteria are assigned acceptable minimum and maximum values and are treated as problem constraints. However, if careful considerations are not given while selecting the acceptable levels, a feasible solution that satisfies all the constraints may not exist. This problem is overcome by *Goal Programming*, which has become a popular practical approach for solving multiple criteria optimization problems.

Goal programming (GP) falls under the class of methods that use completely pre-specified preferences of the DM in solving the multi-criteria mathematical programming (MCMP) problems. In goal programming, all the objectives are assigned target levels for achievement and a relative priority on achieving those levels. Goal programming treats these targets as *goals to aspire for* and not as absolute constraints. It then attempts to find an optimal solution that comes as "close as possible" to the targets in the order of specified priorities. In this section, we shall discuss how to formulate goal programming models and their solution methods.

Before we discuss the formulation of goal programming problems, we discuss the difference between the terms *real constraints* and *goal constraints* (or simply *goals*) as used in goal programming models. The real constraints are absolute restrictions on the decision variables, while the goals are conditions one would like to achieve but are not mandatory. For instance, a real constraint given by, $x_1 + x_2 = 3$ requires all possible values of $x_1 + x_2$ to always equal 3. As opposed to this, a goal requiring $x_1 + x_2 = 3$ is not mandatory, and we can choose values of $x_1 + x_2 \geq 3$ as well as $x_1 + x_2 \leq 3$. In a goal constraint, positive and negative deviational variables are introduced to represent constraint violations as follows:

$$x_1 + x_2 + d_1^- - d_1^+ = 3 \qquad d_1^+, d_1^- \geq 0$$

Note that, if

$$d_1^- > 0, \text{ then } x_1 + x_2 < 3, \text{ and if } d_1^+ > 0, \text{ then } x_1 + x_2 > 3$$

By assigning suitable weights w_1^- and w_1^+ on d_1^- and d_1^+ in the objective function, the model will try to achieve the sum $x_1 + x_2$ as close as possible to 3. If the goal were to satisfy $x_1 + x_2 \geq 3$, then only d_1^- is assigned a positive weight in the objective, while the weight on d_1^+ is set to zero.

6.4.3.1 General Goal Programming Model

A general multiple criteria mathematical programming (MCMP) problem is given as follows:

$$\text{Max} \quad F(\mathbf{x}) = \{f_1(x), f_2(x), \ldots, f_k(x)\}$$

Subject to

$$g_j(\mathbf{x}) \leq 0 \text{ for } j = 1, \ldots, m \tag{6.21}$$

where $\mathbf{x}$ is an n-vector of *decision variables* and $f_i(x)$, $i = 1, \ldots, k$ are the k *criteria/objective functions*.

$$\text{Let } S = \{x/g_j(\mathbf{x}) \leq 0, \text{ for all "}j\text{"}\}$$

$$Y = \{y/F(\mathbf{x}) = \mathbf{y} \text{ for some } \mathbf{x} \in S\}$$

S is called the *decision space* and Y is called the *criteria or objective space* in MCMP.

Consider the general MCMP problem given in Equation 6.21. The assumption that there exists an optimal solution to the MCMP problem involving multiple criteria implies the existence of some preference ordering of the criteria by the DM. The goal programming (GP) formulation of the MCMP problem requires the DM to specify an acceptable level of achievement (b_i) for each criterion f_i and specify a weight w_i (ordinal or cardinal) to be associated with the deviation between f_i and b_i. Thus, the GP model of an MCMP problem becomes:

$$\text{Minimize } Z = \sum_{i=1}^{k} (w_i^+ d_i^+ + w_i^- d_i^-) \tag{6.22}$$

Subject to:

$$f_i(x) + d_i^- - d_i^+ = b_i \quad \text{for } i = 1, \ldots, k \tag{6.23}$$

$$g_j(x) \leq 0 \quad \text{for } j = 1, \ldots, m \tag{6.24}$$

$$x_j, d_i^-, d_i^+ \geq 0 \quad \text{for all } i \text{ and } j \tag{6.25}$$

Equation 6.22 represents the objective function of the GP model, which minimizes the weighted sum of the deviational variables. The system of equations

(Equation 6.23) represents the goal constraints relating the multiple criteria to the goals/targets for those criteria. The variables, d_i^- and d_i^+, in Equation 6.23 are the deviational variables, representing the underachievement and overachievement of the ith goal. The set of weights (w_i^+ and w_i^-) may take two forms as follows:

1. Pre-specified weights (cardinal)
2. Preemptive priorities (ordinal)

Under pre-specified (cardinal) weights, specific values in a relative scale are assigned to w_i^+ and w_i^- representing the DM's "trade-off" among the goals. Once w_i^+ and w_i^- are specified, the goal program represented by Equations 6.22 through 6.25 reduces to a single objective optimization problem. The cardinal weights could be obtained from the DM using any of the methods discussed in Section 6.3 (Rating, Borda count, and AHP). However, for this method to work effectively, criteria values have to be scaled properly and the Ideal value method discussed in Section 6.3.6, is the most commonly used scaling method. In reality, goals are usually incompatible (i.e., incommensurable) and some goals can be achieved only at the expense of some other goals. Hence, preemptive goal programming, which is more common in practice, uses ordinal ranking or preemptive priorities to the goals by assigning incommensurable goals to different priority levels and weights to goals at the same priority level. In this case, the objective function of the GP model (Equation 6.22) takes the form

$$\text{Minimize } Z = \sum_p P_p \sum_i (w_{ip}^+ d_i^+ + w_{ip}^- d_i^-) \tag{6.26}$$

where P_p represents priority p with the assumption that P_p is much larger than P_{p+1} and w_{ip}^+ and w_{ip}^- are the weights assigned to the ith deviational variables at priority p. In this manner, lower priority goals are considered only after attaining the higher priority goals. Thus, preemptive goal programming is essentially a sequential single objective optimization process, in which successive optimizations are carried out on the alternate optimal solutions of the previously optimized goals at higher priority. In addition to Preemptive and Non-Preemptive goal programming models, other approaches (Fuzzy GP, Min-Max GP) have also been proposed. In the next four sections we will illustrate four different variants of goal programming for the supplier selection model discussed in Section 6.4.2

6.4.4 Preemptive Goal Programming

For the three criteria supplier order allocation problem (Section 6.4.2), the preemptive GP formulation will be as follows:

$$\min P_1 d_1^+ + P_2 d_2^+ + P_3 d_3^+ \tag{6.27}$$

Subject to

$$\sum_i \sum_j \sum_k \sum_m l_{ijk}.x_{ijkm} + d_1^- - d_1^+ = \text{Lead-time goal} \qquad (6.28)$$

$$\sum_i \sum_j \sum_k \sum_m p_{ikm}.x_{ijkm} + \sum_k F_k.z_k + d_2^- - d_2^+ = \text{Price goal} \qquad (6.29)$$

$$\sum_i \sum_j \sum_k \sum_m q_{ik}.x_{ijkm} + d_3^- - d_3^+ = \text{Quality goal} \qquad (6.30)$$

$$d_n^-, d_n^+ \geq 0 \quad \forall n \in \{1,...,3\} \qquad (6.31)$$

$$\sum_j \sum_m x_{ijkm} \leq CAP_{ik}.Z_k \quad \forall i,k \qquad (6.32)$$

$$\sum_k \sum_m x_{ijkm} = d_{ij} \quad \forall i,j \qquad (6.33)$$

$$\sum_k z_k \leq N \qquad (6.34)$$

$$x_{ijkm} \leq (b_{ikm} - b_{ik(m-1)}).y_{ijkm} \quad \forall i,j,k \quad 1 \leq m \leq m_k \qquad (6.35)$$

$$x_{ijkm} \geq (b_{ikm} - b_{ik(m-1)}).y_{ijk(m+1)} \quad \forall i,j,k \quad 1 \leq m \leq m_k - 1 \qquad (6.36)$$

$$x_{ijkm} \geq 0 \quad z_k \in \{0,1\} \quad y_{ijkm} \in \{0,1\} \qquad (6.37)$$

Method for solving preemptive goal program is given in Arthur and Ravindran (1980) and Masud and Ravindran (2008, 2009).

6.4.5 Non-Preemptive Goal Programming

In the non-preemptive GP model, the buyer sets goals to achieve for each objective and preferences in achieving those goals expressed as numerical weights. Here, the buyer has three goals as follows:

- Limit the lead-time to lead goal with weight w_1.
- Limit the total purchasing cost to price goal with weight w_2.
- Limit the quality to quality goal with weight w_3.

The weights w_1, w_2, and w_3 can be obtained using the methods discussed in Section 6.3. The non-preemptive GP model can be formulated as

$$\text{Min } Z = w_1.d_1^+ + w_2.d_2^+ + w_3.d_3^+ \qquad (6.38)$$

Subject to the constraints (6.28) through (6.37).

In the aforementioned model $d_1^+, d_2^+,$ and d_3^+ represent the overachievement of the stated goals. Due to the use of the weights the model needs to be scaled. The weights w_1, w_2, and w_3 can be varied to obtain different goal programming optimal solutions.

6.4.6 Tchebycheff (Min–Max) Goal Programming

In this GP model, the DM only specifies the goals/targets for each objective. The model minimizes the maximum deviation from the stated goals. For the supplier selection problem the Tchebycheff goal program becomes:

$$\text{Min Max } (d_1^+, d_2^+, d_3^+) \tag{6.39}$$

$$d_i^+ \geq 0 \; \forall i \tag{6.40}$$

Subject to the constraints (6.28 through 6.37)

Equation 6.39 can be reformulated as a linear objective by setting

$$\text{Max } (d_1^+, d_2^+, d_3^+) = M \geq 0$$

Thus, Equation 6.39 is equivalent to:

$$\text{Min } Z = M \tag{6.41}$$

Subject to:

$$M \geq (d_1^+) \tag{6.42}$$

$$M \geq (d_2^+) \tag{6.43}$$

$$M \geq (d_3^+) \tag{6.44}$$

$$d_i^+ \geq 0, \; \forall i \tag{6.45}$$

Constraints (6.28 through 6.37) stated earlier will also be included in this model.

The advantage of Tchebycheff goal program is that there is no need to get preference information (priorities or weights) about goal achievements from the DM. Moreover, the problem reduces to a single objective optimization problem. The disadvantages of this method are (i) the scaling of goals is necessary (as required in non-preemptive GP) and (ii) outliers are given more importance and could produce poor solutions.

6.4.7 Fuzzy Goal Programming

Fuzzy goal programming uses the ideal values as targets and minimizes the maximum normalized distance from the ideal solution for each objective. An ideal solution is the vector of best values of each criterion obtained by optimizing each criterion independently ignoring other criteria. In this example, ideal solution is obtained by minimizing price, lead-time, and quality independently. In most situations, the ideal solution is an infeasible solution since the criteria conflict with one another.

If M equals the maximum deviation from the ideal solution, then the fuzzy goal programming model is as follows:

$$\text{Min } Z = M \tag{6.46}$$

Subject to

$$M \geq (d_1^+)/\lambda_1 \tag{6.47}$$

$$M \geq (d_2^+)/\lambda_2 \tag{6.48}$$

$$M \geq (d_3^+)/\lambda_3 \tag{6.49}$$

$$d_i^+ \geq 0, \ \forall i \tag{6.50}$$

Constraints (6.28 through 6.37) stated earlier will also be included in this model, except that the target for Equations 6.28 through 6.30 are set to their respective ideal values. In this model λ_1, λ_2, and λ_3 are scaling constants to be set by the user. A common practice is to set the values λ_1, λ_2, λ_3 equal to the respective ideal values. The advantage of Fuzzy GP is that no target values have to be specified by the DM.

For additional readings on the variants of fuzzy GP models, the reader is referred to Ignizio and Cavalier (1994), Tiwari et al. (1986), Tiwari et al. (1987), Mohammed (1997), and Hu et al. (2007). An excellent source of reference for goal programming methods and applications is the textbook by Schniederjans (1995).

We shall now illustrate the four goal programming methods using a supplier order allocation case study. The data used in all four methods is presented next.

6.4.8 Case Study 2: Supplier Order Allocation

To demonstrate the use of goal programming in supplier selection, consider the case where we have two products, one buyer, five suppliers where each supplier offers two price breaks. The problem here is to find which supplier(s) to buy from and how much to buy from the chosen supplier(s). The goal programming problems are solved using the optimization software LINGO 8.0.

The cost of acquiring one unit of demand for product i from supplier k at price level m, p_{ikm}, and the quantity at which price break occurs for product i for supplier k, b_{ikm}, is given in Table 6.29.

Fixed cost associated with supplier k, F_k, is given in Table 6.30.

The demand of product i by buyer j, d_{ij}, is given in Table 6.31.

TABLE 6.29

Unit Price and Price Break for Supplier Product Combination (Case Study 2)

Product	Supplier	Break	Unit Price	Quantity
1	1	1	190	90
1	1	2	175	200
1	2	1	200	80
1	2	2	170	180
1	3	1	185	100
1	3	2	177	180
1	4	1	188	85
1	4	2	180	170
1	5	1	194	90
1	5	2	172	168
2	1	1	360	200
2	1	2	335	350
2	2	1	370	210
2	2	2	330	330
2	3	1	355	220
2	3	2	340	338
2	4	1	365	180
2	4	2	337	400
2	5	1	357	177
2	5	2	350	365

TABLE 6.30

Fixed Supplier Cost (Case Study 2)

Supplier	Cost
1	1000
2	1500
3	800
4	1600
5	1100

TABLE 6.31

Demand Data (Case Study 2)

Product	Buyer	Demand
1	1	320
2	1	230

TABLE 6.32

Lead-Time Data (Case Study 2)

Product	Buyer	Supplier	Lead-Time
1	1	1	6
1	1	2	10
1	1	3	7
1	1	4	14
1	1	5	5
2	1	1	11
2	1	2	6
2	1	3	7
2	1	4	6
2	1	5	9

Lead-time for supplier k to produce and supply product i to buyer j, l_{ijk}, is given in Table 6.32.

Quality that supplier k maintains for product i, (q_{ik}) and production capacity for supplier k for product i, CAP_{ik}, is given in Table 6.33.

The maximum number of suppliers that can be selected is assumed as three.

TABLE 6.33

Supplier Quality Data and Production Capacity Data (Case Study 2)

Product	Supplier	Quality	Capacity
1	1	0.03	450
1	2	0.04	400
1	3	0.08	470
1	4	0.09	350
1	5	0.06	500
2	1	0.06	600
2	2	0.08	550
2	3	0.04	480
2	4	0.03	590
2	5	0.03	640

All four goal programming models (Preemptive, Non-Preemptive, Tchebycheff, and Fuzzy) are used to solve the supplier order allocation problem. Each model produces a different optimal solution. They are discussed next.

6.4.8.1 Preemptive Goal Programming Solution

In Preemptive goal programming, lead-time is given the highest priority, followed by price and quality respectively. The target values for each of the objectives are set at 105% of the ideal value. For example, the ideal (minimum) value for price objective is \$201,590; hence the target value for price is \$211,669 and the goal is to minimize the deviation above the target value. Table 6.34 illustrates the solution using the Preemptive goal programming model.

6.4.8.2 Non-Preemptive Goal Programming

In non-preemptive goal programming, weights w_1, w_2, and w_3 are obtained using AHP. The values of the weights for lead-time, price, and quality are assumed to be 0.637, 0.185, and 0.178 respectively. The target values used are the same that are used in preemptive goal programming. The solution of the non-preemptive goal programming model is shown in Table 6.35. Since non-preemptive GP requires scaling, the target values are used as scaling constants.

TABLE 6.34

Preemptive GP Solution (Case Study 2)

Preemptive GP	Ideal Values	Preemptive Priorities	Target for Preemptive Goal (Ideal+5%)	Actual Achieved	Whether Goal Achieved	Suppliers Chosen
Lead-time	4,610	1	4,840	4,648	Achieved	S1,S4,S5
Price	201,590	2	211,669	203,302	Achieved	
Quality	25.2	3	26.46	30.59	Not-achieved	

TABLE 6.35

Non-Preemptive GP Solution (Case Study 2)

Non-Preemptive GP	Ideal Values	Weights	Scaling Constant	Target for Non-Preemptive Goal (Ideal+5%)	Actual Achieved	Whether Goal Achieved	Suppliers Chosen
Lead-time	4,610	0.637	4,840	4,840	4,840	Achieved	S1,S2,S3
Price	201,590	0.185	211,669	211,669	204,582	Achieved	
Quality	25.2	0.178	26.46	26.46	29.76	Not-achieved	

TABLE 6.36

Tchebycheff GP Solution (Case Study 2)

Tchebycheff GP	Ideal Values	Weights	Scaling Constant	Targets for Tchebycheff GP	Actual Achieved	Whether Goal Achieved	Suppliers Chosen
Lead-Time	4,610	0.637	0.1	4,840	4,932	Not-achieved	S1,S4,S5
Price	201,590	0.185	0.001	211,669	205,196	Achieved	
Quality	25.2	0.178	10	26.46	29.76	Not-achieved	

6.4.8.3 Tchebycheff Goal Programming

In Tchebycheff GP, the target values are the same that are used in preemptive goal programming. Scaling constants are chosen in such a way that all the three objectives have a similar magnitude. For example, lead-time when multiplied by 0.1 gives 461 and quality when multiplied by 10 yields 252; this makes lead-time and quality of similar magnitude. Using the Tchebycheff method, the solution obtained is illustrated in Table 6.36.

6.4.8.4 Fuzzy Goal Programming

Recall that in Fuzzy GP, the ideal values are used as targets for the different goals. The solution obtained using fuzzy goal programming is shown in Table 6.37. The scaling constants are calculated such that all the objectives have similar values.

Here the final set of suppliers is different since the targets are set at ideal values and it is generally not possible to achieve any of them.

6.4.9 Value Path Approach

The presentation of results presents a critical link, in any multi-objective problem. Any sophisticated analysis, just becomes numbers, if they are not presented to the DM in an effective way. In case of multi-objective problems, a lot of information needs to be conveyed which not only includes performance of

TABLE 6.37

Fuzzy GP Solution (Case Study 2)

Fuzzy GP	Ideal Values	Weights	Scaling Constant	Target for Fuzzy GP	Actual Achieved	Whether Goal Achieved	Suppliers Chosen
Lead-time	4,610	0.637	0.1	4,610	4,675	Not-achieved	S1,S2,S5
Price	201,590	0.185	0.001	201,590	206,056	Not-achieved	
Quality	25.2	0.178	10	25.2	27.53	Not-achieved	

various criteria but also their trade-offs. In this section, we discuss how to present the four different optimal solutions obtained by the different GP models.

The value path approach (Schilling et al., 1983) is one of the most efficient ways to demonstrate the trade-offs among the criteria obtained by the different solutions. The display consists of a set of parallel scales; one for each criterion, on which is drawn the value path for each of the solution alternative. Value paths have proven to be an effective way to present the trade-offs in problems with more than two objectives. The value assigned to each solution on a particular axis is that solution's value for the appropriate objective divided by the best solution for that objective. The minimum value will be one if all the objectives were to minimize. Following are some properties of the value path approach (Schilling et al., 1983):

- If two value paths representing solutions A and B intersect between two vertical scales then the line segment connecting A and B in objective space has a negative slope and neither objective dominates other.
- If three or more value paths intersect, then their associated points in the objective space are collinear.
- If two paths do not intersect then one path must lie entirely above the other and is therefore inferior, if the objective were to minimize.

6.4.9.1 Value Path Approach for the Supplier Selection Case Study

The supplier selection case study was solved using four different GP approaches as previously illustrated. A summary of the results is provided in Table 6.38.

To present these results to the DM, value path approach is used as follows:

1. Find the best (minimum) value obtained for each criterion. For the price criterion, preemptive GP has the best value of 203,302; for the lead-time criterion Preemptive GP is best with a value of 4648; and for the quality objective the best value of 27.53 is obtained through Fuzzy GP.

2. For each solution, divide their objective values by the best value for that objective. For example, for the *preemptive GP* solution, the value

TABLE 6.38

Summary of GP Solutions for Case Study 2

Solution	Method	Lead-Time	Price	Quality
1	Preemptive GP	4648	203,302	30.59
2	Non-preemptive GP	4840	204,582	29.76
3	Tchebycheff GP	4932	205,196	29.76
4	Fuzzy GP	4675	206,056	27.53

TABLE 6.39

Criteria Values for Value Path Approach (Case Study 2)

Alternative	Method	Lead-Time	Price	Quality
1	Preemptive GP	1	1	1.1112
2	Non-preemptive GP	1.0413081	1.0063	1.081
3	Tchebycheff GP	1.0611015	1.00932	1.081
4	Fuzzy GP	1.005809	1.01355	1

of lead-time, price, and quality are 4648, 203302, and 30.59 respectively. The best values for lead-time, price, and quality are 4648, 203302, and 27.53. Therefore, the values for the value path approach corresponding to *preemptive GP* are obtained as (4648/4648), (203302/203302), and (30.59/27.53) respectively. Similar values are calculated for the other solutions under the value path approach as shown in Table 6.39.

3. The last step is to plot the results, with lead-time, price and quality on X-axis and ratios of the objective values on the Y-axis.

The graph is shown in Figure 6.7.

6.4.9.2 Discussion of Value Path Results

Based upon the preference of the DM, the preferred suppliers and the quantity ordered from each can change. The value path approach is a useful tool to compare the trade-offs among the suppliers. In some cases, the price of the product may dictate the suppliers who are chosen and in some other cases, the suppliers chosen may be dictated by lead-time or quality. Hence, value path approach can be used to study the trade-offs between different solutions. For example, from Figure 6.7 it can be seen that preemptive GP does

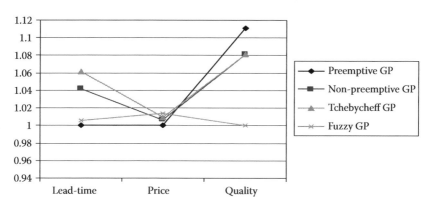

FIGURE 6.7
Graphical representation of value path approach (Case study 2).

1% better on lead-time and 2% better on price compared to Fuzzy GP, but Fuzzy GP is 11% better on quality; such comparisons can easily demonstrate the trade-offs between different solutions.

6.5 Summary and Further Readings

Procurement has become a critical function in recent years in all industries (manufacturing, service, and government). The purchasing cost of raw material in most companies constitutes up to 70% of the total cost. In recent years, the department of defense (DOD) has increased its reliability on commercial suppliers as a way to reduce both purchasing cost and procurement time (Office of Management and Budget, 2008). As of 2006, DOD had a budget of over $400 billion a year, of which 15% was spent on procurement activities. The increased reliance on suppliers requires a better selection process. Hence, supplier selection, which includes identification and evaluation of the right suppliers, has become a critical activity of purchasing.

This chapter illustrated the use of both discrete and continuous multi-criteria decision making techniques to optimize the supplier selection process. We presented the supplier selection problem in two phases. In the first phase called, pre-qualification, we reduced the initial set of large number suppliers to a manageable set. Phase one reduced the effort of the buyer and made the pre-qualification process entirely objective. In the second phase, we analyzed the shortlisted suppliers using the multi-objective technique known as goal programming. We considered several conflicting criteria, including, price, lead-time, and quality. An important distinction of multi-objective techniques is that it does not provide one optimal solution, but a number of solutions known as efficient solutions. Hence, the role of the DM (buyer) is more important than before. By involving the DM early in the process, the acceptance of the model results by the top management becomes easier. The efficient solutions are compared using the value path approach to show the criteria trade-off obtained using different goal programming approaches. In this section, we will summarize the methods discussed in this chapter and suggest further readings about other methods available in the literature.

6.5.1 Ranking Suppliers

In this chapter, we discussed several multiple criteria methods for ranking suppliers—linear weighted point (LWP) method, L_p metric, rating, Borda count, pair-wise comparison, and AHP. It was pointed out that for all the ranking methods (except AHP) to work effectively, the criteria values have to be scaled properly. We then discussed several methods for scaling—simple scaling, simple linearization, and vector scaling by L_p Norm.

The supplier rankings are used in two ways in practice. One is to select the best supplier to negotiate with, under single sourcing. Another is to use the rankings to prequalify or shortlist suppliers under multiple sourcing.

6.5.2 Supplier Order Allocation

Pre-qualification
Pre-qualification reduces a large set of initial suppliers to a smaller set of acceptable suppliers for further assessment. De Boer et al. (2001) have cited many different techniques for pre-qualification. Some of these techniques are categorical methods, data envelopment analysis (DEA), cluster analysis, case-based reasoning (CBR) systems, and multi-criteria decision making method (MCDM). Several authors have worked on pre-qualification of suppliers. Weber and Ellram (1992) and Weber et al. (2000) have developed DEA methods for pre-qualification. Hinkel et al. (1969) and Holt (1998) used cluster analysis for pre-qualification and finally Ng and Skitmore (1995) developed CBR systems for pre-qualification. Mendoza et al. (2008) developed a three phase multi-criteria method to solve a general supplier selection problem. The paper combines analytic hierarchy process (AHP) with goal programming for both pre-qualification and final order allocation.

Multiple sourcing models: Multiple sourcing can offset the risk of supply disruption. In multiple sourcing, a buyer purchases the same item from more than one supplier. Mathematical programming is the most appropriate method for multiple sourcing decisions. Two types of mathematical programming models are found in the literature, single objective and multi-objective models.

- *Single objective models*: In Section 6.2, we presented a single objective linear programming model for order allocation using Example 6.3. The model considered supplier capacities, price discounts, buyers' demand, quality, and lead-time constraints. The objective was to minimize total cost, which included fixed and variable cost of the suppliers. We shall briefly review here, some of the other single objective models that have been discussed in the literature.

Moore and Fearon (1973) stated that price, quality, and delivery are important criteria for supplier selection. They discussed the use of linear programming in decision making.

Gaballa (1974) applied mathematical programming to supplier selection in a real case. He used a mixed integer programming to formulate a decision making model for the Australian Post Office. The objective for this approach is to minimize the total discounted price of allocated items to the suppliers. Anthony and Buffa (1977) developed a single objective linear programming model to support strategic purchasing scheduling (SPS). The linear model minimized the total cost by considering limitations of purchasing budget, supplier capacities, and buyer's demand. Price and storage cost were included in the objective function. The costs of ordering, transportation, and inspection were not included in the model. Bender et al. (1985) applied single objective programming to develop a commercial computerized model for supplier selection at IBM. They used mixed integer programming, to minimize the sum of purchasing, transportation, and inventory costs. Narasimhan and Stoynoff (1986) applied a single objective, mixed integer-programming model to a large manufacturing firm in the midwest, to optimize the allocation procurement for a group of suppliers. Turner (1988) presented a single objective linear programming model for British Coal. This model minimized the total discounted price by considering the supplier capacity, maximum and minimum order quantities, demand, and regional allocated bounds as constraints. Pan (1989) proposed multiple sourcing for improving the reliability of supply for critical materials, in which more than one supplier is used and the demand is split between them. The author used a single objective linear programming model to choose the best suppliers, in which three criteria are considered: price, quality, and service. Seshadri et al. (1991) developed a probabilistic model to represent the connection between multiple sourcing and its consequences, such as number of bids, the seller's profit, and the buyer's price. Benton (1991) developed a nonlinear programming model and a heuristic procedure using Lagrangian relaxation for supplier selection under conditions of multiple items, multiple suppliers, resource limitations, and quantity discounts. Chaudhry et al. (1991) developed linear and mixed integer-programming models for supplier selection. In their model price, delivery, quality, and quantity discount were included. Papers by Degraeve et al. (1999) and Ghodsypour and O'Brien (2001) tackled the supplier selection issue in the framework of TCO or total cost of logistics. Jayaraman et al. (1999) formulated a mixed integer linear programming model for solving the supplier selection problem with multiple products, buyers, and suppliers. Feng et al. (2001) presented a stochastic integer-programming model for simultaneous selection of tolerances and suppliers based on the quality loss function and process capability index.

- *Multi-criteria models*: Among the different multi-criteria approaches for supplier selection, goal programming (GP) is the most commonly used method. Appendix A has a review of goal programming and other approaches for solving multi-criteria optimization problems. We discussed several goal programming models in Section 6.4. They included preemptive, non-preemptive, min–max, and fuzzy goal programs. A case study was used to illustrate the differences among the various goal programming approaches. We then presented the value path approach to compare different solutions visually.

We shall briefly review some of the other mathematical programming models and solutions presented in the literature. Buffa and Jackson (1983) presented a multi-criteria linear goal programming model. In this model two sets of factors are considered: supplier attributes, which include quality, price, service experience, early, late and on-time deliveries, and the buying firm's specifications, including material requirement and safety stock. Sharma et al. (1989) proposed a GP formulation for attaining goals pertaining to price, quality, and lead-time under demand and budget constraints. Liu et al. (2000) developed a decision support system by integrating AHP with linear programming. Weber et al. (2000) used multi-objective linear programming for supplier selection to systematically analyze the trade-off between conflicting factors. In this model aggregate price, quality, and late delivery were considered as goals. Karpak et al. (1999) used visual interactive goal programming for supplier selection process. The objective was to identify and allocate order quantities among suppliers while minimizing product acquisition cost and maximizing quality and reliability. Bhutta and Huq (2002) illustrated and compared the technique of TCO and AHP in supplier selection process. Wadhwa and Ravindran (2007) formulated a supplier selection problem with price, lead-time, and quality as three conflicting objectives. The suppliers offered quantity discounts and the model was solved using goal programming, compromise programming, and weighted objective methods.

Besides goal programming, there are other approaches to solve the multi-criteria optimization model for the supplier selection problem formulated in Section 6.4.2. They include weighted objective method, compromise programming, and interactive approaches. Interested readers can refer to Wadhwa and Ravindran (2007), Masudand Ravindran (2008, 2009) for more details. Masudand Ravindran (2008, 2009) also provide information on the computer software available for the MCDM methods.

Table 6.40 gives a brief summary of some of the other papers not reviewed in this section but have been published since 2000 dealing with supplier selection methods.

6.5.3 Global Sourcing

Firms now recognize international supply management as a key driver of financial performance and overall competitiveness. The main objective of a global sourcing strategy is to exploit both the supplier's competitive advantages and the comparative location advantages of various countries in global competition. As companies turn to global suppliers, they must be

TABLE 6.40

Summary of Supplier Selection Methods

Authors	Method	Brief Description
Yang (2006)	Multi-criteria math model	Developed a five-step multi-criteria strategic supplier selection model incorporating the supply risk
Xia and Wu (2007)	AHP with multi-objective mathematical programming	Formulated a multi-criteria supplier selection problem with supplier price breaks. Incorporates AHP to calculate criteria weights to be used in the model
Saen (2007)	DEA (imprecise DEA)	Used a modified version of DEA to include both qualitative and quantitative data in supplier selection
Chen et al. (2006)	Fuzzy TOPSIS	Used fuzzy numbers to handle linguistic judgments and applied TOPSIS MCDM technique to address the supplier selection problem
Haq and Kannan (2006)	Fuzzy AHP, genetic algorithm	Addressed the supplier selection using fuzzy AHP. Formulates a multi-echelon supply chain configuration problem and solves it using GA
Cachon and Zhang (2006)	Game theory	Formulated a game theoretic supplier selection model under information asymmetry
Ding et al. (2005)	Simulation, genetic algorithm	Proposed a hybrid method where GA is used to search for optimal supplier portfolios and simulation is used to estimate key supplier performance parameters
Piramuthu (2005)	Agent based	Proposed an agent-based model with a learning component
Deng and Elmaghraby (2005)	Tournament, game theory	Proposed a tournament type model where suppliers compete against each other
Liu and Wu (2005)	AHP and DEA	Developed an integrated method by combining AHP and DEA, applied to the supplier selection problem
Valluri and Croson (2005)	Reinforcement learning (RL)	Proposed a RL approach to supplier selection with two separate selection policies
Emerson and Piramuthu (2004)	Agent based	Proposed an agent-based manufacturing supplier selection model with a learning component
Agrell et al. (2004)	Game theory	Applied a game theoretic model to select suppliers in a telecommunication firm
Chan (2003)	AHP	Developed an interactive technique to facilitate data collection prior to AHP implementation in supplier selection problems
Choy et al. (2003)	Neural networks	Used neural networks to benchmark suppliers

aware of opportunities and the barriers. Understanding world markets can be extremely difficult since each country is unique and complex. Therefore, global operations increase uncertainty and reduce control capabilities. Problems might arise from the number of intermediaries, customs requirements, currency fluctuations, political instability, taxes as well as trade restrictions. In addition, uncertainty results from greater distances, longer lead-times, and diverse global market conditions. We shall discuss in detail global supply chain management, including global sourcing, in Chapter 8.

6.5.4 Supplier Risk

In the global economy, firms are continuously seeking their supplier base around the world to find opportunities for reducing supply chain costs. However, singular emphasis on supply chain cost can make the supply chain brittle and more susceptible to the risk of supply disruptions. Supply chain management no longer just involves moving products efficiently, but it also includes mitigating risks along the way. Risks in supply chains are dynamic in nature; the frequency and severity of risk events keep changing. Some risks can be reduced or even eliminated, while new ones may appear at any time. The best way to handle risks is to adopt a proactive strategy and eliminate them before occurring if it is technically and economically feasible. Enterprise and supply chain risk management topics have been extensively treated in the recent books by Zsidisin and Ritchie (2009), Handfield and McCormack (2008), Olson and Wu (2008), and Haimes (2004).

Supplier risk can be defined as the probability and severity of adverse impacts of supply disruptions due to man-made or natural events. Translation of risk to quantitative terms has been a challenging but critical task for successful business applications. Yang (2006) focuses on two dimensions of risk, "severity of impact" and "frequency of occurrence" of risk events and defines supplier risk as a function of these two. Severity of impact is used as an all-embracing term that covers financial loss, reputation loss, market loss, etc. Frequency of occurrence is the probability distribution of the risk event. Different levels of impacts and frequencies lead to two significantly different types of risks (Yang, 2006). *Value-at-Risk* (**VaR**) type risks are used to model less frequent events which disrupt operations at suppliers and can bring severe impact to buyers (e.g., labor strike, terrorist attack, natural disaster, etc.). *Miss-the-target* (**MtT**) type risks, on the other hand, are used to model events that might happen more frequently at suppliers with lesser damage to buyers (e.g., late delivery, missing quality requirements, etc.). Ravindran et al. (2010) recently developed risk-adjusted multiple criteria supplier selection models using **VaR** and **MtT** type risks. We will discuss these risk models in detail in Chapter 7.

Exercises

6.1 Discuss the pros and cons of single sourcing and multiple sourcing.

6.2 How does pre-qualification help the supplier selection process?

6.3 Why is scaling of criteria values necessary in supplier ranking?

6.4 Discuss the similarities and differences between L_p metric and AHP for ranking suppliers.

6.5 What is the difference between a goal and a constraint as used in goal programming?

6.6 What are the drawbacks of using preemptive weights in goal programming?

6.7 A company needs two parts, A and B, for its product. It can either buy them from another company, or can make them in its own plant, or do both. The costs of each alternative and the in-house production rates are as follows:

	Part A	Part B
Make	\$1.00/unit	\$2.00/unit
Buy	\$1.20/unit	\$1.50/unit
In-house production rate	3 units/h	5 units/h

The company must have at least 100 units of part A and 200 units of part B each week. There are 40 h of production time per week and idle time on the machine costs \$3.00/h. Furthermore, no more than 60 units of A can be made each week, and no more than 120 units of B can be made each week. Also, no more than 150 units of B can be bought per week.

The company wants to determine an optimal plan which will minimize the total costs per week. Formulate this as a linear programming problem and solve. What is the optimal make or buy plan for the company?

6.8 Recall Example 6.5 discussed in Section 6.3.6. Determine the ranking of the suppliers using the following methods:

(a) L_1-norm to scale the criteria values and L_1-metric to rank.

(b) L_2-norm to scale the criteria values and L_2-metric to rank.

Do you find any rank reversals including the rankings obtained in Section 6.3.6. Explain any differences.

6.9 Consider a supplier selection problem where the five most important criteria are identified as follows:

C_1—Risk

C_2—Delivery time

C_3—Quality

C_4—Price

C_5—Business Performance

(a) Using the method of paired comparison, rank the criteria and compute their weights.

(b) Suppose 5% of purchasing managers were interviewed and they were asked to rank the aforementioned criteria. The summary of their responses are given in Table 6.41.

TABLE 6.41

Criteria Rankings for Exercise 6.9

Rank Criteria	1	2	3	4	5
C_1	9	4	4	7	6
C_2	3	13	4	8	2
C_3	6	3	11	9	1
C_4	8	7	9	2	4
C_5	4	3	2	4	17

Determine the weights of the criteria and their rankings based on the sample survey, using the Borda count method for multiple decision makers discussed in Section 6.3.9. How does their ranking compare with yours?

6.10 Suppose you are planning to use single sourcing and have narrowed down the choices to four suppliers A, B, C, and D. Your criteria for selection are price (min), company size in market capitalization in millions of dollars (max), and quality in a scale of 1–100 (max). The relevant data is given in the following:

	Price ($) (min)	Size (millions of dollars) (max)	Quality (max)
A	180,000	2800	75
B	160,000	3200	85
C	140,000	2600	80
D	190,000	3600	65

(a) Scale the data using the ideal value method.

(b) Use the L_∞ metric method to rank the four suppliers.

6.11 Consider an order allocation problem under multiple sourcing, where it is required to buy 2000 units of a certain product from three different suppliers. The fixed set-up cost (independent of the order quantity), variable cost (unit price), and the maximum capacity of each supplier are given in the following (two suppliers offer quantity discounts):

Supplier	Fixed Cost	Capacity	Unit Price
1	$100	600 units	$ 10/unit for the first 300 units
			$7/unit for the remaining 300 units
2	$500	800 units	$2/unit for all 800 units
3	$300	1200 units	$6/unit for the first 500 units
			$4/unit for the remaining 700 units

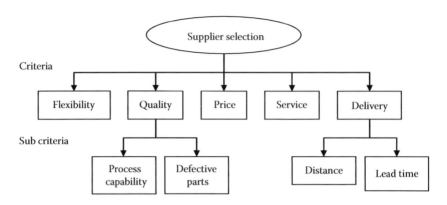

FIGURE 6.8
Supplier selection criteria and sub-criteria (Exercise 6.12).

The objective is to minimize the total cost of purchasing (fixed plus variable cost). Formulate this as a linear integer-programming problem. You must define all your variables clearly, write out the constraints to be satisfied with a brief explanation of each and develop the objective function.

6.12 Case study 3: Supplier Ranking

In this real case study, you will be working on a supplier ranking problem. It is an actual application for a manufacturing company located in Tijuana, Mexico. Because of confidentiality issues, the data given here have been disguised. The supplier selection criteria and sub-criteria have been defined by the Purchasing Manager as shown in Figure 6.8. Note that flexibility, process capability (C_{pk} Index), and service have to be maximized, while the other criteria have to be minimized. The company is considering 21 potential suppliers and the supplier data with respect to the criteria are given in Table 6.42.

Questions

(a) Scale the supplier criteria values using linear normalization (Simple linearization).

(b) Using the scaled values from part (a), apply the L_2 metric method to rank the suppliers.

(c) Determine the criteria/sub-criteria weights using the following methods (use your own judgments).

   (i)   Rating method

   (ii) Pair-wise comparison of criteria/sub-criteria and Borda count

   (iii) AHP. Test the consistency of all pair-wise comparison matrices

(d) Using the weights obtained in part (c) and the scaled supplier data in part (a), determine the supplier rankings by all three methods.

TABLE 6.42

Supplier Criteria Data (Exercise 6.12)

				Criteria Values			
Supplier	Price ($)	C_{pk} (Index)	Defective Parts (ppm)	Flexibility (%)	Service (%)	Distance (km)	Lead-Time (h/Part)
1	50	0.95	105,650	10	75	500	0.25
2	80	2.00	340	0	100	1,500	0.60
3	45	0.83	158,650	25	65	50	0.20
4	60	1.00	66,800	15	85	5,000	0.80
5	40	1.17	22,750	18	90	9,500	0.95
6	60	1.50	1,350	5	99	7,250	0.50
7	65	1.33	6,200	0	100	10	0.10
8	70	1.50	1,350	0	50	15,000	1.50
9	45	1.00	66,800	5	80	7,500	1.75
10	70	1.25	12,225	10	85	12,500	2.00
11	75	0.83	158,650	15	75	1,345	1.25
12	65	1.00	66,800	0	80	6,680	1.15
13	80	1.33	6,200	0	85	5,000	1.00
14	75	1.15	22,750	2	87	16,000	0.90
15	70	1.33	6,200	5	86	17,000	0.95
16	70	1.05	44,500	0	65	1,860	1.50
17	85	1.25	12,225	5	70	1,789	1.45
18	65	0.95	105,650	0	77	1,775	0.90
19	55	0.83	158,650	10	89	2,500	0.75
20	80	1.25	12,225	10	85	12,500	1.50
21	85	0.83	158,650	0	50	17,500	2.00

(*Note*: For AHP, do not perform pair-wise comparison of alternatives (suppliers) with each criterion/sub-criterion. Instead, use directly the scaled supplier data from part (a) as the S matrix)

(e) Compare the supplier rankings obtained by L_2 metric, rating method, Borda count, and AHP. Discuss any rank reversals.

6.13 Case study 4: Supplier Selection and Order Allocation

In this case study, you will solve a supplier order allocation problem with two products, two buyers, and two suppliers. The problem has three conflicting criteria, namely, total cost, lead-time, and quality (measured by rejects). All three objectives have to be minimized.

1. Minimize total cost, which consists of fixed cost and the variable cost. Fixed cost is a one-time cost that is incurred if a supplier is used irrespective of the number of units bought from that supplier.

2. Minimize the weighted average lead-time.

3. Minimize the weighted average quality.

   For objectives 2 and 3, you can use the total demand as the denominator for the weighted average expressions. The constraints in the model include capacity constraints of the supplier and the demand constraint of the buyers.

   The data regarding the various model parameters are given in Tables 6.43 through 6.46. The fixed costs are $1200 and $1325 for suppliers 1 and 2, respectively.

   Questions

   (a) Formulate the supplier order allocation problem as a three criteria mixed integer linear programming (MILP) problem. You must define your variables clearly, write out the

TABLE 6.43

Product Demand (Exercise 6.13)

Product	Buyer	Demand
1	1	150
1	2	175
2	1	200
2	2	180

TABLE 6.44

Supplier Capacities (Exercise 6.13)

Product	Supplier	Capacity
1	1	300
1	2	350
2	1	280
2	2	360

TABLE 6.45

Product Prices (Including Shipping Cost) and the Lead-Time of Products in Days (Exercise 6.13)

Product	Buyer	Supplier	Price/Unit	Lead-Time
1	1	1	$190	15
1	1	2	175	17
1	2	1	194	19
1	2	2	200	18
2	1	1	335	24
2	1	2	360	21
2	2	1	370	11
2	2	2	365	12

TABLE 6.46

Quality of Product (Measured by
Percentage of Rejects) (Exercise 6.13)

Product	Supplier	Quality (%)
1	1	7
1	2	9
2	1	6
2	2	5

constraints explaining their significance, and write down the
objective functions. Specify the problem size: the number of
variables (binary + continuous) and the constraints.

(b) Obtain the ideal values by optimizing each objective separately. Determine the appropriate lower and upper bounds
for the objectives.

(c) Formulate the three criteria MILP model as a goal program.
Set the target values at 110% of the ideal values obtained in
part (b). Solve the MILP goal programming model using the
following methods

   (i) Preemptive goal programming: Preemptive priorities
   are cost, lead-time, and quality.

   (ii) Non-preemptive goal programming: Weights are 0.5, 0.3,
   and 0.2 for cost, lead-time, and quality.

   (iii) Min–max goal programming.

   (iv) Fuzzy goal programming (use the bounds obtained in
   part (b)).

   Notes:

   • Scale the objective functions/goals where necessary.

   • Write the objective function of the GP model for each part
   clearly. Explain any new variable(s) and constraints you
   add to the GP model for each part.

(d) Compare the four GP solutions using the value path approach.

References

Aberdeen Group. 2004. Outsourcing portions of procurement: Now a core strategy.
Supplier Selection and Management Report. 04(07): 4.

Agrell, P. J., R. Lindroth, and A. Norman. 2004. Risk, information and incentives in
telecom supply chains. *International Journal of Production Economics.* 90: 1–16.

Aissaoui, N., M. Haouari, and E. Hassini. 2007. Supplier selection and order lot sizing modeling: A review. *Computers and Operations Research.* 34(12): 3516–3540.

Akarte, M. M., N. V. Surender, B. Ravi, and N. Rangaraj. 2001. Web based casting supplier evaluation using analytical hierarchy process. *Journal of the Operational Research Society.* 52: 511–522.

Anthony, T. F. and F. P. Buffa. 1977. Strategic purchasing scheduling. *Journal of Purchasing and Materials Management.* 13: 27–31.

Arthur, J. L. and A. Ravindran. 1980. PAGP: An efficient algorithm for Linear Goal Programming problems. *ACM Transactions on Mathematical Software.* 6(3): 378–386.

Barbarosoglu, G. and T. Yazgac. 1997. An application of the analytic hierarchy process to the supplier selection problem. *Product and Inventory Management Journal.* 38(1): 14–21.

Bender, P. S., R. W. Brown, M. H. Issac, and J. F. Shapiro. 1985. Improving purchasing productivity at IBM with a normative decision support system. *Interfaces.* 15: 106–115.

Benton, W. C. 1991. Quantity discount decision under conditions of multiple items. *International Journal of Production Research.* 29: 1953–1961.

Bharadwaj, N. 2004. Investigating the decision criteria used in electronic components procurement. *Industrial Marketing Management.* 33(4): 317–323.

Bhutta, K. S. and F. Huq. 2002. Supplier selection problem: A comparison of the total cost of ownership and analytical hierarchy process. *Supply Chain Management.* 7(3–4): 126–135.

Bilsel, R. U. and A. Ravindran. 2011. A Multi-objective chance constrained programming model for supplier selection. *Transportation Research Part B.* 45(8): 1284–1300.

Buffa, F. P. and W. M. Jackson. 1983. A goal programming model for purchase planning. *Journal of Purchasing and Materials Management.* Fall: 27–34.

Cachon, G. P. and F. Zhang. 2006. Procuring fast delivery: Sole sourcing with information asymmetry. *Management Science.* 52(6): 881–896.

Chan, F. T. S. 2003. Interactive selection model for supplier selection process: An analytical hierarchy process approach. *International Journal of Production Research.* 41: 3549–3579.

Chaudhry, S. S., F. G. Forst, and J. L. Zydiak. 1991. Multicriteria approach to allocating order quantities among suppliers. *Production and Inventory Management Journal.* 3rd Quarter: 82–85.

Chen, C.-T., C.-T. Lin, and S.-F. Huang. 2006. A fuzzy approach for supplier evaluation and selection in supply chain management. *International Journal of Production Economics.* 102: 289–301.

Choy, K. L., W. B. Lee, and V. Lo. 2003. Design of an intelligent supplier relationship management system: A hybrid case based neural network approach. *Expert Systems with Applications.* 24: 225–237.

De Boer, L., E. Labro, and P. Morlacchi. 2001. A review of methods supporting supplier selection. *European Journal of Purchasing and Supply Management.* 7: 75–89.

Degraeve, Z., E. Labro, and F. Roodhooft. 2004. Total cost of ownership purchasing of a service: The case of airline selection at Alcatel Bell. *European Journal of Operational Research.* 156: 23–40.

Degraeve, Z. and F. Roodhooft. 1999. Improving the efficiency of the purchasing process using total cost of ownership information: The case of heating electrodes at CockerillSambre S.A. *European Journal of Operational Research.* 112: 42–53.

Degraeve, Z. and F. Roodhooft. 2000. A mathematical programming approach for procurement using activity based costing. *Journal of Business Finance and Accounting*. 27: 69–98.

Deng, S. J. and W. Elmaghraby. 2005. Supplier selection via tournaments. *Production and Operations Management*. 14(2): 252–276.

Dickson, G. W. 1966. An analysis of vendor selection systems and decisions. *Journal of Purchasing*. 2(1): 5–17.

Ding, H., L. Benyoucef, and X. Xie. 2005. A simulation optimization methodology for supplier selection problem. *International Journal of Computer Integrated Manufacturing*. 18(2–3): 210–224.

Ellram, L. M. 1995. Total cost of ownership, an analysis approach for purchasing. *International Journal of Physical Distribution and Logistics Management*. 25(8): 4–23.

Emerson, D. and S. Piramuthu. 2004. Agent – based framework for dynamic supply chain configuration. *In Proceedings of the 37th IEEE International Conference on Systems Science*, pp. 1–8. Big Island, HI: IEEE.

Feng, C., X. Wang, and J. S. Wang. 2001. An optimization model for concurrent selection of tolerances and suppliers. *Computers and Industrial Engineering*. 40: 15–33.

Gaballa, A. A. 1974. Minimum cost allocation of tenders. *Operational Research Quarterly*. 25: 389–398.

Ghodyspour, S. H. and C. O'Brien. 1998. A decision support system for supplier selection using an integrated analytic hierarchy process and linear programming. *International Journal of Production Economics*. 56–57: 199–212.

Ghodsypour, S. H. and C. O'Brien. 2001. The total cost of logistics in supplier selection, under conditions of multiple sourcing, multiple criteria and capacity constraint. *International Journal of Production Economics*. 73(1): 15–27.

Goicoechea, A., D. R. Hansen, and L. Duckstein.1982. *Multiobjective Decision Analysis with Engineering and Business Applications*. Chapter 9. New York: Wiley.

Haimes, Y. 2004. *Risk Modeling, Assessment, and Management*. Hoboken, NJ: Wiley.

Handfield, R. B. and K. McCormack. 2008. *Supply Chain Risk Management: Minimizing Disruptions in Global Sourcing*. Boca Raton, FL: Auerbach Publications.

Haq, A. N. and G. Kannan. 2006. Design of an integrated supplier selection and multi-echelon distribution inventory model in a built-to-order supply chain environment. *International Journal of Production Research*. 44(10): 1963–1985.

Hinkel, C. L., P. J. Robinson, and P. E. Green. 1969. Vendor evaluation using cluster analysis. *Journal of Purchasing*. 5(3): 49–58.

Holt, G. D. 1998. Which contractor selection methodology? *International Journal of Project Management*. 16(3): 153–164.

Hu, C. F., C. J. Teng, and S. Y. Li. 2007. A fuzzy goal programming approach to multi objective optimization problem with priorities. *European Journal of Operational Research*. 176: 1319–1333.

Ignizio, J. M. and T. M. Cavalier. 1994. *Linear Programming*. Chapter 13. Eaglewood cliff, NJ: Prentice Hall.

Jayaraman, V., R. Srivastava, and W. C. Benton. 1999. Supplier selection and order quantity allocation: A comprehensive model. *Journal of Supply Chain Management*. 35(2): 50–58.

Karpak, B., E. Kumcu, and R. Kasuganti. 1999. An application of visual interactive goal programming: A case in supplier selection decisions. *Journal of Multi-Criteria Decision Analysis*. 8(2): 93–105.

Khattree, R. and N. Dayanand. 2000. *Multivariate Data Reduction and Discrimination with SAS Software*. Cary, NC: SAS Institute Inc.

Liu, F., F. Y. Ding, and V. Lall. 2000. Using data envelopment analysis to compare suppliers for supplier selection and performance improvement. *Supply Chain Management: An International Journal*. 5(3): 143–150.

Liu, J. and C. Wu. 2005. An integrated method for supplier selection in SCM. *In Proceedings of the International Conference on Services Systems and Services Management*. 35(1): 617–620. Chongging, China: IEEE.

Mandal, A. and S. G. Deshmukh. 1994. Vendor selection using interpretive structural modeling (ISM). *International Journal of Operations and Production Management*. 14(6): 52–59.

Masud, A. S. M. and A. Ravindran. 2008. Multiple criteria decision making. In *Operations Research and Management Science Handbook*, ed. A. Ravi Ravindran, Chapter 5. Boca Raton, FL: CRC Press.

Masud, A. S. M. and A. Ravindran. 2009. Multiple criteria decision making. In *Operation Research Methodologies*, ed. A. Ravi Ravindran, Chapter 5. Boca Raton, FL: CRC Press.

Mendoza, A. 2007. Effective methodologies for supplier selection and order quantity allocation. PhD dissertation, Department of Industrial Engineering, The Pennsylvania State University, University Park, PA.

Mendoza, A., A. Ravindran, and E. Santiago, E. 2008. A three phase multi-criteria method to the supplier selection problem. *International Journal of Industrial Engineering*. 15(2): 195–210.

Mohammed, R. H. 1997. The relationship between goal programming and fuzzy programming. *Fuzzy Sets and Systems*. 89: 215–222.

Moore, D. L. and H. E. Fearon. 1973. Computer-assisted decision-making in purchasing. *Journal of Purchasing*. 9(4): 5–25.

Muralidharan, C., N. Anantharaman, and S. G. Deshmukh. 2001. Vendor rating in purchasing scenario: A confidence interval approach. *International Journal of Operations and Production Management*. 21(10): 1305–1326.

Narasimhan, R. and K. Stoynoff. 1986. Optimizing aggregate procurement allocation decisions. *Journal of Purchasing and Materials Management*. 22(1): 23–30.

Ng, S. T. and R. M. Skitmore. 1995. CP-DSS: decision support system for contractor pre-qualification. *Civil Engineering Systems: Decision making Problem Solving*. 12(2): 133–159.

Nishiguchi, T. and A. Beaudet. 1998. Case study: The Toyota group and the Aisin fire. *Sloan Management Review*. 40(1): 49–59.

Nydick, R. L. and R. P. Hill. 1992. Using the analytic hierarchy process to structure the supplier selection procedure. *International Journal of Purchasing and Materials Management*. 28(2): 31–36.

Office of Management and Budget, Department of Defense. 2008. Available at http://www.whitehouse.gov/omb/budget/fy2008/defense.html.Accessed1/19/08

Olson, D. L. and D. D. Wu. 2008. *Enterprise Risk Management*. Hackensack, NJ: World Scientific.

Pan, A. C. 1989. Allocation of order quantity among suppliers. *Journal of Purchasing and Materials Management*. 25: 36–39.

Patel, U. R. 2007. Experiments in group decision making in the analytic hierarchy process. MS Thesis, Department of Industrial Engineering, The Pennsylvania State University, University Park, PA.

Piramuthu, S. 2005. Knowledge-based framework for automated dynamic supply chain configuration. *European Journal of Operational Research*. 165: 219–230.

Powdrell, B. J. 2003. Comparison of MCDM algorithms for discrete alternatives. MS Thesis, Department of Industrial Engineering, The Pennsylvania State University, University Park, PA.

Ravindran, A. R., U. Bilsel, V. Wadhwa, and T. Yang. 2010. Risk adjusted multicriteria supplier selection models with applications. *International Journal of Production Research.* 48(2): 405–424.

Ravindran, A. R. and V. Wadhwa. 2009. Multiple criteria optimization models for supplier selection. In *Handbook of Military Industrial Engineering,* eds. A. Badiru and M. U. Thomas, Chapter 4. Boca Raton, FL: CRC Press.

Saaty, T. L. 1980. *The Analytic Hierarchy Process.* New York: McGraw Hill.

Saen, R. F. 2007. Suppliers selection in the presence of both cardinal and ordinal data. *European Journal of Operational Research.* 183: 741–747.

Schilling, D. A., C. Revelle, and J. Cohon. 1983. An approach to the display and analysis of multi-objective problems. *Socio-Economic Planning Sciences.* 17(2): 57–63.

Schniederjans, M. 1995. *Goal Programming: Methodology and Applications.* Boston, MA: Kluwer Academic Publishers.

Sharma, D., W. C. Benton, and R. Srivastava. 1989. Competitive strategy and purchasing decisions. *In Proceedings of the 1989 Annual Conference of the Decision Sciences Institute,* pp. 1088–1090. New Orleans, Louisiana.

Sheshadri, S., K. Chatterjee, and G. L. Lilien. 1991. Multiple source procurement competitions. *Marketing Science.* 10(3): 246–253.

Smytka, D. L. and M. W. Clemens. 1993. Total cost supplier selection model: A case study. *International Journal of Purchasing and Materials Management.* 29(1): 42–49.

Sonmez, M. 2006. A review and critique of supplier selection process and practices. Loughborough University Business School.

Timmerman, E. 1986. An approach to vendor performance evaluation. *Journal of Purchasing and Supply Management.* 22(4): 2–8.

Tiwari, R. N., S. Dharmar, and J. R. Rao. 1986. Priority structure in Fuzzy goal programming. *Fuzzy Sets and Systems.* 19: 251–259.

Tiwari, R. N., S. Dharmar, and J. R. Rao. 1987. Fuzzy goal programming an additive model. *Fuzzy sets and Systems.* 24: 27–34.

Treece, J. 1997. Just-too-much single-sourcing spurs Toyota purchasing review: Maker seeks at least 2 suppliers for each part. *Automotive News.* 3: 3.

Turner, I. 1988. An independent system for the evaluation of contract tenders. *Journal of the Operational Research Society.* 39(6): 551–561.

Valluri, A. and D. Croson. 2005. Agent learning in supplier selection models. *Decision Support Systems.* 39(2): 219–240.

Velazquez, M. A., D. Claudio, and A. R. Ravindran. 2010. Experiments in multiple criteria selection problems with multiple decision makers. *International Journal of Operational Research.* 7(4): 413–428.

Vokurka, R. J., J. Choobineh, and L. Vadi. 1996. A prototype expert system for the evaluation and selection of potential suppliers. *International Journal of Operations and Production Management.* 16(12): 106–127.

Wadhwa, V. and A. Ravindran. 2007. Vendor selection in outsourcing. *Computers and Operations Research.* 34: 3725–3737.

Weber, C. A., J. R. Current, and W. C. Benton. 1991. Supplier selection criteria and methods. *European Journal of Operational Research.* 50: 2–18.

Weber, C. A., J. R. Current, and A. Desai. 2000. An optimization approach to determining the number of suppliers to employ. *Supply Chain Management: An International Journal*. 2(5): 90–98.

Weber, C. A. and L. M. Ellram. 1992. Supplier selection using multi-objective programming a decision support system approach. *International Journal of Physical Distribution and Logistics Management*. 23(2): 3–14.

Wilson, E. J. 1994. The relative importance of supplier selection. *International Journal of Purchasing and Materials Management*. 30: 34–41.

Wind, Y. and P. J. Robinson. 1968. The determinants of vendor selection: Evaluation function approach. *Journal of Purchasing and Materials Management*. 4(8): 29–46.

Xia, W. and Z. Wu. 2007. Supplier selection with multiple criteria in volume discount environments. *Omega*. 35: 494–504.

Yang, T. 2006. Multi objective optimization models for managing supply risks in supply chains. PhD dissertation, Department of Industrial Engineering, The Pennsylvania State University, University Park, PA.

Zenz, G. 1981. *Purchasing and the Management of Materials*. New York: Wiley.

Zhang, Z., J. Lei, N. Cao, K. To, and K. Ng. 2003. Evolution of supplier selection criteria and methods. *E-article retrieved from* http://www.pbsrg.com/overview/downloads/Zhiming%20Zhang_Evolution%20of%20Supplier%20Selection%20Criteria%20and%20Methods.pdf on 8th June 2010.

Zsidisin, G. A. and B. Ritchie. 2009. *Supply Chain Risk: A Handbook of Assessment, Management, and Performance*. New York: Springer.

7

Managing Risks in Supply Chain*

7.1 Supply Chain Risk

The devastating 9.0-magnitude earthquake in northeastern Japan, followed by a massive tsunami, on March 11, 2011, caused more than 16,000 deaths and cost over 300 billion dollars in property damage and economic loss. It also caused major disruptions to the global supply chains of multi-national corporations. Since Japan was a key supplier of electronic components, which go into the production of automobiles, computers, aircrafts and cell phones, companies such as Toyota, Honda, Boeing, GM, and Apple, had to slow down or shut-down their factories due to shortage of parts. It was reported (Martyn, 2011) that Toyota could lose more than $70 million everyday if its factories are idle. In fact, Toyota's quarterly (January–March 2011) net profits plunged 77%. Honda reduced its auto production in half in 2011 in India due to a shortage of parts. It also had to cut production in North America. Its quarterly net profits (January–March 2011) dropped by 38%. Nissan had to ramp up the production of engines in its American plant in Tennessee and ship engines to its assembly plants in Japan and Southeast Asia to produce cars for the Asian market (Powell, 2011).

The vulnerability of supply chains around the world due to a disruptive event in one part of a country is the result of globalization that started during the 1990s. (Chapter 8 discusses globalization and its impacts on supply chain management in detail). During the last two decades, companies have witnessed the emergence of global competitive environment, industrial restructuring, changes in manufacturing, crumbling of international barriers and increased use of information technologies. Many manufacturers seek to expand their supplier base globally. They are moving toward more outsourcing, off shoring, long-term contracts and relationships with just a few suppliers. These strategies have provided the companies an opportunity to significantly reduce supply chain costs. However, overemphasis on supply chain cost can make the supply chain brittle and more susceptible to the risk of disruptions. Hence, *supply chain management calls for balancing the logistics efficiency with risk mitigation.*

---

* The authors gratefully acknowledge the contribution of Tao Yang and Ufuk Bilsel to this chapter.

In this chapter, we will answer the following questions:

1. How to identify supply chain risks?
2. How to classify and prioritize supply chain risks?
3. How to develop appropriate risk mitigation strategies?
4. How to quantify the impacts of supply chain disruptions?
5. How to integrate quantitatively supply chain risks in decision making at both strategic and operational levels?

7.2 Real World Risk Events and Their Impacts

The September 11, 2001 terrorist attacks in the United States, brought home the vulnerabilities of global supply chains. All the U.S. borders (land, sea and air) were closed and it affected the delivery of parts to auto companies. Assembly lines of Toyota and GM were shut down due to lack of parts.

Table 7.1 gives some examples of world's risk events and their impacts on the supply chains of companies. It is important to note that not all world risk events affect the operations of the global supply chains. Some have no impact, some impact the supply chains in the country where the event occurred and some impact the supply chains world-wide.

TABLE 7.1

World Risk Events and Their Impacts

Risk Event	Date	Impact on Supply Chains
Oklahoma city bombing	Apr 1995	None
Toyota brake plant fire	Feb 1997	Global
General motors labor strike	Mar 1996	Mostly the United States
Toyota brake plant fire	Feb 1997	Mostly Japan
Taiwan earthquake	Sep 1999	Global
Phillips New Mexico fire	Mar 2000	Global
9/11 Terrorist attack in the United States	Sep 2001	Global
US west coast ports lockout	Sep 2002	Global
SARS outbreak	Nov 2002–July 2003	Global
Indian ocean Tsunami	Dec 2004	None
London subway bombings	July 2005	None
Hurricane Katrina	Aug 2005	Mostly the United States
Mumbai terrorist attacks	Nov 2008	None
BP Gulf oil spill	Apr 2010	None
Great east Japan earthquake	Mar 2011	Global
Massive Thailand floods	Nov 2011	Global

Toyota practiced 100% JIT supply system with a sole source brake supplier. A major fire in February 1997 destroyed the brake plant and disrupted the production in 20 Toyota assembly plants worldwide costing $1.8 billion in lost sales. A 17-day strike in March 1996 at the Delphi brake plant in Dayton, Ohio shut down 26 GM's North American assembly plants and it cost nearly $1 billion to the company's first quarter earnings (Fitzgerald, 1996). The earthquake that struck Taiwan on September 21, 1999 knocked out the production of memory chips, circuit boards, and other components. It impacted 50% of the world's supplies to major computer manufacturers and their earnings dropped by 5%. Investors reacted more negatively to "Pull-type" supply chains (Dell, Gateway) as opposed to 'Push-type" supply chains (IBM and Compaq).

A lightning strike caused a fire in a Phillips electronics semi-conductor facility in New Mexico in March 2000. Millions of silicon chips used in mobile phones manufactured by Ericsson and Nokia were damaged. Nokia reacted quickly to the disruption by going to a backup supplier and production returned to normal in 3 weeks. Phillips was the sole supplier to Ericsson and Ericsson's production of mobile phones was affected for several weeks, which resulted in $640 million loss in 2000 and the eventual loss of the North American mobile phone market. Nokia's market share increased by 30% (Christopher and Peck, 2004). The US West Coast ports' lockout affected 29 ports and lasted 11 days. The lockout caused an estimated $11–$22 billion in lost sales, cost of airfreight and spoilage.

The SARS (severe acute respiratory syndrome) outbreak in Hong Kong and China during November 2002 to July 2003 affected worldwide travel. A.T. Kearney (Monahan et al., 2003) reported that it affected the global supply chains of HSBC Bank, Motorola and Honda. Technology companies increased their inventories anticipating disruption to the airfreight business.

According to the Kearney report, the frequency of natural disasters has more than *tripled* since the 1960s. Due to globalization, their impacts have increased *tenfold*. Manmade disasters such as terrorist attacks, wars, and computer viruses, have increased dramatically by a factor of 50 during 1970–2011!

7.2.1 Importance of Supply Chain Risk Management

Supply Chain disruptions cost money and affect investor confidence. Hendricks and Singhal (2003) were the first to quantify the financial impacts of supply chain disruptions which they called supply chain "glitches." They focused on glitches that resulted in delays in production or customer fulfillment. Initially, they studied 517 glitches and then extended the results to 885 glitches during 1989–2000 to publicly traded companies (Hendricks and Singhal, 2005). The list of companies included small, medium and large with respect to market capitalization and covered both manufacturing and IT industries. A summary of their findings is given as follows:

- Part shortages, production problems and order changes by customers are the top three reasons for the supply chain glitches
- An average loss of over $250 million in market capitalization
- An average reduction of 10% in stock-market prices
- 92% reduction in Return on Assets
- 7% lower sales
- 11% increase in cost
- 14% increase in inventory

According to Crisis Management International, companies that had prolonged supply chain disruptions of 10 days or more had the following impacts (Mahoney, 2004):

- 73% closed or had significant long-term impact.
- 43% never recovered sufficiently to resume business, and of those that did, only 29% were still operating 2 years later.

In a survey conducted by CFO Research Services and UPS Consulting (Mahoney, 2004) the following points were noted:

- Only 32% of the companies could make major changes to the supply chains when needed by using alternate suppliers, alternate routes and flexible manufacturing capabilities.
- 38% said that their companies were sitting on "too much unmanaged supplier risk".

In a 2003 study, the Gartner Group has predicted (Monahan et al., 2003) the following:

- Two in five businesses will experience a crisis over the next 5 years, anything from a minor fire to a major IT failure.
- 60% of those companies will be forced to shut down within 2 years.

According to a recent survey of 560 companies from 62 countries by Zurich Financial services and U.K. Business Continuity Institute, 85% said that they had suffered at least one disruption to their supply chains during 2011 (Veysey, 2011). Listed next, in order, are the major causes of the supply chain disruptions:

- Weather (51%)
- IT failure (41%)

- Transportation network disruption (21%)
- Earthquake/Tsunami (21%)

It is interesting to note *where* the supply chain disruptions occurred:

- 61% at Tier 1 suppliers
- 30% at Tier 2
- 9% at Tier 3 or lower

In terms of costs, the cost of a single supply chain disruption for most companies (83%) was less than $1.4 million; but, for 14% of the companies surveyed it was between $1.4 million and $14 million. About 1% reported costs of more than $100 million!

In the following sections, we will discuss the sources of supply chain risks, risk identification and classification, risk prioritization, and risk intervention strategies.

7.3 Sources of Supply Chain Risks

Risk is inherent in almost all business operations. However, supply chain risks have different characteristics, impacts, and sources. Hence, the risks need to be categorized and different strategies should be developed to manage them effectively.

Johnson (2001) suggested that when viewed as a whole, risks fall into two major categories: supply risks (including capacity limitations, currency fluctuations, and supply disruptions) and demand risks (including seasonal imbalance, volatility of fads, and new products). Following Chopra and Sodhi (2004) and Yang (2006), we divide the risks that exist in supply chain into two categories: external and internal (Table 7.2). Generally, external risks are from the outside, such as business partners, natural environment, governments, and competitors, and internal risks are from the inside, such as operations, management strategies and activities, and employees. Basically, firms have better control on internal risks than on external risks.

As shown in Table 7.2, risks in supply chain to firms are not only from their business partners, but also from customers, internal operations, new technologies, political issues, natural disasters, etc. Some risks can be reduced or even eliminated, but the others are hard to control. How to successfully manage the risks in supply chain becomes more and more critical to firms. Although many companies have realized their importance, few are well prepared because of the complexity of the risk issues in supply chain and the lack of good risk

TABLE 7.2

Examples of Risks and Their Sources in Supply Chains

Sources		Risks
External	Suppliers	Failures to meet time/quantity/quality requirements
		Price fluctuations
		Outmoded technologies
	Customers	Demand fluctuations in quantity and type for products or service
		Order changes including quantity, type, and time
		Returns
	Global business	Currency exchange rate fluctuations
		Import tax rate changes
		Export restrictions
		Language barriers
		Cultural issues
	Nature	Earthquake, flood, hurricane, blizzard, blackout
		Terrorist attack, war, strike
Internal	Human resource	Key employees' leave
		Short of employees for suddenly increased demands
	Technology	New technologies
		Outmoded product designs
	Management	Inappropriate business strategies
		Forecast errors
	Production	Failures to meet quality goals
		Delivery problems
	Finance	Failed investments
		Stock price fluctuations
	Transportation	Failures to make the time/quantity/quality promises to customers

mitigation techniques. A study completed by FM global, a leading commercial insurance company (www.fmglobal.com), indicated that more than one-third of the financial executives and risk managers surveyed do not feel they are adequately prepared for disruptions to their business. The 2003 Protecting Value study showed that 34% of respondents rated the extent of their preparation for disruptions to their major source of revenue as fair or poor (Bradford, 2003).

7.4 Risk Identification

Risk can be defined as the *uncertainty in the outcome of an event, particularly negative consequences*. People make decisions under uncertainty by minimizing the negative impact of risk. For example, people buy life insurance to manage the negative impact of premature death. Investors diversify their portfolios with different types of stocks (large/small companies, growth/value, local/international) to minimize capital loss.

A *risk portfolio* is defined as the list of all risks a company faces, both internally and externally. Generating a risk portfolio helps the company to assess and prioritize the various types of risks and develop appropriate risk mitigation strategies.

Elkins et al. (2008a) classify the industry portfolio of risks into four categories as follows:

1. Financial risks
2. Strategic risks
3. Hazard risks
4. Operational risks

Examples of *financial risks* include interest rate fluctuations, changes in currency exchange rates, credit rating for company's bonds, changes in accounting and tax laws. *Strategic risks* include new competitors, negative press coverage, customer demand changes, erosion of brand loyalty, poor customer relations, etc. *Hazard risks* are disruptions due to natural disasters (earthquake, foods, lightning, volcano eruptions, etc.) or manmade disasters (terrorism, labor strikes, wars, border closings, etc.). Examples of *operational risks* include supplier problems, IT systems failure, computer viruses, product recalls, and logistics failures.

Elkins et al. (2008a) recommend the use of a cross-functional team of experts to brainstorm and develop the *risk portfolio*. The team should have members from key stakeholders and experts including operations research analysts, statisticians, accountants, manufacturing engineers, risk managers, purchasing staff, logistics, and supply chain managers. Creating a *risk portfolio* also identifies risk owners, who will be responsible for managing those risks, including risk monitoring and risk mitigation.

7.5 Risk Assessment

After developing the risk portfolio, the next step is to narrow the list of risks by assessing their importance to the manufacturing and supply chain operations. This step will be called *risk assessment* and will be accomplished using a two-step procedure as follows:

Step 1: *Risk Mapping* (subjective)

Step 2: *Risk Prioritization* (subjective and objective)

7.5.1 Risk Mapping

Risk mapping is a subjective process where the risks are broadly classified based on *risk occurrence* and *risk impact*. Risk occurrence measures the likelihood of that risk event happening and is subjectively assessed as high or low. Risk impact is used as an all-embracing term that covers financial loss, market share, stock prices, etc. Risk impact is also assessed subjectively as high or low. Based on the assessment of risk occurrence and risk impact, a 2 × 2 matrix of *risk map* can be constructed as shown in Figure 7.1 (Elkins et al., 2008a).

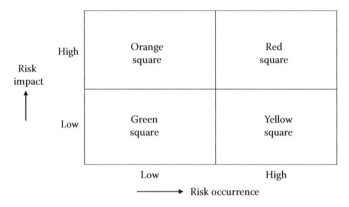

FIGURE 7.1
Risk map. (From Elkins, D. et al., Identifying and assessing supply chain risk, in *Supply Chain Risk Management*, Handfield, R.B. and McCormack, K.B., eds., Auerback Publishers, Boca Raton, FL, 2008a, Chapter 3.)

The risk events that fall into the "red square" have high chances of happening with high impact. Hence, they deserve the most and immediate attention for risk intervention. Risk events that fall into the green square need the least attention in terms of risk management. However, they do require some monitoring.

Given next are examples of risk events and their classifications. (Note: The examples are merely illustrative and should not be construed as hard-and-fast rules for risk mapping classification).

- Tier 1, 2, 3, …, *n* supplier problems (red)
- Loss of critical supplier (red)
- Quality problems and product recalls (red/orange)
- Terrorism and sabotage (orange)
- Flooding/hurricanes/tornados (orange)
- Union and labor problems (orange)
- New competitor in the market (orange)
- Customer demand (red/orange)
- Warehouse fire (yellow/green)

- Blizzard/Ice storm (yellow)
- IT system failure (orange)
- Logistics provider failure (yellow)
- Equipment breakdown (green)
- Product returns from customers (green)
- Temporary work stoppages (yellow)
- Computer virus (yellow/green)
- Interest rate fluctuation (yellow)
- Delivery delays (yellow)
- Defective materials (yellow/green)
- Health and safety violations (green)

7.5.2 Risk Prioritization

Developing a risk map provides a quick and easy way to identify and focus on critical risks (e.g., those in the red square, high occurrence and high impact). The next step is to rank the risks in the various squares at least in red, orange and yellow squares, so that risk intervention strategies can be developed. We present a simple *risk scoring* method that uses *risk priority numbers* (RPNs) to rank the critical risks or at least identify the top 10 or 20 risks that need attention.

7.5.2.1 Risk Priority Numbers

For each risk event, a *Risk Priority Number* (RPN) is computed which is, the product of the numerical scores assigned to the four risk factors, *occurrence, impact, detection* and *recovery*.

$$\text{RPN} = (\text{Occurrence}) \, (\text{Impact}) \, (\text{Detection}) \, (\text{Recovery})$$

RPNs have been used successfully in auto industries and oil refineries as part of the *failure mode and effect analysis* (FMEA) to identify design problems early in the process and provide a risk analysis (Breyfogle, 2003).

We have already discussed the first two risk factors. *Detection* refers to how long it takes to realize that a risk event that needs attention has happened. For example, a fire at a tier-1 supplier will be known to the OEM (original equipment manufacturer) in a relatively short period of time. However, a fire at a tier-3 supplier that will impact tier-1 and eventually the production at OEM may take a longer detection time.

Recovery refers to the time it takes for the company to return to full normal operations after the disruptive event. For example, a small fire in a warehouse may have a negligible or short recovery time, while a major fire in a manufacturing plant may affect normal operations for a longer time. After the earthquake and tsunami in Japan on March 11, 2011, it look Toyota until June 2011 to return to full production globally!

In order to calculate the RPN for each risk event, the experts assign a rating of 1–10 (1 = Low and 10 = High) to quantify each risk factor's importance. For example, a major fire at a tier-1 critical supplier may get ratings of 4, 9, 2, and 8 for occurrence, impact, detection and recovery respectively. Then the RPN is calculated as follows:

$$RPN = 4 \times 9 \times 2 \times 8 = 576$$

Risk events are then prioritized, within each colored square in the risk map, using RPNs, the most important risk with the highest RPN value.

In practice, risk scoring methods are easier to use to get a "quick and dirty" ranking of the risk events. In Sections 7.8–7.13, we discuss mathematical models to quantify the risk factors.

7.6 Risk Management

In 2006, McKinsey did a global survey to understand how companies manage supply chain risks (Muthukrishnan and Shulman, 2006). Some of the key findings from the McKinsey survey are listed below:

- Nearly two-thirds indicated that their supply chain risks had increased
- Nearly one-fourth did not have a formal risk assessment program
- Nearly half did not have company-wide standards to mitigate risk

The most frequently reported risk management actions in the McKinsey survey are as follows:

- Use of performance contracts with suppliers
- Use of backup suppliers
- Currency hedges
- Insurance

Complete details about the global survey and its findings are available in McKinsey article (Muthukrishnan and Shulman, 2006).

A study in FM Global (2011) reported the following.

- Numbers of natural and man-made disasters have tripled since the 1980s.
- Annual losses due to the disasters have increased, on the average, from $10 billion to more than $100 billion in recent years.

- 96% of the companies believe their business operations are exposed to the natural disasters; but less than 20% are really concerned about their financial impact.
- Out of the 25 most costly disasters in the world, between 1970 and 2008, none occurred before 1987, and two-thirds of them occurred since 2001.

7.6.1 Risk Management Strategies

The risk map (Figure 7.1) and the RPNs, discussed in Section 7.5, help to focus on the most important risks. Risks in the red square should be addressed first, followed by those in the orange and yellow squares. Risk management strategies should be developed to reduce the likelihood of the occurrence of the risk or its financial impact or both. Hurricane Katrina showed that companies cannot rely on their governments to mitigate disaster risks. FM Global (2010) recommended that effective risk management should be viewed as a competitive advantage and not simply as a way to reduce financial impacts. The Nokia–Ericsson incident we discussed in Section 7.3 brought this out clearly. With an effective risk management plan, Nokia survived the crisis and increased its market share in the mobile phone market, while Ericsson failed. When outsourcing business operations, companies have to realize that they are not outsourcing the risks; on the contrary they are actually increasing their risks!

Listed next are frequently practiced *risk management strategies* in companies:

1. *Take the risk*: Here the company owns the risk and takes actions to reduce it. Maintaining inventory to manage supply and demand risks is an example of this strategy.
2. *Share the risk*: Here the company takes some risks and asks its supply partners to take the rest. An example of this is the "portfolio strategy" followed by HP to mitigate demand risk. We will discuss HP's portfolio strategy in detail in Section 7.7.
3. *Transfer the risk*: Here the company makes its suppliers to assume all the risk. For example, Dell computer has long required its suppliers to maintain inventory of parts near its assembly plant in Texas.
4. *Reduce the risk*: Company takes action to reduce the risk to an acceptable level by minimizing the impact of a risk occurrence. For example, HP uses multiple suppliers, geographically displaced, to reduce supply disruptions.
5. *Eliminate the risk*: Here the company eliminates the risk by breaking partnerships or finding alternatives to its raw materials and parts needs. This could also involve a redesign of the manufacturing process.

6. *Risk monitoring*: Real-time monitoring of suppliers' performance is done so as to alert for any potential problems. An example of this is the Risk Monitoring System developed by Johnson and Johnson to monitor critical suppliers. Another example is the Global Operations Emergency Control Center at FedEx. We will discuss these in detail in Section 7.7.

These risk management strategies can be broadly grouped into three categories as follows:

1. Risk mitigation
   These are intervention strategies that are built into the product design, manufacturing and supply chain operations, in order to reduce or eliminate risks that can disrupt the company's supply chain. It is very important that the cost of the risk mitigation strategy does not outweigh its benefits.

2. Contingency planning
   These are actions that are preplanned and are set in motion as soon as the risk is identified. Certain resources are identified ahead of time in contingency planning and they have to be available at short notice when a disruption occurs. Having "back up" suppliers is an example of contingency planning.

3. Business insurance
   Business insurance has been used as a method to cover the financial loss due to business interruptions caused by natural disasters. With the increase in the number of risk events, due to natural and man-made disasters, this option is becoming more expensive lately. For some risk events, this may not even be available. Moreover, insurance covers only financial loss. It does nothing to reduce the negative impact on brand name, market share and stock prices.

7.6.2 Developing a Risk Management Plan

The risk map, developed in Figure 7.1 (Section 7.5) helps the management to focus on the critical risks and develop appropriate risk intervention plans. Risks in the red square have a high probability of occurrence and high financial impact. These risks generally require proactive *mitigation strategies* to reduce the risk. For example, if the company has been *single sourcing* a critical part, it should consider using *multiple sourcing*. If, alternate suppliers do not exist, the company should seriously consider changing the product design (if feasible) to eliminate the dependence on a single supplier.

After developing intervention plans to manage the risks in the red square, risks in the yellow and orange squares (Figure 7.1) should be addressed next. Risks in the orange square have low probability of occurrence, but the financial impact is high. Buying business insurance is a good risk management plan for those risks that are insurable. For non-insurable risks, either mitigation strategies or contingency plans should be developed.

Risks that fall in the yellow square have high occurrence but low financial impact. Real-time monitoring and contingency planning are typical risk management plans for these risk events. Risks in the green square have the lowest priority in terms of risk management. Costly risk mitigation strategies are not appropriate for these risk events. Simple emergency plans and real-time monitoring are generally sufficient to manage these risks.

7.6.3 Risk Mitigation Strategies

Braithwaitte (2003) and Chopra and Sodhi (2004) discuss several traditional and flexible risk mitigation strategies to avoid supply chain breakdown.

7.6.3.1 Traditional Strategies

1. *Inventory*—It is commonly used as a buffer to manage unpredictable supply and uncertain demand risks. However, inventory is expensive and may become a risk for high-tech products due to obsolescence.

2. *Capacity planning*—This refers to the use of flexible production lines that can make multiple products and flexible workforce. For example, Toyota makes sure that their plants are flexible enough to supply multiple markets. This reduces idle capacity. HP uses a portfolio approach to workforce planning by maintaining a combination of full time, part time, and temporary workers.

3. *Dual sourcing*—Use of two or more suppliers for critical parts is commonly used to avoid supply disruptions. HP follows the portfolio approach to vendors by maintaining a mix of suppliers from different geographic regions.

4. *Long-term contracts*—These are used as hedge against future price increases. For example, Southwest Airlines used long-term contracts with suppliers for the delivery of aviation fuel to ride out the escalating global oil price increases during 2006.

7.6.3.2 Flexible Strategies

1. *Generic raw material inventory*—Use of generic raw materials helps the company to respond to changes in market demands for its products. Excess inventory of raw materials can be sold easily at the market price.

2. *Product postponement*—Using commonality of parts, the product differentiation is done as late as possible in the manufacturing cycle. Even though the demands for the final products may have a large variance, their component parts demand will have a lower variance and hence less safety stock inventory. For example, paint inventory used to be kept in a multitude of colors. Now, they are kept in basic colors only

and are mixed for different colors after receiving the exact customer orders. Another example is Giant Food stores, which used to have specialized plastic bottles for its different milk products (whole milk, 2%, 1%, no fat, etc.). This resulted in maintaining a huge inventory of empty bottles for each product type and size. In 2008, they switched to using a standard (generic) bottle for each size and specialized labels to differentiate the products. Thus, they only have to maintain inventory of one type of bottles for each size. When the bottle is filled with a certain milk product, the label is affixed. Obviously, it is much cheaper to maintain inventory of different labels than that of different bottles!

Another example is Volkswagen (VW), which is implementing a modular design for its cars. VW's assembly plants, that make subcompact to SUVs, share common car modules. The modular design system is expected to reduce production cost by 20% and the average manufacturing time of each car by 30% (Reed, 2010).

3. *Distribution and logistics alternatives*—These include the use of risk pooling, cross-docking, alternate modes of transport and routing to manage risks posed by suppliers, customers and logistics providers. For example, Dell computers air-lifted parts from its Asian plants to meet its production needs in the United States, during the 2002 west coast ports lockout. Amazon.com maintains inventory in a small number of warehouses, each supplying a large geographical region. This results in a stable forecast of demand and lower inventory due to risk pooling benefits.

4. *Cross training of employees*—This is an effective strategy for maintaining a flexible work force to manage personnel risk.

5. *Supply chain visibility*—According to Blackhurst et al. (2005) supply chain visibility refers to the sharing of information in real-time across the supply chain stages and among their partners. The net effect of visibility is to make the supply chain more responsive, increase availability and reduce inventory risk. For example, Dell shares customer demand information with its suppliers so that they can maintain proper inventory of needed parts. Wal-Mart shares points-of-sales (PoS) data with its suppliers so that they can forecast and plan their replenishment strategies.

7.7 Best Industry Practices in Risk Management

Table 7.3 summarizes the best practices of risk management in industries at both strategic and operational levels (Elkins et al., 2008b, Atkins, 2003).

We shall now discuss some examples of best practices in industries to manage supply chain risks.

TABLE 7.3

Best Supply Chain Risk Management Practices in Industries

Strategic Level	• Monitoring current or potential business partners' performance
	• Requiring key business partners to have risk management plans
	• Considering risks when choosing business partners
	• Improving collaboration
	• Improving visibility in supply chain
	• Using "portfolio" strategy
	• Joining professional programs such as C-TPAT, CSI, CSA and FAST
	• Buying insurance
	• Reviewing current business models
	• Special funding and cash reserve
Operational Level	• Keeping scheduled meetings with key business partners for risk issues
	• Learning the lessons
	• Preparing contingency plans
	• Forming special teams to handle risks with members from different divisions and business partners
	• Considering and controlling risks in all related business activities including product designs, promotions, etc.
	• Getting regular employees involved and conducting training programs

7.7.1 Teradyne Inc. (Atkinson, 2003)

Teradyne, located in Boston, Massachusetts, makes expensive automated systems for testing semi-conductors and circuit boards. Its product development cycle is very long. Since 70%–80% of product cost happens during the design phase, Teradyne focuses on "Pre-crisis Risk Management." It has a color-coded risk management and mitigation plan early in the product design stage. The company considers risks posed by suppliers, parts selection and technology. The objective of the risk management program is to identify and mitigate risks that can adversely impact product cost, time to market, ability to ramp up production, and reliability.

Risks identified as *green* pose no problems. Those identified as *yellow* pose some risk but there is a potential solution. *Orange* risks require a mitigation plan. For example, if a particular part poses a risk that would increase the cost target; the company may go offshore for that part or redesign it to reduce cost. Once the mitigation plan is in place, the color will be changed to yellow. Risks identified as *red* require mitigation plans that have not been developed yet. Hence, they require continuous monitoring until a plan is developed.

It is interesting to note that the U.S. Department of Homeland Security used a similar color code during 2002–2011 to warn Americans of the terrorist threat levels in the country.

7.7.2 Hewlett-Packard (HP) (Billington, 2002)

HP has developed a successful portfolio approach to procurement. It is similar in concept to the portfolio approach in investment, where an investor buys several different stocks in order to reduce the risk of capital loss. HP emphasizes multiple sources for parts and different geographical regions for suppliers in order to reduce the supply and demand risks. Instead of one or two sources with long-term contracts, HP uses a portfolio of several options as follows:

1. A long-term structured contract to meet 90% of the expected demand
2. A short-term contract for "supply-on demand"
3. Spot market purchases

Under option 1, HP *takes* the risk by committing to buy a fixed amount over time for a certain price from the supplier. Under option 2, HP *transfers* the risk to the supplier for supply/demand uncertainty. Of course, HP may have to pay a higher price for the short-term contract. Under option 3, HP buys on the open market at that prevailing price. Naturally HP takes all the risks here.

HP negotiated electricity purchase for its San Diego facility by using the portfolio approach. Under the traditional "free market" approach (short term buying with no contracts and paying market rates), HP's average quarterly cost for electricity was $555,000 with a standard deviation of $49,000. Using the portfolio approach, HP reduced that cost to $352,000 with a standard deviation of $2000. Thus, HP not only reduced its average cost but also its variability!

7.7.3 Federal Express (Florian-Kratz, 2005)

FedEx promises "On-time delivery 100% of the time." Because of its global reach, FedEx has to deal with crisis all the time somewhere in the world. For example, in 2004, FedEx activated its contingency plans for 37 tropical storms alone. It cannot wait for disruptions to happen and then react. It has to be proactive.

Some of the proactive risk management actions practiced by FedEx include the following:

- *Real-time monitoring* of the movement of hundreds of planes and thousands of trucks at the Global Operations Emergency Center in Memphis, Tennessee. When problems develop, contingency plans are activated immediately.
- Eight disaster kits are kept for facility repair in Memphis. Each kit weighs 2 ton and contains fuel, communication gear, water, etc.

- Five empty FedEx planes roam the skies each night as stand-by. They can be used at moment's notice to handle a sudden surge in demand or to replace a broken-down plane.
- Conducting disaster drills several times a year for disruptions due to earthquakes, hurricanes, terrorism, labor strikes, etc.
- FedEx allows the Red Cross to maintain, at the FedEx hubs, shipping containers filled up with disaster relief supplies such as bandages, blankets, batteries, water, etc. These containers can be delivered by FedEx around the world at short notice, when a disaster strikes.

Even the best crisis management plan of FedEx was no match for hurricane Katrina that swept through the gulf coast in 2005. The New Orleans airport was closed for 15 days and FedEx had to return thousands of packages to their senders and stop new shipments to the disaster zones. FedEx shifted its area hub and its employees from New Orleans to Lafayette, Louisiana (135 miles away) in a matter of days. FedEx learned two new things from the Katrina disaster—one is to have temporary housing for displaced employees and second, cell phones cannot be relied on during a disaster since the cell phone networks were down days after Katrina. FedEx has now increased the number of satellite phones that can be deployed.

7.7.4 Wal-Mart (Leonard, 2005)

Wal-Mart has studied points-of-sales (PoS) data of customers in the disaster zones, before and after the natural disaster. Customers usually buy bottled water, flashlights, diesel generators, tarps, chain saws, mops, and non-perishable food items. Based on the PoS data, Wal-Mart has stocked dedicated distribution centers with disaster relief supplies, so that trucks can be dispatched at short notice to the stores in the disaster zones well before the storm hits. Wal-Mart's emergency operations center in Bentonville, Arkansas handles all emergencies that affect the operations of the Wal-Mart stores. The center uses data from the National Weather Service and other meteorologists to track the paths of hurricanes and tropical storms. For example, Wal-Mart dispatched disaster relief supplies to its stores in the gulf coast 6 *days* before Katrina hit New Orleans. Wal-Mart's "Loss-Prevention" teams, whose jobs were to protect the stores from vandalism and theft, arrived in New Orleans days before the Federal emergency management agency and Red Cross officials. In many cases, the "loss-prevention" team members became the sole providers of relief supplies free to the disaster victims.

Hurricane Katrina tested Wal-Mart's emergency operations system to its extreme limits. Katrina shut down 126 Wal-Mart stores in the disaster areas. Even though all but 13 were up and running in 2 weeks, it broke Wal-Mart's typical recovery time of one day after a natural disaster.

Wal-Mart's corporate mantra was to make available day-to-day things to customers efficiently, at competitive prices. During Katrina, Wal-Mart delivered on that promise and more.

7.7.5 Johnson and Johnson (Atkinson, 2003)

Johnson and Johnson (J & J), a leading manufacturer of consumer products, pharmaceuticals and medical devices, has one of the best real-time supplier monitoring programs. Called "Continuity of Supply," it is one of J & J's strategic initiatives. First, the supply risk management team identifies J & J's critical suppliers. For example, these could be suppliers falling in the red square in the risk map (Figure 7.1). Then, it continuously monitors each supplier's performance in six key areas, either directly by site visits, or indirectly from published data. J & J conducts the "process vitality checks" in six key areas as given in the following:

1. *Process operations*: This area includes the supplier's normal manufacturing capabilities and stability of operations. Stability will check whether a supplier has more than one production facility or the supplier has a good disaster management plan if his supply is disrupted.

2. *Quality*: Product quality, Quality assurance plan, ISO certifications, and Six sigma practices are monitored here.

3. *Financial vitality*: Supplier's financial strength is monitored with published data from Dun and Bradstreet (www.dnb.com) and other sources.

4. *Engineering/technical Expertise*: Here the depth of engineering and technical support for the supplier in manufacturing and information technology (IT) is evaluated and monitored.

5. *Delivery*: Meeting time windows on delivery and supplier reliability are monitored here.

6. *Leadership*: The vision and mission of supplier's top management and its commitment and support to J & J's business are monitored. Changes in top leadership may result in a site visit to the supplier by the J & J risk management team.

By monitoring the performance of the critical suppliers in the six key areas, J & J has built a successful risk management program to minimize supply chain disruptions.

7.8 Risk Quantification Models

The risk quantification models presented in the remaining sections of this chapter are based on the dissertations of the author's doctoral students (Yang, 2006; Bilsel, 2009) and the resulting publications of their doctoral work (Yang and Ravindran, 2007; Bilsel and Ravindran, 2011, 2012).

The risk quantification models discussed in this chapter will take a broader view of supply chain risk and model it as a function of *occurrence, impact, detectability,* and *recovery*. Methods to quantify each risk component will be developed. We will begin with the development of a *basic risk quantification model* as a function of impact and occurrence. Separate mathematical models will then be developed for risk detectability and risk recovery time. All the models will be integrated and illustrated with a case study on risk adjusted multi-criteria supplier selection model at the end of the chapter.

7.8.1 Basic Risk Quantification Models

Following Yang (2006), we classify risks, natural or man-made, that can cause supply chain disruptions into two types for the purpose of quantification:

1. Value-at-Risk (VaR) type risks
2. Miss-the-Target (MtT) type risks

The basic model will quantify risk as a function of *severity of impact* and *frequency of occurrence* as follows:

$$\text{Risk} = f(\text{impact, occurrence})$$

It is worthwhile to note a subtle difference between the definitions of impact and occurrence in VaR and MtT type risk functions. In the case of VaR type risk function, impact is a probability distribution of loss due to a risk event and occurrence is the probability distribution of the number of risk events during a period. For the MtT type risk function, impact represents the loss due to deviation from a performance target value and occurrence is the distribution of that performance measure.

Table 7.4 illustrates the differences between VaR and MtT type risks. VaR risks are rare events that can disrupt severely supply chain operations. MtT risks are more common with less severe impact. Yang (2006) used *extreme value theory* (EVT) to model the impact of VaR type risk event and *Taguchi's Loss Functions* to model the impact of MtT type risk. Supply disruptions due

TABLE 7.4

Quantification of Risk

Type	Occurrence	Impact	Examples
Value-at-risk (VaR)	Rare	Severe	Hurricane, strike, fire, terrorist attack
Miss-the-target (MtT) risk	Frequent	Mild to moderate	Late delivery of raw materials, low quality replenishment

Source: Yang, T., Multi-objective optimization models for managing supply risks in supply chains, Ph.D dissertation, Pennsylvania State University, University Park, PA, 2006.

to natural/manmade disasters will be modeled as VaR type risks. Supplier delivery problems will be modeled using the MtT type risks.

In the following sections, we shall discuss each type of risk in detail and present mathematical models to quantify them.

7.9 Value-at-Risk (VaR) Models

Generally, VaR type risks are caused by rare events, such as earthquakes, floods, fire, wars, terrorism, etc. They can result in catastrophic disruptions to the company's supply chain. Recall how a lightning bolt hitting a semi-conductor plant in New Mexico in 2000, resulted in over $640 million revenue loss to Ericsson and changed the landscape of the mobile phone industry.

VaR models were first developed for the financial industry in the early 1990s. They are considered as a standard measure for market risk and used extensively in portfolio risk management. From the financial point of view, VaR measures the maximum possible loss in the market value of a given portfolio. Considering the characteristics of VaR type risks caused by rare events, the concept of VaR can be applied to risk quantification in supply chain management also.

7.9.1 VaR Type Impact Function

Extreme value theory (EVT) is used to model and predict the severity of impact for VaR type events. EVT is chosen for its characteristics and successful applications in several industries including insurance, telecommunications, and finance to model rare events. EVT approaches random variables from a completely different perspective than classical statistics. Traditionally, the main purpose of statistics is to find properties (mean, variance, etc.) about the central characteristics of a given population. In EVT, the focus is to find similar traits related to the extremes (maximum or minimum) of the population. For example, EVT models are used in ocean engineering (e.g. maximum expected wave heights for building offshore oil rinks), civil engineering (e.g., including wind, earthquake resistance in calculations), meteorology (e.g., estimating very high–low temperatures) and transportation engineering (e.g., estimating congestion). Both discrete and continuous models are used in EVT. Additionally, univariate and multi-variate models are also available.

Impact of disruptive events is the financial loss due to those events. In general, the impact of such extreme events is modeled using heavy-tailed distributions such as Weibull, Gumbel, and Frechet distributions. Heavy tails of these distributions are appropriate for maximum or minimum extreme values. A more general family of distributions called the *generalized extreme value distributions* (GEVD) arises as the limit distributions of Weibull, Gumbel and Frechet distributions (Castillo et al., 2005). There exist two GEVDs: one for maximum extremes and the other for minimum extremes. *Probability density functions* (PDF) and *cumulative distribution functions* (CDF) for the maximum GEVD are given in Equations 7.1 through 7.4. PDF and CDF of the minimum GEVD can be found in Castillo et al. (2005).

$$\text{for } K \neq 0, \quad f_\kappa(x;\lambda,\delta) = \frac{1}{\delta}\exp\left(-\left[1-K\left(\frac{x-\lambda}{\delta}\right)\right]^{\frac{1}{K}}\right)\left[1-K\left(\frac{x-\lambda}{\delta}\right)\right]^{\frac{1}{K}-1} \quad (7.1)$$

$$\text{for } K = 0, \quad f_0(x;\lambda,\delta) = \frac{1}{\delta}\exp\left[-\exp\left(\frac{\lambda-x}{\delta}\right)\right]\exp\left(\frac{\lambda-x}{\delta}\right) \quad (7.2)$$

$$\text{for } K \neq 0, \quad F_K(x;\lambda,\delta) = \exp\left(-\left[1-K\left(\frac{x-\lambda}{\delta}\right)\right]^{\frac{1}{K}}\right) \quad (7.3)$$

$$\text{for } K = 0, \quad F_0(x;\lambda,\delta) = \exp\left[-\exp\left(\frac{\lambda-x}{\delta}\right)\right] \quad (7.4)$$

Table 7.5 provides definitions of the GEVD parameters.

The mean of the GEVD can be calculated using Equation 7.5 if the parameter $K \neq 0$

$$E(X) = \lambda - \frac{\delta}{K} + \frac{\delta}{K}\Gamma(1-K) \quad (7.5)$$

TABLE 7.5

GEVD Parameters

Parameter	Interpretation
K	Shape parameter
	$K > 0$, corresponds to a Frechet distribution
	$K = 0$, corresponds to a Gumbel distribution
	$K < 0$, corresponds to a Weibull distribution
δ	Scale parameter
λ	Location parameter

where $\Gamma(.)$ is the Gamma function. In the case when $K = 0$, GEVD boils down to a Gumbel distribution and the mean can be approximated as $\lambda + \xi\delta$, where $\xi = 0.57721$ is the Euler–Mascheroni constant.

7.9.2 Generalized Extreme Value Distribution (GEVD) Functions for Risk Impact

Plots of the density functions for GEVD impact functions (dgev) for different parameter values are given in Figures 7.2 through 7.4. Effect of changing the shape parameter K is presented in Figure 7.5.

7.9.3 Estimating GEVD Parameters

Several methods have been proposed to estimate the GEVD parameters in Table 7.5 as given in the following:

- *Maximum likelihood*: Gives parameter values that maximize a given likelihood function.
- *Method of moments*: Set the first k moments of a random variable equal to the corresponding sample moments and then solve the resulting system of k equations.
- *Probability weighted moments* (PWM): Set the first k weighted moments of a random variable equal to the corresponding weighted sample moments and then solve the resulting system of k equations.

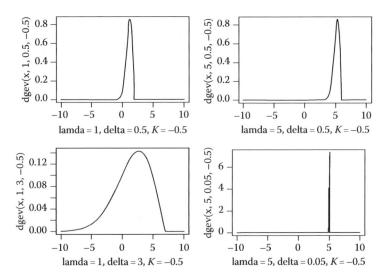

FIGURE 7.2
Case $K = -0.5$.

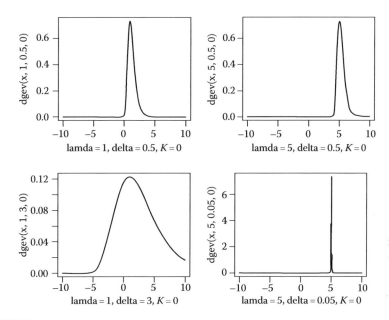

FIGURE 7.3
Case $K = 0$.

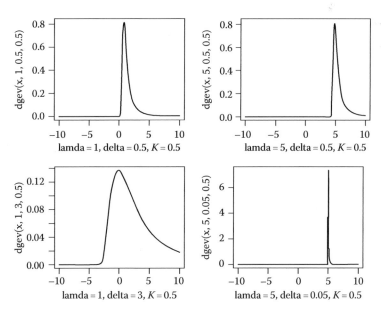

FIGURE 7.4
Case $K = 0.5$.

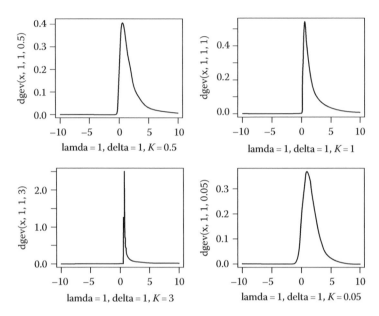

FIGURE 7.5
Different K values ($K > 0$).

- *Element percentile method* (EPM): It is used as a substitute to maximum likelihood estimators (MLEs). MLEs are more suitable to estimate moments in models where the central tendency is important, rather than extremes. Moreover, in some cases MLEs may not exist (e.g., in GEVD, MLEs do not exist for $K \leq -1$). EPM can be used in such cases.

Although the *method of moments* is the simplest way to estimate the parameters, it is also one with the highest bias values and hence, is not recommended. Normally, only a small number of data points will be available for each VaR event. Since *PWM* is less biased, compared to MLE, for small samples (Coles and Dixon, 1998), PWM method is used in the illustrative Example 7.1. Sherman test and Kolmogorov–Smirnov statistic can be used to check the goodness-of-fit after estimating the GEVD parameters. Users should be aware that although both tests tend to over-fit, they still provide a rough basis for accepting/rejecting a model.

7.9.4 VaR Occurrence Functions

After the function of the VaR event is determined, the occurrence function of the risk event has to be modeled. Occurrence of risk is the number of times a particular type of VaR event that happens over a period of time. The time period can be a year, 2 years or several years depending on the event of interest. If historical data were available, then the occurrence function could be modeled using an empirical distribution or by fitting it to a known probability distribution.

For example, *occurrence* can be modeled as a Poisson distribution under the assumption that the disruptive events are not correlated. Equation 7.6 gives the probability mass function of the Poisson distribution:

$$P(X = k) = \frac{e^{-\lambda}\lambda^k}{k!} \quad \text{for } k = 0, 1, 2, \ldots \tag{7.6}$$

The random variable X is the number of occurrences over a given period. The mean and variance of Poisson distribution are both equal to λ. Note that several references, including Olsen et al. (1998) and Bogachev et al. (2008) cite that the independence assumption may not hold for all disruptive events, since some events show long-term correlation. However, we do not consider long-term correlations in the models presented in this chapter.

7.9.5 VaR Disruption Risk Function

The final form of the disruption risk function is obtained by combining the impact and occurrence functions described in Sections 7.9.3 and 7.9.4. Aggregation can be carried out in two ways:

1. By Monte Carlo simulation (Yang, 2006)
2. By analytical methods (Bilsel, 2009)

We will present the simulation approach presented by Yang (2006). The analytical approach is available in Bilsel and Ravindran (2011a).

7.9.5.1 Simulation Approach

Under this approach, the number of extreme events during a given period, and their impacts are simulated and aggregated. Thus, each simulation run will consist of the following steps:

Step 1: Sample from the occurrence function to generate the number of risk events during a given period

Step 2: For each risk event, sample from the GEVD function to generate its impact, in terms of financial loss.

Step 3: Sum the impacts to determine the total loss due to all the events.

By repeating steps 1, 2, and 3 several hundred simulation runs can be made to determine the total impact (financial loss) due to VaR type risk. From these values, one can compute the mean, variance and the entire distribution of the disruption risk function. Using Example 7.1, we shall illustrate the estimation of GEVD parameters for the impact function, the occurrence function and the aggregate disruption risk function by the simulation approach.

Example 7.1 (Yang and Ravindran, 2007)

An original equipment manufacturer (OEM) in the United States wants to evaluate the risks from its supplier in Southern China. Flood is one of the major problems affecting the supplier due to its location. In the last 10 years, floods forced the supplier to shut down at least part of its production lines from several days to 2 weeks. In other years, although the supplier managed to continue its production, finished components could not be shipped out at scheduled time with desired quantity to the OEM because either the outgoing road was blocked by flood or the nearby airport was forced to close. Correspondingly, the OEM suffers severe losses since the supplier is a major supplier. Senior managers of the OEM finally decide to solve this problem. In order to make the final decision, the OEM wants to evaluate the risk of flood from the supplier. Then, the OEM can buy either business insurance or renegotiate contracts with the supplier. Table 7.6 shows the OEM's losses due to floods at the supplier location in the last 10 years.

Solution

Quantifying the Impact Function
To estimate the parameters of the generalized extreme value distribution (GEVD), we use the PWM method, described in detail in Hosking et al. (1985). The PWM method requires moment estimates, which can be

TABLE 7.6

Losses Caused by Floods from 1997 to 2007 (Example 7.1)

Year	Number of Floods	Loss
1997	1	$734,900
1998	2	$580,070
		$354,180
1999	0	
2000	1	$457,820
2001	4	$258,410
		$1,250,000
		$780,540
		$243,000
2002	0	
2003	3	$1,358,110
		$981,250
		$548,270
2005	4	$254,170
		$158,970
		$987,410
		$578,940
2006	0	
2007	1	$875,240

computed using plotting positions. A plotting position is a distribution free estimate of the empirical CDF of a data set. For a data set of size n, the plotting position p_i of a data point i is given by:

$$p_i = \frac{i-a}{n}$$

with $-0.5 < a < 0.5$.

Then the estimate of the rth moment is given by

$$\hat{\beta}_r(p_i) = \frac{1}{n}\sum_{i=1}^{n} p_i^r \times i$$

The estimates of moments 0, 1, and 2 denoted by b_0, b_1, and b_2, and an intermediary parameter c, are necessary to estimate the GEVD parameters from empirical data.

By using the approximation formulas suggested by Hosking et al. (1985), we get:

$$c = \frac{2b_1 - b_0}{3b_2 - b_0} - \frac{\log 2}{\log 3} = 0.012 \quad \hat{\kappa} = 7.859c + 2.9554c^2 = 0.095$$

$$\hat{\delta} = \frac{(2b_1 - b_0)\hat{\kappa}}{\Gamma(1+\hat{\kappa})(1-2^{-\hat{\kappa}})} = 314478.674 \quad \hat{\lambda} = b_0 + \frac{\hat{\delta}}{\hat{\kappa}}(\Gamma(1+\hat{\kappa})-1) = 495901.049$$

Since $K \neq 0$, the following form of the GEVD will fit the VaR type impact function,

$$F_{\lambda,\delta,\kappa}(x) = \exp\left\{-\left[1 - \kappa\left(\frac{x-\lambda}{\delta}\right)\right]^{1/\kappa}\right\} = \exp\left\{-\left[1 - 0.095\left(\frac{x - 495901.049}{314478.674}\right)\right]^{1/0.095}\right\}$$

Kolmogorov–Smirnov Statistics are used in this example for the fitness test. Table 7.7 shows the $F(x_i)$, $\frac{i}{n} - F(x_i)$, and $F(x_i) - \frac{i-1}{n}$ values from the original data, where $F(x_i)$ is the GEVD CDF using the estimated parameters.

Then,

$$\sqrt{n}D^+ = \sqrt{16}\max_i\left\{\frac{i}{n} - F(x_i)\right\} = 0.497 \quad \sqrt{n}D^- = \sqrt{16}\max_i\left\{F(x_i) - \frac{i-1}{n}\right\} = 0.313$$

$$\sqrt{n}D = \sqrt{16}\max(D^+, D^-) = 0.497 \quad \sqrt{n}V = \sqrt{16}(D^+ + D^-) = 0.810$$

From percentage points of Kolmogorov–Smirnov statistics table (Chandra et al., 1981), we conclude that there is no evidence to reject the GEVD model.

TABLE 7.7

Kolmogorov–Smirnov Statistics

Index	Loss x	$F(x_i)$	$\dfrac{i}{n} - F(x_i)$	$F(x_i) - \dfrac{i-1}{n}$
1	158,970	0.062431	6.92826E-05	0.062430717
2	243,000	0.11414	0.010859683	0.051640317
3	254,170	0.122483	0.065016977	−0.002516977
4	258,410	0.125738	0.124261775	−0.061761775
5	354,180	0.211264	0.101236413	−0.038736413
6	457,820	0.323697	0.051303025	0.011196975
7	548,270	0.429355	0.008145022	0.054354978
8	578,940	0.465174	0.034825995	0.027674005
9	580,070	0.466486	0.096014067	−0.033514067
10	734,900	0.634863	−0.009863226	0.072363226
11	780,540	0.678354	0.009145668	0.053354332
12	875,240	0.757554	−0.007554354	0.070054354
13	981,250	0.8283	−0.015800234	0.078300234
14	987,410	0.831854	0.04314599	0.01935401
15	1,250,000	0.936387	0.001113057	0.061386943
16	1,358,110	0.959168	0.040831608	0.021668392

7.9.5.2 VaR Type Occurrence Function

Based on the number of floods in each year, given in Table 7.6, the OEM adopts the Poisson distribution and estimates the parameter for the VaR type occurrence function as $\hat{\lambda} = \bar{x} = 1.6/\text{year}$.

7.9.5.3 VaR Type Disruption Risk Function

In order to get the final VaR type disruption distribution function, we aggregate the impact function $F_{495901.049,314478.674,0.095}(x)$ and the occurrence function $P(1.6)$. The following steps can be used to get the final risk distribution using an Excel spreadsheet (Table 7.8).

Step 1: Generate Poisson random numbers with $\lambda = 1.6$ in the second column of the spreadsheet (the first column is used to count the number of runs)

Step 2: Generate as many uniform random variables as demanded by the frequency (numbers in the second column) and use them as the probabilities

Step 3: Based on generated probabilities, find out the corresponding x value from the GEVD distribution

Step 4: Sum up the results in the column Total.

Table 7.8 shows the first few rows of the spreadsheet only. It illustrates 15 simulation runs.

TABLE 7.8

Excel Spreadsheet for 15 Simulation Runs (Example 7.1)

Run #	Frequency	1	First Event	2	Second Event	3	Third Event	4	Fourth Event	5	Fifth Event	6	Sixth Event	Total
1	0													0
2	3	1	489,250	1	508,046	1	797,556							1,794,852
3	1	1	867,520											867,520
4	0													0
5	0													0
6	0													0
7	1	1	1,493,229											1,493,229
8	1	1	1,069,418											1,069,418
9	1	1	773,856											773,856
10	3	1	566,584	1	832,807	1	382,004							1,781,395
11	0													0
12	3	1	1,257,954	1	419,163	1	846,171							2,523,289
13	2	1	1,444,241	1	889,247									2,333,488
14	0													0
15	2	1	1,143,845	1	752,208									1,896,053

TABLE 7.9

Distribution of Loss (in Dollars) due to Floods (Example 7.1)

Percentile (%)	Loss	Percentile (%)	Loss	Percentile (%)	Loss	Percentile (%)	Loss
99	5,132,621	74	1,794,852	49	1,069,418	24	180,416
98	4,331,860	73	1,781,395	48	1,032,636	23	157,818
97	3,898,051	72	1,753,989	47	902,777	22	25,071
96	3,843,777	71	1,732,442	46	901,385	21	0
95	3,520,809	70	1,634,668	45	885,153	20	0
94	3,235,638	69	1,583,325	44	867,520	19	0
93	2,800,572	68	1,509,111	43	821,346	18	0
92	2,690,496	67	1,505,201	42	805,942	17	0
91	2,655,008	66	1,497,237	41	779,759	16	0
90	2,587,499	65	1,493,229	40	773,856	15	0
89	2,523,289	64	1,432,587	39	769,862	14	0
88	2,483,131	63	1,330,984	38	768,509	13	0
87	2,366,726	62	1,314,882	37	757,585	12	0
86	2,333,488	61	1,303,204	36	706,869	11	0
85	2,311,285	60	1,295,468	35	674,269	10	0
84	2,126,165	59	1,288,498	34	602,216	9	0
83	2,059,962	58	1,281,809	33	550,667	8	0
82	2,017,960	57	1,251,826	32	523,134	7	0
81	2,015,964	56	1,229,143	31	491,213	6	0
80	2,006,971	55	1,203,298	30	476,911	5	0
79	1,941,751	54	1,195,368	29	429,708	4	0
78	1,938,188	53	1,120,779	28	413,228	3	0
77	1,926,715	52	1,109,613	27	320,535	2	0
76	1,896,053	51	1,091,191	26	200,436	1	0
75	1,838,048	50	1,072,811	25	191,941	0	0

The final disruption distribution function for the annual loss due to floods is given in Table 7.9. Note that the values in Table 7.9 are associated with a certain time unit (1 year in this example) and correspond to a certain percentile. An entry in Table 7.9 can be read as follows: "With 95% confidence, the *annual loss* from flood will not be larger than $3,520,809." In some cases, firms may be more interested with the expected value instead of the maximum possible loss. The expected value can be approximated by averaging the loss values shown in Table 7.9.

The information in Table 7.9 may be valuable to the OEM in justifying contingency plans. Suppose the OEM can get a contingency insurance plan of $6 million to cover all its losses due to flood disruption. From Table 7.9, we note that with 99% probability, the loss will not exceed $5,132,621. The company may then take the contingency insurance to cover their disruption losses due to floods.

Note also that Table 7.9 indicates that there will be no loss with 21% probability. This information is also valuable to a firm that does not want to spend money for a contingency plan. The intermediate loss values and their probabilities can be used to justify investments for different contingency plans.

7.10 Miss-the-Target (MtT) Risk Models (Yang, 2006; Yang and Ravindran, 2007)

The presentation in this section is based on the doctoral work of Yang (2006). As discussed in Section 7.8.1 (Table 7.4), supply chain risks are classified into two types:

1. Value-at-Risk (VaR) type risks
2. Miss-the-Target (MtT) type risks

VaR type risks are rare events that can cause supply chain disruptions due to man-made or natural disasters. We discussed quantitative models for VaR type risks, using EVT in Section 7.9. In this section, we will present quantitative models for MtT type risks.

Most MtT type risks are primarily due to suppliers not conforming to the buyers' requirements with respect to cost, quality and delivery. Suppliers deviating from buyers' targets cause cost increases and profit losses to the buyers and OEMs. MtT type risks can also happen when cost targets and launch/ramp-up dates are missed for new product lines.

Compared to VaR type risks, MtT type risks happen more frequently. They can be controlled, but may not be totally eliminated. The impacts to the supply chains are usually not as dramatic as VaR type risks. However, in the long run, the accumulated impacts can also be significant.

7.10.1 MtT Type Impact Function

Genichi Taguchi is considered the father of quality engineering. We use Taguchi's loss functions (Taguchi, 1986) to model MtT type risks. *S-type* (*Smaller the better*) loss function is used in cases where values from real performance smaller than the target value are preferred, such as, defective rate with target value of 0.001%. *N-type* (*Nominal the better*) loss function is used when values from real performance should be within the target range. For example, delivery time could be earlier or later than the target time by a small amount. Finally *L-type* (*Larger the better*) loss function is used when values from real performance larger than the target value are preferred, such as, customer satisfaction rate with target value of 95%. Assuming that the buyers have past performance data on the suppliers, the MtT type impact function can be calculated as the difference between the supplier's actual performance and the target, $\Delta x = x - X_0$. The x variable is the actual performance of a supplier with respect to a certain measure and X_0 is the target value imposed by the buyer. The mathematical functions for the three types of Taguchi's loss functions, assuming a quadratic form, are given as follows:

S-type impact function

$$H_{\text{MtT-S}} = \begin{cases} 0 & X_0 \le x < r_1^+ \\ a^+(x - r_1^+)^2 + c^+ & r_1^+ \le x < r_2^+ \quad \text{(safe zone)} \\ M^+ & r_2^+ < x \end{cases} \qquad (7.7)$$

where
  a^+ and c^+ are hazard parameters
  M^+ is the maximum possible loss in the worst case

N-type impact function

$$H_{\text{MtT-N}} = \begin{cases} M^- & x < r_2^- \\ a^-(x - r_1^-)^2 + c^- & r_2^- \le x \le r_1^- \\ 0 & r_1^- < x < r_1^+ \quad \text{(safe zone)} \\ a^+(x - r_1^+)^2 + c^+ & r_1^+ \le x \le r_2^+ \\ M^+ & r_2^+ < x \end{cases} \qquad (7.8)$$

where
  a^+, a^-, c^+, and c^- are hazard parameters
  M^+ and M^- are the maximum possible loss values in the worst case

L-type impact function

$$H_{\text{MtT-L}} = \begin{cases} M^- & x < r_2^- \\ a^-(x - r_1^-)^2 + c^- & r_2^- \le x \le r_1^- \quad \text{(safe zone)} \\ 0 & r_1^- < x \end{cases} \qquad (7.9)$$

where
  a^- and c^- are hazard parameters
  M^- is the maximum possible loss in the worst case.

Plots of Taguchi's loss functions are given in Figure 7.6.

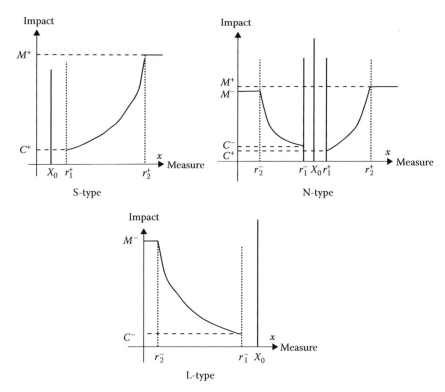

FIGURE 7.6
Taguchi's loss functions.

7.10.2 MtT Type Occurrence Function

Taguchi's loss functions represent the impact of the risk. Since the risk is a function of both impact and occurrence, we need the occurrence function of the risk event as well. MtT type occurrence function is actually the distribution of the performance measure from historical data and it can be used as the probability function to predict risk in the future. Firms can use past data to fit an appropriate occurrence function, or use some widely adopted distributions such as Gamma distribution for S-type occurrence function, Beta distribution for L-type occurrence function, and Generalized Hyperbolic distribution for N-type occurrence function.

7.10.2.1 Gamma Distribution for S-Type

The PDF and CDF for the Gamma distribution are:

$$f_{\text{Gamma}}(\Delta x; \lambda, \theta) = \frac{\lambda^{\theta} \Delta x^{\theta-1} e^{-\lambda \Delta x}}{(\theta-1)!} = \frac{\lambda^{\theta} \Delta x^{\theta-1} e^{-\lambda \Delta x}}{\Gamma(\theta)} \quad 0 \leq \Delta x < \infty$$

$$F_{\text{Gamma}}(\Delta x; \lambda, \theta) = \int_0^{\Delta x} f_{\text{Gamma}}(\Delta x; \lambda, \theta) d\Delta x = \int_0^{\Delta x} \frac{\lambda^\theta \Delta x^{\theta-1} e^{-\lambda \Delta x}}{\Gamma(\theta)} d\Delta x$$

where $\Delta x = x - x_0$.

7.10.2.2 Beta Distribution for the L-Type

The PDF function for Beta distribution is:

$$f_{\text{Beta}}(x; \lambda, \theta) = \frac{x^{\lambda-1}(1-x)^{\theta-1}}{\beta(\lambda, \theta)} = \frac{\Gamma(\lambda + \theta)}{\Gamma(\lambda)\Gamma(\theta)} x^{\lambda-1}(1-x)^{\theta-1} \quad \lambda > 0, \theta > 0$$

where $0 \le x \le 1$. However, the actual x value may not be restricted to range $[0, 1]$. Hence, for variable X at any range $[a, b]$, the following transformation has to be done:

$$Y = \frac{X-a}{b-a}$$

Then, Y follows a beta distribution. The CDF function is:

$$F_{\text{Beta}}(y; \lambda, \theta) = \int_0^y f_{\text{Beta}}(y; \lambda, \theta) dy = \int_0^y \frac{y^{\lambda-1}(1-y)^{\theta-1}}{\beta(\lambda, \theta)} dy = I_{\text{Beta}}(y; \lambda, \theta)$$

7.10.2.3 Generalized Hyperbolic Distribution for N-Type

Normal distribution is probably the most commonly used distribution in practice. However, due to it symmetric shape, its application in some areas, such as the financial industry, is limited. In a lot of cases, other asymmetric distributions are used such as lognormal distribution. As a much more flexible distribution, generalized hyperbolic distribution can be used for N-type occurrence function.

Generalized hyperbolic distribution is introduced by Barndorff-Nielsen (1978). Comparing to traditional normal distribution with two parameters μ and σ, hyperbolic distribution has five parameters λ, α, β, δ, and μ, and it is much more flexible. The density function is:

$$f_{\text{GH}}(\Delta x; \lambda, \alpha, \beta, \delta, \mu) = a(\lambda, \alpha, \beta, \delta)\left(\delta^2 + (\Delta x - \mu)^2\right)^{\left(\lambda - \frac{1}{2}\right)/2}$$

$$\times K_{\lambda - \frac{1}{2}}\left(\alpha\sqrt{\delta^2 + (\Delta x - \mu)^2}\right) \exp(\beta(\Delta x - \mu))$$

where $\Delta x = x - x_0$ and

$$a(\lambda, \alpha, \beta, \delta) = \frac{(\alpha^2 - \beta^2)^{\frac{\lambda}{2}}}{\sqrt{2\pi}\alpha^{\lambda - \frac{1}{2}}\delta^{\lambda}K_{\lambda}\left(\delta\sqrt{\alpha^2 - \beta^2}\right)}$$

is a normalized constant. The values that parameters can take are:

$$\delta \geq 0, \quad |\beta| < \alpha, \quad if \quad \lambda > 0$$
$$\delta > 0, \quad |\beta| < \alpha, \quad if \quad \lambda = 0$$
$$\delta > 0, \quad |\beta| \leq \alpha, \quad if \quad \lambda < 0$$

K_v denotes the modified Bessel function of the third kind with index v. An integral representation of K_v is

$$K_v(z) = \frac{1}{2}\int_0^{\infty} y^{v-1}\exp\left(-\frac{1}{2}z(y + y^{-1})\right)dy$$

In the density function, α determines the shape, β determines the skewness, μ determines the location, and λ determines the heaviness of the tails. δ is the scaling parameter, which is comparable to σ in the normal distribution.

Roughly, about 250 data points are required to fit the generalized hyperbolic distributions. However, about 100 data points can offer reasonable results. Although maximum-likelihood estimation method can be used to estimate the parameters, it is very difficult to solve such a complicated nonlinear equation system with five equations and five unknown parameters. Therefore, numerical algorithms are suggested such as modified Powell method (Wang, 2005). Kolmogorov–Smirnov statistics can also be used here for the fitness test.

7.10.3 MtT Type Risk function

Mathematical expressions for S-type, N-type, and L-type risks can be derived (for a desired α level of confidence) by aggregating the impact function of Section 7.10.1 and the occurrence function of Section 7.10.2. Since Taguchi's loss functions are relatively simple, the aggregation can be done analytically.

7.10.3.1 S-Type Risk Function

Assuming a Gamma distribution for the occurrence function and a quadratic form for the impact function (Equation 7.7), the final S-type risk function will be as follows.

$$R_{MtT-S}(\alpha) = \begin{cases} c^+ & 0 \le \alpha < F_{Gamma}(r_1^+ - X_0; \lambda, \theta) \\ a^+ (F_{Gamma}^{-1}(\alpha; \lambda, \theta))^2 + c^+ & \\ & F_{Gamma}(r_1^+ - X_0; \lambda, \theta) \le \alpha \le F_{Gamma}(r_2^+ - X_0; \lambda, \theta) \\ M^+ & F_{Gamma}(r_2^+ - X_0; \lambda, \theta) < \alpha \le 1 \end{cases} \quad (7.10)$$

where $F_{Gamma}^{-1}(\alpha; \lambda, \theta)$ is the value of $x - X_0$ which makes $F_{Gamma}(x - X_0; \lambda, \theta) = \alpha$. Equation 7.10 is used to calculate the maximum loss with respect to a given confidence level α, i.e., decision maker provides a confidence level α and plug it in Equation 7.10 to determine the maximum dollar loss due to the MtT type event.

7.10.3.2 L-Type Risk

Assuming a Beta distribution for the occurrence function and a quadratic form for the impact function (Equation 7.9), the final form of the L-type risk function is as follows:

Suppose that the possible value range for x is [a, b], then, let

$$y = \frac{x - a}{b - a}$$

Note that, for L-type risks, $X_0 = b$ and, $y_0 = 1$. The L-type risk function is then

$$R_{MtT-L}(\alpha) = \begin{cases} 0 & 0 \le \alpha < 1 - F_{Beta}\left(\dfrac{r_1^- - a}{b - a}; \lambda, \theta\right) \\ a^-(b-a)^2(1 - F_{Beta}^{-1}(1 - \alpha; \lambda, \theta))^2 + c^- & \\ & 1 - F_{Beta}\left(\dfrac{r_1^- - a}{b - a}; \lambda, \theta\right) \le \alpha \le 1 - F_{Beta}\left(\dfrac{r_2^- - a}{b - a}; \lambda, \theta\right) \\ M^- & 1 - F_{Beta}\left(\dfrac{r_2^- - a}{b - a}; \lambda, \theta\right) < \alpha \le 1 \end{cases} \quad (7.11)$$

where $F_{Beta}^{-1}(1 - \alpha; \lambda, \theta)$ is the value of $\dfrac{x - a}{b - a}$, which makes $F_{Beta}(\alpha; \lambda, \theta) = 1 - \alpha$.

7.10.3.3 N-Type Risk Function

Since N-type impact exists on both sides of the target value, its mathematical form is more complicated than S-type and L-type. There are 11 normal cases and 13 special cases that need to be considered for the final closed-form

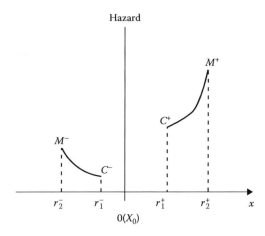

FIGURE 7.7
Case 1 of N-type risk function.

expression for the N-type risk function, depending on the relationship among M^+, M^-, c^+, c^- in Equation 7.8. Given next is the final form of the N-type risk function derived for case 1.

Case 1: If $M^+ > c^+ \geq M^- > c^-$ (Figure 7.7)

$$R(\alpha)_{MtT-N} = \begin{cases} 0 & 0 \leq \alpha < F(r_1^+) - F(r_1^-) \\ a^-(F^{-1}(F(r_1^+) - \alpha))^2 + c^- & F(r_1^+) - F(r_1^-) \leq \alpha \leq F(r_1^+) - F(r_2^-) \\ M^- & F(r_1^+) - F(r_2^-) \leq \alpha < F(r_1^+) \\ a^+(F^{-1}(\alpha))^2 + c^+ & F(r_1^+) \leq \alpha \leq F(r_2^+) \\ M^+ & F(r_2^+) \leq \alpha \leq 1 \end{cases}$$

$$(7.12)$$

The final forms of the risk functions for the remaining 23 cases are given in Yang (2006).

Normal Cases

Case 2	$M^+ > M^- > C^+ > C^-$
Case 3	$M^+ > M^- > C^+ = C^-$
Case 4	$M^+ > M^- > C^- > C^+$
Case 5	$M^+ = M^- > C^- > C^+$
Case 6	$M^+ = M^- > C^+ > C^-$
Case 7	$M^+ = M^- > C^- = C^+$
Case 8	$M^- > C^- \geq M^+ > C^+$
Case 9	$M^- > M^+ > C^- > C^+$
Case 10	$M^- > M^+ > C^- = C^+$
Case 11	$M^- > M^+ > C^+ > C^-$

Special Cases

Case 12	$M^- > C^- > M^+$
Case 13	$M^- > C^- = M^+$
Case 14	$M^- > M^+ > C^-$
Case 15	$M^- = M^+ > C^-$
Case 16	$M^+ > M^- > C^-$
Case 17	$M^- > M^+ > C^+$
Case 18	$M^- = M^+ > C^+$
Case 19	$M^+ > M^- > C^+$
Case 20	$M^+ > M^- = C^+$
Case 21	$M^+ > C^+ > M^-$
Case 22	$M^- > M^+$
Case 23	$M^+ = M^-$
Case 24	$M^+ > M^-$

Example 7.2 (Yang and Ravindran, 2007)

Computer manufacturer XYZ in the US wants to evaluate the risks from its supplier ABC in Southern China. Inaccurate delivery time is one of the major problems affecting XYZ's supply chain. This represents an N-type risk for XYZ. After analyzing the historical data, XYZ found that ABC's delivery time fits a generalized hyperbolic distribution with parameters $\lambda = -0.5$, $\alpha = \beta = 0$, $\delta = 1$, and $\mu = 0.1$. This actually is a special case of generalized hyperbolic distribution called the *Cauchy distribution*. The PDF and CDF functions of a Cauchy distribution are as follows:

$$f(x) = \frac{1}{\delta\pi\left[1+\left(\dfrac{x-\mu}{\delta}\right)^2\right]} = \frac{1}{\pi(1+(x-0.1)^2)}$$

and

$$F(x) = 0.5 + \frac{1}{\pi}\tan^{-1}(x-0.1)$$

A plot of the Cauchy PDF is presented in Figure 7.8.

N-type impact function, in dollars, has the following expression (target value is 0 with ±0.05 allowance):

$$H_{\text{MtT-N}} = \begin{cases} 0 & -0.05 \le x \le 0.05 \\ 4{,}000x^2 + 5{,}000 & 0.05 < x \le 2 \\ 1{,}000x^2 + 500 & -2 \le x < -0.05 \\ 4{,}500 & x < -2 \\ 21{,}000 & x > 2 \end{cases} \qquad (7.13)$$

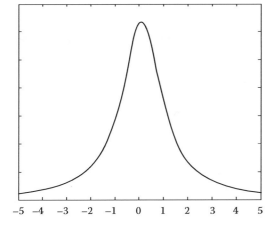

FIGURE 7.8
Plot of the Cauchy PDF.

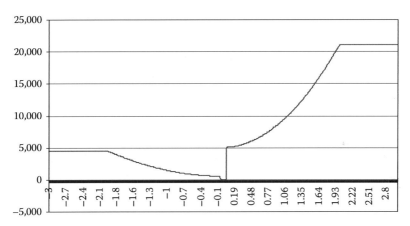

FIGURE 7.9
Plot of the N-type impact function (Example 7.2).

The plot of the N-type impact function is given in Figure 7.9.

Note that, using the notation of Figure 7.7, we can identify $M^- = 4,500$, $M^+ = 21,000$, $r_2^- = -2$, $r_1^- = -0.05$, $r_1^+ = 0.05$, $r_2^+ = 2$, $C^- = 502.5$ and $C^+ = 5,010$.

The parameters of the impact function fit into case 1 presented in Equation 7.12. Thus the MtT type risk function is given by:

$$R(\alpha)_{MtT-N} = \begin{cases} 0 & 0 \le \alpha < 3.15\% \\ 1,000(\tan(-\alpha - 0.015902)\pi + 0.1)^2 + 500 & 3.15\% \le \alpha \le 34.26\% \\ 4,500 & 34.26\% \le \alpha < 48.41\% \\ 4,000(\tan(\alpha - 0.5)\pi + 0.1)^2 + 5,000 & 48.41\% \le \alpha \le 84.58\% \\ 21,000 & 84.58\% \le \alpha \le 100\% \end{cases}$$

(7.14)

where α is the confidence percentile value. For example, if XYZ chooses $\alpha = 0.80$, using the aforementioned MtT type risk formula, it can be calculated that 80% of time, XYZ's loss will not be larger than $9595.

7.11 Risk Measures

So far in our discussion of VaR and MtT type risks, we have used the risk measure as the *maximum loss at confidence level* α, denoted by $R(\alpha)$. We can also derive other risk measures to support managerial decision making in risk management and mitigation. Given next are examples of several risk measures that can be obtained directly as a by-product of the VaR and MtT type disruption risk functions developed earlier.

1. $R(\alpha)$: In $\alpha \times 100\%$ of the times, the risk value will not be larger than $R(\alpha)$.
2. $PLe(r)$: The probability of risk value being smaller than r. Naturally, $0 \le PLe(r) \le 1$.
3. $PLa(r)$: The probability of risk value being larger than r. Note that, $0 \le PLa(r) \le 1$ and $PLe(R) + PLa(R) = 1$ for continuous distributions.
4. ER: The expected loss from a certain risk.

These risk measures can be used for both VaR type and MtT type risks. Values of $R(\alpha)$, $PLe(r)$, and $PLa(r)$ can be read from the risk table for VaR type risks or calculated using the risk function for MtT type risks. When the risk table does not have the desired value, an interpolation method can be used to estimate the real value. Numerical search methods can be used to find out the corresponding α value for a certain risk value from the MtT type risk functions. In order to get ER, extra calculations need to be done based on the VaR type risk table or the MtT type risk function.

We shall illustrate the various risk measures for both VaR and MtT risks using Examples 7.3 and 7.4 respectively.

Example 7.3

Consider the disruption risk due to floods in China discussed in Example 7.1. Using the VaR disruption risk function developed in Table 7.9, compute the following risk measures:

  (i)   Maximum loss in a year due to floods at 90% confidence level
  (ii) Probability that the annual loss due to floods will not exceed $2 million
  (iii) Probability that the annual loss due to floods exceeds $3 million
  (iv) Expected value of the annual loss due to floods

Solution

Using the distribution of annual loss ($) due to floods, given in Table 7.9, we get:

(i) $R(0.9) = \$2,587,499$ (90% chance that the annual loss due to floods will not exceed this amount)

(ii) To calculate $PLe(\$2M)$, we note that the loss ($2M$) falls between $R(0.79)$ and $R(0.8)$. Using extrapolation we get

$$PLe(2M) = 0.79 + \left(\frac{0.8 - 0.79}{2,006,971 - 1,941,751} \right)(2,000,000 - 1,941,751) = 0.799$$

There is a 79.9% chance that annual loss will not exceed $2M$.

(iii) To calculate $PLa(\$3M)$, we note that it will be between 6% and 7%. Using extrapolation again, we get:

$$PLa(3M) = 0.06 + \left(\frac{0.07 - 0.06}{3,235,638 - 2,800,572} \right)(3,000,000 - 2,800,572) = 0.0646$$

There is a 6.46% chance that the annual loss will exceed $3M$.

(iv) To compute the expected annual loss due to floods (ER), we can simply average the annual losses given in the last column of Table 7.8 over all the simulation runs. We can also use the following analytical result for compound random variable:

$$ER = E(N)E(X) \tag{7.15}$$

where

N is the number of occurrences of floods in a year

X is the loss per occurrence

For a proof of this result, refer to Bilsel (2009).

In Example 7.1, N is distributed Poisson with parameter $\lambda = 1.6$. The impact function (loss per occurrence) follows a GEVD with parameters $K = 0.095$, $\delta = 314,478.674$, and $\lambda = 495,901.049$. Using Equation 7.5,

$$E(X) = \lambda - \frac{\delta}{K} + \left(\frac{\delta}{K} \right) \Gamma(1 - K)$$

where $\Gamma \cdot$ is the Gamma function.

Substituting the values of K, δ, and λ, we get

$$E(X) = 709,839.686$$

Using Equation 7.16, we get

$$ER = (1.6)(709,839.686) = \$1,135,743.5/\text{year}$$

Example 7.4 (MtT type risk measures)

Consider the delivery time problems faced by the computer manufacturer XYZ from its supplier ABC in Example 7.2. This was modeled as an MtT–N type risk. Using the disruption risk function given by Equation 7.14, compute the following risk measures for missing delivery time window:

 (i) Cost due to inaccurate delivery time at 90% confidence level
 (ii) Probability that the delivery time cost will not exceed $15,000
 (iii) The expected value of delivery time cost

Solution

Before we begin the solution, we would like to reemphasize the subtle difference between the definitions of impact and occurrence in VaR and MtT type risk functions. For VaR functions, impact is a probability distribution of loss due to a risk event and occurrence is the probability distribution of the number of risk events during a planning period. For MtT risk functions, impact function represents the loss (cost) due to deviation from a performance target value and the occurrence function is the probability distribution of that performance measure.

In Example 7.2, the performance measure was delivery time which followed a *Cauchy Distribution*, given by Figure 7.8. The impact function was an N-type Taguchi loss function, representing the cost due to deviation from the target window for delivery time, given by Equation 7.13. The MtT–N type risk function was given by Equation 7.14.

 (i) At 90% confidence level, the delivery time cost will not exceed $21,000. In fact, we can say *with certainty* that the cost will never exceed $21,000.
 (ii) In order to compute the probability that the delivery time cost will not exceed $15,000, we have to solve the following equation for α:

$$4,000(\tan(\alpha - 0.5)\pi + 0.1)^2 + 5,000 = 15,000$$

Solving for α, we get

$$\alpha = 0.81$$

The delivery time cost will not exceed $15,000 approximately 81% of the times.
(*Note*: The tan inverse has to be calculated in radians).
 (iii) The expected value of delivery time cost = $7445.

7.12 Combining VaR and MtT Type Risks

Supply chains are constantly exposed to multiple VaR and MtT type risks. Hence, these risks have to be combined and assessed for risk management and mitigation. For example, an OEM may be interested in assessing the total VaR or MtT type risk posed by a particular overseas supplier or in a given geographic region. This could be used as additional criteria for supplier selection.

In this section, we shall discuss how to combine the various risks for managerial decision making, in the context of *supplier risk management*. In particular, we shall discuss three different ways to combine VaR and MtT type risk of suppliers as follows:

1. Combine different VaR or MtT type risks from the same supplier
2. Combine the same VaR and MtT type risks from different suppliers
3. Combine the total VaR or MtT type risk from all suppliers

We will use $R_{VaR}(\alpha)$ and $R_{MtT}(\alpha)$ in this section to denote the combined risk values. Note that the term "supplier" could be easily replaced by "geographic region" in the context of supply chain network design optimization, where risk can be explicitly included as part of the criteria functions for locating plants and warehouses.

7.12.1 Combining Different VaR Type or MtT Type Risks from the Same Supplier

7.12.1.1 VaR Type Risk Combination

Combining VaR type risks can be done fairly easily by summing all VaR type risks occurring in the same time period assuming that the different VaR type risks are not correlated. For example, if an earthquake caused a fire, and then the fire destroyed a supplier's manufacturing facility, we consider this risk is from earthquake and not from the fire. In other words, we categorize VaR type risks by their fundamental sources.

$$R_{VaR}(\alpha; S_i) = \sum_{k=1}^{K} R_{VaR}(\alpha; S_i)_k$$

where
  $R_{VaR}(\alpha; S_i)_k$ represents the k-th VaR type risk value (at α percentile) at supplier S_i
  $R_{VaR}(\alpha; S_i)$ is the total VaR type risk value from supplier S_i at α percentile
  k is the index of VaR type risks

Since $R_{VaR}(\alpha; S_i)_k$ represents the maximum loss at confidence level α, the aforementioned summation represents the worst case scenario, when all the VaR type risk events happen to the supplier S_i.

7.12.1.2 MtT Type Risk Combination

Different MtT type risks from the same supplier are usually not independent. For example, suppose that $R_{MtT}(\alpha; S_i)_1$ is caused by defective components and $R_{MtT}(\alpha; S_i)_2$ is caused by late delivery. Then, the total $R_{MtT}(\alpha; S_i)$ value most likely will not be the sum of $R_{MtT}(\alpha; S_i)_1$ and $R_{MtT}(\alpha; S_i)_2$, and in most cases, it will actually be smaller. For example, both defective components and late delivery can slow down or even stop production. However, if production has slowed down or halted due to late delivery, defective components cannot cause the same level of loss because the production has already been impacted. Since a large number of parameters are involved, it is very difficult to find out the exact correlations among the different MtT type risks. In this section, an approximation method is introduced to estimate the final result in order to support decision making. Suppose that a total of j different kinds of MtT type risks from supplier S_i need to be combined, and the risk values are denoted as $R_{MtT}(\alpha; S_i)_j$ for $j = 1,\ldots, J$ at α percentile. For each MtT type risk, suppose M_j is the maximum possible impact by risk j alone. That is, for L-type risks, $M_j = M_j^-$, for S-type risks, $M_j = M_j^+$; for N-type, $M_j = \max\left(M_j^-, M_j^+\right)$. Assume M_{total} is the estimate of the maximum possible impact in the worst case, from all the different MtT type risks for supplier S_i. Then, the combined $R_{MtT}(\alpha; S_i)$ is

$$M^* = \sum_{j=1}^{J} M_j \tag{7.16}$$

$$R_{MtT}(\alpha) = \sum_{j=1}^{J} R_{MtT}(\alpha; S_i)_j \times \frac{M_{total}}{M^*} \tag{7.17}$$

**Example 7.5 (Combining the MtT type risks;
　　　　　Yang and Ravindran, 2007)**

Company A is concerned with five different MtT type risks from its supplier, B. The first and second risks are L-type risks and the corresponding M^- values are \$540,000 and \$390,000. The third and fourth risks are S-type risks and the corresponding M^+ values are \$760,000 and \$560,000. The last one is an N-type risk and the maximum possible loss is \$1,200,000. In the worst case, impact to company A caused by supplier B is estimated as \$2,500,000. A wants to know the total possible loss at 95% confidence level.

Solution

Company A calculates the risk value for each MtT type risk by setting $\alpha = 0.95$ as \$510,000, \$350,000, \$710,000, \$480,000, and \$1,000,000 respectively. Then, following the approximation given by Equations 7.16 and 7.17, A can get the combined risk as follows:

$$M^* = \$540,000 + \$390,000 + \$760,000 + \$560,000 + \$1,200,000$$

$$= \$3,450,000$$

$$\frac{M_{total}}{M^*} = \frac{\$2,500,000}{\$3,450,000} = 0.725$$

$$R_{MtT}(95\%) = (\$510,000 + \$350,000 + \$710,000 + \$480,000 + \$1,000,000) \times 0.725$$

$$= \$2,211,250$$

This means that in 95% of all cases, the loss will not be larger than \$2,211,250 due to all the MtT type risks.

7.12.2 Combining the Same VaR Type or MtT Type Risks from Different Suppliers

7.12.2.1 VaR Type Combination

When aggregating the same risk over multiple suppliers, the combined VaR type risk is given by the sum of the same VaR type risk values from all suppliers.

$$R_{VaR}(\alpha;S)_k = \sum_{i=1}^{I} R_{VaR}(\alpha;S_i)_k \tag{7.18}$$

where

$R_{VaR}(\alpha; S)_k$ is the total k type VaR type risk value from all suppliers at α confidence level

$R_{VaR}(\alpha; S_i)_k$ denotes the k type VaR type risk value from supplier i at α confidence level.

7.12.2.2 MtT Type Risk Combination

MtT type risks are mainly decided by internal operations; therefore, it is reasonable to assume that there is no correlation among the same MtT type risks from different suppliers. Then the combined MtT type risk can be calculated using

$$R_{MtT}(\alpha;S)_j = \sum_{i=1}^{I} R_{MtT}(\alpha;S_i)_j \tag{7.19}$$

where

$R_{MtT}(\alpha; S)_j$ is the total j type MtT type risk value from all suppliers at α confidence level

$R_{MtT}(\alpha; S_i)_j$ denotes the j type MtT type risk value from supplier i at α confidence level.

7.12.3 Combining Total VaR Type or MtT Type Risks from All Suppliers

7.12.3.1 VaR Type Combination

The total VaR type risk from all suppliers is the sum of the VaR type risks from each supplier.

$$R_{VaR}(\alpha;S) = \sum_{i=1}^{l} R_{VaR}(\alpha;S_i) \tag{7.20}$$

NOTE: Equation 7.20 assumes that the suppliers are in different geographic regions and their risks are not correlated.

7.12.3.2 MtT Type Combination

The overall MtT type risk from all suppliers is the sum of the MtT type risks from each supplier.

$$R_{VaR}(\alpha;S) = \sum_{i=1}^{l} R_{MtT}(\alpha;S_i) \tag{7.21}$$

The total VaR and MtT type risks from all the suppliers, given by Equations 7.20 and 7.21 assume that all the suppliers are equally important. Hence, their individual risks are simply added up to give the total risk. However, not all the suppliers are the same. Companies do different volume of business with their suppliers. This can be incorporated by assigning a weight W_i to supplier S_i. The weight of a supplier can be defined as the ratio of the total value of the contracts with that supplier to the value of all the supply contracts for the company. For example, if company A plans to spend $50 million for all the components and material purchases in the coming fiscal year and the supplier B's contract for components is worth $10 million, then supplier B's overall weight in the supply portfolio of A is 0.2. Thus, risks from a supplier change when the weight of that supplier changes. Once all the supplier weights are determined, then the total risk can be computed by replacing Equations 7.20 and 7.21 by the weighted sum of the risks.

7.13 Risk Detectability and Risk Recovery (Bilsel, 2009)

Global supply chains have large and complex structures that pose additional difficulties to managers. One particular challenge is the transfer of information among the various stages (nodes) in a supply chain. Although information sharing has proven to enhance effective supply chain practice (see for instance Zhou and Benton (2007) for recent survey results of U.S. companies), it also has undesirable effects and brings additional risks, including loss of bargaining power and sharing of business secrets, as listed in Yuan and Qiong (2008). Information sharing becomes even more complex, when nodes in a supply chain do not belong to the same organization, due to trust issues and unwillingness to share corporate information (Fawcett et al., 2007) and possible software incompatibilities among parties forming the supply chain. This is often the situation in supply chains, where some activities, such as services, manufacturing, transportation and logistics, are outsourced to different companies. The aim of this section is to model the propagation of disruption information from a failed supplier to the buyer of the end product. The failed supplier can be located at any tier in the supply chain.

7.13.1 Detectability of Disruptive Events

Detectability is defined as the time it takes for the buyer to realize that a disruption to supply network has occurred. Modeling the propagation of disruption information in a supply chain depends on the way nodes communicate with one another. Under the best-case scenario, each supplier, independent of its tier, communicates directly with the buyer. This is a very optimistic scenario since it requires every supplier to know the end destination of its product and to establish a direct connection with the buyer. The next best scenario assumes that every disrupted supplier reports to its immediate buyer. This situation is more realistic; however, it does not ensure that the information will reach the primary buyer promptly since the immediate buyer of the disrupted supplier may not share the information with his customers or with the main buyer. Another scenario would be to assume that the information flows randomly in the supply chain network. This scenario may be perceived as pessimistic at first, but the method can be customized to better model the reality.

Bilsel (2009) developed a risk detectability model based on Markov Chain theory. His work was motivated by Petri-net based model to analyze propagation of disruption information (Wu et al., 2007). We will present the Markov Chain model of Bilsel and Ravindran (2012) next.

Information flow is related to the *mean first passage time* (MFPT) concept used in Markov chain analysis. MFPT represents the mean time it takes for a signal released at a node to reach another node in the Markov chain.

In the following subsections, we will present some Markov chain properties required to hold in order to compute the MFPT values, provide formulas to compute the MFPT matrix of a Markov chain and discuss how detectability can be modeled for improving supply chain risk management.

7.13.1.1 Some Basic Properties of Markov Chains

Consider a stochastic process $\{X_n, n = 0, 1, 2, \ldots\}$ that takes a finite or countable number of possible values. Let $X_n = i$ indicate that the process is at state i at time (or stage) n. This stochastic process is called a Markov chain if the probability of being in a certain state at a stage depends only on the state at the immediately previous stage. That is, $\Pr[X_{n+1} = i_{n+1} | X_n = i_n, X_{n-1} = i_{n-1}, \ldots, X_0 = i_0] = \Pr[X_{n+1} = i_{n+1} | X_n = i_n]$.

Two states accessible from each other in a Markov chain are said to communicate. A class of states is a group of states that communicate with each other. If a Markov chain has only one class, then it is said to be *irreducible*. The period in a Markov chain is the minimum number of transitions required to return to a state upon leaving it. A Markov chain of period one is called *aperiodic*.

Let $P = (p_{ij})$ be the transition probability matrix of a Markov chain, where p_{ij} is the probability of moving to a state j in the next stage while the process is in state i at the current stage; that is, $p_{ij} = \Pr[X_{t+1} = j | X_t = i]$. A Markov chain is said to have steady state probabilities if the transition probability matrix converges to a constant matrix. Note that the term *steady state probability* is used here in a rather loose sense since only aperiodic recurrent Markov chains admit this property.

Every Markov chain with a finite state set has a unique stationary distribution. In addition, if the Markov chain is aperiodic, then it admits steady state probabilities. Given the transition probability matrix, steady state probabilities of a Markov chain can be computed using the methods detailed in Kulkarni (1995).

Note that aperiodicity is often a strong assumption in Markov chains. One can still carry out the steady state probability computations even if a Markov chain has a period greater than one. Components of the steady state probability vector Π can then be interpreted as long run proportions of time that the underlying stochastic process would be in a given state.

7.13.1.2 Computing the MFPT Matrix

Once Π is calculated, we can compute the MFPT matrix M as described in Kemeny and Snell (1976). Several auxiliary matrices have to be computed to reach M. Let $Z = (1 - P + e\Pi^T)^{-1}$ be the *fundamental matrix*, where I is the identity matrix, P the transition probability matrix, e a vector of all ones and Π the vector of steady state probabilities. Let also Z_{dg} be the matrix that has the same elements as Z in the diagonal and zeros elsewhere; and

D be the matrix that has $1/\pi_j$ in the diagonal and zeros elsewhere. Then, M can be computed as

$$M = (I - Z + EZ_{dg})D$$

where E is a matrix of all ones. We can also compute the variance of the first passage time once we have computed the MFPT matrix M, using

$$V = M(2Z_{dg}D - I) + 2(ZM - E(ZM_{dg}))$$

where ZM_{dg} is a matrix that agrees with the product matrix ZM on the diagonal and has zeros elsewhere.

7.13.1.3 Using MFPT in Disruption Risk Quantification

There are several ways to include detectability of disruption risks in supply chain risk models. One way is to directly use the values in the MFPT matrix, the m_{ij} values, and create an objective function to minimize the number of transitions between suppliers and the buyer. Otherwise, the MFPT values may not be suitable to use directly in risk quantification since values in the MFPT matrix are in transitions and need to be transformed to actual time units (e.g., hours, days, or weeks) for proper use in disruption quantification. This transformation to time units is supply chain specific, since the speed with which the information spreads through the nodes depends on the information technology systems implemented at each node and the availability and strength of connection among the nodes. For instance, if a buyer has implemented an ERP system that allows communication with all tiers of his supply chain, he would have much better connectivity to any supplier and the transition times would be much shorter than buyers that do not have a similar visibility. We call the time it takes any disruption news to reach from node i to j as the *disruption delay between nodes i and j* and denote it as Δ_{ij}.

Detectability of disruption risks can be included in the general risk model that we developed earlier in the chapter. We can assume that the magnitude of disruption will increase with the time it takes for the news to reach the buyer, but only up to a certain level. That is, the damage from delay of disruption news should have a maximum level, which will indeed be the point where the company goes bankrupt and cannot be further hurt by the disruption. In mathematical terms, we can use an increasing function of the disruption delay with a flat horizontal tail. One can use the natural logarithm function, $f_1(\Delta) = \ln(\Delta)$, even though it tends toward infinity as $\Delta \to \infty$. We believe that the function $f_1(\Delta)$ is appropriate for a resilient supply chain which can reconfigure itself under heavy disruptions without breaking down. Another option is to use a function with a horizontal asymptote,

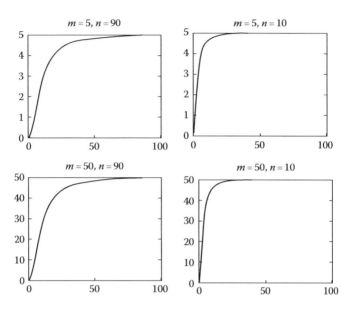

FIGURE 7.10
Plots of $f_2(\Delta)$ for different n and m values.

such as $f_2(\Delta) = \dfrac{m\Delta^2}{\Delta^2 + n}$. This function has a horizontal asymptote at m which is reached at n. Several plots of $f_2(\Delta)$ with different m and n parameter values are given in Figure 7.10.

Function $f_2(\Delta)$ is more suitable for robust supply chains, which can resist disruptions up to a certain degree, but then break down. As in Figure 7.10, the parameter n actually acts as a measure of supply chain robustness. We shall illustrate the computation of detectability function $f_1(\Delta)$ and $f_2(\Delta)$, through Example 7.6 later in the Section.

7.13.2 A Conceptual Model for Risk Recovery

Resurgence from a disruption is a vital phase in supply chains. Risk recovery plays a key role in a company's survival together with the aspects of risk discussed throughout this chapter. It is important to note that recovery can start only after the occurrence of the risk event has been detected. Hence, *risk recovery is the time it takes to neutralize a risk event completely, after it has been detected. We call the total time it takes from the occurrence of disruption to its neutralization as the risk time, which is the sum of the detection time and recovery time.*

This section proposes an abstract model of risk recovery based on concepts borrowed from queuing theory and reliability engineering (Bilsel and Ravindran, 2012). In queuing theory, it is customary to model a system using a Poisson process and assume Exponential inter-arrival times. Similar methods are used in reliability engineering, especially for modeling the reliability of component systems. For instance, Mohammed et al. (2002) model multi-component serial and parallel systems and use exponentially distributed random variables to include system recovery time from a failure. We adopt a similar model in this section with an emphasis on the parameter of the exponential distribution. Here, the exponential distribution is used to model the recovery time from disruptive events, that is, the time between the disruptive events is detected and is completely neutralized.

Exponential distribution is used to model the time between occurrences of two events and has a single parameter, μ, which represents the rate of occurrence of events. In our models, μ is the recovery rate at the failed node and $1/\mu$ is the recovery time at the failed node. The Exponential CDF reaches faster to 1 as the μ parameter gets larger, i.e., a higher μ leads to a faster recovery time. Therefore, there is a direct relationship between μ and recovery time. Note that the recovery time, hence μ, is a function of at least the inventory level (I), availability of mitigation plans (δ), and impact of the risk event (S). The higher the inventory levels along the supply chain, the better chances a company has in meeting the customer demand through available inventory and alleviates the effects of disruption. Therefore, μ increases with the inventory level. Risk mitigation plans can greatly alleviate the effects of disruption risks. A mitigation plan can be a backup supplier plan as developed in Bilsel (2009). The buyer can immediately contact the backup supplier upon disruption, if a backup supplier is available to replace a failed supplier. The backup supplier may still need some time to react to the buyer's request, but the recovery will happen much quicker compared to waiting for the failed supplier to recover. We therefore define δ as a continuous variable in [0,1] to capture different degrees of mitigation. A δ parameter close to 1 should be used for buyers that have good mitigation plans. CDF of a probability distribution or a fuzzy number can be used to model δ. Impact of the risk event (S), on the other hand, increases the recovery time, that is, the higher the impact, the longer it takes to recover from the risk event. Within this scheme, the recovery rate μ of the disruption risk can be modeled as shown in Equation 7.22.

$$\mu(I, \delta, S) \propto (\delta)\frac{I}{S} \tag{7.22}$$

Risk recovery can be included in supply chain optimization models. One can first create risk scenarios based on previously computed risk values (using techniques described in Section 7.9). Then, assuming certain inventory levels, the μ parameter for each scenario can be calculated and the sum of μ parameters of each supplier can be minimized.

7.13.3 Illustrative Example of Risk Detectability and Recovery

We shall now illustrate the computation of Risk Detectability and Risk Recovery functions with a numerical example.

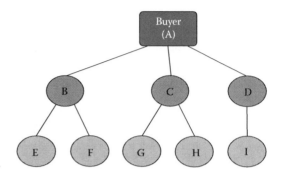

FIGURE 7.11
A 3-tier supply chain (Example 7.6).

**Example 7.6 (Risk Detectability and Recovery;
Bilsel and Ravindran, 2012)**

Consider a small supply chain with three stages, a buyer, three tier-1 suppliers and five tier 2 suppliers as shown in Figure 7.11. Assume that the buyer wants to select a single supplier and that all tier-1 and their tier-2 suppliers have enough capacity to meet the buyer's demand.

To model detectability, let us assume that with 0.8 probability, disruption information will be shared with nodes upstream (toward the buyer) and with 0.2 probability information will be shared with nodes downstream in the supply chain of Figure 7.11. Our computational experiments showed that allocating these probabilities yield much more reasonable outputs than assuming that information flows randomly. Then, the transition probability matrix of the supply chain in Figure 7.11 is given in Table 7.10 and the relevant MFPT matrix is given in Table 7.11.

Let us assume that the transition time (in days) between a tier-1 supplier and the buyer follows a Uniform (1,2) distribution and transition time between a tier-2 and a tier-1 supplier follows a Uniform (2,4) distribution. Sampling from these uniform distributions, let $t_{tran}(B) = 2$, $t_{tran}(C) = 2$,

TABLE 7.10

Transition Probability Matrix of the Supply Chain
Network in Figure 7.11 (Example 7.6)

	A	B	C	D	E	F	G	H	I
A	0	0.33	0.33	0.33	0	0	0	0	0
B	0.8	0	0	0	0.1	0.1	0	0	0
C	0.8	0	0	0	0	0	0.1	0.1	0
D	0.8	0	0	0	0	0	0	0	0.2
E	0	1	0	0	0	0	0	0	0
F	0	1	0	0	0	0	0	0	0
G	0	0	1	0	0	0	0	0	0
H	0	0	1	0	0	0	0	0	0
I	0	0	0	1	0	0	0	0	0

TABLE 7.11

MFPT Matrix of the Supply Chain Network
in Figure 7.11

	A	B	C	D	E	F	G	H	I
A	2.5	6.0	6.0	6.0	65.0	65.0	65.0	65.0	35.0
B	1.5	6.0	7.5	7.5	59.0	59.0	66.5	66.5	36.5
C	1.5	7.5	6.0	7.5	66.5	66.5	59.0	59.0	36.5
D	1.5	7.5	7.5	6.0	66.5	66.5	66.5	66.5	29.0
E	2.5	1.0	8.5	8.5	60.0	60.0	67.5	67.5	37.5
F	2.5	1.0	8.5	8.5	60.0	60.0	67.5	67.5	37.5
G	2.5	8.5	1.0	8.5	67.5	67.5	60.0	60.0	37.5
H	2.5	8.5	1.0	8.5	67.5	67.5	60.0	60.0	37.5
I	2.5	8.5	8.5	1.0	67.5	67.5	67.5	67.5	30.0

TABLE 7.12

Transition Times of the Supply Chain
Network in Figure 7.11

Source (i)	Path	Δ_{iA}	$f_1(\Delta_{iA})$
B	$B \to A$	3	1.098
C	$C \to A$	3	1.098
D	$D \to A$	3	1.098
E	$E \to B \to A$	$4 + 3 = 7$	1.946
F	$F \to B \to A$	$2 + 3 = 5$	1.609
G	$G \to C \to A$	$2 + 3 = 5$	1.609
H	$H \to C \to A$	$4 + 3 = 7$	1.956
I	$I \to D \to A$	$3 + 3 = 6$	1.609

$t_{tran}(D) = 1$, $t_{tran}(E) = 4$, $t_{tran}(F) = 2$, $t_{tran}(G) = 2$, $t_{tran}(H) = 4$, $t_{tran}(I) = 3$. We finally have the Δ_{iA} disruption delay values in days between suppliers and the buyer as in Table 7.12.

Values in the Δ_{iA} column of Table 7.12 can be used in supply chain decision making. For example, a buyer may consider minimizing the total transition time of his supply chain as a supplier selection criterion. Taking the analysis one step further, one of the functions $f_1(\Delta)$ or $f_2(\Delta)$ introduced earlier in the section can be used to characterize losses to the supply chain due to disruption delays. The last column in Table 7.12 shows the magnitude of losses the buyer would suffer from supplier disruptions under the $f_1(\Delta)$ function, where $f_1(\Delta) = \ln(\Delta)$. As previously noted, buyers can use the $f_1(\Delta)$ values (or $f_2(\Delta)$ values given that they estimate the required parameters) to select among different suppliers.

Another important aspect when dealing with disruptions is the risk recovery time discussed in the Conceptual Model for Risk Recovery

Section (Section 7.13.2). We use an Exponential model to compute recovery time where the parameter μ of the exponential distribution can be computed as in Equation 7.22. Let us assume that tier-1 suppliers have good backup plans and set $\delta = 1$. Assume that supplier B has $3000 worth of inventory, supplier C has $5000 worth of inventory and supplier D has $1000 worth of inventory.

Then, $\mu_B = \dfrac{3000}{3000} = 1$; $\mu_C = \dfrac{5000}{3000} = 1.67$ and $\mu_D = \dfrac{1000}{3000} = 0.33$. Therefore, it would take supplier B $\dfrac{1}{1} = 1$ day, supplier C $\dfrac{1}{1.67} = 0.6$ days and supplier D $\dfrac{1}{0.33} = 3$ days to recover from disruptions. Recovery time can be used as a standalone criterion to evaluate suppliers. Also, the total risk time, which is the sum of the detection and recovery times, can be calculated and used as a separate criterion. In our example, risk time for supplier B is $3 + 1 = 4$ days. For supplier C the risk time is $3 + 0.6 = 3.6$ days and supplier D's risk time is $3 + 3 = 6$ days.

An extreme case would arise if some of the suppliers work in a just-in-time framework and practically carry no inventory. In such situations, Equation 7.22 would yield $\mu = 0$ for which $1/\mu$ is not defined. It may be assumed that recovery time at those suppliers would equal to the maximum lead time between that supplier and his own suppliers, given that the disrupted supplier would be able to manufacture, and hence recover, once all required items reach his facility.

7.14 Multiple Criteria Optimization Models for Supplier Selection Incorporating Risk

The objective of this section is to develop optimization models for supplier selection incorporating disruption and operational risks of the supplier and apply them to a real company. The presentation is based on the work of Ravindran et al. (2010). We model the risk-adjusted supplier selection problem as a Multiple Criteria Decision Making (MCDM) problem and solve it in two phases.

The two-phase method follows the sequence presented in Figure 7.12. Phase 1 is the pre-qualification step, where a large set of initial suppliers is reduced to a smaller set of manageable suppliers using various multi-criteria ranking methods discussed in Chapter 6, Section 6.3. In Phase 2, order quantities are allocated among the shortlisted suppliers using a multi-objective optimization model. In the multi-objective formulation, price, lead time, VaR type risk of disruption loss due to hurricanes and MtT type risk of quality are considered as four conflicting objectives. The multi-objective optimization problem is solved using the four variants of goal programming methodology discussed in Chapter 6, Section 6.4.

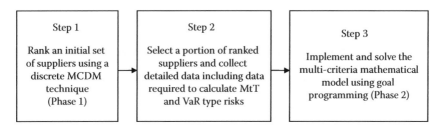

Step 1	Step 2	Step 3
Rank an initial set of suppliers using a discrete MCDM technique (Phase 1)	Select a portion of ranked suppliers and collect detailed data including data required to calculate MtT and VaR type risks	Implement and solve the multi-criteria mathematical model using goal programming (Phase 2)

FIGURE 7.12
Steps of the two-phase method.

The models are illustrated with an actual vendor management application to a global IT company (Ravindran et al., 2010).

7.14.1 Phase 1 Model (Short-Listing Suppliers)

The objective of Phase 1 is to shortlist suppliers by ranking them under conflicting criteria. Optimal order quantities are allocated among short-listed suppliers in Phase 2. A set of criteria to be used in the evaluation of the candidate suppliers is required to perform Phase 1 calculations. These criteria then have to be presented to the decision makers (DMs) to capture their preferences in two ways: first, the importance of each criterion and attribute with respect to one another (input 1) and second, the evaluations of each candidate supplier with respect to each attribute (input 2). These input data from the DMs are then analyzed using several MCDM methods, which provide the ranking of the suppliers as an output. Figure 7.13 illustrates the flow of activities that take place in Phase 1.

For Phase 1, 20 candidate suppliers were considered for the global IT Company. Each supplier was evaluated using 7 criteria measured by 14 attributes as shown in Table 7.13. The supplier base was a mixture of domestic and international suppliers.

Four DMs from the company, two from procurement and two from R&D, participated in the study. The interaction with the DMs (Activity 2 of Figure 7.13) was carried out using survey sheets. Each DM provided

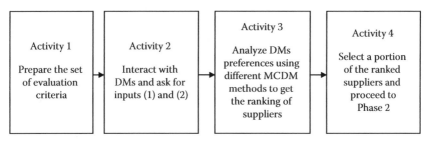

Activity 1	Activity 2	Activity 3	Activity 4
Prepare the set of evaluation criteria	Interact with DMs and ask for inputs (1) and (2)	Analyze DMs preferences using different MCDM methods to get the ranking of suppliers	Select a portion of the ranked suppliers and proceed to Phase 2

FIGURE 7.13
Flow of activities in phase 1.

TABLE 7.13

Attribute Weights for DM-1

No	Criterion	Attribute	Definition
1	Delivery (C1)	Accuracy (A1)	Accuracy in meeting the promised delivery time
2		Capacity (A2)	Capacity of the supplier
3		Lead time (A3)	Promised delivery lead time
4	Business performance (C2)	Financial status (A4)	Financial performance
5		Compatibility of business strategy (A5)	Compatibility of strategic plans of the suppliers with buyer's long-term plans
6	Quality (C3)	Defective rate (A6)	Rate of defective items among shipped
7		Responsiveness (A7)	Reaction time of supplier to correct defects and other supply related issues
8	Cost (C4)	Unit cost (A8)	Cost per item
9		Order change and cancellation charges (A9)	Fees associated with modifying or cancelling orders after placement
10	Information technology (C5)	Online (A10)	Availability of online ordering and order tracking
11		EDI (A11)	Availability of EDI systems at the supplier
12	Long-term improvement (C6)	Improvement programs (A12)	Improvement of customer service related activities at the supplier
13		R&D activities (A13)	Incentive in pursuing R&D
14	Risk (C7)	Risk score (A14)	Risk due to supply disruption

the following data on the importance of the different supplier attributes (Table 7.13):

- Rating of each attribute (1–10 scale)
- Pair-wise comparison of attributes
- Strength of preference (1–9 scale) for pair-wise comparisons

Once the survey forms were received electronically, they were analyzed using Rating, Borda Count, and AHP methods (Activity 3 of Figure 7.13). Detailed descriptions of these three Ranking methods are available in Chapter 6, Section 6.3.

7.14.2 Results of Phase 1 Experiments

7.14.2.1 Ranking of the Criteria

Based on the DM's preferences on the supplier criteria, obtained through survey data, the criteria rankings were determined for each DM using the three MCDM methods (Rating, Borda Count, and AHP). Comparison of criteria rankings were made under three categories:

1. Criteria rankings for the same DM by the three MCDM methods
2. Criteria rankings across DMs using the same MCDM method
3. Group rankings of criteria by combining the inputs of all DMs

7.14.2.2 Comparison across Methods for the Same DM

Table 7.14 compares criteria rankings by the three methods for DM-2. Note that cost, quality, and delivery are the top three criteria for all three methods. However, the Rating method failed to break several ties. For example, three criteria are tied for second place and two for the last place, in the Rating method; other methods do not suffer from this problem. Also, the Rating method ranks IT above risk (Table 7.14), whereas risk is rated higher by both Borda Count and AHP. In general, the Rating method would have difficulty in differentiating criteria that are close to one another in importance. Borda Count requires less DM input compared to AHP and is easier to implement.

For DM-2, both Borda Count and AHP provide the same rankings. Similar results were obtained for the other DMs also. It was found that cost, quality and delivery were the three most important supplier criteria and Borda Count and AHP methods yielded similar rankings.

7.14.2.3 Comparison across DMs for the Same Method

It is also interesting to compare the criteria rankings of all DMs using each method. Table 7.15 illustrates the rankings by all the four DMs by AHP.

The objective of this comparison is to see whether there exist any differences in criteria importance between Procurement (P) and R&D staff. We note from Table 7.15 that the preference structures of DM-2, DM-3 and DM-4 are very similar, with cost, quality and delivery as the top three criteria in

TABLE 7.14

Criteria Rankings for DM-2

	Rank Using		
Criterion	Rating	Borda	AHP
Delivery	2	3	3
Business performance	2	4	4
Quality	2	2	2
Cost	1	1	1
IT	5	6	6
Long-term improvement	6	7	7
Risk	6	5	5

TABLE 7.15

Criteria Rankings by AHP for All DMs

	Rank According to			
Criterion	DM-1 (P)	DM-2 (P)	DM-3 (R&D)	DM-4 (R&D)
Delivery	2	3	3	3
Business performance	3	4	5	4
Quality	1	2	2	2
Cost	4	1	1	1
IT	6	6	7	7
Long-term improvement	7	7	6	6
Risk	5	5	4	5

supplier selection. In addition, no significant differences were found between Procurement and R&D staff rankings.

7.14.2.4 Individual Supplier Rankings

To determine the supplier rankings, we compute the attribute weights using Rating, Borda Count, and AHP methods for each DM. This results in a total of 12 different weights (one for each DM, by each method) for each of the 14 attributes and potentially 12 different supplier rankings. Table 7.16 gives the actual attribute weights for DM-1.

Note that additional data on the evaluation of suppliers with respect to each attribute is needed to determine the supplier rankings. This is given in Table 7.17. (Data is scaled such that all attributes are maximized)

TABLE 7.16

Attribute Weights for DM-1

No	Attribute	Rating	Borda Count	AHP
1	Accuracy (A1)	0.059	0.107	0.173
2	Capacity (A2)	0.052	0.071	0.083
3	Lead time (A3)	0.059	0.036	0.040
4	Financial status (A4)	0.079	0.119	0.120
5	Compatibility of business strategy (A5)	0.071	0.060	0.040
6	Defective rate (A6)	0.085	0.083	0.273
7	Responsiveness (A7)	0.085	0.167	0.091
8	Unit cost (A8)	0.051	0.048	0.059
9	Order change and cancellation charges (A9)	0.081	0.095	0.020
10	Online (A10)	0.057	0.054	0.018
11	EDI (A11)	0.057	0.054	0.018
12	Improvement programs (A12)	0.066	0.018	0.011
13	R&D activities (A13)	0.047	0.018	0.011
14	Risk score (A14)	0.151	0.071	0.044

TABLE 7.17

Evaluation of Candidate Suppliers with respect to Attributes

Supplier	Attribute													
	1	2	3	4	5	6	7	8	9	10	11	12	13	14
1	80	40	70	76	53	15	30	90	20	65	55	80	60	45
2	55	30	20	70	36	45	90	45	70	78	75	69	78	50
3	34	70	40	40	67	70	67	60	45	24	35	40	45	56
4	75	65	35	60	60	30	78	76	45	65	78	70	65	68
5	90	55	76	80	57	13	23	65	37	75	76	80	65	50
6	87	29	70	70	27	24	34	30	73	38	43	70	76	25
7	39	78	30	20	75	75	70	30	40	40	51	40	20	80
8	85	60	55	70	62	20	50	67	60	92	87	65	67	72
9	45	43	30	39	50	50	79	41	68	76	68	47	40	52
10	77	56	60	57	50	30	45	80	34	85	84	72	60	60
11	50	60	30	65	60	45	80	76	40	50	55	89	45	70
12	76	70	40	73	73	30	70	45	76	65	60	85	74	65
13	20	34	70	55	30	75	34	40	63	30	40	40	65	45
14	86	43	35	75	48	26	70	67	35	76	65	65	75	50
15	65	56	86	50	50	40	20	60	40	55	48	60	55	60
16	70	17	35	45	22	35	75	34	55	54	65	40	25	30
17	80	71	60	45	77	36	45	30	69	78	68	70	50	75
18	35	65	74	89	70	69	30	56	50	34	40	55	70	60
19	96	90	13	90	93	6	90	94	19	90	88	95	89	85
20	11	68	85	15	50	85	10	50	70	30	30	18	10	50

To determine the supplier rankings, the score for each supplier is obtained by multiplying their attributes (row elements of Table 7.17) by their respective weights (Table 7.16). In Table 7.18 we present the *top five suppliers* for each DM using their Borda Count weights as an illustration.

It can be observed from Table 7.18 that supplier 19 is always in the first place. Also, observe that suppliers 4, 8 and 12 are almost always in the top 5. If the ultimate Phase 1 decision would be based on individual preferences, it

TABLE 7.18

Shortlist of Suppliers Using Borda Count Weights

Supplier Rank	Rank According to			
	DM-1 (P)	DM-2 (P)	DM-3 (R&D)	DM-4 (R&D)
1st	19	19	19	19
2nd	12	8	4	18
3rd	8	4	11	4
4th	4	12	8	8
5th	2	11	12	11

would be logical to select these suppliers to advance to the Phase 2. For the fifth place, one of the other suppliers (e.g., Supplier 11) that showed up more frequently in the short lists might be selected. In situations, where individual DM rankings differ significantly, weights can be assigned to each DM based on their positions, knowledge and skills.

7.14.2.5 Group Supplier Rankings

Most of the supplier selection practices involve the participation of multiple DMs and the ultimate decision is based on the aggregation of DMs individual judgments to arrive at a group decision. We have aggregated the preferences of the four DMs using different group decision making techniques discussed in Chapter 6, Section 6.3.9:

- *Multiple DM rating*: Rating of a given attribute is the average of the ratings for that particular attribute across all DMs. Ratings are then normalized to obtain the group weights associated with each attribute.

- *Multiple DM Borda Count*: Points are assigned based on the number of DMs that assign a particular rank for an attribute. These points are then totaled for each attribute and normalized to get weights. (This is similar to how AP college polls are done to get the top 25 teams.)

- *Multiple DM AHP*: Strength of preference scores assigned by DMs is aggregated using geometric means and then used in the AHP calculations.

- *Multiple DM AHP with Borda Count*: At the beginning, each DM is considered individually and an AHP weight calculation is conducted for each DM to come up with attribute weights and supplier rankings. The individual supplier rankings are then aggregated using Borda Count to get the group ranking.

The top five suppliers identified by the aforementioned group decision-making methods are presented in Table 7.19.

It is to be noted that the exact ranking of the suppliers can vary by different methods. However, the objective of Phase 1 is to short-list the promising suppliers. Hence, the list can be expanded when there are conflicts in the rankings.

TABLE 7.19

Results of the Group Decision Making Analysis

	Rating	Borda Count	AHP (Geometric Mean)	AHP (with Borda Count)
Supplier rankings	19, 8, 12, 4, 17	19, 4, 8, 12, 11	19, 4, 11, 8, 12	19, 8, 12, 4, 11

7.14.2.6 Conclusions from Phase I Results

- Three out of four DMs indicated cost, quality and delivery as the most important criteria in supplier selection for the company.
- Supplier 19 came in the first place in both individual and group decision making rankings.
- Supplier 19 was followed by suppliers 8, 4, and 12 (not necessarily in that order) in all analyses.
- Supplier 11 was included in the shortlist of five suppliers for Phase 2, since three out of four methods selected it.
- In most cases, Borda Count results were in line with AHP results. Given the increased cognitive burden and expensive calculations for AHP, Borda Count might be selected as the appropriate MCDM method for supplier ranking. This observation is consistent with the results obtained previously in several research projects done by Industrial Engineering students to compare different multiple criteria ranking methods in the graduate level Multi-criteria Optimization class taught by the author at Penn State.

7.14.3 Risk Adjusted Multi-Criteria Optimization Model for Supplier Sourcing (Phase 2)

In Phase 1, we had 14 attributes. In Phase 2, we used only four of them: lead time, quality, cost and disruption risk, based on the large relative importance derived from the DM's inputs in Phase 1. The remaining attributes were deemed significantly less important and hence, were eliminated from further analyses. As a result of Phase 1, the initial set of 20 suppliers was reduced to 5 suppliers (19, 12, 11, 8, and 4). In the next step, detailed quantitative data on price, capacity, quality and risk were generated for these 5 suppliers and were used in a multi-objective optimization model. For the Phase 2 model, we consider multiple buyers, multiple products, multiple suppliers and price discounts. The scenario of multiple buyers is possible in case of a central purchasing department, where different divisions of an organization buy through one purchasing department. Here the number of buyers will be equal to number of divisions buying through the central purchasing. In all other cases, the number of buyers is equal to one. We consider the least restrictive case for modeling where any of the buyers can acquire one or more products from any suppliers.

In Phase 2, an organization will make the following decisions:

- To choose the most favorable suppliers for raw materials and parts to meet its supplier selection criteria.
- To order desired quantities from the chosen suppliers to meet its production plan or demand.

TABLE 7.20

Phase 2 Model Notations

Indices	Description
I	Set of products available to purchase
J	Set of buyers who procure multiple units to fulfill demand
K	Potential set of suppliers
M	Set of incremental price breaks for different price levels
Parameters	**Description**
p_{ikm}	Cost of acquiring one unit of product i from supplier k at price level m
b_{ikm}	Quantity at which incremental price breaks occur for product i at supplier k
F_k	Fixed ordering cost of supplier k
D_{ij}	Demand for product i for buyer j
l_{ijk}	Lead time of supplier k to supply product i to buyer j
CAP_{ik}	Production capacity at supplier k for product i
N	Maximum number of suppliers that can be selected
VaR_k	Quantified VaR type risk for supplier k
MtT_k	Quantified MtT type risk for supplier k
Decision Variables	**Description**
x_{ijkm}	Number of units of product i that buyer j procures from supplier k at price level m
z_k	Binary variable indicating whether supplier k is selected or not
y_{ijkm}	Binary variable used in separating price levels m for item i in a transaction between buyer j and supplier k

The model presented here is an extension of the multiple criteria model for supplier selection, developed by Wadhwa and Ravindran (2007). The notations used in the multi-criteria models are given in Table 7.20.

7.14.3.1 Model Objectives

1. Total purchasing cost

$$Min\, Z_1 = \sum_i \sum_j \sum_k \sum_m p_{ijkm} \cdot x_{ijkm} \qquad (7.23)$$

2. MtT risk for quality (loss in dollars)

$$Min\, Z_2 = \sum_i \sum_j \sum_k \sum_m MtT_k \cdot x_{ijkm} \qquad (7.24)$$

3. Average lead time

$$Min\, Z_3 = \frac{\sum_i \sum_j \sum_k \sum_m l_{ijk} \cdot x_{ijkm}}{\sum_i \sum_j D_{ij}} \qquad (7.25)$$

4. VaR risk (loss in dollars)

$$Min\, Z_4 = \sum_i \sum_j \sum_k \sum_m VaR_k \cdot x_{ijkm} \qquad (7.26)$$

The first objective function (7.23) minimizes the total purchasing cost. Purchasing costs has two components, *total variable cost* (TVC) and *fixed cost* (FC). TVC constitutes the first part and FC is the second part of Equation 7.23. Note that FC depends on whether a supplier is selected or not; therefore, we use the binary variable z_k in the expression of FC. The second objective (7.24) represents MtT risk for quality resulting from past transactions with the suppliers. The third objective refers to the minimization of average lead time (7.25). Note that $\sum_i \sum_j D_{ij}$ is a constant and can be removed during the optimization process. The fourth objective in (7.26) minimizes VaR risk of supply disruptions due to hurricanes. We model the quality objective as an *S-type* MtT risk (see Figure 7.6). The calculations of exact mathematical expressions for VaR and MtT type risks for the different suppliers are given in Section 7.14.5.

7.14.3.2 Model Constraints

$$\sum_j \sum_m x_{ijkm} \le CAP_{ik} \cdot z_k \quad \forall i, k \qquad (7.27)$$

$$\sum_k \sum_m x_{ijkm} = D_{ij} \quad \forall i, j \qquad (7.28)$$

$$\sum_k z_k \le N \qquad (7.29)$$

$$x_{ijkm} \le (b_{ikm} - b_{ik(m-1)}) \cdot y_{ijkm} \quad \forall i, j, k, 1 \le m \le m_k \qquad (7.30)$$

$$x_{ijkm} \geq (b_{ikm} - b_{ik(m-1)}) \cdot y_{ijk(m+1)} \quad \forall i, j, k, \ 1 \leq m \leq m_k - 1 \qquad (7.31)$$

$$x_{ijkm} \geq 0, z_k \in \{0,1\} \qquad (7.32)$$

Constraints (7.27) represent the capacity limits at each supplier. The total order placed with a supplier should be less than or equal to the capacity available at the supplier. Note that the binary variable z_k is used to activate the constraint for a supplier k only if k is chosen in the model. Constraints (7.28) introduce the demand constraints. Demand for product i at buyer j must be satisfied. Constraint (7.29) limits the number of selected suppliers to N. Constraints (7.30) and (7.31) are used to linearize the original nonlinear cost function that arises due to price discounts. The sequence $0 = b_{ik0} < b_{ik1} < \cdots < b_{ikm_k}$ is the sequence of quantities where price discounts occur. If the $y_{ijkm} = 1$ for a discount interval m, then all quantities in intervals 1 to $m - 1$ should be at the maximum of those ranges. Finally, constraints (7.32) force non-negativity and binary restrictions in the model.

Numerical complexity of the model can be expressed as a function of problem size. Overall, the model has $i(2jk + k + j) + 1$ constraints and $k(2ijm + 1)$ variables, of which $k(ijm + 1)$ are binary. Model complexity increases with the number of variables. Binary variables, in particular, are known to cause computational difficulty. Moreover, adding more price breaks will make the model harder to solve because of the additional y_{ijkm} variables required in the formulation. Constraints, on the other hand, generally reduce the complexity of integer programs by cutting suboptimal regions of the feasible set.

7.14.4 Solution Methodology

As discussed in Chapter 6, *goal programming* (GP) is a practical method for handling multiple objectives. GP falls under the class of methods that use completely pre-specified preferences of the DM in solving the multi-objective optimization problem. In GP, each objective is assigned a target level for achievement and relative priority on achieving the target. GP treats these targets as *goals to aspire for* and not as absolute constraints. It attempts to find an optimal solution that comes as close as possible to the targets in the order of specified priorities. We solve the supplier selection problem, given by Equations 7.23 through 7.32 using four variants of GP; preemptive GP, non-preemptive GP, Tchebycheff (Min–Max) GP, and fuzzy GP. Detailed descriptions of the GP methods are given in Chapter 6, Section 6.4.

7.14.4.1 Preemptive GP Model

In preemptive GP, priority is assigned for each incommensurable goal and weights to goals at the same priority. Goals at higher priority have to be satisfied before lower priority goals are considered. The preemptive GP formulation for the supplier selection problem is as follows:

$$\min z = P_1 d_1^+ + P_2 d_2^+ + P_3 d_3^+ + P_4 d_4^+ \tag{7.33}$$

subject to

$$\left(\sum_i \sum_j \sum_k \sum_m p_{ikm}.x_{ijkm} + \sum_k F_k.z_k \right) + d_1^- - d_1^+ = \text{Price goal} \tag{7.34}$$

$$\sum_i \sum_j \sum_k \sum_m \text{MtT}_k.x_{ijkm} + d_2^- - d_2^+ = \text{MtT type risk goal} \tag{7.35}$$

$$\sum_k \sum_m l_{ijk}.x_{ijkm} + d_3^- - d_3^+ = \text{Lead time goal} \tag{7.36}$$

$$\sum_i \sum_j \sum_k \sum_m \text{VaR}_k.x_{ijkm} + d_4^- - d_4^+ = \text{VaR type risk goal} \tag{7.37}$$

$$d_n^-, d_n^+ \geq 0 \quad \forall n \in \{1,\ldots,4\} \tag{7.38}$$

Constraints (7.27 through 7.32) stated earlier will also be included in this model.

Based on Phase 1 results, price is the most important goal followed by MtT goal on quality, lead time, and VaR in that order. d_n^- and d_n^+ are the deviation variables representing how far the solution deviates from each goal. In general, goal values are to be provided by the DMs. To determine the optimal solution for a preemptive GP model, a sequence of optimization problems has to be solved.

7.14.4.2 Non-Preemptive GP Model

In non-preemptive GP, the buyer sets goals to achieve for each objective and preference in achieving those goals are expressed as numerical weights. The buyer has the following four goals:

- Limit the total purchasing cost to Price goal with weight W_1
- Limit the MtT type risk to MtT goal with weight W_2
- Limit the lead time to Lead goal with weight W_3
- Limit the VaR type risk to VaR goal with weight W_4

The weights W_1, W_2, W_3, and W_4 are obtained from Phase 1. The criteria weights used in the model are, cost (0.343), quality (0.338), lead time (0.246) and VaR risk (0.073). The non-preemptive GP model can be formulated as follows:

$$\min z = W_1 * d_1^+ + W_2 * d_2^+ + W_3 * d_3^+ + W_4 * d_4^+ \tag{7.39}$$

Subject to the constraints (7.27 through 7.32) and (7.34 through 7.38). Because of the use of numerical weights, the objectives have to be scaled properly. However, only a single objective optimization problem had to be solved here.

7.14.4.3 Tchebycheff (Min–Max) GP Model

The Tchebycheff GP model minimizes the maximum weighted deviation from the stated goals. The objective function of the Tchebycheff GP model is as follows:

$$\text{Min Max}\left(W_1 * d_1^+, W_2 * d_2^+, W_3 * d_3^+, W_4 * d_4^+\right) \tag{7.40}$$

where W_1, W_2, W_3, and W_4 are the weights used in the non-preemptive GP. Objective function (7.40) is not linear but can be reformulated as a linear objective by setting

$$\max W_1 * d_1^+, W_2 * d_2^+, W_3 * d_3^+, W_4 * d_4^+ = M, \quad \text{Where } M \geq 0.$$

Thus, the Tchebycheff GP model can be rewritten as follows:

$$\text{Min } Z = M \tag{7.41}$$

subject to

$$M \geq \left(W_1 * d_1^+\right) \tag{7.42}$$

$$M \geq \left(W_2 * d_2^+\right) \tag{7.43}$$

$$M \geq \left(W_3 * d_3^+\right) \tag{7.44}$$

$$M \geq \left(W_4 * d_4^+\right) \tag{7.45}$$

Constraints (7.27 through 7.32) and (7.34 through 7.38) are also included. Tchebycheff GP method is more focused on minimizing outliers than satisfying goals and may lead to poor solutions.

7.14.4.4 Fuzzy GP Model

Fuzzy GP uses the ideal values as targets and minimizes the maximum weighted normalized distance from the ideal solution for each objective. An *ideal solution* is the vector of best values of each objective obtained by optimizing each objective independently, ignoring other objectives. In this application, ideal solutions were obtained by minimizing price, lead time, MtT and VaR type risks independently. If M equals the maximum deviation from the ideal solution, then the fuzzy GP model is as follows:

$$\text{Min } z = M \tag{7.46}$$

subject to

$$M \geq W_1 \left(\frac{Z_1 - Z_1^*}{Z_1^*} \right) \tag{7.47}$$

$$M \geq W_2 \left(\frac{Z_2 - Z_2^*}{Z_2^*} \right) \tag{7.48}$$

$$M \geq W_3 \left(\frac{Z_3 - Z_3^*}{Z_3^*} \right) \tag{7.49}$$

$$M \geq W_4 \left(\frac{Z_4 - Z_4^*}{Z_4^*} \right) \tag{7.50}$$

Constraints (7.27 through 7.32) stated earlier will also be included in this model.

Z_1^*, Z_2^*, Z_3^* and Z_4^* in the aforementioned formulation represent individual minima, that is, the ideal values for the respective objectives. Weight values W_1, W_2, W_3, and W_4 are the same that are used in the non-preemptive GP model. The advantage of fuzzy GP is that no target values have to be specified by the DM and there are no deviational variables.

7.14.5 Data Description

As outlined in Section 7.14.3, data regarding the price, leadtime, quality, capacity, VaR and MtT type risk from different suppliers for different products are required to model the multi-objective optimization problem. As an illustration, consider the case where we have two products, two buyers, five suppliers and each supplier offers two price breaks. The problem here is to find which suppliers to buy from and how much to buy from each supplier.

7.14.5.1 MtT Type Risk Calculations

Suppliers are the source of several MtT type operational risks; e.g., late delivery, low service rate, high defective rate, etc. We assume buyers suffer mainly from defective parts; hence, we focus on the defective rate (i.e., quality) issue. Naturally, buyers ask for the lowest defective rate possible. For MtT type risks of *smaller-the-better* type, we proposed using an S-type impact function (see Figure 7.6). We present the general form of the S-type impact function in Equation 7.51

$$
I(\alpha) = \begin{cases} 0 & x_0 \leq x < r_1^+ \\ a^+(x - x_0)^2 c^+ & r_1^+ \leq x \leq r_2^+ \\ M^+ & r_2^+ < x \end{cases} \tag{7.51}
$$

We assume $M^+ = 6960$, $a^+ = 4000$, $c^+ = 5000$, $x_0 = 0$, $r_1^+ = 10^{-5}$ and $r_2^+ = 0.7$ for the first candidate supplier S19. A low r_1^+ value indicates that buyers will be adversely affected as soon as the supplier delivers defective items. A high r_2^+ value, on the other hand, models a relatively high tolerance to the amount of defective items before the maximum damage is faced by the buyers. In other words, r_2^+ models the level up to which buyers can absorb the particular MtT type risk. Parameters a^+, c^+, and x_0 are used to model the impact of defective items within the range $[r_1^+, r_2^+]$. M^+ is the maximum damage that S19 can do to its buyers for this particular risk measure and it is reached once the defective level exceeds r_2^+.

As mentioned in Section 7.10, Gamma distribution can be used to model the S-type occurrence function. Combining the impact and occurrence functions, we obtain the following S-type risk function (see Yang (2006) for the calculations of other types of risks functions).

$$
R(\alpha) = \begin{cases} c^+ & 0 \leq \alpha < F_{Gamma}(r_1^+ - x_0; \lambda, \theta) \\ a^+ \left(F_{Gamma}^{-1}(\alpha; \lambda, \theta) \right)^2 + c^+ & F_{Gamma}(r_1^+ - x_0; \lambda, \theta) \leq \alpha \leq F_{Gamma}(r_2^+ - x_0; \lambda, \theta) \\ M^+ & F_{Gamma}(r_2^+ - x_0; \lambda, \theta) < \alpha \leq 1 \end{cases}
$$

$$\tag{7.52}$$

TABLE 7.21

MtT Type Risk Values for Suppliers with 80% Confidence

Supplier	19	8	11	4	12
MtT risk value	5426.62	6080.00	5176.70	8470.00	8240.00

where α represents the confidence level chosen by the decision maker. For illustration, we choose Gamma (0.15, 2) as the cumulative distribution function to be used in Equation 7.52 and obtain the risk function in Equation 7.53 for supplier S 19.

$$R(\alpha) = \begin{cases} 5000 & 0 \le \alpha < 0.172 \\ 4000(F_{Gamma}^{-1}(\alpha; 0.15, 2))^2 + 5000 & 0.172 \le \alpha \le 0.877 \\ 6960 & 0.877 < \alpha \le 1 \end{cases} \quad (7.53)$$

MtT type risk calculations for the remaining suppliers selected for Phase 2 are done in a similar way. A summary of calculations is provided in Table 7.21. The data in Table 7.21 are calculated with 80% confidence level. For example, losses due to MtT risk from supplier S19 will not exceed $5426.62 with 80% confidence. Values in Table 7.21 are used in Equation 7.24 of the multi-objective model.

7.14.5.2 VaR Type Risk Calculations

Suppliers may face several different disruptive events. In our case, hurricanes are assumed to cause the most significant supply disruptions; hence, buyers initially want to focus on VaR type risk brought by hurricanes. We present next the steps to calculate VaR type risk values for supplier S19.

Step 1: Collect historical data about supplier S19 (see Table 7.22). Note that the method does not require any increasing or decreasing pattern in data.

Step 2: Use Table 7.22 data to fit an appropriate GEVD; that is, find the parameter values to be used in the GEVD function. The values are calculated as $K = 0.15621$, $\delta = 197{,}835$, and $\lambda = 250{,}632$. (Refer to Section 7.9.5 for estimating the parameters of GEVD.)

TABLE 7.22

Loss ($) due to Hurricanes

Year	1997	1998	1999	2000	2001	2002	2003	2004	2005	2006
Loss (1000' $)	50	65	150	210	350	365	400	435	600	755

TABLE 7.23

VaR Type Risk Values for Suppliers with 80% Confidence

Supplier	19	8	11	4	12
VaR risk value	694,308.35	924,786.55	759,941.40	1,231,267.23	956,771.57

Step 3: Estimate the mean parameter of the Poisson distribution to model the occurrence. In this example, we assumed, on the average, only one hurricane impacted the operations at S19 every year.

Step 4: Run a simulation. We generate Poisson random variables to model occurrence. For instance a Poisson random variable of value 1 means that for the simulated year, there will be one hurricane affecting S19. Then we generate as many uniform random variables as the realization of the occurrence random variable for a given year and use these to sample from the GEVD distribution. (Refer to Section 7.9.5 for the simulation approach.)

The same routine was executed for the remaining suppliers selected for Phase 2. VaR type risk values to be used in the multi-objective model with 80% confidence are given in Table 7.23.

Values in Table 7.23 can be interpreted similar to the MtT type risk values. For example yearly losses from hurricanes at S19's region would not exceed $694,308.35 with 80% confidence. The use of small sample size in VaR type risk calculations is justified by the infrequency of these events. PWM estimators and the approximate estimators used in this section were shown to yield bias smaller than other estimators (Holsking et al., 1985).

7.14.6 Phase 2 Model Results

Four GP models discussed in Section 7.14.4 were used to solve the multi-objective supplier selection problem. Ideal solutions, obtained by optimizing each objective separately in the model, given by Equations 7.23 through 7.32, are given in Table 7.24. Optimal solutions obtained by each GP model are discussed next.

7.14.6.1 Preemptive GP Solution

In the preemptive GP model, price is given the highest priority, followed by MtT risk of quality, lead time and VaR risk respectively. This is consistent with the criteria rankings obtained from the company staff (Section 7.14.2).

TABLE 7.24

Ideal Solutions to the Supplier Selection Problem

	Price (Z_1)	MtT (Z_2)	Lead Time (Z_3)	VaR (Z_4)
Ideal solution	508,680	11,907,800	9.56	1,477,343,000

TABLE 7.25

Preemptive GP Solution

Objectives	Preemptive Priorities	Objective Targets	Objective Values Obtained
Price (Z_1)	1	559,548	520,029
MtT (Z_2)	2	13,098,580	13,098,580
Lead-time (Z_3)	3	10.516	10.516
VaR (Z_4)	4	1,625,077,300	1,796,133,000

Target values for each of the objectives are set at 110% of the ideal value. For example, the ideal (minimum) value for price objective is $ 508,680; hence the target value for price is $559,548 and the goal is to minimize the deviation above the target value. Table 7.25 illustrates the optimal solution using the preemptive GP model. Suppliers S8, S11, and S19 are selected and target values for objectives Z_1, Z_2, and Z_3 are achieved in the optimal solution. The optimal solution also provides the optimal order quantities for each product and supplier.

7.14.6.2 Non-Preemptive GP Solution

In the non-preemptive GP model, weights W_1, W_2, W_3, and W_4, obtained from Phase 1, are used. Same target values as in the preemptive GP model are used. Target values are also used as scaling constants since non-preemptive GP requires scaling of objectives. That is, each objective function is divided by its target value during optimization. The optimal solution of the non-preemptive GP model is shown in Table 7.26. The same suppliers S8, S11, and S19 are chosen. However, a lower value for the price criterion is achieved in comparison to the preemptive GP solution, but at the expense of quality and lead time goals.

7.14.6.3 Tchebycheff GP Solution

Same target values are used in the Tchebycheff GP model also. Objectives are scaled as in the non-preemptive GP. Using the Tchebycheff method, the optimal solution obtained is illustrated in Table 7.27. The same three suppliers S8, S11, and S19 are chosen and only the price goal is achieved.

TABLE 7.26

Non-preemptive GP Solution

Objective	Weights	Objective Targets	Objective Values Obtained
Price (Z_1)	0.343	559,548	518,330
MtT (Z_2)	0.338	13,098,580	13,156,000
Lead-time (Z_3)	0.246	10.516	10.68
VaR (Z_4)	0.073	1,625,077,300	1,731,573,000

TABLE 7.27

Tchebycheff GP Solution

Objective	Weights	Objective Targets	Objective Values Obtained
Price (Z_1)	0.343	559,548	520,270
MtT (Z_2)	0.338	13,098,580	13,455,030
Lead-time (Z_3)	0.246	10.516	10.623
VaR (Z_4)	0.073	1,625,077,300	1,754,038,000

TABLE 7.28

Fuzzy GP Solution

Objectives	Weights	Objective Targets	Objective Values Obtained
Price (Z_1)	0.343	508,680	518,763
MtT (Z_2)	0.338	11,907,800	14,011,130
Lead-time (Z_3)	0.246	9.56	10.611
VaR (Z_4)	0.073	1,477,343,000	1,687,676,000

7.14.6.4 Fuzzy GP Solution

In the fuzzy GP model, ideal values (Table 7.24) are used as targets for the four objectives. The optimal solution obtained using fuzzy GP is shown in Table 7.28.

Here, the final set of suppliers is different since targets are set at the ideal values which are not possible to achieve. As a result, none of the targets are achieved and suppliers S4, S11, and S19 are chosen.

7.14.7 Comparison of Phase 2 Results

The four GP models provided different optimal solutions and goal achievements. In order to compare the four solutions and their trade-offs, we use the value path approach (VPA) discussed in Chapter 6, Section 6.4.9. VPA is one of the most efficient ways to demonstrate the trade-offs among conflicting criteria. The display consists of a set of parallel scales, one for each criterion, on which is drawn the value path for each optimal solution.

The value assigned to each optimal solution on a particular axis is the ratio of that solution's value for the appropriate objective to the best value obtained for that objective. Since we are minimizing all the four objectives, the minimum value for each axis is 1 and all the ratios are greater than or equal to 1.

To present these results to the DMs, the following steps are used in VPA. First, the best solution achieved for each objective is found. For the price objective, non-preemptive GP has the best value of 518,330 (see Table 7.26); for lead time, preemptive GP is best with a value of 10.516; for MtT the best value of 13,098,580 is obtained through preemptive GP and for the VaR, the best value of 1,687,676,000 is obtained through fuzzy GP. Then, for all of the remaining methods, divide their respective objective values by the best value for that objective to get their ratios. For example, for the preemptive GP method, the values for price, lead time, MtT and VaR are 520,029, 10.516, 13,098,580, and 1,796,133,000, respectively. The best values for price, lead time, MtT and VaR are 518,330, 10.516, 13,098,580, and 1,687,676,000. Therefore, the ratios corresponding to the preemptive GP solution are obtained as (520,029/518,330), (10.516/10.516), (13,098,580/13,098,580) and (1,796,133,000/1,687,676,000) respectively. Similar ratios are calculated for the other GP solutions under the VPA. Table 7.29 shows the objective values and VPA values for each GP method with their ratios in parenthesis.

The results are then plotted with price, lead time, MtT, and VaR on the horizontal axis and ratios of the objective values on the vertical axis as shown in Figure 7.14.

TABLE 7.29

Summary of GP Solutions

GP Method	Price	Lead Time	MtT	VaR
1. *Preemptive*	520,029	10.516	13,098,580	1,796,133,000
	(1.003)	(1.000)	(1.000)	(1.064)
2. *Non-preemptive*	518,330	10.68	13,156,000	1,731,573,000
	(1.000)	(1.016)	(1.004)	(1.026)
3. *Tchebycheff*	520,270	10.623	13,455,030	1,754,038,000
	(1.004)	(1.010)	(1.069)	(1.039)
4. *Fuzzy*	518,763	10.611	14,011,130	1,687,676,000
	(1.001)	(1.009)	(1.027)	(1.000)

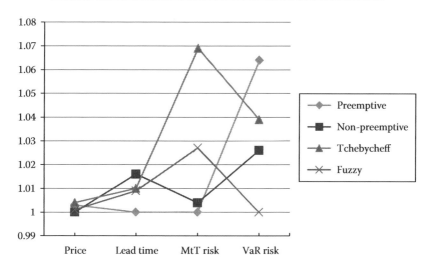

FIGURE 7.14
Comparison of phase 2 results by value path approach.

7.14.8 Discussion of the Results

The VPA is a valuable tool to compare objective trade-offs among suppliers. In some cases, price of the product may dictate the chosen suppliers and in some other cases, the suppliers chosen may be dictated by lead time or quality. The output of the VPA can be used to determine trade-offs among different optimal solutions. For example, it can be seen from Figure 7.14 that preemptive GP solution does 1.6% better than non-preemptive GP on the lead time objective and 0.4% better on MtT but it does 0.3% worse on price and 3.8% worse on VaR objectives. Such comparisons can lead to the choice of an optimal solution that is acceptable to the management. VPA also allows assessing the sensitivity of optimal solutions with respect to each criterion. It can be seen from Figure 7.14 that all the GP methods perform very close with respect to the price objective; hence the optimal supplier selection is very sensitive to the price criterion, which is the most important one. For the VaR objective, on the other hand, GP methods provide a scattered performance pattern, which indicates that the optimal solution is not very sensitive to the least important criterion.

7.15 Summary and Further Readings

Most companies are continuously seeking their supplier base around the world. They are moving more toward outsourcing, single sourcing, long-term contracts and relationships. These developments have provided the companies an opportunity to significantly reduce their supply chain costs. However, overemphasis on costs can make the supply chain brittle and more susceptible to the risk of disruptions. Supply Chain Management is no longer moving products efficiently but also mitigating risks along the way. In this Chapter, we developed models and methods to better manage and mitigate supply chain risks.

7.15.1 Summary

Disruptions are among the most significant threats to supply chain performance. We began by giving examples of several disruptive world events, such as earthquakes, fires, hurricanes, labor strikes and terrorist attacks and their impacts on the supply chain networks. We then discussed methods for risk identification, risk classification, and risk assessment. In risk classification, we grouped risks under different categories depending on the source, impact, internality and manageability of the risk events. Depending on the type of risk, different risk management and mitigation strategies were discussed. Best practices in supply chain risk management in industries were presented as examples.

Despite the seriousness of supply chain risks, the literature is still short of research on risk quantification, especially for the quantification of disruption risks. For the purpose of quantifying supply chain risks, we classified them into two major categories *disruption risks* and *operational risks*. Disruption risks are rare events that can bring catastrophic outcomes. (e.g., floods, earthquakes, terrorist attacks, etc.). Operational risks are more frequent events but with less severe impact on the supply chain. (e.g., missing delivery dates, missing quality standards, etc.). *Value-at-Risk* (VaR) models based on Extreme Value Theory were used to build mathematical models to quantify disruption risks. On the other hand, operational risks were quantified using *Miss-the-Target* (MtT) type risk models based on Taguchi's loss functions. Methods for aggregating VaR and MtT type risk functions were also presented with examples. *Detectability* of risk events was discussed next and a model based on Markov Chain theory was presented for *risk detectability*. A conceptual model of *risk recovery* was presented next. Risk recovery time represents the total time it takes to neutralize a risk event completely after it has been detected.

The chapter concluded with the development of optimization models for supplier selection incorporating VaR and MtT type risks. The risk adjusted supplier selection problem was modeled as a multiple criteria decision making (MCDM) model and solved in two phases. Phase 1 was the pre-qualification step, where a large set of initial suppliers was reduced to a smaller set of manageable suppliers, using MCDM ranking methods. In Phase 2, order quantities were allocated optimally among the shortlisted suppliers, by considering price, lead-time, MtT type risk of quality, and VaR type risk of disruption loss due to hurricanes as four conflicting criteria. The multi-objective model was solved using four different variants of goal programming. The models were illustrated with an actual application to a global IT company for supplier selection and supply risk management.

7.15.1.1 Extensions

The VaR and MtT type risk models presented in this chapter were based on the work of Yang (2006). Bilsel and Ravindran (2012) extended Yang's VaR model by breaking disruption risks down into four components: impact, occurrence, detectability and recovery. They developed an analytical framework to model each component and methods to combine the individual components of risks. Bilsel and Ravindran (2011) have also developed a multi-objective stochastic programming model for supplier selection under uncertainty. They use *chance constrained programming* to formulate a multi-objective *multiple sourcing model* (MSM) to assign one or more suppliers for each product for a single buyer. Suppliers are assigned as either *primary* or *backups*. Primary suppliers are used in order fulfillment, whereas backup suppliers are used as reserve suppliers, in case a primary supplier faces disruptions. MSM is a risk mitigation model and proposes an optimal strategy to follow when a disruption occurs. We shall now review some of the other supply chain risk management models available in the literature.

7.15.2 Literature on Supply Chain Risk Quantification and Management

The emphasis in this section is on publications on risks related to events that can cause disruptions in a supply chain. We review those models and methods that have not been discussed in detail in this chapter. To date, most of the work on supply chain risk considers demand and supply risks rather than survivability of supply chains. Therefore, we also review recent papers addressing risks related to uncertainty in demand and supply.

7.15.2.1 *Mathematical Models for Supply Chain Risk Quantification and Management*

Hopp and Yin (2006) used a nonlinear mixed integer programming formulation to account for supply disruption due to catastrophic failure. They solved the model for minimizing total cost defined as the sum of the inventory cost and protection cost. They also analyzed the effects of several different protection policies to mitigate disruption risks. In short, they considered three policies: holding inventory, using high capacity (which is only valid for service supply chains) and a mixture of both. Snyder and Shen (2006a) developed a simulation model to determine the optimal order frequency, inventory placement and supply chain structure under demand uncertainty and supply disruptions. Tomlin (2006) analyzed the supplier selection problem under supply disruptions through stochastic optimization. He considered two different suppliers, a reliable but more expensive and a less reliable but cheaper and argued that supplier's percentage uptime and the nature of the disruptions are key determinants of supplier selection. Wu et al. (2006) came up with a classification of inbound supply chain risks: internal/external controllable, internal/external partially controllable and internal/external uncontrollable. They used AHP to calculate weights describing the significance of a particular type of risk. These weights are multiplied by probability of occurrence for each type of risk provided by the manager and a simple LP model is obtained. The LP model yields an uncertainty factor that is used in the analysis. Xiao and Yu (2006) analyzed the effect of supply disruption on retailers in a supply chain. Two strategies for retailers, namely profit maximization and revenue maximization, are considered in the paper. They used the evolutionary game theory approach to model the problem and derive the optimal strategy the retailer should follow in case of a disruption. Babich (2005) used the *theory of financial options* to tackle the uncertainty problem.

In his paper, the firm has the option to defer ordering decisions until uncertainty is unfolded and the supplier has the option to defer the pricing decision. Supply chain disruption is introduced in the model as a Bernoulli random variable. Chopra et al. (2005) separated two sources of uncertainty/risk, supply risk and supply disruption risk in their work. They argued that combining these two causes underutilization of reliable supplier and over utilization of unreliable supplier. Based on this separation, they proposed a stochastic extension to the economic order quantity (EOQ) model. Snyder and Daskin (2005) studied a facility location problem. They built a multi-objective mixed integer program considering disruption risks and costs related to disruption (e.g., relocation cost) and solved the model through a Lagrangian relaxation algorithm. Babich et al. (2004) proposed a game theoretic model to study supply disruption and incorporated demand uncertainty in the model. The study assumed that suppliers were subject to correlated random defaults. The effects of correlation in random defaults were analyzed under deterministic and stochastic demand and the paper proposed a method to determine optimal timing for supplier payments and optimal wholesale price under supply disruption risk. Bundschuh et al. (2003) proposed a nonlinear mixed integer programming model to address the issues of reliability and robustness in supply chains. They created robust and reliable supply chain models under several different scenarios.

7.15.2.2 Conceptual Models for Supply Chain Risk Management

Hendricks and Singhal (2005) ran an empirical study using data from 885 publicly traded firms and reported performance changes. They used inventory level and economic growth as performance indicators and statistically showed that independent of the causes, disruptions (referred to as a *glitch*) have negative effects on the performance of the studied firms. Kleindorfer and Saad (2005) introduced a framework to be used in managing disruption risks in supply chains. Their conceptual methodology consists of two steps, "specifying sources of risk and vulnerability" and "risk assessment and mitigation." The proposed framework was based upon 10 principles including risk diversification, preventive actions, information sharing and total quality management practices.

7.15.2.3 Surveys and Case Studies on Supply Chain Risk Management

Snyder and Shen (2006b) transformed their Snyder and Shen (2006a) paper into an executive summary emphasizing most significant parts of their work. Tang (2006) presented an extensive literature survey on risk management in supply chains, including manmade and natural disasters. Blackhurst et al. (2005) conducted an empirical study within a number of different industries to depict the manner in which supply chain disruption is understood and mitigated in practice. Hau (2004) investigated ways to improve security in supply chains. Based on the case study, it was suggested that implementing RFID technologies can be a viable option. Rice and Caniato (2003a,b) presented review reports for various risks associated with supply chains. They put forward supply risk, transportation risk, facilities risk, communication risk and human resources risk as different types of risks and concentrated on supply chain disruptions due to terrorist attacks. They also proposed a framework for risk assessment.

A good tutorial on planning for disruptions in supply chain networks is given in Snyder et al. (2006). Sodhi et al. (2012) present a survey on the perspectives of the researchers working on supply chain risk management. They found the following:

- Varied definitions of supply chain risk management among the researchers
- Inadequate coverage of responses to supply chain disruptions
- Inadequate use of empirical methods

Brief descriptions of all the publications mentioned in this section are given in Table 7.30.

We conclude this section by referring readers to four recent books published in the area of *Supply Chain Risk Management*. The *Supply Chain Risk Management* book edited by Handfield and McCormack (2008), discusses strategies for minimizing disruptions in global sourcing. The use of *risk mapping* for prioritizing risk, discussed in Section 7.5 is based on the work of Elkins et al. (2008) and is discussed in detail in this book. It also contains two case studies from the automotive and healthcare industries for managing supplier risks. The book by Olson and Wu (2008) is on Enterprise Risk Management. In addition to supply chain risks, it discusses risk management in accounting and finance. It addresses tools to assess risk such as balanced score card, multiple criteria analysis and financial risk measures. It also contains four case studies on risk managementon. The *Supply Chain Risk Management* book by Walters (2007) has a good discussion of how to identify, analyze, and respond to supply chain risks. The principles of designing a resilient supply chain are also included. Finally, the book by Haimes (2004) is a general reference book on risk modeling, assessment and management. It has a good mix of both qualitative and quantitative methods of risk management, covering such topics as, decision-tree analysis, multi-objective trade-off analysis, risk filtering, fault trees, Bayesian analysis, and project risk management. It also contains several case studies in risk management applied to engineering, business and public policy.

Dun and Bradstreet (www.dnb.com) have a "Supplier Risk Manager" product that is offered to help their customers "effectively manage suppliers and mitigate supply risks."

TABLE 7.30

Survey of Supply Chain Risk Management Literature

Authors	Method	Addressed Risk(s)/ Uncertainties	Brief Description
Tang (2006)	None	Uncertain demand, supply yields/ capacity, lead times and supply cost	Literature review
Wu et al. (2006)	AHP	Multiple types of inbound risks, grouped under six categories	Provides a 6-level (high level: internal/external; low level; controllable/partially controllable/uncontrollable) risk classification and identifies up to 25 sources of risk. Ranks them using AHP
Hopp and Yin (2006)	Non linear mixed integer programming	Supply disruption, catastrophic failure	Analyzed effects of protection policies to mitigate disruption risks. Considers three cases: holding inventory, high capacity (only for service supply chains) and a combination of the two methods; solves for minimizing total cost (expressed as the sum of inventory cost and protection cost)
Xiao and Yu (2006)	Evolutionary game theory	Demand uncertainty, supply disruption	Analyzed effects of demand uncertainty and supply disruption on retailer strategies in a duopoly setting. Proposes a recovery model in case of demand disruptions
Snyder and Shen (2006a)	Simulation	Demand uncertainty, supply disruptions	Determines the optimal order frequency, inventory placement and supply chain structure under demand uncertainty and supply disruptions
Snyder and Shen (2006b)	None	Demand uncertainty, supply disruptions	Presents executive summary of the Snyder and Shen (2006a) working paper.
Tomlin (2006)	Stochastic optimization	Supply disruption	Analyzes supplier selection under supply disruption. Assumes that there are two suppliers, a reliable but more expensive and a less reliable but cheaper. Argues that supplier's percentage uptime and the nature of the disruptions are key determinants of supplier selection

(continued)

TABLE 7.30 (continued)

Survey of Supply Chain Risk Management Literature

Authors	Method	Addressed Risk(s)/ Uncertainties	Brief Description
Hendricks and Singhal (2005)	Empirical study, hypothesis testing	Supply disruption (referred as *supply chain glitch* in the paper)	Conducts empirical study on over 885 publicly traded firms, reports changes in performance metrics as inventory level and economic growth in cases of supply chain glitches. Argues that no matter what caused the glitch there were negative effects on the performance of the studied firms
Chopra et al. (2005)	Stochastic extension of EOQ	Supply risk, supply disruption risk	Argues that combining supply risk and supply disruption risk causes underutilization of reliable supplier and overutilization of unreliable supplier. Proposes separation of these two
Kleindorfer and Saad (2005)	Conceptual model proposition, probability assessment	Supply disruption risk	Analyzes sources of disruption risk, proposes a model for disruption risk analysis based on probability assessment
Babich (2005)	Financial—real options, continuous time stochastic processes	Supply risk, supply disruption, lead time uncertainty	Assumes firm has an option of deferring ordering decisions due to uncertainty in lead times. Also suppliers have the option of deferring pricing decisions. Analyzes a single-period two-echelon multi-stage supply chain with competing suppliers
Blackhurst et al. (2005)	Real life applications report/ literature review		Proposes directions for research on supply chain disruption risk based on interactions with industry. Focuses on how to discover potential sources of disruption, how to mitigate the risks and how to redesign the supply chain
Snyder and Daskin (2005)	Multi-objective mixed integer programming, Lagrangian relaxation	Facility disruption risk	Considers facility location problems with disruption risks and costs related to disruption (e.g., relocation cost)

TABLE 7.30 (continued)

Survey of Supply Chain Risk Management Literature

Authors	Method	Addressed Risk(s)/ Uncertainties	Brief Description
Babich et al. (2004)	Game theory (Stackelberg game)	Supply disruption, demand uncertainty	Suppliers are subject to correlated random defaults. The paper analyzes the effect of correlation under deterministic and stochastic demand, determines optimal timing for supplier payments and optimal wholesale price under supply disruption risk
Hau (2004)	Case study	Supply chain security	Discusses implementation of RFID to enhance supply chain security in international shipping
Bundschuh et al. (2003)	Nonlinear mixed integer programming	Supplier risk, supply disruption, reliability and robustness in supply chains	Addresses the issues of reliability and robustness in supply chains. Builds robust and reliable supply chain models under several different scenarios
Rice and Caniato, (2003a)	Report	Supply risk, transportation risk, facilities risk, communication risk, human resources risk	Reviews and discusses risks associated with supply chains and concentrates on supply chain disruptions due to terrorist attacks. Proposes a framework for risk assessment
Rice and Caniato, (2003b)	Review	Supply risk, transportation risk, facilities risk, communication risk, human resources risk,	Proposes actions to mitigate risk factors and discusses their advantages/disadvantages. Discusses and groups actions to take for risk mitigation at four different levels

Exercises

7.1 What is *Risk Mapping*? How is it used in risk management in supply chains?

7.2 What is the meaning of a *risk priority number*? How is it used in supply chain risk management?

7.3 Discuss the key differences among the following risk management strategies.

- Risk mitigation
- Contingency planning
- Business insurance

Which is more appropriate under what conditions?

7.4 What is the meaning of *product postponement*? Discuss its effectiveness as a risk mitigation strategy.

7.5 Explain the portfolio strategy in risk management. Where and how is it used for risk mitigation? Name one company that uses this strategy effectively and explain how it is done.

7.6 Give examples of both the traditional and flexible risk mitigation strategies used by companies for controlling risk.

7.7 Discuss the risk management strategies practiced by the following companies

- HP
- Federal Express
- Wal-Mart
- Johnson & Johnson

7.8 Explain the major differences between value-at-risk and miss-the-target risk. Give examples of each.

7.9 What are the pros and cons of using extreme value distributions to model supply chain risks?

7.10 Discuss the differences among the following terms in risk management.

- Risk occurrence
- Risk impact
- Risk detection
- Risk recovery

7.11 How do the following goal programming (GP) methods differ in solving multiple criteria optimization problems?

   i. Preemptive GP

   ii. Non-preemptive GP

   iii. Min–max (Tchebycheff's GP)

   iv. Fuzzy GP

7.12 What is Value Path Analysis? How is it used in multiple criteria decision making?

7.13 Consider the disruption risk due to floods in China discussed in Example 7.1. Using the VaR disruption risk function developed in Table 7.9, compute the following risk measures:

(a) Probability that the annual loss due to floods will not exceed $3 million.

(b) Maximum loss in a year due to floods at 80% confidence level.

(c) Probability that the annual loss due to floods exceed $4 million.

(d) Probability that the annual loss is between $2 and $3 million.

7.14 Consider the delivery time problem faced by the computer manufacturer from its supplier in Example 7.2. This was modeled as an MtT-N type risk. Using the disruption risk function given by Equation 7.14, compute the following risk measures for missing the delivery time window:

(a) Cost of missing delivery window at 80% confidence level.

(b) Probability that the delivery time cost will not exceed $12,000

(c) Probability that the delivery time cost exceeds $18,000

7.15 Consider the problem of ranking suppliers discussed in Section 7.14. Using the supplier attribute weights given in Table 7.16 and their attribute values given in Table 7.17, determine the complete ranking of the 20 suppliers using the following methods:

(a) Weights obtained by rating.

(b) Weights obtained by Borda Count.

(c) Weights obtained by AHP.

(d) Do you find any *rank reversals*? Explain.

Notes:

(i) You can use a spreadsheet to answer this.

(ii) For a discussion of the different ranking methods, refer to Appendix A or Chapter 6, Section 6.3 of this text.

References

Atkins, E. 2003. Top 10 tips to secure your supply chain. *Materials Management and Distribution*. 13(1): 56–60.

Atkinson, W. 2003. Supply chain risk management: Riding out global challengers. *Purchasing*. 132(14): 43–46.

Babich, V. 2005. Vulnerable options in supply chains: Effects of supplier competition. *Naval Research Logistics*. 53(7): 656–673.

Babich, V., A. N. Burnetas, and P. H. Ritchken. 2004. Competition and diversification effects in supply chains with supplier default risk. Working paper. Ann Arbor, MI: Department of Industrial Operations and Engineering, University of Michigan.

Barndorff-Nielsen, O. E. 1978. Hyperbolic distributions and distributions on hyperbolae. *Scandinavian Journal of Statistics*. 5: 151–157.

Billington, C. 2002. HP cuts risk with portfolio approach. *Purchasing*. 131(3): 43–45.

Bilsel, R. U. 2009. Disruption and operational risk quantification models for outsourcing operations. PhD dissertation. University Park, PA: Pennsylvania State University.

Bilsel, R. U. and A. Ravindran. 2012. Modeling disruption risk in supply chain risk management. *International Journal of Operations Research and Information Systems.* 3(3): 15–39.

Bilsel, R. U. and A. Ravindran. 2011. A multiobjective chance constrained programming model for supplier selection under uncertainty. *Transportation Research Part B.* 45(8): 1284–1300.

Blackhurst, J., C. W. Craighead, D. Elkins, and R. B. Handfield. 2005. An empirically derived agenda of critical research issues for managing supply chain disruptions. *International Journal of Production Research.* 43(19): 4067–4081.

Bogachev, M., J. Eichner, and A. Bunde. 2008. On the occurrence of extreme events in long-term correlated and multifractal data sets. *Pure and Applied Geophysics.* 165(6): 1195–1207.

Bradford, M. 2003. Keeping risks from breaking organizations' supply chains. *Business Insurance.* 37(31): 9–10.

Braithwaite, A. 2003. The supply chain risk of global sourcing. http://www.lcp-consulting.com/thought-leadership/supply-chain.

Breyfogle, F. W. 2003. *Implementing Six Sigma,* 2nd edn. Hoboken, NJ: John Wiley & Sons.

Bundschuh, M., D. Klabjan, and D. L. Thurston. 2003. Modeling robust and reliable supply chains. Working paper. Urbana and Champaign, IL: University of Illinois at Urbana–Champaign.

Castillo, E., A. Hadi, N. Balakrishnan, and J. Sarabia. 2005. *Extreme Value and Related Models with Applications in Engineering and Science.* Hoboken, NJ: Wiley.

Chandra, M., N. Singpurwalla, and M. A. Stephens. 1981. Kolmogorov statistics for test of fit for the extreme value and Weibull distributions. *Journal of American Statistical Association.* 76: 729–731.

Chopra, S., G. Reinhardt, and U. Mohan. 2005. The importance of decoupling recurrent and disruption risks in a supply chain. Working paper. Evanston, IL: Northwestern University.

Chopra, S. and M. S. Sodhi. 2004. Managing risk to avoid supply chain breakdown. *MIT Sloan Management Review.* 46(1): 53–61.

Christopher, M. and H. Peck. 2004. Building the resilient supply chain. *International Journal of Logistics Management.* 15(2): 1–14.

Coles, S. and M. Dixon. 1998. Likelihood-based inference for extreme value models. *Extremes.* 2(1): 5–23.

Elkins, D., R. B. Handfield, J. Blackhurst, and C. W. Craighead. 2008b. A to-do list to improve supply-chain risk management capabilities. In *Supply Chain Risk Management,* ed. R. B. Handfield and K. B. McCormack, Chapter 4. Boca Raton, FL: Auerback Publishers.

Elkins, D., D. Kulkarni, and J. Tew. 2008a. Identifying and assessing supply chain risk. In *Supply Chain Risk Management,* ed. R. B. Handfield and K. B. McCormack, Chapter 3. Boca Raton, FL: Auerback Publishers.

Fawcett, S., P. Osterhaus, G. Magnan, J. Brau, and M. McCarter. 2007. Information sharing and supply chain performance: The role of connectivity and willingness. *Supply Chain Management: An International Journal.* 12(5): 358–368.

Fitzgerald, G. M. 1996. Weak links. *Ward's Auto World.* 32(9): 109.

Florian-Kratz, E. 2005. For FedEx, it was time to deliver. *Fortune.* 2005(October 3): 83–84.

FM Global. 2010. Managing risk in the next decade. http://www.fmglobal.com/assets/pdf/P07001_0310c.pdf

FM Global. 2011. Risk: Flirting with disaster. http://www.fmglobal.com/assets/pdf/ P07001_0311e.pdf

Haimes, Y. 2004. *Risk Modeling, Assessment and Management*, 2nd edn. Hoboken, NJ: John Wiley & Sons.

Hau, L. L. 2004. Supply chain security: Are you ready? *Stanford Global Supply Chain Management Forum.* September 3, 2004: 1–16.

Hendricks, K. B. and V. R. Singhal. 2003. The effect of supply chain glitches on shareholder wealth. *Journal of Operations Management.* 21: 501–552.

Hendricks, K. B. and V. R. Singhal. 2005. Association between supply chain glitches and operating performance. *Management Science.* 51(5): 695–711.

Hopp, W. J. and Z. Yin. 2006. Protecting supply chain networks against catastrophic failures. Working paper. Evanston, IL: Northwestern University.

Hosking, J., J. Wallis, and E. Wood. 1985. Estimation of generalized extreme value distribution by the method of probability weighted moments. *Technometrics.* 27(3): 251–261.

Johnson, M. E. 2001. Learning from toys: Lessons in managing supply chain risks from the toy industry. *California Management Review.* 43(3): 106–124.

Kulkarni, V. 1995. *Modeling and Analysis of Stochastic Systems.* New York: Chapman & Hall.

Kemeny, J. and J. Snell. 1976. *Finite Markov Chains.* Princeton, NJ: Springer-Verlag.

Kleindorfer, P. R. and G. H. Saad. 2005. Managing disruption risks in supply chains. *Production and Operations Management.* 14(1): 53–68.

Leonard, D. 2005. The only lifeline was the Wal-Mart. *Fortune.* October 3, 2005: 74–80.

Mahoney, C. D. 2004. Making your supply chain risk-resistant. *Inside Supply Management.* 15(10): 6.

Martyn, P. 2011. Supply chain lessons from Japan. *Forbes.* http://blogs.forbes.com/ ciocentral/2011/04/06/supply-chain-lessons-from-japan

Mohammed, A., A. Ravindran, and L. Leemis. 2002. An interactive multicriteria availability allocation algorithm. *International Journal of Operations and Quantitative Management.* 8(1): 1–19.

Monahan, S., P. Laudicina, and D. Atlis. 2003. Supply chains in a vulnerable, volatile world. *Executive Agenda.* Vol. 6, No. 3. Chicago, IL: A.T. Kearney.

Muthukrishnan, R. and J. A. Shulman. 2006. Understanding supply chain risk: A McKinsey global survey. *The McKinsey Quarterly.* 2006: 1–9.

Olsen, R., J. Lambert, and Y. Haimes. 1998. Risk of extreme events under nonstationary conditions. *Risk Analysis.* 18(4): 497–510.

Olson, D. L. and D. D. Wu. 2008. *Enterprise Risk Management.* Singapore: World scientific publishing Co.

Powell, B. 2011. The global supply chain: So very fragile. *CNN Money.* Accessed December 12, 2011, http://tech.fortune.cnn.com

Ravindran, A., R. U. Bilsel, V. Wadhwa, and T. Yang. 2010. Risk adjusted multi-criteria supplier selection models with applications. *International Journal of Productions Research.* 48(2): 405–424.

Reed, J. 2010. Carmaking: A drive to Lego land. *Financial Times.* November 30, 2010: 1–6.

Rice, J. B. and F. Caniato. 2003a. Supply chain response to terrorism: Creating resilient and secure supply chains. Report by MIT Center for Transportation and Logistics, Cambridge, MA: MIT.

Rice, J. B. and F. Caniato. 2003b. Building a secure and resilient supply network. *Supply Chain Management Review.* 2003(September/October): 22–30.

Snyder, L. V. and M. S. Daskin. 2005. Reliability models for facility location: The expected failure cost case. *Transportation Science.* 39(3): 400–416.

Snyder, L. V., M. P. Scaparra, M. S. Daskin, and R. L. Church. 2006. Planning for disruptions in supply chain networks. In *Tutorials in Operations Research, INFORMS Practice Conference 2006*, Baltimore, MD, eds. M. P. Johnson, B. Norman, and N. Secomandi, pp. 234–257. Hanover, MD: Institute for Operations Research and the Management Sciences.

Snyder, L. V., Z.-J. M. Shen. 2006a. Supply and demand uncertainty in multi-echelon supply chains. Working paper. Bethlehem, PA: Lehigh University.

Snyder, L. V., Z.-J. M. Shen. 2006b. Managing disruptions to supply chains. Working paper. Bethlehem, PA: *Lehigh University*.

Sodhi, M. S., B. Son, and C. S. Tang. 2012. Researchers perspectives on supply chain risk management. *Production and Operations Management*. 21(1): 1–13.

Taguchi, G. 1986. *Introduction to Quality Engineering*. Tokyo, Japan: Asian Productivity Organization.

Tang, C. 2006. Perspectives in supply chain risk management. *International Journal of Production Economics*. 103(2): 451–488.

Tomlin, B. 2006. On the value of mitigation and contingency strategies for managing supply chain disruption risks. *Management Science*. 52(5): 639–657.

Veysey, S. 2011. Majority of companies suffered supply chain disruptions in 2011: Survey. *Business Insurance*. Accessed November 2, 2011, www.businessinsurance.com

Wadhwa, V. and A. Ravindran. 2007. Vendor selection in outsourcing. *Computers & Operations Research*. 34(12): 3725–3737.

Walters, D. 2007. *Supply Chain Risk Management*. London, UK: Kogan Page Limited.

Wang, C. 2005. On the numerics of estimating generalized hyperbolic distributions. Masters thesis. Berlin, Germany: Universität Zu Berlin.

Wu, T., J. Blackhurst, and V. Chidambaram. 2006. A model for inbound supply risk analysis. *Computers in Industry*. 57(4): 350–365.

Xiao, T. and G. Yu. 2006. Supply chain disruption management and evolutionarily stable strategies of retailers in the quantity-setting duopoly situation with homogeneous goods. *European Journal of Operational Research*. 173(2): 648–668.

Yang, T. 2006. Multi-objective optimization models for managing supply risks in supply chains. PhD dissertation. University Park, PA: Pennsylvania State University.

Yang, T. and A. Ravindran. 2007. Supply risk quantification in supply chain. Working paper. University Park, PA: Department of Industrial Engineering, Pennsylvania State University.

Yuan, Q. and Z. Qiong. 2008. Research on information risk sharing in supply chain management. In *Proceedings of the International Conference on Wireless Communications and Mobile Computing*, IWCMC 2011, Istanbul, Turkey, pp. 1–6.

Zhou, H. and W. Benton. 2007. Supply chain practice and information sharing. *Journal of Operations Management*. 25(6): 1348–1365.

8

Global Supply Chain Management

In the 1600s, Galileo was excommunicated for stating that earth was not flat and was not the center of the universe. In the twenty-first century, Friedman (2005) sold millions of copies of his book *The World is Flat*, by arguing that the "flattening" of the world is due to the growth in global trade and the Internet economy.* In the 1960s factories in the United States made 95% of the clothes and 98% of the shoes sold in the United States. Today only 5% of the clothes and 10% of the shoes are made in the United States. The emerging theme is the rapid growth in world trade beginning with the last quarter of the twentieth century. Even though the world gross domestic product (GDP) increase has been modest, the increase in world trade in manufacturing has been exponential during the last 50 years.

In this chapter, we will discuss globalization and its impacts on supply chain management. We will begin with the history of globalization and illustrate the trends in globalization with examples. We will discuss the key issues in managing global supply chains. Trends in outsourcing due to globalization will also be discussed. Next, criteria for international supplier selection will be presented, along with issues in international logistics. Finally, multi-objective optimization models for designing global supply chains will be discussed and illustrated with an actual case study.

8.1 History of Globalization

For centuries, international trade was the pursuit of many kings and emperors. The pace increased dramatically in the twenty-first century due to advances in air travel, telecommunication, digital revolution, and the Internet. Sourcing of raw materials and parts is done globally; products are manufactured and assembled in many countries and sold to customers around the world.

According to the U.N. Economic and Social Council, globalization in an economic context, refers to the "reduction and removal of barriers between national borders in order to facilitate the flow of goods, capital, services and labor." Globalization was facilitated by free trade agreements between selected nations through the *Global Agreement on Trade and Tariff (GATT)*

---

* We owe this remark, with slight alterations, to our friend, Partha V. Iyer, of Reading, Pennsylvania.

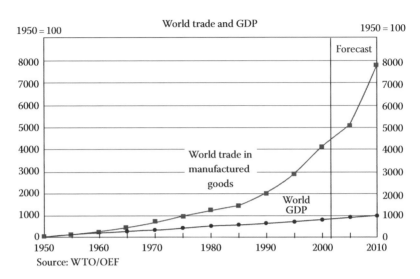

FIGURE 8.1
Globalization trend.

in 1948. At that time, there were 23 countries in GATT. The *World Trade Organization (WTO)* replaced GATT in 1995 to further lower the barriers to global trade. Now, there are 153 countries, who are members of WTO. According to the Global Policy Forum, world exports rose from 8.5% in 1970 to 16.2% of total gross world product in 2001. Figure 8.1 illustrates the growth in World GDP and the world trade in manufactured goods during 1950–2000. The exponential growth in world trade is due to the formation of WTO in 1995 and China's entry into WTO in 2001.

After an explosive and sustained growth during the twenty-first century, global trade suffered for the first time a 12% decline in 2009, due to the global recession.

8.2 Impacts of Globalization

8.2.1 Changes to World Economies

According to the World Bank, the gross national incomes of the world's leading economies in 2008 are as shown in Figure 8.2.

The United States leads the world economies with over 14 trillion dollars. In fact, during the third quarter of 2010, China surpassed Japan as the second largest economy in the world. According to FM Global (2010), globalization is leading the way to a significant shift in the economies of *developed* and *developing nations*. The economies of Brazil, Russia, India, and China, known as the *BRIC countries*, will become larger than the economies of the

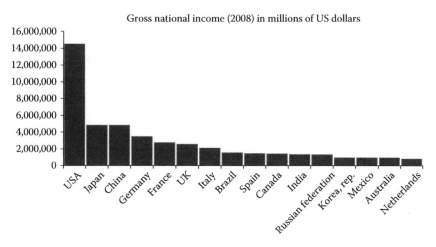

FIGURE 8.2
World economies (2008). (From World Bank, 2008 World Development Indicators.)

TABLE 8.1

Growth Rates around World (2005–2020)

	World	United States	Argentina	Brazil	India	China
Annual growth rate (%)	3.1	3.2	3.6	3.6	5.5	6.6

Source: The World Bank.

industrialized G7 countries (United States, Canada, Japan, Germany, France, United Kingdom, and Italy) during the decade of the 2030s. The number of companies, in the Global Fortune 500, headquartered in the emerging markets rose from 20 in 1995 to 91 in 2010.

According to the World Bank forecasts, China's economy will grow 6.6% annually, while the global GDP is expected to grow at 3.1% from 2005 to 2020. Table 8.1 gives the growth rates for selected countries around the world.

China's economy actually grew at the average annual rate of 10% from 2003 to 2008. A recent report from Goldman Sachs predicted that China may become the world's largest economy by 2020.

8.2.2 Global Products

Globalization has resulted in the collaboration of the best minds and talents in the world to produce innovative global products. For example, Apple iPod was conceived and designed in the United States, the engineering work was done in India and all the components were manufactured in China.

Another example of globalization is the production of an automobile. According to Balasubramanian and Tewary (2005), nine countries were involved in some aspect of production, marketing and selling (WTO, 1998) of an "American car." The value added by different countries was as follows:

- 30% in Korea for assembly
- 17.5% in Japan for advanced technology components
- 7.5% in Germany for design
- 4% in Taiwan and Singapore for minor parts
- 2.5% in the United Kingdom for advertising and marketing
- 1.5% in Ireland and Barbados for data processing

Thus only 37% of the car's value was generated in America!

Mangan et al. (2008) cite another example of globalization. The all-American Barbie doll's hair comes from Japan, the doll's plastic body from Taiwan, and her cotton clothing comes from China. Only the molds and pigments are made in the United States!

A final example is the production of a very popular smart phone. According to the Asian Development Bank, five countries were involved in manufacturing 14 key parts and the final assembly. Germany made six parts (e.g., camera modules, GPS receiver), Korea made the application processor, United States made three parts (e.g., memory chip, Bluetooth) and Japan made four parts (e.g., display module, flash memory, touch screen). The final assembly of the smart phone was done in China!

According to Friedman (2005), the Internet revolution, WTO and the fall of the Berlin wall contributed to the globalization of the market place. He claims that anyone with high speed internet connection can compete for nearly any job on the global market!

During the last 10 years, offshore manufacturing has increased significantly, particularly in countries like China, Mexico and India, where the cost of labor is very low. China's share of manufactured goods in the world market increased significantly after it joined WTO in 2001. According to the Asian Development Bank, China's share of the world market in 2004 was as follows:

- 50% of all cameras
- 25% of all air-conditioners and TVs
- 25% of all washing machines

China's share of the U.S. market in 2004 was as follows:

- 81% of toys
- 72% of shoes
- 54% of consumer electronics
- 9% of apparels

China's share of apparel in the United States and Europe has increased dramatically now, after the expiration of the quotas on textiles in 2005, under the WTO agreement.

In a study done by Accenture and Northeastern University (Mulani, 2005), it was found that *Business Process Offshoring* (BPO) doubled from 40% to 80% of companies surveyed during 1994–2004. As a percent of the budget, BPO increased from less than 20% in 1999 to 40% in 2004, nearly 15% annual growth. The same study also found that transportation and related services were the most supply chain management functions to be outsourced. Another growth area was the outsourcing of procurement function.

In the beginning of globalization, mostly manufacturing services were outsourced. Then came the outsourcing of call centers to India. Now, many multi-national companies are moving their R&D functions to the developing countries. Companies have realized that "knowledge work" is easier to outsource compared to manufacturing. There is no transportation cost or long lead times and no "customs" to clear at the international borders! This had a negative impact on the IT salaries in the United States and the developed countries. Multinational IT companies (e.g., IBM, Accenture, and SAP) found that they were not able to compete globally with software providers in India (e.g., Tata, Infosys, and Wipro), which employed highly skilled Indian IT personnel at lower wages. To overcome the cost differential, multinational IT companies have now established offices in India and hire the same Indian IT personnel to compete for global projects. This has resulted in a big increase in IT salaries in India.

A McKinsey article in 2008 raised the question whether it is time to rethink offshore manufacturing in Asia (Goel et al., 2008). The authors argued that big increases in oil prices and wages, made offshore manufacturing not as attractive as it used to be. For example, the average annual wage inflation from 2003 to 2008 was 3% in the United States, whereas in Brazil it was 21%, China 19%, Malaysia 8%, and Mexico 5%.

8.2.3 Impact of Globalization in U.S. Manufacturing

The top three countries with the highest percent of world labor force are China (21%), India (17%), and the United States (5%), a distant third! Table 8.2

TABLE 8.2

Employment by Sector

Sector	China (%)	India (%)	United States (%)
Agriculture	50	60	3
Manufacturing	15	17	27
Service	35	23	70

gives the percent of labor employed in agriculture, manufacturing, and services sectors in the three countries in 2005.

Consider the following facts about U.S. Manufacturing (*Source:* U.S. Bureau of Labor Statistics):

- In 1980, 40% of US GDP was in manufacturing and 50% in service. In 2008, only 10% was in manufacturing, while GDP in service increased to 70%.

- During World War II, manufacturing accounted for 4 out of 10 jobs in the United States; now it is reduced to 1 out of 10.

- Employment in the manufacturing sector went from 17.6 million in 1998 to 13.4 million in 2008. By 2018, it is projected to decrease to 12.2 million!

- The United States produced 90% of what it consumed 30 years ago; in 2008, it was 65%.

- In the 1960s, America *made* 98% of its shoes; now, it *imports* 90% of its shoes.

However, based on this data, it is *incorrect* to conclude that U.S. manufacturing is dead or dying. In fact, U.S. manufacturing has moved *upscale*. American companies have moved to manufacturing "high-end" products, as low value goods have moved offshore. Until 2011, the United States was the leading manufacturer in the world based on the *value of goods* produced. For example, in 2007 it produced $1.6 trillion doubling the $811 billion produced in 1987, with an annual growth rate of more than 7%. The United States generated $2.50 in value of goods, for every $1 produced in China! Examples of the largest manufacturers by revenue are Boeing (aircraft), General Electric (gas turbines), Intel (computer chips), Lockheed Martin (military planes), John Deere (farm equipment) and General Motors (autos).

8.2.4 Risks in Globalization

Accenture conducted a survey of U.S. companies that operated globally to identify the top risk factors in global business operations (Mulani, 2005). The companies included IT and manufacturing industries (small and big). The top seven risk factors from the study were as follows:

1. Time zone (27%)
2. Language barrier (25%)
3. Cultural differences (23%)
4. Unknown suppliers (9%)

5. Distance (8%)

6. Unknown legal rights (6%)

7. Geopolitical instability (2%)

It was interesting to note that time zone, language barrier, and cultural differences were cited as the top three risk factors, while "distance" was a distant fifth! It should be noted that some multinational companies use time zone difference to their advantage.

In the global economy, most manufacturers are continuously seeking their supplier base around the world and looking for opportunities to significantly reduce supply chain costs. Singular emphasis on supply chain cost, however, can make the supply chain brittle and more susceptible to the risk of disruptions. In Section 8.4, we will discuss the risks and hidden costs in global sourcing.

8.3 Managing Global Supply Chains

8.3.1 Global Risk Factors

Globalization provides companies with opportunities to enter new markets, find better suppliers and take advantage of cheaper labor available in other countries. Inherent in global supply chains is the risk due to disruptions. This was clearly demonstrated by the earthquake, tsunami and the nuclear disaster in Japan in March 2011. Since Japan was a key supplier of electronic components, that go into the production of automobiles, computers, aircrafts and cell phones, companies such as Toyota, Honda, Boeing, GM, Sony, and Apple had to slow down or shut down their factories due to shortage of parts.

In a McKinsey survey conducted in 2008, more than three-fourth of companies responded that their supply chain risks have increased significantly (33%) or increased slightly (44%) during the past 5 years (Paulonis and Norton, 2008). A 2007 Accenture survey categorized the sources of risk that impact the performance of the global supply chains (Ferre et al., 2007). The top five risk factors were natural disasters, performance of supply chain partners, changing fuel prices, currency fluctuations and logistics. The 2008 McKinsey survey also found that the three big challenges in managing global supply chains are the following:

1. Required resources

2. Recruitment and retention of local talent

3. Integration of IT systems

8.3.2 Global Supply Chain Strategies

According to Simchi-Levi et al. (2003), global supply chains can have many forms and shapes and they fall into one of the following types:

- *Global distribution*: Manufacturing is done mainly in one country and the products are distributed worldwide.
- *Global suppliers*: International vendors supply raw materials and parts, but the final assembly and manufacturing are done domestically.
- *Offshore manufacturing*: The product is manufactured in a foreign location, imported back and distributed locally to customers.
- *Integrated global supply chain*: Here raw materials are sourced from some countries, sub-assemblies and final assemblies are done in other countries and final products are distributed using distribution centers (DCs) in several countries.

Depending on the type of global supply chain, the issues and complexities in managing it will vary widely. Cohen and Mallik (1997) published one of the early papers to address the differences between domestic and global supply chains. They cite numerous issues that global supply chains should consider, such as duties and tariffs, currency exchange rate fluctuations, international taxes, longer lead times, lower costs, potential for economies of scale in manufacturing and sourcing, new markets, local content rules, counter trade and quotas, transfer prices and cultural and language differences.

The key flows in any supply chain are the following:

1. *Products*: Include raw materials, work-in-progress, sub-assemblies, finished goods, etc.
2. *Funds*: Include invoices, payments, credits, consignments, etc.
3. *Information*: Include orders, deliveries, marketing promotions, plant capacities, inventory, etc.

Because products and funds flow across international boundaries, adopting a global supply chain strategy becomes very complex. According to Cohen and Huchzermeier (1999), a successful strategy has to reduce the cost, risks, and pay-off to the company. It has to involve both operational and financial decisions in the long term. Mangan et al. (2008) state that a successful corporate mantra should be to *"Think globally, but act locally."* In other words, product development decisions should be based on global market conditions, as if the world is one large market. At the same time, some products should be region-specific, to meet the local conditions and needs.

Fawcett et al. (2007) state that companies often fail to recognize the differences between domestic and global operations in designing their supply chains. They recommend that companies consider the following four issues for managing global supply chains:

1. *Political conditions*: When companies include foreign locations for their facilities, political stability should be an important consideration. A sudden change in government may alter the policies and agreements of the previous government and could result in the sudden increase in cost and potential disruptions to the global supply chain network. A recent example is what happened in Venezuela after Hugo Chavez's election as president.

2. *Legal conditions*: Rules and regulations in what is legal for business activities could vary widely from country to country. For example, laws regulating product liability, sustainability, bribery, labor conditions, local area content in manufacturing product and advertising are country dependent, making decisions on global supply chain strategies very complex.

3. *Finance*: Fluctuating currency exchange rate is one of the most critical issues facing a global supply chain. *Hedging* and *real options* are considered good strategies to manage currency risk. Taxation policy of each country is another financial issue that should be included in the global strategy.

4. *Cultural conditions*: Culture affects people's behavior. Hence, companies need to pay special attention to the people's culture, habits and other traits prevailing in the foreign countries they plan to do business in. Without an established personal relationship, a business relationship may not succeed. This is particularly true in many developing countries in Asia.

8.3.3 Examples of Globalization Strategies

Cohen and Huchzermeier (1999) point out that adopting a global supply chain strategy will be costly and risky, but has enormous potential benefits. They cite Ford Motor Company's "World car" design as an example of a successful globalization strategy that restructured car manufacturing from a regional to a global organization. Such a strategy resulted in $3 billion savings, reduction in order-to-delivery times to less than 15 days, and new car development time from 37 to 24 months. A similar strategy was used by Whirlpool Corporation for its "World Washers."

Cohen and Huchzermeier (1999) also cite a different global supply chain strategy called "Production Switching," adopted by Toyota Motor Company in Southeast Asia. The strategy leverages its assembly plants in different countries to make those products that are currently in high demand. For example, when automobile demand drops in one country, the Toyota plant in that country will produce parts that can be exported to other countries, where the demands are high. Such a strategy requires training workers for multiple jobs.

Mangan et al. (2008) cite McDonald's as an example of the "Think Global, but act local" strategy. In addition to selling its world products (burgers, French fries, coca-cola, etc.) in all its world-wide outlets, McDonald's also sells locally desired products. McDonald's India has established close cooperation with hundreds of Indian farmers to feed its "French fries" supply chain; because cows are sacred in India, "beef patties" have been replaced by "lamb patties!"

Global companies have also used their presence in one country to expand into other markets in that geographic region. For example, Kimberly-Clark and Black & Decker have successfully used their presence in central America to expand into other South American markets. Fawcett et al. (2007) point out that competing directly in the competitor's home market is another globalization strategy that is used to slow down the competitor from moving into its home market. Wal-Mart used this strategy in Germany and England against Carrefour. This strategy also provides competitive intelligence with regard to advanced technology and suppliers. GM and Ford have also adopted this strategy by selling cars in Japan to counter the rapid growth of Japanese cars in the United States.

Finally, another important issue in the globalization strategy is *centralization*, namely in what areas supply chain decisions should be made centrally and in what should be delegated to local management. A 2008 McKinsey survey (Paulonis and Norton, 2008) found that most companies prefer centralization, compared to local management for their supply chains and this trend had increased in recent years. For example, majority of the companies surveyed, indicated they managed centrally both sourcing and logistics, even though both functions were outside the country. Moreover, 56% of the large companies (annual revenues over $1 billion) stated that the centralization trend had increased during 2003–2008. Primary reason for centralization was to maximize economies of scale to reduce cost.

8.4 Global Sourcing*

Most firms now recognize that global sourcing is a key driver of financial performance and overall competitiveness. In a survey of 170 companies by Aberdeen group (Enslow, 2005), half of the companies reported having at least 25% international suppliers with the expectation that it will grow to 50% in 10 years. However, international supplier selection is more complicated and risky than domestic supplier selection. In this section, we begin with the benefits and barriers to global sourcing and their effects on supplier selection. Next, we analyze the financial, logistics, manufacturing and strategic issues inherent in global sourcing. Finally, we review the different approaches available for strategic and tactical supplier selection decisions.

---

* The material in this section is based on the working paper by Tisminisky and Ravindran (2005).

8.4.1 Benefits and Barriers to Global Sourcing

8.4.1.1 Reasons for Global Sourcing

Globalization and the introduction of international time-based competition have led many companies to use global sourcing. Customers expect items that fulfill international standards, in turn forcing companies to look for new sources and products. Firms now recognize that international supply management is essential for their survival in the global competition and thereby try to develop a global sourcing strategy.

The main objective of a global sourcing strategy is to exploit both the supplier's competitive advantages and the comparative location advantages of various countries in global competition. Handfield (1994) argues that global sourcing does not need to be implemented by every firm, and it should only be developed by those that have to deal with competition using continuous performance improvements. Quality improvements, meeting schedule requirements and importing technologies are some of the reasons why a buyer might pursue international purchasing. Monczka et al. (2005) and Peterson et al. (2000) discuss the following reasons why companies purchase materials and services from other countries.

- Exploiting cost-price benefits
- Having access to quality goods and process technologies
- Introducing competition to domestic suppliers
- Increasing the number of available suppliers
- Establishing a presence in the foreign market
- Reacting to sourcing practices of competitors

8.4.1.2 Barriers to Global Sourcing

As companies turn to global suppliers, they must be aware of both the opportunities and the threats present. Understanding world markets can be extremely difficult since each country is unique and complex. Therefore, global operations increase uncertainty and reduce control capabilities. Problems might arise from the number of intermediaries, customs requirements as well as trade restrictions. In addition, uncertainty could result from greater distances, longer lead times, and less knowledge of the market conditions.

Monczka et al. (2005) have pointed out several barriers that are implicit in global sourcing:

- Cultural and communication differences
- Longer lead times and increased uncertainty
- Political and financial risks

- Trade regulations and agreements
- Lack of knowledge about foreign business practices
- Trade risks when sharing new technologies

Nevertheless, education, training and measurements systems are management strategies that can help overcome these obstacles. Technological improvements like globally linked computer aided design (CAD), electronic mail, bar codes, and RFID tags, are also key features in finding solutions to purchasing problems.

8.4.2 Issues in Global Sourcing

When determining the supplier sourcing strategy and identifying potential supplier sources, purchasing personnel should consider the issues that are particularly important in international sourcing and are sometimes overlooked in domestic sourcing. The following are some examples:

- Significant cost difference between domestic and foreign
- Supplier's prices over time
- Effects of longer material pipelines and higher inventories in the supply chain
- Consistency in delivery times
- Trustworthiness and legal system to follow
- Supplier's payment terms
- Currency exchange management

Another important factor in the supplier sourcing strategy and supplier selection stages is the total cost of doing business with an international source. The different costs involved in global sourcing are classified into logistic and hidden costs. Logistic costs are generated from all the activities that involve transporting the goods from the supplier to the buyer's facilities. The inland and overseas transportation, warehousing, extra tooling, import processing fees are few examples of these costs.

8.4.2.1 Hidden Costs in Global Sourcing

Item price is not the sole criterion in global sourcing. There are a number of hidden costs that should be included in the decision making process. They include the following:

- Higher inventory cost
- Cost of extra packaging
- Loss due to damaged items

- Cost of production disruption
- Cost of handling returns due to quality problems
- Currency risks

The *total cost of ownership* (TCO) approach developed at General Electric Wiring Devices (Smytka and Clemens, 1993) can be used to include the hidden costs of global sourcing. As discussed in Chapter 6, (Section 6.2.4.1) TCO looks beyond the unit price of the item and includes the cost of other factors, such as quality, delivery, supply disruption, safety stock, etc. TCO then assigns a cost to each factor and computes the total cost of ownership with respect to each supplier. Finally the business is awarded to the supplier with the lowest total cost.

8.4.3 Factors Affecting International Supplier Selection

An important review of criteria and analytical methods in the area of domestic supplier selection has been done by Weber et al. (1991), who reviewed 74 articles that discuss supplier selection. Further, they provide a comprehensive list of the criteria that academicians and purchasing practitioners have considered as important in the supplier selection decision (Table 8.3). It is interesting to note that most articles consider more than one criterion when developing supplier selection models, emphasizing the inherent multiobjective nature of the supplier selection process.

Some of the criteria mentioned in Table 8.3 have also been used for international supplier selection. They are classified by Choy and Lee (2002) according to technical capabilities, quality assessments, and organization profiles, while Atkinson (2003) divides them into political, currency, and logistics risks. In Table 8.4, we propose an alternative classification for international supplier selection criteria. Political issues are not considered in this classification. In most cases, they are rarely a concern since only a major event could paralyze the exports of a country. We shall discuss the four major international categories of criteria in detail next.

8.4.3.1 Financial Issues

Currency exchange fluctuations are the biggest financial issues when implementing a global sourcing strategy. In global sourcing, it is common that the parties involved acquire long term debts supported by a signed contract. The currency exchange rates might fluctuate significantly during the contract period.

Carier and Vickery (1989) state two primary strategies to manage currency exchange rates-*macro* and *micro strategies*. A *macro level strategy* focuses on the sourcing decision itself, tries to determine the best time to purchase and in what volume. On the other hand, a *micro level strategy* focuses on protecting the buyer from the risk of exchange rate fluctuations and occurs mostly after the sourcing decision has been made.

TABLE 8.3

Supplier Selection Criteria (Weber et al., 1991)

Rank	Criteria
1	Net price
2	Delivery
3	Quality
4	Production facilities and capability
5	Geographical location
6	Technical capability
7	Management and organization
8	Reputation and position in industry
9	Financial position
10	Performance history
11	Repair service
12	Attitude
13	Packaging ability
14	Operational controls
15	Training aids
16	Bidding procedural compliance
17	Labor relations record
18	Communication system
19	Reciprocal arrangements
20	Impression
21	Desire for business
22	Amount of past business
23	Warranties and claim policies

TABLE 8.4

Criteria for International Supplier Selection

Category	Criteria
Financial issues	Currency fluctuations, inflation, deflation, forecast availability, market volatility
Logistic issues	Delivery lead time, supply lots, flexibility in changing the order, delivery in good condition, capacity to meet demand
Manufacturing practices	Lean manufacturing, just-in-time, low inventories, zero defects, flexible production, small batches
Strategic issues	Relationship type, quality, culture, capabilities, reputation, position in the sector, financial strength, management skills, trust, codifiability

A common macro level strategy is *volume-timing technique*. This approach determines a forward buying policy or a short term purchasing policy according to the stability of the currency. If the currency at which the item is being sold is going to be devalued in comparison with the buyer's currency, then it might be worthwhile purchasing a bigger quantity than usual and store the item in inventory until needed. The volume-timing technique might also affect whether the company sources domestically or internationally. For example, if the dollar appreciates, U.S. companies will find it easy to procure items from abroad. On the contrary, if the dollar depreciates, U.S. companies will find it difficult to rely on global suppliers because they will have to pay higher prices in dollars.

Apart from the volume-timing technique used in the macro level strategy, there are several other micro level strategies that can be used to reduce financial risk. Handfield (1994) suggests that currency fluctuations can be controlled using contracts that define terms in the buyer's home country's currency, provide currency options, and allow for future contracts, periodic negotiations or re-negotiation. Branch (2001) states that currency options, forward exchange rates, as well as average rate options, can be used to overcome disadvantages when purchasing internationally. Nevertheless, these contracts usually have a cost to the buyer and require a valuation analysis. Atkinson (2003) states that sourcing with a fixed dollar value is not an effective strategy because "if there is a major weakening of the buyer's currency, the supplier will almost always have to ask for a price increase." A better solution is to structure a plan before the supplier is selected in case the supplier's currency gets stronger. This plan could include the use of forward contracting in case there is a wrong forecasting of the exchange rate trend.

8.4.3.2 Logistic Issues

In addition to the financial issues, logistic factors might also be significantly important in the international supplier selection decision. Cebi and Bayraktar (2003) state that delivery lead time, supply lots, flexibility in changing the order, delivery in good condition, and capacity to meet demand are just some of the criteria considered in this category. We will discuss international logistics in detail in Section 8.5.

8.4.3.3 Manufacturing Practices

Apart from the effect of financial and logistic issues concerning the vendors in global sourcing, the buyer's manufacturing practices and the ability of the supplier to adapt to them can also have a profound impact in the supplier selection decision. As a consequence of the increased reliance on lean manufacturing techniques, it is more important to consider a number of intangible factors beyond those used in tactical decisions. According to Sarkis and Talluri (2002), some examples of these intangible factors are: emphasis on quality at the source, design competency, process capabilities, operators' cross-training, concurrent design, and dedicated capacity.

Lean manufacturing covers tools such as just-in-time (JIT), small inventory, zero defects, flexible production, small batches, and close technical cooperation with suppliers. Some of its advantages, such as reduction of defects and engineering changes, can facilitate implementation of a global purchasing strategy.

Despite the benefits that JIT provides to global purchasing, it might become more difficult to maintain a JIT environment as a consequence of the geographical dispersion of the supply chain when using a global sourcing strategy. Levy (1997) argues that the shipping time, lower frequency of freight connections, high rates for numerous small shipments and unpredictable delays (weather, customs and documentation) implicit in global purchasing can reduce the chance of a successful supply chain. Longer lead times might also increase the volatility of inventory levels over time. In addition, other important attributes of lean manufacturing like the close relationship between the supplier and customer might also be affected due to the distance and communication barriers inherent in global sourcing.

Firms are increasingly using lean manufacturing techniques. An important issue to be considered in international supplier selection is the ability of a supplier to work with the buyer in maintaining these manufacturing practices. Although there are several tools from a lean environment that will benefit global sourcing, international purchasing can become an obstacle to sustaining a lean manufacturing strategy.

8.4.3.4 Strategic Issues

In a global economy, firms do not frequently face situations in which different suppliers look similar, but only vary on few criteria like price and lead time. Especially for strategic commodities, it is hard to find two suppliers that have exactly the same advantages. Kannan and Tan (2002) point out that "Soft non-quantifiable selection criteria, such as supplier's strategic commitment to a buyer, have a greater impact on performance than hard quantifiable criteria."

Strategic factors are important in situations where the buyer is looking to develop a long-term partnership with the supplier. Generally, firms are interested in this type of relationship when they want to reduce the risk and transaction costs associated with the product/service procured, or they are interested in restructuring the supply base. They might also be looking for relationship-specific investments, improvement in codifiability, prevention of knowledge loss and the control of scarce resources.

Intangible supplier characteristics also become more important in a global sourcing environment compared to a domestic one. Sarkis and Talluri (2002) mentioned that the organization profile analysis should cover factors such as culture, technology, and relationship. It should focus less on the operations side of the supplier, but more on capabilities of the organization. For Cebi and Bayraktar (2003), business criteria are related to the supplier's reputation and position in the sector, financial strength, as well as management skills and compatibility. When purchasing internationally, companies usually do not have a good understanding of suppliers' background and environment thus increasing the importance of its reputation. For instance, Handfield (1994) claims that when purchasing a critical component, American firms place great importance on trust, schedule reaction, supplier's ability to react to schedule changes, and the fact that the company was established in the United States.

Transaction codifiability determines the relationship type between the buyer and the supplier. Levi (2002) defines perfect codifiability as "the ability to convey all transaction related knowledge in an easily understood document." A product becomes highly uncodifiable when it is difficult to set its requirements. The less codifiable the product is, the less likely it can be bought using structured methods that rely on quantitative criteria. For instance, an emerging technology is harder to standardize due to confidentiality and information issues. A buyer that is looking to increase its component's codifiability will definitely value the fact that a supplier is willing to cooperate to achieve this goal. Therefore, the supplier's ability to make knowledge explicit using written documents will affect the relationship type and should become another criterion in the global supplier selection decision.

8.4.4 Tools for Global Sourcing

Shore et al. (2004) classify supplier selection methods into data-based methodologies and experience-based methodologies.

Data-based approaches can only handle defined measurable information and are only adequate for well-structured problems. They use analytical techniques which are frequently used in tactical decisions. On the other hand, experience-based approaches apply hybrid techniques that combine both quantitative and qualitative criteria. They generally employ expert systems, neural networks, and case-based reasoning to define the problem and determine the criteria in the supplier selection decision. The experience-based approaches use brainstorming, cognitive mapping as well as interpretative structural modeling.

In terms of analytical methods, Weber et al. (1991) present an important review of the most relevant articles in supplier selection. Out of these articles, only 10 of them used mathematical programming. Ghodsypour and O' Brien (2001) also present a comprehensive review of the supplier selection methods. These methods vary in their level of complexity, from simple sourcing and matrix methods to advanced mathematical programming approaches.

The most common quantitative supplier selection methods employ mathematical programming, weighted average method, and statistical approaches. Linear weighted models subjectively assign a weight for each criterion, evaluate the vendor's performance in each criterion, and compute a weighted score for each vendor. The mathematical programming techniques include linear programming, mixed integer programming, and goal programming. They typically try to minimize the total cost by considering different criteria and constraints. Finally, the statistical approaches incorporate probabilistic measurements into the supplier selection problem.

The quantitative methods for supplier selection were discussed in detail in Chapter 6 of this textbook. In Table 6.43, we presented a brief summary of some of the recent supplier selection methods published since 2000. Most of the supplier selection methods in the literature have been developed primarily for domestic supplier selection. A few of these mathematical methods have been used in global supplier selection. We believe that most of the domestic approaches can be extended to global sourcing by incorporating the global selection criteria discussed in Section 8.4.3.

8.5 International Logistics

International logistics is the movement of goods across national boundaries. In a survey of 170 companies conducted by Aberdeen Group (Enslow, 2005), it was found that logistics cost as a percentage of revenue for products bought or sold offshore varied between 6% and 11%. In comparison, the logistics cost was typically less than 3% for a domestically sourced product. Half the companies reported that 25% of their suppliers and customers are located in a foreign country. They expected this percentage to double in a few years. More than three-fourths of the respondents indicated China as the most common foreign location.

Managing international logistics is a complex process. It involves multiple parties handling the goods (at least seven intermediaries for many items). Increased border security and inspections such as C-TPAT certification (Customs-Trade Partnership Against Terrorism) and other U.S. Homeland Security programs that have put border security on the corporate radar. Cost benefits of voluntary security initiatives, such as Green Lane Program, are not easy to quantify. Green Lane program uses RFID-enabled seals to fast-track containers at the border. However, companies have reported quantifiable benefits by using FAST (Free And Secure Trade lane) for moving goods by trucks across the borders in Mexico, Canada, and the United States.

The Aberdeen Group survey also found that companies were increasing the flexibility of the logistics networks for transporting their products. They reduced the number of carriers, logistics providers and IT systems. They also increased visibility at the suppliers, used flexible transportation modes, practiced co-mingling of shipments with other companies and product-postponement strategies. Postponement involves the practice of making the product differentiation as close to the customer as possible.

The Aberdeen Group (Enslow, 2005) recommended the following strategic and tactical logistics strategies to reduce cost and increase responsiveness of the global supply chain.

8.5.1 Steady Demand

For products with predictable shipment volume, low inventory turns and multiple transportation companies, an effective strategic initiative is to use flexible transportation networks. For tactical planning, buying transportation capacity and using comingled freight are effective strategies.

8.5.2 High Demand Variability

For products with fluctuating demand, *postponement* is an effective long term strategy. Tactical strategies include improving forecast accuracy, increasing raw material and product visibility and document automation.

A winning procurement strategy for products with high demand variability is to buy globally from the lowest cost country and handle demand variability through "spot purchases" from local sources. This is similar to HP's portfolio strategy to handle supply risk we discussed in Section 7.7.2.

8.6 Designing a Resilient Global Supply Chain: A Case Study*

Chopra and Meindl (2010) recommend that global supply chain design decisions should consider the following:

1. The network design should include both strategic planning and financial planning.
2. The network should be evaluated using multiple supply chain performance metrics that include supply chain profit, risk, and responsiveness.
3. The final decision should include both quantitative and qualitative factors.
4. Sensitivity analysis on the model parameters should be performed to check the robustness of the solution.

The case study presented in this section incorporates all these elements by developing a supply chain footprint model and applying it to a complex multinational manufacturing enterprise in the health and hygiene industry, listed in the Fortune 500. The footprint model focuses on the development of methods for designing a resilient and responsive global supply chain network that considers manufacturing and distribution facilities that service multiple markets in several countries. The emphasis is on developing a

---

* The material in this section is based on the dissertation of the author's doctoral student (Portillo, 2009).

multi-criteria optimization model to help determine optimal supply chain design to support specific competitive strategies within the complex multi-national environment. The model considers manufacturing and distribution facilities location/allocation selection, capacity and expansion requirements, production and distribution network variables, international issues, exchange rates, lead-times, and transfer prices. The model also includes a set of supply chain design selection criteria that integrates financial, customer service, risk and strategic factors based on multi-criteria selection techniques. In this case study, we will discuss how international issues have been incorporated in decision criteria and how they are assessed for inclusion in the model. Detailed definition of the decision variables and the development of the constraints and objectives will not be presented. Interested readers are referred to the original publications (Portillo, 2009; Portillo et al., 2009). Results and managerial insights obtained from the case study will be discussed, such as capital investment plans, sourcing alternatives and contingency plans, customer oriented supply chain strategies, among others.

8.6.1 Problem Background

The Company's global brands are sold in more than 150 countries and used by approximately 1.3 billion people, holding first or second positions in majority of its markets. The company has operations in 37 countries, employs approximately 55,000 people and sells close to $20 billion a year. This case study focuses on the company's largest international division, selling products across a continent functioning with several offices, distribution centers, and with manufacturing facilities located in 20 countries. To support its competitive strategy, its supply chain is formed by a spectrum of customers ranging from multi-national chains and large distributors to thousands of small "mom and pop" stores. A series of mergers and acquisitions in the last 15 years across the continent, led to a supply chain with a highly complex internal structure. Currently, the supply chain comprises of 45 distribution locations, spread across the continent, and 21 manufacturing facilities located in 10 of these countries. Most of the products sold in the region are supplied by these 21 manufacturing facilities. However, some products are imported from other companies located all around the world. In addition, cross sourcing activity within the continent has increased significantly in the last few years. Today, more than 60% of the production facilities manufacture finished and semi-finished products that are distributed to different countries, beside the local market. At least three facilities are continent-wide facilities, sourcing all markets within the division. Moreover, these facilities export products to other company divisions worldwide. Until now, asset rationalization and streamlining efforts have focused on distribution supply chain designs for specific business units or division-wide designs considering manufacturing facilities for particular product. In order to better support and enhance the company's competitive strategy, a robust, flexible

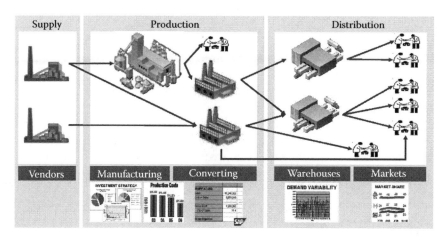

FIGURE 8.3
Supply chain network echelon.

and efficient global supply chain design was required to assure achievement of desired customer service levels and financial goals while considering related risk factors.

The strategic and tactical model, for optimization of manufacturing and distribution networks, incorporates more than 100 customer zones or markets and dozens of product categories, manufactured in more than 250 production lines. Because of the diverse international nature of the problem, many international factors (such as domestic and international freights, transfer prices, taxes and duties) are considered. Some product categories have multi-stage production process and are processed in multiple echelons (Figure 8.3). Multiple productivity rates are considered for multi-product machines and products. The costs (facility overheads, production line fixed and variable costs and raw material consumption costs) are appropriately included in the optimization model.

To address the problem of optimizing the global supply chain network, *a multi-criteria mixed integer linear programming (MILP)* model is developed. The multi-criteria model is solved using *preemptive and non-preemptive goal programming* methods discussed in Appendix A. Goal programming has become a practical method for handling multiple criteria (Masud and Ravindran, 2008). Goal programming is selected due to the fact that complete information from the decision makers about their preferences is available. Multiple goals are defined along-with its corresponding target levels of achievement. The objective is to develop solutions that are as close to the targets, based on the pre-specified priorities/weights. The preemptive priorities and non-preemptive criteria weights are obtained using Rating, Ranking (Borda count), and AHP methods (discussed in Appendix A and Chapter 6, Section 6.3).

8.6.2 Model Features

The multi-criteria global supply chain network design model integrates customer service levels, strategic factors and disruption risk criteria along with the financial measure of performances. Customer service level is measured using two factors: (1) the ability to completely fulfill customer demand and (2) the speed of delivering products to the customers. Demand fulfillment is defined as portion of the demand that is effectively delivered to the customers from any stage in the supply chain. The ability to completely fulfill customer demand is modeled as a goal constraint by specifying demand fulfillment targets for combinations of products and customer zones. The speed of delivering products to the customers is modeled as a goal, minimizing weighted speed measure, based on volume and lead times.* Weighted speed goal targets, for each customer zone, are explicitly considered in the model. In addition, the multi-criteria model takes into account the minimization of *risk associated with supply chain disruptions.* Different measures of risk for domestic and global sourcing are estimated for each manufacturing, converting, and distribution location. These measures incorporate both facility and country specific risk factors. Facility-based risk factors are determined based on assessments performed by the decision makers. Country-specific risk factors are obtained by considering the weighted average cost of capital rates of each country. A more detailed description of the risk measure estimation is given in Section 8.6.3. The objective of minimizing the risk measure is modeled as a goal constraint, by setting the overall risk target value for the entire supply chain.

Decisions related to supply chain network design may also require the modeler to consider *strategic factors* to open new markets, to increase market share, and to strengthen relationships with customers. This model includes strategic significance measures for each facility, based on the ratings provided by the decision makers. A goal constraint is set to achieve the maximum possible overall strategic measure for the entire supply chain network. Finally, the objective function of the goal programming model is to minimize the weighted deviations from the criteria targets: profit, demand fulfillment, speed, risk, and strategic factors.

8.6.3 Decision Criteria and Risk Assessment

This section presents a detailed discussion of the supply chain design criteria used in the case study (Figure 8.4) and the procedure to determine the criteria weights and preferences. Assessment of risk and of strategic factors is also discussed.

The financial objective of maximizing the gross profit considers revenue from sales less production costs, distribution expenses, and freights.

---

* Corresponding to each arc of the network that links a DC and a customer zone.

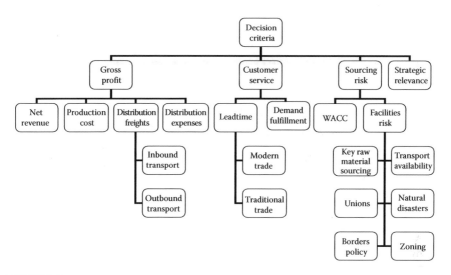

FIGURE 8.4
Decision criteria hierarchy.

The production costs include facility-specific overheads, fixed and variable machine overheads, and the raw material cost. Distribution expenses include all the fixed and variable costs for operating the DCs. The transportation cost is divided into inbound and outbound freight. Inbound transportation cost is incurred in moving the goods among the stages/facilities within the firm's supply chain network. Outbound transportation costs are the freight paid for the shipment of goods to the customers.

Customer service criterion has two sub-criteria: Lead time and Demand fulfillment. Lead time is defined as the time between receipt of a customer sales order and its delivery at the customers facility (used as a proxy for speed). Demand fulfillment is the proportion of the customer order that is effectively delivered.

Supply chain disruption risk criterion includes facility-specific and country-specific disruption risk factors. For each facility, the facility specific risk factor was assessed using an expert opinion-based risk rating method. It is a qualitative assessment done by the decision makers considering factors such as key raw material sourcing, transportation availability, existence of unions, risk of zoning, and occurrence of natural disasters. The decision makers assign a score on a 1–5 scale (see Table 8.5) for each facility considering its impact on the domestic and export market. An example of facility risk rating is given in Table 8.6. Country specific risk factor uses the firm's *weighted average cost of capital* (WACC) as the risk measure, according to the country in which the facility is located. This is an estimate of the domestic cost of capital excluding inflation. WACC is based on U.S. dollar denominated bonds, which are used to determine the spread between the United States and foreign countries. These rates are then adjusted based on *political risk* (50%),

TABLE 8.5

Scale for Facility Risk Rating

Verbal Preference	Score to Assign
No risk	1
Very low risk	2
Medium risk	3
High risk	4
Very high risk	5

TABLE 8.6

Example of Facility Risk Rating

	Open		Closed	
Facilities	Local	Exports	Local	Exports
DC 1	2	2	5	5
Plant 2	2	2	5	5
	2	2	5	5

economic risk (25%), and *financial risk* (25%). *Political risk* considers government stability, corruption, bureaucracy, socioeconomic conditions, involvement of military in politics, investment profile, religious tensions, internal conflicts, law and order, external conflict, ethnic tensions, and democratic accountability. *Economic risk* considers the current account balance and budget balance as a percent of gross domestic product (GDP), GDP per capita, annual inflation, and real annual GDP growth. *Financial risk* includes the percentage of exports of goods and services (XGS), foreign debt as a percent of the GDP, net liquidity as months of import cover, foreign debt service as percent of XGS, and exchange rate stability.

Three methods have been used to obtain the weights of the criteria: *simple rating, pairwise comparison using Borda count*, and *AHP* (see Appendix A and Chapter 6, Section 6.3, for a discussion of these methods). In the rating method, a panel of experts assigned a score between 1 and 10 (1 being the least important and 10 being the most important) for each criterion and attributes under the criterion. The weights for the criteria are computed by normalizing the ratings. Criteria ratings given by one decision maker (DM) are shown in Table 8.7. In the pair wise comparison method, preferences between criteria are specified by each DM. For AHP, pairwise comparison of criteria is performed using a 1–9 strength of preference scale (1—equally important, 9—absolutely more important).

Based on the three methods, three sets of individual weights have been determined for each DM. The main criteria weights obtained (in percentages) for one decision maker is given in Table 8.8. A total of 11 decision makers were involved in the criteria assessment. The decision makers were then

TABLE 8.7

Sample Criteria Ratings for a Decision Maker

	Main Criteria	Rating Score (1–10)
1	Gross profit	10
2	Lead time	6
3	Demand fulfillment	9
4	Sourcing risk	6
5	Strategic factor	6
	Lead Time Attributes	**Rating Score (1–10)**
1	Modern trade	10
2	Traditional trade	7
	Risk Attributes	**Rating Score (1–10)**
1	WACC	7
2	Facility risk	5

TABLE 8.8

Sample Criteria Weights for a Decision Maker

Criteria	Simple Rating (%)	Borda Count (%)	AHP (%)
Gross profit	27	33	43
Speed	16	20	15
Demand fulfillment	24	27	26
Sourcing risk	16	7	11
Strategic factor	16	13	4

asked to indicate the method which best reflected their preferences. Fifty percent of the DMs indicated that simple rating method represented their preferences, 30% chose pairwise comparison method, and 20% chose AHP.

The 11 individual criteria weights are then aggregated to obtain the group weights using group decision making methods (simple averaging, Borda count, and AHP) discussed in Section 6.3.9. Table 8.9 gives the group weights obtained by the three methods.

TABLE 8.9

Group Weights for the Criteria

Criteria	Simple Averaging (%)	Borda Count (%)	AHP (%)
Gross profit	34	31	38
Speed	15	14	12
Demand fulfillment	21	22	22
Sourcing risk	10	13	10
Strategic factor	19	20	18

In all three methods, the ordinal preferences of the criteria are the same—gross profit, demand fulfillment, strategic relevance, speed, and risk. This preference order is used in the preemptive goal programming model. The cardinal weights obtained by each method are reasonably close. For the case study, the set of weights obtained by simple averaging is used in the non-preemptive goal programming model.

The complete formulation of the optimization model (decision variables, constraint equations and objective functions) will not be presented here. References (Portillo, 2009; Portillo et al., 2009) provide complete details of the Mixed Integer Linear Programming (MILP) model constraints and the objective functions.

Briefly, the decision variables include continuous variables that represent the flow of goods through the network, for both production and distribution volume. Binary variables are used to indicate opening or closing decisions associated with the manufacturing plants, converting facilities, distribution facilities, and specific production lines.

The model considers constraints on the distribution and production capacities, flow balance constraints at the facilities, and constraints on customer demands.

8.6.4 Model Results and Managerial Insights

The model has been applied to perform the analysis of the manufacturing and distribution network of the company based on its competitive strategy. Customer demand projections for a 5-year time horizon are used in the analysis. Initially, the MILP model is executed as a single objective problem, with the objective of maximizing the profits. The solution of the profit maximization model is then used as baseline results for comparison against the results from the multi-criteria model. The multi-criteria model considers profit, demand fulfillment, speed, sourcing risk and strategic factors as the conflicting objectives.

8.6.4.1 Results of Profit Maximization Model

The single objective MILP model to maximize profits comprised of three stages:

Stage 1: Determine the ability of the company to fulfill present and projected sales, based on the current supply chain network.

Stage 2: Evaluate how the company's ability would be improved, considering the potential expansions of plants and DCs in the current supply chain.

Stage 3: Optimize the global supply chain network that would deliver the best results for the entire 5 year time horizon.

The MILP model had 7000 constraints and a total of 7500 variables with 300 binary variables. The mathematical model was coded in ILOG and solved using the CPLEX solver (www.ILOG.com). The solver reached optimality in 2 min.

The stage 1 analysis showed that the capacities of the existing facilities were utilized completely under current demand levels. However, considering projected future demand, the results showed that the existing supply chain could fulfill only 75% of the total projected demand, primarily restricted by production and distribution capacities.

The stage 2 analysis included capacity expansions in the production and distribution facilities already under management consideration. Within a 2 year horizon, the management had planned to open nearly a dozen new production lines. Although the management had already decided on their locations, multiple options were allowed in the model to confirm their choices. The results showed that the majority of the chosen locations were optimal. Even though the locations were optimal under the profit objective, the supply chain network was capable of meeting only 85% of the projected demand even after the expansion.

In stage 3 analysis, additional levels of facility expansions were considered. The model determined the optimal levels for production, distribution and sales. The optimal demand fulfillment ratio reached only 96%, emphasizing that certain product-market combinations were not profitable. The model gave optimal results on the magnitude of the required expansions at the current locations, and opening of new facilities for different technologies and product categories. These results could be used to define future investment plans.

Based on the optimal supply chain design, optimal distribution strategy including cross sourcing and distribution for each product category were obtained. The distribution strategy gave the optimal flow of products through the supply chain from the manufacturing facilities to the customers. This involved decisions on the selection of production lines to produce products, sourcing strategy for converting facilities, allocation of distribution centers to production facilities, and allocation of customer zones to either distribution centers or production facilities. The model suggested a more centralized distribution strategy for customers in two groups of countries, where favorable geographic conditions and trade agreements exist. This implied closing the local warehouses and opening an expanded regional distribution center. The aforementioned results are critical, both for determining the strategic actions and for defining the tactical business plans.

8.6.4.2 Multi-Criteria Analysis

In order to enhance the initial analysis of the profit maximization model, a multi-criteria supply chain model was solved by adding objectives on demand fulfillment, speed, sourcing risk and strategic factors. The weights and preferences obtained earlier from the decision makers were used in solving the goal programming models. The results obtained in this analysis are similar to the ones generated for the single objective model. However, important differences in the supply chain network design occurred primarily due to the inclusion of the speed and risk criteria. Also, the magnitude of the expansions was affected by the demand fulfillment criterion. In addition, the results led to a trade-off analysis among the decision criteria. Significant differences in the optimal supply chain network design between the multi-criteria model and the profit maximization model are discussed below:

1. The profit maximization model suggested adopting a centralized distribution strategy for some countries. The multi-criteria model suggested a decentralized structure. This change was primarily due to the relative importance given to the speed criterion. Evidently, a trade-off exists between profit and speed criteria. Based on the model results, the marginal benefits in profits may not justify a reduction in customer service levels. More specifically, the savings from efficiencies gained from economies of scale and risk pooling are partially offset by higher transportation costs, making the marginal benefit not large enough to justify a decrease in the responsiveness to customers. The risk to serve local markets increased with the centralized approach, since the probability of supply chain disruptions is higher due to increased foreign sourcing, longer inter-company routes and lead times.

2. The multi-criteria model suggested keeping specific facilities open in countries where having local production facilities is crucial to minimize risks of supply chain disruptions to serve local customer zones. Previously, results from the single criterion optimization (focused on maximizing profit) considered closing facilities in these countries. However, when considering the high values of the weighted average cost of capital of the firm and the risk assessments obtained from the decision makers primarily driven by the current political instability in those countries, the optimal solution included keeping the facilities open regardless of profit implications.

3. Strategic factors had less impact on the supply chain network design, since there were no significant differences in the assessed values among facilities. However, the results from the multi-criteria models were different in two cases. In case 1, the multi-criteria model resulted in opening a production facility with a very high strategic rating. The profit maximization model considered closing that facility, due to cost benefits. In case 2, one of the new production facilities opened by the profit maximization model, was closed in the optimal solution of the multi-criteria model. This was due to the fact that its strategic rating was very low compared with the second sourcing option for that specific country, which had a larger regional plant with a higher strategic rating. The profit maximization model focused only on marginal benefit in profit.

4. In the multi-criteria model, the overall demand fulfillment ratio increased to 99% with some reduction in profits. A valuable outcome of this analysis was the trade-off evaluations for specific product-market combinations, where increases in sales reduced profitability. These insights led to valuable managerial implications for the evaluation of the firms' competitive strategy.

8.7 Summary and Further Readings

8.7.1 Summary

In this chapter, we discussed globalization and its impacts on supply chain management. We began with the history of globalization and the emerging trends in globalization. The international trade among countries for goods, capital, services and labor increased dramatically in the twenty-first century due to advances in air travel, telecommunication, computer usage, and the Internet technology. Globalization is also facilitated by free trade agreements between countries through World Trade Organization, with a current membership of 153 countries. China has emerged as the most common destination for contract manufacturing. This has resulted in China emerging as the second largest world economy, behind the United States. It is predicted that due to globalization, the economies of developing countries (Brazil, Russia, India, and China) will become larger than the economies of industrialized G7 countries (the United States, Canada, Japan, Germany, France, United Kingdom, and Italy) by 2030.

Next we discussed the impacts of globalization to the consumer. It has resulted in the collaboration of the best minds and talents in the world to produce innovative global products, such as iPods, smart phones, world cars, etc. Globalization has also resulted in the offshore outsourcing of not only manufacturing services to China, but also knowledge services (IT, R&D, etc.) to India. It is also pointed out that manufacturing in the United States is not dead or dying. Instead, it has moved *upscale* to the manufacturing of "high-end" products.

Next, we discussed risks in globalization and the key issues in managing global supply chains. Strategies for managing global supply chains were discussed, including supplier criteria under global sourcing and international vendor management. Several examples of globalization strategies were also presented.

We then discussed issues in international logistics. The logistics cost as a percentage of revenue for products bought or sold offshore (6%–11%) was more than double compared to domestically sourced products (typically 3%). With multiple parties (at least seven) involved in handling globally sourced product and increased border security, managing international logistics has become very complex. Best-in-class multinational companies are moving towards more flexibility to their logistics network. Flexibility is achieved through increased visibility at the suppliers, use of flexible transportation routes, comingling of shipments and product postponement strategies.

Finally, we presented a real-world application for designing a resilient global supply chain for a multinational consumer products company. A multi-criteria optimization model explicitly considered conflicting criteria that integrated financial, customer service, supply chain risk and strategic factors of the company. Model results and managerial implications were also presented.

8.7.2 Further Readings

To learn more about the history of globalization and its impacts, the book *World is Flat* by Thomas Friedman is an excellent resource (Friedman, 2005). It explains how "flattening" of the world happened in the twenty-first century due to globalization, WTO, and the Internet revolution. It also discusses the impacts of globalization on countries, companies, and the average consumer. The author also explains how governments and societies can adapt to the globalization trends.

The book *Outsourcing for Radical Change* by Thomas Davenport discusses outsourcing strategies for enterprise transformation (Davenport, 2004). It is based on a major study by Accenture and in-depth interviews by the author. It discusses workable action plans for transformational outsourcing, navigating its challenges and the leadership skills necessary to succeed.

The book by Robinson and Kalakota on *Offshore Outsourcing* discusses business models, return on investment and best practices for business process outsourcing (Robinson and Kalakota, 2004). It gives step-by-step guide for implementing a successful offshore outsourcing strategy and overcoming the challenges in selecting overseas countries and the suppliers. It illustrates the best practices for offshore outsourcing followed by companies like GE, Dell, IBM, American Express, Delta Airlines, British Airways, and others. It also discusses offshoring benefits in reducing costs, fine-tuning business processes, and increasing profits.

The textbook on *Outsourcing and Insourcing in an International Context* by Schniederjans focuses on the concepts, processes and methodologies for companies interested in outsourcing (Schniederjans, 2005). It includes an extensive discussion of the international risk factors associated with an offshore outsourcing strategy. The author gives a good mix of both pros and cons of outsourcing and provides both qualitative and quantitative methods that can be used for outsourcing decisions. It also contains several numerical examples to illustrate the methods. The book also contains the right ways and the wrong ways to undertake offshore outsourcing.

Exercises

8.1 What are the impacts of *globalization* on the following?
- Supply chain network design
- Transportation strategies
- Inventory decisions
- Cost

8.2 Discuss how *globalization* has impacted the manufacturing and service sectors in the developed economies of Europe and the United States.

8.3 What is the *World Trade Organization* (WTO)? How WTO has impacted *globalization*? What are the *pros* and *cons* of WTO membership?

8.4 Explain *global sourcing*. What are the benefits and barriers to global sourcing?

8.5 Name the *supplier selection criteria* that would be common to both *domestic sourcing* and *global sourcing*. Name the criteria that would be different.

8.6 What is *lean manufacturing*? How does globalization impact the lean manufacturing practice?

8.7 What are the key characteristics of a *resilient global supply chain*? How does one design such a supply chain under globalization?

8.8 (Adapted from Portillo, 2009) Recall the case study on designing a resilient global supply chain discussed in Section 8.6. The five main criteria considered are as follows:

GP: Gross profit

LT: Lead time

DF: Demand fulfillment

RD: Risk of disruption

SF: Strategic factor

(a) Using a *rating method*, the Director of supply chain, assigned a score from 1 to 10 (1 being the least important and 10 being the most important) for the five criteria as shown in Table 8.10.

Determine the criteria weights using the rating method.

(b) Suppose the Director of supply chain ranks the criteria from most important to least important as follows:

(Most important) GP, DF, LT, SF, RD (Least important)

Using *Borda count*, determine the criteria weights.

(*Hint*: Borda count method is discussed in Chapter 6, Section 6.3.4)

TABLE 8.10

Criteria Ratings for Exercise 8.8

Criteria	Rating Score (1–10)
GP	10
LT	6
DF	9
RD	6
SF	6

TABLE 8.11

Pairwise Comparison and the
Strength of Preference (Exercise 8.9)

Criteria Pair	Which Is More Important	Strength of Preference
GP-LT	GP	6
GP-DF	GP	5
GP-RD	GP	3
GP-SF	GP	5
LT-DF	DF	7
LT-RD	LT	7
LT-SF	LT	3
DF-RD	DF	5
DF-SF	DF	5
RD-SF	SF	5

8.9 In Exercise 8.8, assume that the Director of supply chain, not only provides a pair wise comparison of her preference, but also her strength of
preference in a scale of 1–9. (1—equally important, 5—more important,
9—absolutely important) as given in Table 8.11.

(a) Using *analytical hierarchy process* (AHP), determine the criteria
weights.

(b) Test the consistency of the pairwise comparison matrix obtained
from Table 8.11 using AHP.

(c) Compare the AHP criteria weights with those obtained in
Exercise 8.8 by the rating method and Borda count.

(*Hint*: For a discussion of AHP, refer to Chapter 6, Section 6.3.6.)

8.10 Suppose 11 *decision makers* (DMs) are asked to rate the importance of
the five criteria in Exercise 8.8. Based on their responses, the criteria weights in percent have been computed for 11 DMs as shown in
Table 8.12. (*Note*: The weights add up to 100% for each DM.)

TABLE 8.12

Criteria Weights of 11 Decision
Makers (Exercise 8.10)

	Criteria Weights (%)				
	GP	LT	DF	RD	SF
DM-1	22	19	21	18	20
DM-2	36	18	28	10	8
DM-3	22	21	19	18	20
DM-4	27	20	13	7	33
DM-5	40	5	9	14	32
DM-6	29	13	27	6	25
DM-7	53	14	24	6	3
DM-8	33	12	14	10	31
DM-9	44	17	25	6	8
DM-10	33	20	27	7	13
DM-11	33	15	27	12	13

(a) Determine the group criteria weights by simple averaging over the 11 decision makers and the group ranking of the criteria.

(b) Determine the ranking of the criteria for each decision maker using the weights in Table 8.12. Then use the Borda count method, to determine the group ranking and the group weights for the criteria.

(c) Compare the group rankings and group weights obtained in parts (a) and (b).

(*Hint*: See Chapter 6, Section 6.3.9 for a discussion of group decision making methods.)

8.11 (Adapted from Attai, 2003) This is a case study dealing with the global supply chain design for the *Banana company*, whose supply chain spans four countries. Banana's main product is sold in two countries and the company's three *customer zones* (C) are located in countries 2 and 4. Banana's raw material *suppliers* (S) are located in countries 1, 2, and 3. The company is looking into three potential sites in countries two and three for the manufacturing *plants* (P). Four potential sites in countries 2 and 4 are under consideration for *warehouses* (W) for storing finished products. Figure 8.5 illustrates Banana's global supply chain network. Even though the supplier and retailer locations are fixed in the network, the plant and warehouse sites will be chosen optimally in order to maximize the net profit. Selecting a particular site incurs a fixed cost and a variable cost.

Tables 8.13 and 8.14 give the relevant data on the fixed cost, variable production cost, holding cost and capacities at the plants and warehouses respectively. Country 2 uses U.S. dollars and all local currencies in other countries have been converted to U.S. dollars. Table 8.15 gives

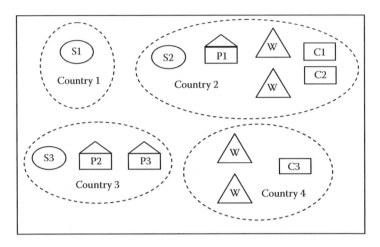

FIGURE 8.5
Banana's global supply chain network.

TABLE 8.13

Plant Data

	Plants		
	P1	**P2**	**P3**
Fixed cost ($)	2,100,000	1,900,000	1,800,000
Holding cost ($)	0.02/unit	0.02/unit	0.03/unit
Production cost ($)	7/unit	6/unit	6/unit
Production capacity	100,000	50,000	50,000
Holding capacity	500,000	500,000	500,000

TABLE 8.14

Warehouse Data

	Warehouses			
	W1	**W2**	**W3**	**W4**
Fixed cost ($)	800,000	750,000	700,000	750,000
Holding cost ($)	0.05/unit	0.05/unit	0.01/unit	0.03/unit
Capacity	500,000	500,000	500,000	500,000

TABLE 8.15

Customer Zone data

	Customer Zones		
	C1	**C2**	**C3**
Holding cost ($)	0.05/unit	0.06/unit	0.07/unit
Sales price ($)	$20/unit	$20/unit	$21/unit

the data on inventory holding cost and selling price for the main product at the customer zones in countries 2 and 4.

The planning horizon covers 15 time periods. The forecasted demands at the customer zones for each time period are given in Table 8.16.

The shipping costs and lead times are summarized in Tables 8.17 through 8.19. Raw material costs are included in the shipping cost. The lead times (in number of periods) are given in parentheses.

TABLE 8.16

Customer Demands (Exercise 8.11)

| Time Period | Customer Zones | | |
	C1	C2	C3
1	104,000	33,000	101,000
2	102,000	35,000	106,000
3	93,000	35,000	115,000
4	102,000	35,000	106,000
5	93,000	39,000	115,000
6	100,000	20,000	60,000
7	200,000	10,000	70,000
8	30,000	10,000	80,000
9	40,000	10,000	10,000
10	50,000	20,000	10,000
11	105,000	10,000	20,000
12	20,000	20,000	20,000
13	130,000	20,000	30,000
14	140,000	20,000	40,000
15	150,000	15,000	150,000

TABLE 8.17

Supplier to Plant Shipping Costs and Lead Times in Periods

| Supplier | Plant | | |
	P1	P2	P3
S1	$3 (2)	$3.5 (5)	$3.5 (5)
S2	$2 (1)	$3.5 (3)	$3.5 (3)
S3	$4 (3)	$2 (1)	$2 (1)

TABLE 8.18

Plant to Warehouse Shipping Costs
and Lead Times

Plant	Warehouse			
	W1	**W2**	**W3**	**W4**
P1	$3 (1)	$3.5 (2)	$4.5 (3)	$4.5 (3)
P2	$4.5 (4)	$4 (3)	$4 (4)	$4 (4)
P3	$4.5 (4)	$4 (3)	$4.5 (4)	$4 (4)

TABLE 8.19

Warehouse to Customer Shipping Costs
and Lead Times

Warehouse	Customer Zone		
	C1	**C2**	**C3**
W1	$3 (1)	$3 (1)	$4.5 (3)
W2	$3 (1)	$3 (1)	$4.5 (3)
W3	$4.5 (3)	$4.5 (3)	$4 (1)
W4	$4.5 (3)	$4.5 (3)	$4 (1)

Questions

(a) Formulate a *mixed integer linear programming* (MILP) model to determine
   the following:

   - Location of plants and production quantity
   - Location of warehouses
   - Raw material procurement plan
   - Distribution of main products from plant to warehouses and ware-
     houses to customer zones
   - Demand fulfillment at the customer zones (percent of demand met)

   The objective is to maximize the net profit over the planning horizon
   subject to the following constraints:

   - Production capacity at the plants
   - Storage capacity at the plants for raw material and finished product
   - Holding capacity at the warehouses
   - Product sales at customer zones cannot exceed the forecasted demands
   - Material balance constraints at the plants and warehouses

Model assumptions

1. One unit of raw material is needed to produce one unit of final product
2. No production lead time at the plants; hence, items produced in period t can be shipped in period t.
3. No capacity limitations at the supplier for raw materials.
4. No capacity restrictions at the customer zones.
5. Unfilled customer demands at any period are considered *lost sales*.
6. Plant storage can be used for both raw materials and finished goods and they require the same amount of space.
7. Initial inventories at the customer zones are as follows: C1 = 250,000; C2 = 100,000; C3 = 250,000.

(b) Solve the *MILP* model formulated in part (a) by any optimization software and answer the following

- Supply chain network design (location of plants and warehouses)
- Raw material procurement plan
- Production quantities at the plants
- Product distribution plans—plants to warehouses and warehouses to customers
- Total net profit for the planning horizon
- Lost sales at the customers

References

Atkinson, W. 2003. Riding out global challenges. *Purchasing*. 132(12): 43–47.

Attai, T. D. 2003. A multiple objective approach to global supply chain design. MS thesis. Department of Industrial Engineering, The Pennsylvania State University, University Park, PA.

Balasubramanian, P. and A. K. Tewary. 2005. Design of supply chains: Unrealistic expectations on collaboration. *Sadhana*. 30(2): 1–11.

Branch, A. 2001. *International Purchasing and Management*. London, U.K.: Thomson Learning.

Carier, J. R. and S. K. Vickery. 1989. Currency exchange rates: Their impact on global sourcing. *Journal of Purchasing Management*. 25(3): 19–25.

Cebi, F. and D. Bayraktar. 2003. An integrated approach for supplier selection. *Logistics Information Management*. 16(6): 395–400.

Chopra, S. and P. Meindl. 2010. *Supply Chain Management*, 4th edn. Upper Saddle River, NJ: Prentice, Hall.

Choy, K. L. and W. B. Lee. 2002. On the development of a case-based supplier management tool for multinational supplier selection. *Measuring Business Excellence*. 6(1): 15–22.

Cohen, M. A. and A. Huchzermeier. 1999. Global supply chain management: A survey of research and applications. In *Quantitative Models of Supply Chain Management*, eds. S. Tayur, R. Ganeshan, and M. Magazine, pp. 671–702. Norwell, MA: Kluwer Academic Publishers.

Cohen, M. A. and S. Mallik. 1997. Global supply chains: Research and applications. *Production and Operations Management*. 6(2): 193–210.

Davenport, T. H. 2004. *Outsourcing for Radical Change*. New York: American Management Association.

Enslow, B. 2005 (March). *New Strategies for Global Trade Management*. Boston, MA: Aberdeen Group.

Fawcett, S. E., L. M. Ellram, and J. A. Ogden. 2007. *Supply Chain Management*. Upper Saddle River, NJ: Pearson Prentice Hall.

Ferre, J., J. Karlberg, and J. Hintlian. 2007 (March). Integration: The key to global success. *Supply Chain Management Review*. 11(2): 24–30.

FM Global. 2010. The new global economy. *Reason Magazine*. 2. (www.fmglobal.com)

Friedman, T. L. 2005. *The World Is Flat: A Brief History of the Twenty-First Century*. New York: Farrar, Strauss and Giroux.

Ghodsypour, S. H. and C. O. O' Brien. 2001. The total cost of logistics in supplier selection, under conditions of multiple sourcing, multiple criteria and capacity constraint. *International Journal of Production Economics*. 73: 15–27.

Goel, A., N. Moussavi, and V. N. Srivatsan. 2008. Time to rethink offshoring? *The McKinsey Quarterly*. 2008(4): 108–111.

Handfield, R. B. 1994. US global sourcing: Patterns of development. *International Journal of Operations & Production Management*. 14(6): 40–52.

Kannan, R. and K. C. Tan. 2002. Supplier selection and assessment: Their impact on business performance. *Journal of Supply Chain Management*. 38(4): 11–21.

Levi, M. 2002. Supplier management, investments in information systems and codifiability. PhD dissertation. The Wharton School, Pennsylvania.

Levy, D. 1997. Lean production in an international supply chain. *Sloan Management Review*. 28(2): 94–101.

Mangan, J., C. Lalwani, and T. Butcher. 2008. *Global Logistics and Supply Chain Management*. Hoboken, NJ: John Wiley.

Masud, A. S. M. and A. Ravindran. 2008. Multiple criteria decision making. In *Operations Research and Management Science Handbook*, ed. A. Ravi Ravindran, Chapter 5. Boca Raton, FL: CRC Press.

Monczka, R. M., R. Trent, and R. Handfield. 2005. *Purchasing and Supply Chain Management*. Cincinnati, OH: ITP.

Mulani, N. 2005 (May). High performance supply chain. In *Production and Operations Research Conference*. Chicago, IL.

Paulonis, D. and S. Norton. 2008. Managing global supply chains. *McKinsey Quarterly*. 2008(4): 1–9.

Petersen, K. J., D. J. Frayer, and T. V. Scannell. 2000. An empirical investigation of global sourcing strategy effectiveness. *Journal of Supply Chain Management*. 36: 29–38.

Portillo, R. C. 2009. Resilient global supply chain network design optimization. PhD dissertation. The Pennsylvania State University, University Park, PA.

Portillo, R. C., A. Ravindran, and R. A. Wysk. 2009 (September). Resilient global supply chain network design: A case study. Working Paper. Department of Industrial and Manufacturing Engineering, The Pennsylvania State University.

Robinson, M. and R. Kalakota. 2004. *Offshore Outsourcing*. Alpharetta, GA: Mivar Press.

Sarkis, J. and S. Talluri. 2002. A model for strategic supplier selection. *Journal of Supply Chain Management*. 38(1): 18–28.

Schniederjans, M. 2005. *Outsourcing and Insourcing in an International Context*. Armonk, NY: M. E. Sharpe Inc.

Shore, B., A. R. Venkatachalam, L. Iandoli, and G. Zollo. 2004. Toward a learning organization perspective to supplier selection for global supply chain management: An integrated framework. *Journal of Information Science and Technology*. 1: 1.

Simchi-Levi, D., P. Kaminisky, and E. Simchi-Levi. 2003. *Designing and Managing the Supply Chain*, 2nd edn. New York: McGraw Hill.

Smytka, D. L. and M. W. Clemens. 1993. Total cost and supplier selection model: A case study. *International Journal of Purchasing and Materials Management*. 29(1): 42–49.

Tisminisky, S. and A. Ravindran. June 2005. A review of global supplier selection criteria, methods and applications. Working Paper. Department of Industrial and Manufacturing Engineering, The Pennsylvania State University.

Weber, C. A., J. R. Current, and W. C. Benton. 1991. Vendor selection criteria and methods. *European Journal of Operational Research*. 50: 2–18.

WTO Annual Report. 1998. Accessed November 5, 2011, www.wto.org/english/res_e/booksp_e/anrep_e/anre98_e.pdf

Appendix A:* Multiple Criteria Decision Making: An Overview

Managers are frequently called upon to make decisions under multiple criteria that conflict with one another. For example, supply chain engineers have to consider conflicting criteria—supply chain costs, customer responsiveness, and supply chain risk, in making decisions. The general framework of a multiple criteria optimization problem is to simultaneously optimize several criteria, usually conflicting, subject to a system of constraints that define the feasible alternatives. Multiple criteria decision making (MCDM) problems are categorized on the basis of whether (1) the constraints are implicit, that is, the feasible alternatives are *finite and known* or (2) the constraints are explicit given by a set of linear and nonlinear inequalities or equations; in this case, the feasible alternatives are *infinite and unknown*. MCDM problems with finite (known) alternatives are called *multiple criteria selection problems* (MCSP). MCDM problems with infinite (unknown) alternatives are called *multiple criteria mathematical programming* (MCMP) problems. In this appendix, we will give an overview of some of the methods that are available for solving both MCSP and MCMP problems.

A.1 Multiple Criteria Selection Problems (MCSP)

For MCSP, the alternatives are finite and their criteria values are known a priori in the form of a *pay-off matrix*. Table A.1 illustrates the pay-off matrix for an MCSP with n alternatives (1, 2,..., n) and p criteria ($f_1, f_2,..., f_p$). The matrix elements f_{ij} denote the value of criterion j for alternative i.

A.1.1 Concept of "Best Solution"

In a single objective optimization problem, the *best solution* is defined in terms of an "optimal solution" that maximizes (or minimizes) the objective, compared to all other feasible solutions (alternatives). In MCSP, due to the conflicting nature of the objectives, the optimal values of the various criteria do not usually occur at the same alternative. Hence, the notion of an optimal

---

* Portions of this appendix have been adapted with permission from Masud and Ravindran (2008).

TABLE A.1

Pay-off Matrix of MCSP

	Objectives (Max)				
	f_1	f_2	f_3	...	f_p
Alt. 1	f_{11}	f_{12}	f_{13}		f_{1p}
2	f_{21}	f_{22}	f_{23}		f_{2p}
.					
.					
.					
n	f_{n1}	f_{n2}	f_{n3}		f_{np}
Max f_i	f_1^*	f_2^*	f_3^*		f_p^*

or best alternative does not exist in MCSP. Instead, decision making in MCSP is equivalent to choosing the *most preferred* alternative or the *best compromise solution* based on the preferences of the *decision maker* (DM). Thus, the objective of the MCSP method is to rank order the alternatives from the *best* to the *worst*, based on the DM's preference structure.

We begin the discussion with some key definitions and concepts in solving MCSP. We will assume that *all the criteria in Table A.1 are to be maximized*.

A.1.2 Dominated Alternative

Alternative i is dominated by alternative k if and only if $f_{kj} \geq f_{ij}$, for all $j = 1, 2, ..., p$ and for at least one j, $f_{kj} > f_{ij}$. In other words, the criteria values of alternative k are as good as those of alternative i and for at least one criterion, alternative k is better than i.

A.1.3 Non-Dominated Alternative

An alternative that is not dominated by any other feasible alternatives is called *non-dominated, Pareto optimal*, or *efficient* alternative. For a non-dominated alternative, an increase in the value of any one criterion is not possible without a decrease in the value of at least one other criterion.

A.1.4 Ideal Solution

It is the vector of the best values achievable for each criterion. In other words, if $f_j^* = \max_i f_{ij}$, then the ideal solution $= \left(f_1^*, f_2^*, ..., f_p^*\right)$. Since the f_j^* values may correspond to different alternatives, the *ideal solution is not achievable* for MCSP. However, it provides good target values to compare against for a tradeoff analysis.

A.2 Multi-Criteria Ranking Methods

Weighted methods are commonly used to rank the alternatives under conflicting criteria. Based on the DM's preferences, a weight w_j is obtained for criterion j such that

$$w_j \geq 0 \quad \text{and} \quad \sum_{j=1}^{p} w_j = 1 \tag{A.1}$$

Next, a weighted score of the criteria values is calculated for each alternative as follows:

$$\text{Score}(i) = \sum_{j=1}^{p} w_j f_{ij} \quad \text{for } i=1, 2, \ldots, n \tag{A.2}$$

The alternatives are then ranked based on their scores. The alternative with the highest score is ranked at the top.

There are two common approaches for determining the criteria weights based on the DM's preferences as discussed next.

A.2.1 Rating Method

Here the DM is asked to provide a rating for each criterion on a scale of 1–10 (with 10 being the most important and 1 the least important). The ratings are then normalized to determine the weights as follows:

$$w_j = \frac{r_j}{\sum_{j=1}^{p} r_j} \tag{A.3}$$

where r_j is the rating assigned to criterion j for $j = 1, 2, \ldots, p$.

Note:

$$w_j \geq 0 \quad \text{and} \quad \sum_{j=1}^{p} w_j = 1$$

A.2.2 Ranking Method (Borda Count)

Under the ranking method devised by Jean Charles de Borda, (an eighteenth century French physicist) the DM is asked to rank the p criteria from the most important (ranked first) to the least important (ranked last). The criterion that is ranked first gets p points, ranked second gets $(p - 1)$ points, and the last place criterion gets 1 point. The sum of all the points for the p criteria is given by

$$S = \frac{p(p+1)}{2} \tag{A.4}$$

The criteria weights are then calculated by dividing the points assigned to criterion j by the sum S, given by Equation A.4.

Let us illustrate the basic definitions (dominated, non-dominated, and ideal solutions) and the two weighting methods with a numerical example.

Example A.1 (Faculty recruiting)

An industrial engineering (IE) department has interviewed five PhD candidates for a faculty position and has rated on a scale of 1–10 (10 being the best and 1 being the worst) on three key criteria—Research, Teaching, and Service. The criteria values of the candidates are given in Table A.2

(a) Determine the ideal solution to this problem. Is the ideal solution achievable?
(b) Identify the dominated and non-dominated candidates.
(c) Determine the ranking of the candidates using
  (i) Rating method
  (ii) Ranking (Borda) method

Solution

(a) Ideal solution represents the best values achievable for each criterion. Since all the criteria are to maximize, the ideal solution is given by (8, 8, 5). The ideal solution is not achievable since the criteria conflict with one another and no candidate has the ideal values.
(b) Candidate A dominates E since A has higher values for all three criteria. Similarly, candidate C dominates B. On the other hand, candidates A, C, and D are non-dominated.

TABLE A.2

(Example A.1) Faculty Recruiting

Candidate	Criteria		
	Research	Teaching	Service
A	8	4	3
B	4	5	3
C	6	6	5
D	2	8	4
E	7	3	2

(c) (i) *Rating method*: Assume that the ratings for research, teaching, and service are 9, 7, and 4 respectively. Then, the weights for research, teaching, and service are computed using Equation A.3 as follows:

$$w_R = 9/(9+7+4) = 0.45$$

$$w_T = 7/20 = 0.35$$

$$w_S = 4/20 = 0.25$$

Note that the sum of the weights equals to one. Using the criteria weights, the weighted score (S) for candidate A is computed using Equation A.2:

$$\text{Score (A)} = (0.45)\ 8 + (0.35)\ 4 + (0.2)\ 3 = 5.6$$

Similarly, the scores for the other candidates are computed and are given below:

Score (B) = 4.15
Score (C) = 5.8
Score (D) = 4.5
Score (E) = 4.6

The five candidates are then ranked, using their scores, from the highest to the lowest. Thus, candidate C is ranked first followed by candidates A, E, D, and B respectively.

(ii) *Ranking (Borda) method*: Assume that the three criteria are ranked as follows:

Rank	Criteria
1st	Research
2nd	Teaching
3rd	Service

Thus, research gets 3 points, teaching 2 points, and service 1 point. Their sum is 6 and the weights are as follows:

$$w_R = 3/6 = 0.50$$

$$w_T = 2/6 = 0.33$$

$$w_S = 1/6 = 0.17$$

Using these weights and Equation A.2, the scores for candidates A, B, C, D, and E are 5.83, 4.17, 5.83, 4.33, and 4.83 respectively. Thus, both candidates A and C are tied for first place, followed by candidates E, D, and B. Note that the rankings are not exactly the same by the two methods. It does happen in practice.

Weighted methods for ranking alternatives, such as the rating method and Borda method, require that the criteria values (f_{ij}) are properly scaled so that they are of similar magnitudes. Otherwise, the rankings will be incorrect. There are several methods available for scaling criteria values, such as *ideal value method, linear normalization,* and *vector normalization using L_p norms.*

A detailed description of these scaling methods is given in Chapter 6 (Section 6.3.6) of this text, as well as in the book chapters by Masud and Ravindran (2008, 2009).

A.2.3 Analytic Hierarchy Process

The *analytic hierarchy process* (AHP), developed by Saaty (1980), is another method used for ranking alternatives. AHP can accommodate both quantitative and qualitative criteria and does not require scaling of criteria values. It uses pair-wise comparison of criteria and alternatives with strength of preference. Hence, its cognitive burden on the DM and computational requirements are much higher. A commercial software for AHP, called *expert choice,* is also available. For a detailed discussion of AHP and its applications, the reader is referred to Chapter 6, Section 6.3.7, of this text and the references (Masud and Ravindran, 2008, 2009). Experiments done with real DMs have shown that the rankings by Borda method are as good as the AHP rankings and require less cognitive burden on the DM and fewer computations (Powdrell, 2003; Ravindran et al., 2010; Velazquez et al., 2010).

A.3 Multiple Criteria Mathematical Programming Problems

In the previous sections, our focus was on solving MCDM problems with *finite* number of alternatives, where each alternative is measured by several conflicting criteria. These MCDM problems were called *multiple criteria selection problems* (MCSP). The ranking methods we discussed earlier helped in identifying the best alternative and rank order all the alternatives from the best to the worst.

In this and the subsequent sections, we will focus on MCDM problems with *infinite number of alternatives.* In other words, the feasible alternatives are not known a priori but are represented by a set of mathematical (linear/nonlinear) constraints. These MCDM problems are called *multicriteria mathematical programming* (MCMP) problems.

A.3.1 MCMP Problem

$$\text{Max } F(\mathbf{x}) = \left\{ f_1(\mathbf{x}), f_2(\mathbf{x}), \ldots, f_k(\mathbf{x}) \right\}$$

$$\text{Subject to, } g_j(\mathbf{x}) \le 0 \quad \text{for } j = 1, \ldots, m \tag{A.5}$$

where $\mathbf{x}$ is an n-vector of *decision variables* and $f_i(\mathbf{x})$, $i = 1, \ldots, k$ are the k *criteria/ objective functions*. All the objective functions are to maximize.

Let $S = \{\mathbf{x}/g_j(\mathbf{x}) \leq 0,$ for all $j\}$

$Y = \{\mathbf{y}/F(\mathbf{x}) = \mathbf{y}$ for some $\mathbf{x} \in S\}$

S is called the *decision space* and Y is called the *criteria or objective space* in MCMP.

A solution to MCMP is called a *superior solution* if it is feasible and maximizes all the objectives simultaneously. In most MCMP problems, superior solutions do not exist as the objectives conflict with one another.

A.3.2 Efficient, Non-Dominated, or Pareto Optimal Solution

A solution $\mathbf{x}^o \in S$ to MCMP is said to be *efficient* if $f_k(\mathbf{x}) > f_k(\mathbf{x}^o)$ for some $\mathbf{x} \in S$ implies that $f_j(\mathbf{x}) < f_j(\mathbf{x}^o)$ for at least one other index j. More simply stated, an efficient solution has the property that an improvement in any one objective is possible only at the expense of at least one other objective.

A *dominated solution* is a feasible solution that is not efficient.

Efficient set: Set of all efficient solutions is called the *efficient set* or *efficient frontier*.

NOTE: Even though the solution of MCMP reduces to finding the efficient set, it is not practical because there could be an infinite number of efficient solutions.

Example A.2

Consider the following bi-criteria linear program:

$$\text{Max } Z_1 = 5x_1 + x_2$$

$$\text{Max } Z_2 = x_1 + 4x_2$$

Subject to

$$x_1 \leq 5$$

$$x_2 \leq 3$$

$$x_1 + x_2 \leq 6$$

$$x_1, x_2 \geq 0$$

Solution

The *decision space* and the *objective space* are given in Figures A.1 and A.2 respectively. Corner Points C and D are efficient solutions while corner points A, B, and E are dominated. The set of all efficient solutions is given by the line segment CD in both figures.

Ideal solution is the vector of individual optima obtained by optimizing each objective function separately ignoring all other objectives.

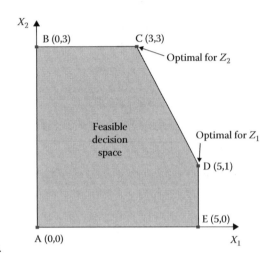

FIGURE A.1
Decision space (Example A.2).

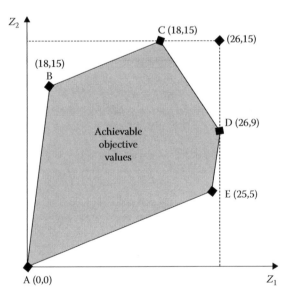

FIGURE A.2
Objective space (Example A.2).

In Example A.2, the maximum value of Z_1, ignoring Z_2, is 26 and occurs at point D. Similarly, maximum Z_2 of 15 is obtained at point C. Thus the ideal solution is (26, 15) but is *not* feasible or achievable.

NOTE: One of the popular approaches to solving MCMP problems is to find an efficient solution that comes "as close as possible" to the ideal solution. We will discuss this approach later in Section A.7.

A.3.3 Determining an Efficient Solution (Geoffrion, 1968)

For the MCMP problem given by Equation A.5, consider the following single objective optimization problem, called the P_λ *problem*.

$$\text{Max } Z = \sum_{i=1}^{k} \lambda_i f_i(x) \tag{A.6}$$

Subject to $x \in S$

$$\sum_{i=1}^{k} \lambda_i = 1$$

$$\lambda_i \geq 0$$

Theorem A.1 (Sufficiency)

Let $\lambda_i > 0$ for all i be specified. If x^o is an optimal solution for the P_λ problem (Equation A.6), then x^o is an efficient solution to the MCMP problem.

In Example A.2, if we set $\lambda_1 = \lambda_2 = 0.5$ and solve the P_λ problem, the optimal solution will be at D, which is an efficient solution. The P_λ problem is also known as the *weighted objective problem*.

Warning: Theorem 1 is only a sufficient condition and is *not* necessary. For example, there could be efficient solutions to MCMP which could not be obtained as optimal solutions to the P_λ problem. Such situations occur when the objective space is not a convex set. However, for MCMP problems, where the objective functions and constraints are linear, Theorem A.1 is both necessary and sufficient.

A.3.4 Test for Efficiency

Given a feasible solution $\bar{x} \in S$ for MCMP, we can test whether or not it is efficient by solving the following single objective problem.

$$\text{Max } W = \sum_{i=1}^{k} d_i$$

Subject to

$$f_i(x) \ge f_i(\bar{x}) + d_i \quad \text{for } i = 1, 2, \ldots, k$$

$$x \in S$$

$$d_i \ge 0$$

Theorem A.2

   (i) If Max $W > 0$, then $\bar{x}$ is a dominated solution.
  (ii) If Max $W = 0$, then $\bar{x}$ is an efficient solution.

NOTE: If Max $W > 0$, then at least one of the d_i's is positive. This implies that at least one objective can be improved without sacrificing the other objectives.

A.4 Classification of MCMP Methods

In MCMP problems, there are often an infinite number of efficient solutions and they are not comparable without the input from the DM. Hence, it is generally assumed that the DM has a real-valued *preference function* defined on the values of the objectives, but it is not known explicitly. With this assumption, the primary objective of the MCMP solution methods is to find the *best compromise solution*, which is an efficient solution that maximizes the DM's preference function.

In the last three decades, most MCDM research have been concerned with developing solution methods based on different assumptions and approaches to measure or derive the DM's preference function. Thus, the MCMP methods can be categorized by the basic assumptions made with respect to the DM's preference function as follows:

1. When *complete* information about the preference function is available from the DM

2. When *no* information is available

3. Where *partial* information is obtainable progressively from the DM

In the following sections, we will discuss the MCMP methods—*goal programming, compromise programming,* and *interactive methods,* as examples of categories 1, 2, and 3 type approaches.

A.5 Goal Programming (Ravindran et al., 2006)

One way to treat multiple criteria is to select one criterion as primary and the other criteria as secondary. The primary criterion is then used as the optimization objective function, while the secondary criteria are assigned acceptable minimum and maximum values and are treated as problem constraints. However, if careful considerations were not given while selecting the acceptable levels, a feasible design that satisfies all the constraints may not exist. This problem is overcome by *goal programming*, which has become a practical method for handling multiple criteria. Goal programming falls under the class of methods that use completely pre-specified preferences of the DM in solving the MCMP problem.

In goal programming, all the objectives are assigned target levels for achievement and relative priority on achieving these levels. Goal programming treats these targets as *goals to aspire for* and not as absolute constraints. It then attempts to find an optimal solution that comes as *close as possible* to the targets in the order of specified priorities.

Before we discuss the formulation of goal programming models, we should discuss the difference between the terms *real constraints* and *goal constraints* (or simply *goals*) as used in goal programming models. The real constraints are absolute restrictions on the decision variables, while the goals are conditions one would like to achieve but are not mandatory. For instance, a real constraint given by

$$x_1 + x_2 = 3$$

requires all possible values of $x_1 + x_2$ to always equal 3. As opposed to this, a goal requiring $x_1 + x_2 = 3$ is not mandatory, and we can choose values of $x_1 + x_2 \geq 3$ as well as $x_1 + x_2 \leq 3$. In a goal constraint, positive and negative deviational variables are introduced as follows:

$$x_1 + x_2 + d_1^- - d_1^+ = 3 \quad d_1^+, d_1^- \geq 0$$

Note that, if

$$d_1^- > 0, \quad \text{then } x_1 + x_2 < 3, \quad \text{and} \quad \text{if } d_1^+ > 0, \quad \text{then } x_1 + x_2 > 3$$

By assigning suitable weights w_1^- and w_1^+ on d_1^- and d_1^+ in the objective function, the model will try to achieve the sum $x_1 + x_2$ as close as possible to 3. If the goal were to satisfy $x_1 + x_2 \geq 3$, then only d_1^- is assigned a positive weight in the objective, while the weight on d_1^+ is set to zero.

A.5.1 Goal Programming Formulation

Consider the general MCMP problem given in Section A.3 (Equation A.5). The assumption that there exists an optimal solution to the MCMP problem involving multiple criteria implies the existence of some preference ordering of the criteria by the DM. The goal programming (GP) formulation of the MCMP problem requires the DM to specify an acceptable level of achievement (b_i) for each criterion f_i and specify a weight w_i (ordinal or cardinal) to be associated with the deviation between f_i and b_i. Thus, the GP model of an MCMP problem becomes

$$\text{Minimize } Z = \sum_{i=1}^{k} \left(w_i^+ d_i^+ + w_i^- d_i^- \right) \tag{A.7}$$

Subject to

$$f_i(x) + d_i^- - d_i^+ = b_i \quad \text{for } i = 1, \dots, k \tag{A.8}$$

$$g_j(x) \le 0 \quad \text{for } j = 1, \dots, m \tag{A.9}$$

$$x_j, d_i^-, d_i^+ \ge 0 \quad \text{for all } i \text{ and } j \tag{A.10}$$

Equation A.7 represents the objective function of the GP model, which minimizes the weighted sum of the deviational variables. The system of equations (Equation A.8) represents the *goal constraints* relating the multiple criteria to the goals/targets for those criteria. The variables, d_i^- and d_i^+, in Equation A.8 are called *deviational variables*, representing *the under achievement* and *over achievement* of the ith goal. The set of weights (w_i^+ and w_i^-) may take two forms as given below:

1. Pre-specified weights (cardinal)
2. Preemptive priorities (ordinal)

Under pre-specified (cardinal) weights, specific values in a relative scale are assigned to w_i^+ and w_i^- representing the DM's "trade-off" among the goals. Once w_i^+ and w_i^- are specified, the goal program represented by Equations A.7 through A.10 reduces to a single objective optimization problem. The cardinal weights could be obtained from the DM using any of the methods discussed in Sections A.2. However, in order for this method to work, the criteria values have to be scaled properly.

In reality, goals are usually *incompatible* (i.e., incommensurable) and some goals can be achieved only at the expense of some other goals. Hence, *preemptive goal programming*, which is more common in practice, uses *ordinal ranking* or *preemptive priorities* to the goals by assigning incommensurable goals to different priority levels and weights to goals at the same priority level. In this case, the objective function of the GP model (Equation A.7) takes the following form:

$$\text{Minimize } Z = \sum_p P_p \sum_i \left(w_{ip}^+ d_i^+ + w_{ip}^- d_i^- \right) \tag{A.11}$$

where P_p represents priority p with the assumption that P_p is much larger then P_{p+1} and w_{ip}^+ and w_{ip}^- are the weights assigned to the ith deviational variables at priority p. In this manner, lower priority goals are considered only after attaining the higher priority goals. Thus, *preemptive goal programming* is essentially a sequence of single objective optimization problems, in which successive optimizations are carried out on the alternate optimal solutions of the previously optimized goals at higher priority.

In both preemptive and non-preemptive GP models, the DM has to specify the targets or goals for each objective. In addition, in the preemptive GP models, the DM specifies a preemptive priority ranking on the goal achievements. In the non-preemptive case, the DM has to specify relative weights for goal achievements.

To illustrate, consider the following bi-criteria linear program (BCLP):

Example A.3 (BCLP)

$$\text{Max } f_1 = x_1 + x_2$$

$$\text{Max } f_2 = x_1$$

Subject to

$$4x_1 + 3x_2 \leq 12$$

$$x_1, x_2, \geq 0$$

Maximum f_1 occurs at $x = (0, 4)$ with $(f_1, f_2) = (4, 0)$. Maximum f_2 occurs at $x = (3, 0)$ with $(f_1, f_2) = (3, 3)$. Thus the ideal values of f_1 and f_2 are 4 and 3 respectively and the bounds on (f_1, f_2) on the efficient set will be as follows:

$$3 \leq f_1 \leq 4$$

$$0 \leq f_2 \leq 3$$

Let the DM set the goals for f_1 and f_2 as 3.5 and 2 respectively. Then the GP model becomes

$$x_1 + x_2 + d_1^- - d_1^+ = 3.5 \tag{A.12}$$

$$x_1 + d_2^- - d_2^+ = 2 \tag{A.13}$$

$$4x_1 + 3x_2 \leq 12 \qquad\qquad\text{(A.14)}$$

$$x_1, x_2, d_1^-, d_1^+, d_2^-, d_2^+ \geq 0 \qquad\qquad\text{(A.15)}$$

Under the preemptive GP model, if the DM indicates that f_1 is much more important than f_2, then the objective function will be

$$\text{Min } Z = P_1 d_1^- + P_2 d_2^-$$

subject to the constraints (A.12 through A.15), where P_1 is assumed to be much larger than P_2.

Under the non-preemptive GP model, the DM specifies relative weights on the goal achievements, say w_1 and w_2. Then the objective function becomes

$$\text{Minimize } Z = w_1 d_1^- + w_2 d_2^-$$

subject to the same constraints (A.12) through (A.15).

A.6 Partitioning Algorithm for Preemptive Goal Programs

A.6.1 Linear Goal Programs

Linear goal programming problems can be solved efficiently by the *partitioning algorithm* developed by Arthur and Ravindran (1978, 1980a). It is based on the fact that the definition of preemptive priorities implies that higher order goals must be optimized before lower order goals are even considered. Their procedure consists of solving a series of linear programming sub-problems by using the solution of the higher priority problem as the starting solution for the lower priority problem. Care is taken that higher priority achievements are not destroyed while improving lower priority goals.

A.6.2 Integer Goal Programs

Arthur and Ravindran (1980b) show how the partitioning algorithm for linear GP problems can be extended with a modified branch and bound strategy to solve both pure and mixed integer GP problems. They demonstrate the applicability of the branch and bound algorithm by solving a multiple objective nurse scheduling problem (Arthur and Ravindran, 1981).

A.6.3 Nonlinear Goal Programs

Saber and Ravindran (1996) present an efficient and reliable method called the partitioning gradient based (PGB) algorithm for solving nonlinear

GP problems. The PGB algorithm uses the partitioning technique developed for linear GP problems and the generalized reduced gradient (GRG) method to solve single objective nonlinear programming problems. The authors also present numerical results by comparing the PGB algorithm against a modified pattern search method for solving several nonlinear GP problems. The PGB algorithm found the optimal solution for all test problems proving its robustness and reliability, while the pattern search method failed in more than half the test problems by converging to a non-optimal point.

Kuriger and Ravindran (2005) have developed three intelligent search methods to solve nonlinear GP problems by adapting and extending the simplex search, complex search, and pattern search methods to account for multiple criteria. These modifications were largely accomplished by using partitioning concepts of goal programming. The chapter also includes computational results with several test problems.

A.7 Method of Global Criterion and Compromise Programming

Method of global criterion (Hwang and Masud, 1979) and *compromise programming* (Zeleny, 1982) fall under the class of MCMP methods that do not require any preference information from the DM.

Consider the MCMP problem given by Equation (A.5). Let

$$S = \{x/g_j(x) \le 0, \text{ for all } j\}$$

Let the ideal values of the objectives $f_1, f_2, \ldots, f_k$ be $f_1^*, f_2^*, \ldots, f_k^*$. The method of global criterion finds an efficient solution that is "closest" to the ideal solution in terms of the L_p distance metric. It also uses the ideal values to normalize the objective functions. Thus the MCMP reduces to

$$\text{Minimize } Z = \sum_{i=1}^{k} \left(\frac{f_i^* - f_i}{f_i^*} \right)^p$$

subject to $x \in S$.

The values of f_i^* are obtained by maximizing each objective f_i subject to the constraints $x \in S$, *but* ignoring the other objectives. The value of p can be 1, 2, 3, ... Note that $p = 1$ implies equal importance to all deviations from the ideal. As p increases larger deviations have more weight.

A.7.1 Compromise programming

Compromise Programming is similar in concept to the method of global criterion. It finds an efficient solution by minimizing the weighted L_p distance metric from the ideal point as given next:

$$\text{Min } L_p = \left[\sum_{i=1}^{k} \lambda_i^p \left(f_i^* - f_i \right)^p \right]^{1/p} \tag{A.16}$$

subject to $x \in S$ and $p = 1, 2,..., \infty$.

where λ_i's are weights that have to specified or assessed subjectively. Note that λ_i could be set to $1/(f_i^*)$.

Theorem A.3

Any point x* that minimizes L_p (Equation A.16) for $\lambda_i > 0$ for all i, $\sum \lambda_i = 1$ and $1 \le p < \infty$ is called a *compromise solution*. Zeleny (1982) has proved that these compromise solutions are non-dominated. As $p \to \infty$, Equation A.16 becomes

$$\text{Min } L_\infty = \text{Min } \underset{i}{\text{Max}} \ [\lambda_i \ (f_i^* - f_i)]$$

and is known as the *Tchebycheff metric*.

A.8 Interactive Methods

Interactive methods for MCMP problems rely on the progressive articulation of preferences by the DM. These approaches can be characterized by the following procedures:

Step 1: Find a solution, preferably feasible and efficient.

Step 2: Interact with the DM to obtain his or her reaction or response to the obtained solution.

Step 3: Repeat steps 1 and 2 until satisfaction is achieved or until some other termination criterion is met.

When interactive algorithms are applied to real-world problems, the most critical factor is the functional restrictions placed on the objective functions, constraints, and the *unknown* preference function. Another important factor is *preference assessment styles* (hereafter, called *interaction styles*). According to Shin and Ravindran (1991), the typical interaction styles are as follows:

(a) *Binary pair-wise comparison*—the DM must compare a pair of two-dimensional vectors at each interaction.

(b) *Pair-wise comparison*—the DM must compare a pair of p-dimensional vectors and specify a preference.

(c) *Vector comparison*—the DM must compare a set of p-dimensional vectors and specify the best, the worst, or the order of preference (note that this can be done by a series of pair-wise comparisons).

(d) *Precise local tradeoff ratio*—the DM must specify precise values of local tradeoff ratios at a given point. It is the *marginal rate of substitution* between objectives f_i and f_j: in other words, tradeoff ratio is how much the DM is willing to give up in objective j for a unit increase in objective i at a given efficient solution.

(e) *Interval tradeoff ratio*—the DM must specify an interval for each local tradeoff ratio.

(f) *Comparative tradeoff ratio*—the DM must specify his preference for a given tradeoff ratio.

(g) *Index specification and value tradeoff*—DM must list the indices of objectives to be improved or sacrificed, and specify the amount.

(h) *Aspiration levels* (or reference point)—the DM must specify or adjust the values of the objectives which indicate his or her optimistic wish concerning the outcomes of the objectives.

Shin and Ravindran (1991) also provide a detailed survey of MCMP interactive methods. Their survey includes the following:

- A classification scheme for all interactive methods.
- A review of methods in each category based on functional assumptions, interaction style, progression of research papers from the first publication to all its extensions, solution approach, and published applications.
- A rating of each category of methods in terms of the DM's cognitive burden, ease of use, effectiveness, and handling inconsistency.

A.9 MCDM Applications

One of the most successful applications of multi-criteria decision making has been in the area of portfolio selection, an important problem faced by individual investors and financial analysts in investment companies.

A portfolio specifies the amount invested in different securities which may include bonds, common stocks, mutual funds, bank CDs, treasury notes, and others. Much of the earlier investment decisions were made by seat-of-the-pant approaches. Markowitz (1959) pioneered the development of the *modern portfolio theory*, which uses bi-criteria mathematical programming models to analyze the portfolio selection problem. By quantifying the tradeoffs between risks and returns, he showed how an investor can diversify portfolios such that the portfolio risk can be reduced without sacrificing returns. Based on Markowitz's work, Sharpe (1963) introduced the concept of the *market risk*, and developed a bi-criteria linear programming model for portfolio analysis. For their pioneering work in modern portfolio theory, both Markowitz and Sharpe shared the 1990 Nobel prize in economics. The Nobel award was the catalyst for the rapid use of modern portfolio theory by Wall Street firms in the 1990s. A detailed discussion of the mean-variance model of Markowitz is available in Heching and King (2008, 2009).

Among the MCDM models and methods, the goal programming models have seen the most applications in industry and government. Chapter 4 of the text book by Schniederjan (1995) contains an extensive bibliography (666 Citations) on goal programming applications categorized by areas—accounting, agriculture, economics, engineering, finance, government, international, management, and marketing. Zanakis and Gupta (1995) also have a categorized bibliographic survey of goal programming applications.

For an application of goal programming models in supplier selection, see Chapter 6, Section 6.4 and Chapter 7, Section 7.14, of this text book.

A.10 MCDM Software

One of the problems in applying MCDM methods in practice is the lack of commercially available software implementing these methods. There is some research software available. Two good resources for these are as follows:

1. http://www. mcdmsociety.org/
2. http://www.sal.hut.fi/

The first is the web page of the International Society on Multiple Criteria Decision Making. It has links to MCDM software and bibliography. A number of this software is available free for research and teaching use. The second link is to the research group at Helsinki University of Technology. It has links to some free software, again for research and instructional use.

References

Arthur, J. L. and A. Ravindran. 1978. An efficient goal programming algorithm using constraint partitioning and variable elimination. *Management Science*. 24(8): 867–868.

Arthur, J. L. and A. Ravindran. 1980a. PAGP: An efficient algorithm for linear goal programming problems. *ACM Transactions on Mathematical Software*. 6(3): 378–386.

Arthur, J. L. and A. Ravindran. 1980b. A branch and bound algorithm with constraint partitioning for integer goal programs. *European Journal of Operational Research*. 4: 421–425.

Arthur, J. L. and A. Ravindran. 1981. A multiple objective nurse scheduling model. *Institute of Industrial Engineers Transactions*. 13: 55–60.

Geoffrion, A. 1968. Proper efficiency and theory of vector maximum. *Journal of Mathematical Analysis and Applications*. 22: 618–630.

Heching, A. R. and A. J. King. 2008. Financial engineering. In *Operation Research and Management Science Handbook*, ed. A. R. Ravindran, Chapter 21. Boca Raton, FL: CRC Press.

Heching, A. R. and A. J. King. 2009. Financial engineering. In *Operation Research Applications*, ed. A. R. Ravindran, Chapter 7. Boca Raton, FL: CRC Press.

Hwang, C. L. and A. Masud. 1979. *Multiple Objective Decision Making-Methods and Applications*. New York: Springer-Verlag.

Kuriger, G. and A. Ravindran. 2005. Intelligent search methods for nonlinear goal programs. *Information Systems and Operational Research*. 43: 79–92.

Markowitz, H. 1959. *Portfolio Selection: Efficient Diversification of Investments*. New York: Wiley.

Masud, A. S. M. and A. Ravindran. 2008. Multiple criteria decision making. In *Operations Research and Management Science Handbook*, ed. A. R. Ravindran, Chapter 5. Boca Raton, FL: CRC Press.

Masud, A. S. M. and A. Ravindran. 2009. Multiple criteria decision making. In *Operation Research Methodologies*, ed. A. R. Ravindran, Chapter 5. Boca Raton, FL: CRC Press.

Powdrell, B. J. 2003. Comparison of MCDM algorithms for discrete alternatives. MS thesis, Department of Industrial Engineering, The Pennsylvania State University, University Park, PA.

Ravindran, A., K. M. Ragsdell, and G. V. Reklaitis. 2006. *Engineering Optimization: Methods and Applications*, 2nd edn., Chapter 11. Hoboken, NJ: John Wiley.

Ravindran, A., U. Bilsel, V. Wadhwa, and T. Yang. 2010. Risk adjusted multicriteria supplier selection models with applications. *International Journal of Production Research*. 48(2): 405–424.

Saaty, T. L. 1980. *The Analytic Hierarchy Process*. New York: McGraw Hill.

Saber, H. M. and A. Ravindran. 1996. A partitioning gradient based (PGB) algorithm for solving nonlinear goal programming problem. *Computers and Operations Research*. 23: 141–152.

Schniederjans, M. 1995. *Goal Programming: Methodology and Applications*. Boston, MA: Kluwer Academic Publishers.

Sharpe, W. F. 1963. A simplified model for portfolio analysis. *Management Science*. 9: 277–293.

Shin, W. S. and A. Ravindran. 1991. Interactive multi objective optimization: Survey I-continuous case. *Computers and Operations Research.* 18: 97–114.

Velazquez, M. A., D. Claudio, and A. R. Ravindran. 2010. Experiments in multiple criteria selection problems with multiple decision makers. *International Journal of Operational Research.* 7(4): 413–428.

Zanakis, S. H. and S. K. Gupta. 1995. A categorized bibliographic survey of goal programming. *Omega: International Journal of Management Science.* 13: 211–222.

Zeleny, M. 1982. *Multiple Criteria Decision Making.* New York: McGraw Hill.

Index

A

ABC analysis
  CBA, 100
  divide and conquer approach, 99
  inventory
    plot, 100–101
    results, 100
  Pareto principle, 99
The Aberdeen Group, 466
Analytic hierarchy process (AHP), 494
  model steps
    consistency index, 325
    consistency ratio, 325
    degree of importance scale, 323
    final criteria weights, 324–325
    pair-wise comparison of
      criteria, 324
    sub-criteria weights, 325–326
    supplier ranking, 326–327
    total score, 326
  principles, 322
  supplier criteria, 323
Annual order cost (AOC), 258
Auto regressive integrated moving
    average (ARIMA) method,
    81–82

B

Bi-criteria linear program (BCLP),
    501–502
Borda count, 315, 423
  individual supplier rankings,
    421–422
  MCDM, 315, 423
  ranking method
    criteria weights, 491–492
    faculty recruiting, 492–494
  supplier selection models and
    methods, 314–315
BPO, *see* Business Process Offshoring
    (BPO)

Bullwhip effect
  ad hoc, 157
  base-stock ordering policy, 159
  base-stock policy, 162
  centralized *vs.* decentralized
    control, 140
  consumer demand, 161–163
  CPFR, 163
  definition, 154
  demand forecast, 157, 160
  example, 156–157
  four-stage supply chain, 155
  manufacturer and base-stock order,
    161, 164–165
  scale-driven costs, 155
  super-linear function, 158
  supply chain management, 155
  two-period lead time, 158–159
  updated base-stock orders, 159–161
Business Process Offshoring (BPO), 453

C

Charlie's Bavarian Automotive (CBA), 100
Cluster analysis (CA)
  hierarchical clustering algorithms, 327
  partitional clustering algorithms, 327
  procedure, 328–329
  research, 326
Collaborative planning, forecasting and
    replenishment (CPFR), 85, 163
Constant-work-in-process (WIP)
    systems, 97
CPFR, *see* Collaborative planning,
    forecasting and replenishment
    (CPFR)

D

Decision criteria and risk assessment
  binary variables, 474
  criteria weights and preferences, 470
  customer service criterion, 471

decision criteria hierarchy, 470–471
facility risk rating, 471–472
group weights for criteria, 473
MILP model, 474
preemptive goal programming model, 474
sample criteria ratings, 472–473
sample criteria weights, 472–473
WACC, 471
Disruptive events detectability
Markov chain properties, 410
MFPT
concept, 409–410
disruption risk quantification, 411–412
matrix computation, 410–411
Distribution of demand over lead time plus review period (DLTR) distribution, 123, 126
Distribution requirements planning (DRP)
adjustments, 131–132
average per-period cost, 131, 133
distribution centers (DCs), 128
dynamic-programming solution, 131
EOQ computation, 131
human intervention, 131
make-to-stock product setting, 127
ordering and holding costs, 131, 133
records for MeltoMatic Snow Blowers, 128–130
ROP-OQ method, 127–128

E

Economic order quantity (EOQ)
computation, 124, 131
formula, 255, 260
model, 19, 112–114
sawtooth pattern, 141
Element percentile method (EPM), 386
Extreme value theory (EVT)
MtT type risks, 393
risk quantification models, 381
VaR type impact function, 382

F

Free and secure trade lane (FAST), 466

G

GDP, *see* Gross domestic product (GDP)
Generalized extreme value distributions (GEVD)
impact function, 383
PWM, 384, 386
risk impact functions, 384–386
risk measures, 403
VaR
disruption risk function, 387–389
type impact function, 383–384
type risk calculations, 431–432
GEVD, *see* Generalized extreme value distributions (GEVD)
Global products
"American car," 452
automobile production, 451
BPO, 453
China's share, 452–453
offshore manufacturing, 453
smart phone, 452
Global supply chain management, 21–22
enterprise transformation, outsourcing strategies, 478
GDP (*see* Gross domestic product (GDP))
globalization history, 449–450
globalization impacts
globalization risks, 454–455
global products (*see* Global products)
U.S. manufacturing, 453–454
world economies, 450–451
globalization strategies, 457–458
global risk factors, 455
global sourcing
benefits and barriers, 459–460
international supplier selection (*see* International supplier selection)
issues, 460–461
tools, 465–466
global supply chain strategies, 456–457
international logistics, 466–467
logistics cost, 477
logistics network flexibility, 478
outsourcing companies, 478

resilient global supply chain
design
decision criteria and risk
assessment (*see* Decision
criteria and risk assessment)
design decision, 467
model features, 470
multi-criteria analysis, 475–477
multi-criteria selection
techniques, 468
problem background, 468–469
profit maximization model
results, 474–475
Goal programming (GP)
formulation, 500–502
methodology
general goal programming
model, 335–336
MCMP problems, 334–335
multiple criteria optimization
problems, 334
real constraints and goal
constraints, 334
Goods in transit
annual in-transit inventory cost, 183
consignee, 181
decision-making firm, 182
freight charges, 181
freight transportation
cost, 180
terms, 181
in-transit holding cost
computations, 183
re-supply lead time, 182
TAC expression, 182
Gravity model, 266–267
facility location, 270–271
iterations, 270
limitations, 271
three-stage supply chain,
268–269
total distribution cost, 269
Green Lane program, 466
Gross domestic product (GDP)
decision criteria and risk
assessment, 472
globalization history, 450
U.S. manufacturing, 454
world economy, 451

H

High demand variability, 467

I

Individual supplier rankings
Borda count weights, 421–422
candidate suppliers evaluation,
420–421
decision makers attribute
weights, 420
Industrial engineers (IEs) design, 1
International supplier selection
criteria, 461–462
financial issues
currency exchange fluctuations, 461
financial risk reduction, 463
macro and micro strategies, 461
volume-timing technique, 462–463
logistic issues, 463
manufacturing practices, 463–464
strategic issues, 464–465
Interpretive structural modeling (ISM), 298
Inventory management, 19–20
cost minimization, 152
decision framework
administrative costs, 98
constant-WIP systems, 97
dependent-demand items, 96
independent-demand items, 95, 97
stationary/non-stationary
demand, 97
multi-echelon inventory systems
arborescent system, 137–138
assembly system, 136–137
centralized *vs.* decentralized
control, 139–140
distribution system, 136–137
echelon inventory, 138
echelons, 137–138
four-stage serial system, 136
general arborescent network,
137–138
management, 137
serial supply chain (*see* Serial
supply chain)
two-stage serial system (*see*
Two-stage serial system)

multi-item inventory models
coordinated replenishment, 135
joint-ordering TAC model, 133
joint replenishment, 134–135
"must-order" and "can-order"
levels, 134
ordering cost, 133
preliminary modeling issues
ABC analysis (*see* ABC analysis)
critical tasks, 98–99
production scheduling, 154
single-item, multi-period problems
base-stock system, 126–127
constant-demand setting, 116
continuous-review system, 109
cycle service level, 118
DLTR, 126
in-stock probability, 118
inventory level, 115
inventory system optimization, 120
lead-time demand, 117
newsvendor problem, 109
non-stationary demand
(*see* Distribution requirements
planning (DRP))
optimal inventory policy, 119
optimization software
packages, 119
order-up-to system, 125–126
penalty cost, 117
perpetual-demand model,
115, 118
probability distribution of
demand, 108
reorder point-order quantity
model (*see* Reorder point-order
quantity model)
reorder-point-order-up-to models
(*see* Reorder-point-order-up-to
models)
replenishment lead time, 116
safety stock, 115
sample inventory profile, 125
single-point/-season demand
cost tradeoffs, 105
critical ratio, 105
demand and cost, 101
marginal analysis, 104
newsvendor demand, 103

newsvendor model application
and problem, 102
optimal order quantity, 104
safety stock, 108
service measures, 105–106
shortage costs service impact,
106–108
supply chain safety
inventory, 153
ISM, *see* Interpretive structural
modeling (ISM)

L

Less-than-truckload (LTL) mode
inventory decision model
advantage, 195
linear approximation, 195
linear regression, 197
lower per-unit rates, 195
LTL effective rates, 196–197
non-linear curve-fitting
process, 196
OAK-ATL rate tariff, 194–195
power function estimate, 198
published tariff discount
annual transportation cost,
200, 202
CzarLite-based rate tariff, 199
CzarLite rate tariff block, 206
logistics-related costs, 204
OAK-ATL *vs.* NYC-ATL
shipments, 209–210
optimal inventory decision
model, 201, 204
optimization problem, 201
optimization results, 207–208
power function estimate,
206–207
producer price index, 200
rate estimation function, 199
rate function, 205
replenishment lead time, 205
shipment decision, 207
TL optimization, 199
total annual cost function, 199
transportation cost, 202–203
traveling salesman problem, 205
vehicle routing problem, 205

service general rate model
  data-intensive and
    computationally-intensive
    approach, 211
  Kay-Warsing estimation function,
    213–214
  Kay-Warsing LTL rate model, 212
  rate alternative estimation,
    213–214
  rate function, 212
  rate tariffs, 211
  transportation cost analysis, 211
Linear programming model
  "chase" and "level" strategies, 65
  constraints, 66
  decision variables, 66
  demand data, 65
  demand/inventory balance, 67–68
  production capacity, 68
  training, 67
  workforce assignment, 67
  workforce size, 66
Linear weighted point (LWP) method
  ranking suppliers, 346
  rating/scoring method, 312
  single sourcing methods, 300–301
Location and distribution decisions
  binary variables
    capital budgeting problem,
      230–231
    fixed charge problem, 231–232
    nonlinear integer programs,
      236–238
    quantity discounts (*see* Quantity
      discounts)
    right-hand-side constants,
      232–233
    set covering models (*see* Set
      covering models)
    set partitioning models, 240
    warehouse location, 240
  case studies, 282
  continuous location models
    gravity model (*see* Gravity
      model)
    iterative method, 267–268
    multiple facility location, 271–272
  decision making, 229
  facility location decisions, 281

integer programming models, 230
network design multiple criteria
  models, 279–280
real-world applications
  AT&T, 277
  BMW, 276–277
  Ford motor company, 275–276
  Hewlett–Packard (HP), 276
  multi-national consumer products
    company, 273–274
  Procter and Gamble (P&G),
    274–275
  UPS, 277–278
risk pooling (*see* Risk pooling)
supply chain network
  management, 229
supply chain network optimization
  (*see* Supply chain network
  optimization)
LTL mode, *see* Less-than-truckload
  (LTL) mode
LWP, *see* Linear weighted point (LWP)
  method

M

MCDM, *see* Multiple criteria decision
  making (MCDM)
MCSP, *see* Multiple criteria selection
  problems (MCSP)
Mean first passage time (MFPT)
  disruption risk quantification,
    411–412
  Markov chain analysis, 409–410
  matrix, 410–411
  risk detectability and recovery,
    414–415
Miss-the-target (MtT) risk models
  L-type
    beta distribution, 396
    impact function, 393–394
    risk function, 398
  N-type
    Cauchy distribution, 400
    Cauchy PDF, 400–401
    generalized hyperbolic
      distribution, 396–397
    impact function, 393–394, 401
    mathematical form, 398–399

normal cases, 399
parameters, 401–402
special cases, 400
S-type
gamma distribution, 395–396
impact function, 393–394
risk function, 397–398
Taguchi's loss functions, 394–395
Multi-criteria mathematical
programming (MCMP)
methods classification, 498
problems, 494–495
bi-criteria linear program, 495–496
decision space and objective
space, 496–497
dominated solution, 495
efficiency test, 497–498
efficient set/efficient frontier, 495
GP, 334–335
solution determination, 497
Multi-criteria mixed integer linear
programming
decision criteria and risk
assessment, 474
resilient global supply chain
design, 469
single objective problem, 474
Multiple criteria decision making
(MCDM), 2
applications, 505–506
Borda count (*see* Borda count)
compromise programming, 504
data analysis, 417
decision criteria and risk
assessment, 472
distance metric, 503
goal programming
formulation (*see* Goal
programming (GP),
formulation)
real constraints and goal
constraints, 499
interactive methods, 504–505
MCMP (*see* Multi-criteria
mathematical programming
(MCMP))
MCSP
'best solution' concept, 489–490
dominated alternative, 490

ideal solution, 490
non-dominated alternative, 490
methods, 419
multi-criteria ranking methods
AHP, 494
rating method, 491
preemptive goal programs
partitioning algorithm
integer goal programs, 502
linear goal programs, 502
nonlinear goal programs, 502–503
ranking methods, 437
rating/scoring method, 313
resilient global supply chain
design, 469
risk-adjusted supplier selection
problem, 416, 437
software, 506
supplier order allocation, 348
supply chain optimization, 13
Multiple criteria selection problems
(MCSP)
'best solution' concept, 489–490
dominated alternative, 490
ideal solution, 490
MCMP problem, 494
non-dominated alternative, 490

N

National Motor Freight Transportation
Association (NMFTA), 190
Nonlinear programming model
changing production cost, 70
inventory/shortage cost, 71–72
production cost, 70

P

Planning production
aggregate planning
greedy algorithm, 78–80
linear programming (*see* Linear
programming model)
nonlinear programming
(*see* Nonlinear programming
model)
problem, 64–65
strategies, 80

transportation problem
(*see* Transportation problem)
ARIMA method, 81–82
Bullwhip effect, 84–85
CPFR, 85
Croston's method, 82
demand forecasting, 81
demand management, 83–84
forecast accuracy monitoring, 57–59
forecasting errors, 54
   bias, 55
   forecasting method selection, 57
   mean absolute deviation, 55
   mean absolute percentage
      error, 56
   mean squared error, 55
   standard deviation, 55
   uses, 57
forecasting process, 28–29
multiple periods forecasting
   constant level, 51
   seasonality, 52
   seasonality and trend, 53–54
   trend, 52–53
production planning
   decisions, 83
   process, 63–64
qualitative forecasting methods
   customer surveys, 31
   Delphi method, 30
   executive committee consensus,
      29–30
   sales force survey, 30–31
quantitative forecasting methods
   averaging method, 34–35
   constant level forecasting
      methods, 33–34
   exponential smoothing method,
      38–39
   last value method, 34
   linear programming model,
      36–37
   simple moving average
      method, 35
   time series forecasting (*see* Time
      series forecasting)
   weighted moving average
      method, 35
real world applications, 61–62

regression and Box-Jenkin's ARIMA
   methods, 82
SCM demand forecasting, 27–28
seasonality in forecasting
   deseasonalized demand data, 40
   deseasonalized forecast, 41–42
   exponential smoothing
      forecasting method, 41
   naïve method, 42
   4-quarter moving average, 42
   seasonality index, 39–40
   seasonality indices, 41
software, forecasting
   automatic software type, 59–60
   manual software type, 60–61
   semi-automatic software type, 60
   user experience, 61
static seasonality indices, 47–48
survey results, 62–63
trend, forecasting
   Holt's method, 45–46
   simple linear trend model, 43–44
Winters' method
   computations, 50–51
   Holt's method, 49
   periodicity, 49
   seasonality index, 50
Probability weighted moments (PWM)
   GEVD parameters estimation,
      384, 386
   moment estimates, 388–389
   VaR type risk calculations, 432

Q

Quantity discounts
   "all-unit" quantity discounts,
      233–235
   application, 236
   graduated quantity discount,
      235–236

R

Reorder point-order quantity (ROP-OQ)
      model
   control methods, 127–128
   cost of holding inventory, 111–112
   cycle stock, 113

EOQ
  cost curve, 114–115
  cost estimation, 114
  per-unit holding cost, 112
  reorder point-order quantity
      inventory policy, 113
  total annual cost, 113
  infinite-horizon and perpetual
      demand setting, 111
  inventory position, 110
  inventory system levels, 110
  per-unit holding cost, 112
  weighted average cost of capital, 111
Reorder-point-order-up-to models
  ABC analysis, 121
  approximate periodic-review
      solution, 124
  computational approximations, 121
  DLTR distribution, 123
  inventory system parameters
      computation, 125
  less intensive management, 121
  ordering and holding inventory
      cost, 123
  periodic-review reorder point, 123
  sample inventory profile, 121–122
  stochastic dynamic
      programming, 121
Risk management
  actions, 372
  chance constrained
      programming, 437
  disruption and operational risks., 437
  industry practices
    Federal express, 378–379
    Hewlett–Packard (HP), 378
    Johnson and Johnson, 380
    strategic and operational levels,
        376–377
    Teradyne inc., 377
    Wal-Mart, 379–380
  MCDM model, 437
  MtT risk models (*see* Miss-the-target
      (MtT) risk models)
  multiple criteria optimization
      models (*see* Supplier selection
      incorporating risk)
  multiple sourcing model, 437–438
  plan development, 374–375

real world risk events and impacts
  market capitalization, 366
  pull- and push-type supply
      chains, 365
  supply chain disruptions, 366–367
  supply chain "glitches," 365
  supply chain of companies, 364
risk assessment
  risk mapping, 370–371
  risk prioritization, 371–372
risk detectability and recovery
  disruptive events detectability (*see*
      Disruptive events detectability)
  exponential model, 416
  just-in-time framework, 416
  MFPT matrix, 414–415
  risk recovery conceptual model,
      412–413
  3-tier supply chain, 413–414
  transition probability matrix, 414
  transition times, 415
risk identification, 368–369
risk measures
  maximum loss, 402
  MtT type risk measures, 404
  VaR disruption risk function,
      402–404
risk quantification models
  impact severity and occurrence
      frequency, 380–381
  VaR and MtT type risks, 381–382
strategies, 373–376
supply chain risk, 363–364
supply chain risk quantification and
    management
  conceptual models, 439
  disruption planning, 440
  manmade and natural
      disasters, 439
  mathematical models, 438–439
  risk managementon, 440
  survey, 440–443
supply chain risk sources, 367–368
VaR models (*see* Value-at-risk (VaR)
    models)
Risk pooling
  advantages and drawbacks, 259–260
  AOC, 258
  average inventory, 255, 257

definition, 230
demand correlation matrix, 263
demand uncertainty
    completely deconsolidated
        distribution network,
        260–261
    regional demands, 262–263
    risk pooled/consolidated
        network, 261–262
distribution options, 264–265
IHC, 257
number of orders, 256–257
optimal order quantity, 255
practical applications, 265–266
3-stage supply chain network,
    253–254
supply chain
    metrics impact, 254
    network design, 254
    safety stock, 265
time between orders, 256
weekly demands, 263

S

Scaling criteria values
    ideal value method, 317–318
    ideal value scaling, 320
    L_p norm, 321–322
    L_p norm usage/vector scaling, 318
    simple linearization/linear
        normalization, 318, 320–321
    simple scaling, 317, 319–320
    supplier criteria matrix, 317
    suppliers, 318–319
Serial supply chain
    deterministic demand and fixed
        ordering costs, 140–141
    fixed costs and stochastic demand,
        151–152
    stochastic demand and negligible
        fixed ordering costs
        echelon holding costs, 146
        echelon j-truncated system,
            148–149
        installation inventory
            position, 150
        inventory holding and backorder
            penalty costs, 146

multi-echelon base-stock policies,
    149–150
newsvendor-type fractiles, 147
optimal echelon base-stock
    level, 148
Set covering models
    integer programming, 239
    set covering matrix definition, 238
    set covering problem, 239–240
Steady demand, 467
Supplier order allocation
    demand data, 340–341
    fixed supplier cost, 340
    fuzzy goal programming, 343
    goal programming, 339–340
    lead-time data, 341
    non-preemptive goal
        programming, 342
    preemptive goal programming
        solution, 342
    supplier quality data and production
        capacity data, 341
    Tchebycheff goal programming, 343
    unit price and price break, 340
Supplier selection incorporating risk
    data description
        MtT type risk calculations,
            430–431
        VaR type risk calculations, 431–432
    phase 1 experiment result
        criteria ranking, 418–419
        decision makers comparison,
            419–420
        group supplier rankings, 422
        individual supplier rankings
            (*see* Individual supplier
            rankings)
        methods comparison, 419
        multiple criteria ranking
            methods, 423
    phase 2 model results
        fuzzy GP solution, 434
        non-preemptive GP
            solution, 433
        preemptive GP solution, 432–433
        Tchebycheff GP solution,
            433–434
    phase 1 model/short-listing
        suppliers, 417–418

phase 2 results comparison, 434–435
solution methodology
  fuzzy GP model, 429
  non-preemptive GP model,
    427–428
  preemptive GP model, 426–427
  Tchebycheff (min–max) GP model,
    428–429
supplier sourcing, risk adjusted
  multi-criteria optimization
  models
  model constraints, 425–426
  model objectives, 424–425
  multi-criteria models, 424
  organization decisions, 423
two-phase method, 416–417
value path approach, 435–436
VaR and MtT type risks (*see*
  VaR and MtT type risks
  combination)
Supplier selection models and
  methods, 21
global sourcing
  global competition, 349
  uncertainty, 350–351
multi-criteria ranking methods
  AHP (*see* Analytic hierarchy
    process (AHP))
  Borda count, 314–315
  cluster analysis (*see* Cluster
    analysis (CA))
  group decision making, 329–330
  L_p metric, 311–312
  pair-wise criteria comparison, 316
  ranking methods comparision, 330
  rating/scoring method, 312–314
  scaling criteria values (*see* Scaling
    criteria values)
  suppliers ranking (*see* Suppliers
    ranking)
multi-objective supplier allocation
  model
  fuzzy goal programming, 339
  goal programming methodology
    (*see* Goal programming (GP),
    methodology)
  non-preemptive goal
    programming, 337–338
  notations, 331

order allocation problem, 332–333
preemptive goal programming,
  336–337
supplier order allocation (*see*
  Supplier order allocation)
Tchebycheff (min–max) goal
  programming, 338
value path approach (*see* Value
  path approach)
multiple sourcing methods
average lead-time
  requirements, 305
constraints, 306–307
decision variables, 306
mathematical programming
  models, 303
objective function, 306
optimal order allocation, 308
product demand, 303–304
product prices, 305
products lead-time, 303–304
products quality, 303–304
supplier capacities, 303–304
total cost minimization, 303
ranking suppliers, 346–347
selection criteria
ISM, 298
key criteria, 298–299
purchasing agents and managers
  survey, 297
relative importance, 297–298
supplier selection process, 297
single sourcing methods
LWP, 300–301
TCO, 301–303
sourcing strategy, 296–297
supplier order allocation
goal programming, 349
linear programming, 349
multi-criteria linear goal
  programming model, 349
multi-criteria models, 348–349
pre-qualification, 347
single objective models, 347–348
supplier risk, 351
supplier selection problem
in-house/outsource, 295–296
supplier base pre-qualification/
  pre-screening, 296

supplier selection factors, 293–294
supplier selection process,
    294–295
suppliers pre-qualification, 299
Suppliers ranking
    cost criteria, 310
    experience criteria, 310
    miscellaneous criteria, 310
    multiple criteria ranking
        methods, 310
    organizational criteria, 309
    performance criteria, 310
    quality criteria, 310
    selection criteria, 309
    supplier criteria values, 310–311
Supply chain and financial metrics
    business financial measures
        cash-to-cash cycle, 15–16
        return on assets, 15
        working capital, 15
    inventory measures
        days of inventory, 14
        inventory capital, 14–15
        inventory turns, 13–14
Supply chain decisions
    operational decisions, 6–7
    strategic decisions
        information technology, 6
        network design, 5
        production and sourcing, 6
    tactical decisions, 6
Supply chain engineering
    decisions (*see* Supply chain
        decisions)
    definition, 2
    drivers
        facilities, 9
        inventory, 8
        suppliers, 9
        transportation, 8–9
    enablers, 7–8
    flows, 4
    location and distribution decisions,
        20–21 (*see also* Location and
        distribution decisions)
    network, 3
    performance assessment and
        management
        efficiency, 10–11

optimization, 13
responsiveness-efficiency tradeoff
    frontier, 10
supply chain responsiveness, 12
supply chain risk, 10, 12
planning production, 19 (*see also*
    Planning production)
risk management, 21 (*see also* Risk
    management)
SCM definition, 4
Supply chain management (SCM)
    definition, 4
    operations research models, 18
    supply chain disruption, 17
    supply chain efficiency, 16
    supply chains top 25, 17–18
    transportation decisions, 20
        (*see also* Transportation
        decisions)
Supply chain network optimization
    distribution planning
        strategic decision, 242
        tactical decision, 242
        transportation problem,
            243–244
    location-distribution, dedicated
        warehouses
        dedicated warehouse
            problems, 247
        formulation, 248–249
        multiple deliveries, 247
        supply constraint, 248
    location-distribution problem
        location decisions and
            distribution decisions, 244
        mixed integer program, 247
        optimal distribution plan, 245
        supply chain network, 245
        supply constraint, 246
        warehouse data, 245–246
    supply chain network design
        annual demand, 249–250
        binary and continuous
            variables, 250
        constraints, 251
        multi-state supply chain
            network design problem, 249
        objective function, 251–252
        optimal solution, 252

3-stage supply chain, 249
unit shipping cost, 249–250
warehouse capacities and
investment cost, 249–250
warehouse location, 241–242
Supply chain operations reference
(SCOR) model, 22–23

T

TCO, *see* Total cost of
ownership (TCO)
Time series forecasting
constant level
and seasonality, 31–32
and trend, 31–32
systematic and random
component, 31
trend and seasonality, 31, 33
Total annual inventory-related cost
(TAC) expression, 182, 219
Total cost of ownership (TCO)
ideal value scaling, 320
L_p norm scaling, 321–322
scaling criteria values, 318–319
simple scaling, 319
single sourcing methods,
301–303
supplier order allocation, 348–349
Trailer-on-flatcar/container-on-flatcar
(TOFC/COFC), 215
Transportation decisions
customer-supplier negotiations, 219
freight rates general models
distance- *vs.* -transit time
relationship, 190
line-haul costs, 189
shipment quantity, 188
shipping distance *vs.* lead
time, 189
terminal/accessorial costs, 189
truck-based freight rates, 188
truckload service, 187
freight transportation
airfreight, 185
intermodal, 186–187
LTL quantity, 184
motor freight, 185
pipeline, 186

private carriers and for-hire
carriers, 185
railroads, 185
service characteristics and cost, 184
shippers and carriers, 184
water-borne freight, 185–186
goods in-transit risk, 219
lead-time demand, 220
LTL service
"class-based" rating system, 190
freight charge formalization, 194
general rate model and mode
(*see* Less-than-truckload (LTL)
mode)
NMFTA, 190
OAK-ATL lane, 191
rate tariff, 191
rate weight breaks, 192–193
shipment freight charge, 193–194
total shipment charge, 192
motor carrier freight
continuous-review cost, 178
cost-based formulation, 176
goods in transit (*see* Goods in transit)
truckload computations, 177–178
truckload freight transportation
cost, 179
truckload optimization, 180
truckload service, 176
rail and air cargo
NYC-ATL lane, 217–219
rail carload, 214
rail to motor freight results, 215–216
solver-driven solution, 215
TOFC/COFC, 215
truck-based shipments, 215
shipper, 175
VRP, 205, 220
Transportation problem
cost, 72–73
decision variables, 75
greedy algorithm, 77
linear programming formulation, 73
standard transportation problem, 73
standard transportation table, 74–75
transportation formulation, 77
transportation table, 75–78
unbalanced transportation problem,
73–74

Two-stage serial system
  centralized control
    Axsater's algorithm, 144–145
    continuous first-order
      condition, 144
    echelon holding costs, 143
    inventory cycle lengths, 143
  decentralized control
    average installation inventory, 141
    installation and echelon stock, 142
    inventory cycle length, 141
    order cycle, 143
    stage-optimal policy, 141
    total cost, 142

U

United Parcel Service (UPS), 277–278

V

Value-at-risk (VaR) models
  GEVD parameters estimation
    EPM, 386
    maximum likelihood, 384
    method of moments, 384
    PWM, 384
    Sherman test and Kolmogorov–
      Smirnov statistic, 386
  GEVD risk impact functions, 384–386
  impact function
    disruptive events, 382
    Euler–Mascheroni constant,
      383–384
    EVT, 382
    GEVD parameters, 383
    Gumbel distribution, 383–384
    VaR disruption risk function (*see* VaR
      disruption risk function)
  VaR occurrence functions, 386–387

Value path approach
  multi-objective problems, 343
  properties, 344
  supplier selection case study,
    344–345
  trade-offs demonstration, 344
  value path results, 345–346
VaR and MtT type risks combination
  all suppliers
    MtT type combination, 408
    VaR type combination, 408
  different suppliers
    MtT type risk combination,
      407–408
    VaR type combination, 407
  same supplier
    MtT type risk combination,
      406–407
    VaR type risk combination,
      405–406
VaR disruption risk function
  annual loss disruption distribution
    function, 392–393
  excel spreadsheet, 390–391
  occurrence function, 390
  risk distribution, 390
  simulation approach
    GEVD, 388–389
    Kolmogorov–Smirnov statistics,
      389–390
    plotting positions, 389
    PWM method, 388–389
    steps, 387
Vehicle routing problem (VRP),
  205, 220

W

Weighted average cost of capital
  (WACC), 471

Made in the USA
Las Vegas, NV
04 November 2020

10559498R00299